AF899406

Solar and Stellar Flares

Lyndsay Fletcher • Petr Heinzel
Editors

Solar and Stellar Flares

Observations, Simulations, and Synergies

Previously published in *Solar Physics* Volume 290,
Issue 12, 2015

Springer

Editors
Lyndsay Fletcher
School of Physics and Astronomy
University of Glasgow School of Physics and Astronomy
Glasgow, UK

Petr Heinzel
Academy of Sciences
Astronomical Institute Academy of Sciences
Ondrejov, Czech Republic

ISBN 978-94-024-0934-5
Springer New York Heidelberg Dordrecht London

Library of Congress Control Number: 2016960757

© Springer Science+Business Media B.V. 2017
This work is subject to copyright. All rights are reserved by the Publisher, whether the whole or part of the material is concerned, specifically the rights of translation, reprinting, reuse of illustrations, recitation, broadcasting, reproduction on microfilms or in any other physical way, and transmission or information storage and retrieval, electronic adaptation, computer software, or by similar or dissimilar methodology now known or hereafter developed.
The use of general descriptive names, registered names, trademarks, service marks, etc. in this publication does not imply, even in the absence of a specific statement, that such names are exempt from the relevant protective laws and regulations and therefore free for general use.
The publisher, the authors and the editors are safe to assume that the advice and information in this book are believed to be true and accurate at the date of publication. Neither the publisher nor the authors or the editors give a warranty, express or implied, with respect to the material contained herein or for any errors or omissions that may have been made.

Cover Image: X6.9 flare of August 9, 2011 (08:37:50 UTC) observed by NASA SDO/AIA.
Image processing by means of PM-NAFE algorithm was done by Miloslav Druckmüller (Brno University of Technology) using 21.1, 17.1 and 30.4 nm bands.
Courtesy of NASA/SDO and AIA science team.

Printed on acid-free paper

Springer is part of Springer Science+Business Media (www.springer.com)

Contents

Reprinted from *Solar Physics* Journal, DOI 10.1007/s11207-015-0818-5

Preface

L. Fletcher[1] · P. Heinzel[2] · L. van Driel-Gesztelyi[3,4,5] · C.H. Mandrini[6,7] · F. Fárník[2]

Published online: 24 November 2015
© Springer Science+Business Media Dordrecht 2015

Solar flares and flares on other cool stars are currently extensively studied using new ground- and space-based instruments, together with highly sophisticated numerical simulations. This was reflected well in the many contributions to the conference "Solar and Stellar Flares: Observations, simulations and synergies", held in Prague during 23 – 27 June 2014, from which this Topical Issue (TI) is drawn. The conference was organised in honour and memory of Prof. Zdeněk Švestka (1925 Prague – 2013 Bunschoten), a leading expert in the field of solar flares and the co-founder and editor in chief of the journal *Solar Physics* (see his memoirs,

Solar and Stellar Flares: Observations, Simulations, and Synergies
Guest Editors: Lyndsay Fletcher and Petr Heinzel

✉ L. Fletcher
lyndsay.fletcher@glasgow.ac.uk

P. Heinzel
pheinzel@asu.cas.cz

L. van Driel-Gesztelyi
Lidia.vanDriel@obspm.fr

C.H. Mandrini
mandrini@iafe.uba.ar

F. Fárník
frantisek.farnik@asu.cas.cz

1 SUPA School of Physics and Astronomy, University of Glasgow, Glasgow G12 8QQ, UK

2 Astronomical Institute, The Czech Academy of Sciences, 251 65 Ondřejov, Czech Republic

3 Mullard Space Science Laboratory, University College London, Holmbury St Mary, Dorking, Surrey RH5 6NT, UK

4 Observatoire de Paris, LESIA-CNRS UMR 8109, 9295 Meudon Cédex, France

5 Konkoly Observatory of Hungarian Academy of Sciences, Budapest, Hungary

6 Instituto de Astronomía y Física del Espacio, CONICET-UBA, Buenos Aires, Argentina

7 Facultad de Ciencias Exactas y Naturales (FCEN), UBA, Buenos Aires, Argentina

Švestka, 2010, and obituary in the June 2013 issue of the journal). Almost 100 participants came to Prague from all over the world to attend the meeting, and they presented many new interesting results on flares. This TI represents a selection of articles (both reviews and original research papers) based on presentations at the meeting. The Prague conference was held 25 years after a similar TI was published in *Solar Physics*, collecting papers presented during IAU Colloquium 104 on solar and stellar flares (Haisch and Rodono, 1989). At that meeting, Zdeněk Švestka presented an invited review on the gradual phase of solar flares (Švestka, 1989).

This TI starts with Hugh Hudson's reminiscences on Zdeněk Švestka's research, coupled with current achievements, problems, and perspectives (Hudson, 2015), followed by four reviews based on invited talks. Janvier, Aulanier, and Démoulin (2015) give an overview of coronal observations and magnetohydrodynamic (MHD) simulations, the building blocks for 3D models of solar flares. Flare energetics is currently a hot topic (*e.g.* Milligan *et al.*, 2014; Kerr and Fletcher, 2014; Kleint *et al.*, 2015) that requires multiwavelength spectral observations, and Milligan (2015) reviews the extreme ultra-violet (EUV) spectroscopy of the lower solar atmosphere during flares, from which the properties of flare plasmas can be deduced, using observations from the *Extreme ultraviolet Variability Experiment* (EVE) experiment onboard the *Solar Dynamics Observatory* (Woods *et al.*, 2012) and the *Extreme ultraviolet Imaging Spectrometer* (EIS, Culhane *et al.*, 2007). In the optical range, broadband spectra of stellar flares are being obtained by large telescopes, and Kowalski *et al.* (2015) present comprehensive insights into the white-light flare emission from stellar flares. These data are hard to obtain, but are crucial also in the solar case. In particular, the authors analyse the strongly enhanced hydrogen Balmer continuum, which was always rather difficult to detect in solar flares (*e.g.* Švestka, 1966), but see the recent findings from the *Interface Region Imaging Spectrograph* (IRIS, De Pontieu *et al.*, 2014) near-ultraviolet (NUV) flare spectra by Heinzel and Kleint (2014). The article by Kowalski *et al.* (2015) also shows how the technique of numerical radiation hydrodynamics can be applied to both solar and stellar flare studies. Finally, the broad topic of the relation of flares and coronal mass ejections (CMEs) is covered by Schmieder, Aulanier, and Vršnak (2015). These four articles are then followed by ten contributed articles that cover a broad range of interesting and current topics. From among these, we highlight the exciting and new topic of "super flares" discussed during the conference, which also has implications for exoplanetary habitability (*e.g.* Katsova and Livshits, 2015).

The scientific organising committee (SOC) of the conference was co-chaired by P. Heinzel and L. Fletcher, with members B. Dennis, L. van Driel-Gesztelyi, F. Fárník, S.L. Hawley, H.S. Hudson, M. Karlický, E.R. Priest, K. Shibata, J. Sylwester, and A. Veronig. The local organising committee (LOC) was chaired by F. Fárník, with members P. Heinzel, J. Kašparová, M. Varady, and Z. Pecková. Since Z. Švestka was an honorary member of the Czech Learned Society, the conference was held under the auspices of this Society and was opened by its president, Professor J. Bičák. The main organising body was the Astronomical Institute of the Czech Academy of Sciences, together with CBT Prague. We would like to acknowledge that the conference was considered to be part of the activities of seven European research centres working within the EC-funded FP7 project "F-CHROMA" – Flare Chromospheres: Observations, Models and Archives. And last but not least, we also thank all of the referees for their helpfulness and careful attention in improving the quality of the articles in this volume.

We were delighted to dedicate this conference and now its related TI of *Solar Physics* to the memory of Zdeněk Švestka, our colleague and friend. We were also very happy to welcome his wife Lída to the meeting, remembering the early years at Ondřejov Observatory, where she worked with Zdeněk on optical flare spectroscopy. We hope that Zdeněk would have enjoyed the meeting, and also leafing through its Topical Issue.

Acknowledgements Sponsorship of the conference by Springer is gratefully acknowledged.

References

Culhane, J.L., Harra, L.K., James, A.M., Al-Janabi, K., Bradley, L.J., Chaudry, R.A., *et al.*: 2007, *Solar Phys.* **243**, 19.

De Pontieu, B., Title, A.M., Lemen, J.R., Kushner, G.D., Akin, D.J., Allard, B., *et al.*: 2014, *Solar Phys.* **289**, 2733.

Haisch, B.M., Rodono, M.: 1989, *Solar Phys.* **121**, 1. DOI.

Heinzel, P., Kleint, L.: 2014, *Astrophys. J. Lett.* **794**, L23. DOI.

Hudson, H.S.: 2015, *Solar Phys.* this issue. DOI.

Janvier, M., Aulanier, G., Démoulin, P.: 2015, *Solar Phys.* this issue. DOI.

Katsova, M.M., Livshits, M.A.: 2015, *Solar Phys.* this issue. DOI.

Kerr, G.S., Fletcher, L.: 2014, *Astrophys. J.* **783**, 98. DOI.

Kleint, L., Heinzel, P., Judge, P., Krucker, S.: 2015, *Astrophys. J.* arXiv.

Kowalski, A.F., Hawley, S.L., Carlsson, M., Allred, J.C., Uitenbroek, H., Osten, R.A., Holman, G.: 2015, *Solar Phys.* this issue. DOI.

Milligan, R.O.: 2015, *Solar Phys.* this issue. DOI.

Milligan, R.O., Kerr, G.S., Dennis, B.R., Hudson, H.S., Fletcher, L., Allred, J.C., Chamberlin, P.C., Ireland, J., Mathioudakis, M., Keenan, F.P.: 2014, *Astrophys. J.* **793**, 70. DOI.

Schmieder, B., Aulanier, G., Vršnak, B.: 2015, *Solar Phys.* this issue. DOI.

Švestka, Z.: 1966, *Space Sci. Rev.* **5**, 388. DOI.

Švestka, Z.: 1989, *Solar Phys.* **121**, 399. DOI.

Švestka, Z.: 2010, *Solar Phys.* **267**, 235. DOI.

Woods, T.N., Eparvier, F.G., Hock, R., Jones, A.R., Woodraska, D., Judge, D., Didkovsky, L., *et al.*: 2012, *Solar Phys.* **275**, 115. DOI.

Reprinted from *Solar Physics* Journal, DOI 10.1007/s11207-015-0669-0

Švestka's Research Then and Now

Invited Review

H.S. Hudson[1,2]

Received: 16 November 2014 / Accepted: 10 March 2015 / Published online: 25 March 2015
© Springer Science+Business Media Dordrecht 2015

Abstract Zdeněk Švestka's research work influenced many fields of solar physics, especially in the area of flare research. In this article I take five of the areas that particularly interested him and assess them in a "then and now" style. His insights in each case were quite sound, although of course in the modern era we have learned things that he could not readily have envisioned. His own views about his research life have been published recently in this journal, to which he contributed so much, and his memoir contains much additional scientific and personal information (Švestka in *Solar Phys.* **267**, 235, 2010).

1. Introduction

In a quasi-historical, quasi-scientific review such as this, one must expect a fairly ignorant account that is heavily biased to the author's own knowledge base and working concepts. Since the author's own research career only began some two decades after Zdeněk's, and since we had not actually met until the late 1960s, the literature trail provides the main material for the background, along with the author's speculations about the scientific environment before that time. Thus the reader should be wary of a certain degree of mismatch (knowledge and opinion) between Zdeněk and myself, but in general, we have certainly been interested in the same topics, and in particular in the still-fascinating white-light flare phenomenon.

I have structured this as a "then and now" summary of specific topics to which Zdeněk made important early contributions, and recommend that any interested reader also read Zdeněk's own memoir (Švestka, 2010) for further insight especially into the "then" aspects. The five specific topics (in Sections 2 – 6: optical spectroscopy, proton flares, loop-prominence systems, the Flare Build-up Study, and "giant arches") overlap considerably and

Solar and Stellar Flares: Observations, Simulations, and Synergies
Guest Editors: Lyndsay Fletcher and Petr Heinzel

✉ H.S. Hudson
hhudson@ssl.berkeley.edu

[1] School of Physics and Astronomy, University of Glasgow, Glasgow G12 8QQ, UK

[2] Space Sciences Laboratory, University of California, Berkeley, CA 94720, USA

mainly provide a framework in time for the discussion. No doubt one thing led to another for Zdeněk's research interest, as it will, and the topics have considerable overlap because of the broadly interconnected dynamics of the components of the overall flare process. He had a prescient interest in "proton flares," which today we would call Solar Particle Events (jargonized to SPEs, which consist of Solar Energetic Particles, or SEPs). This interest predated the discovery of coronal mass ejections (CMEs), now known to play a vital role in particle acceleration, and we will pick up the historical trail below.

A solar flare is a complicated set of phenomena that reflect sudden energy release in the solar atmosphere (*e.g.* Švestka, 1976; Fletcher *et al.*, 2011). The effects range from the photosphere ("white-light flares") to the distant heliosphere ("Interplanetary Coronal Mass Ejections," or ICMEs). Flares can occur without detectable CMEs, and sometimes *vice versa*, with the "stealth CMEs" exemplifying the latter (*e.g.* Webb *et al.*, 1998; Robbrecht, Patsourakos, and Vourlidas, 2009). The latter typically come from filament channels remote from active regions. The white-light emission of a solar flare reflects radiation emitted at the photosphere itself, or even from slightly below it (Xu *et al.*, 2004). Flare effects indeed extend into the deep interior, as known via seismic signatures, see Kosovichev and Zharkova (1998). The energy that is to be released during one of these sudden transients comes from slowly-accumulated stresses in the magnetic field, according to general consensus, and indeed alternative possible sources for this energy seem implausible (*e.g.* Hudson, 2007).

High-resolution spectroscopic observations of solar flares (Section 2), Zdeněk's main interest in the early decades, has returned to the forefront now, both because technology has advanced considerably beyond photographic techniques and because the limitations imposed by Earth's atmosphere have disappeared. As revealed with crystal clarity by the very first observations of a solar flare (Carrington, 1859; Hodgson, 1859) because of the high radiated energy, the lower solar atmosphere plays a fundamental role in flare development. Today, not only do we have the broader coverage afforded by observatories in space, but we are also seeing a remarkable development of high-resolution ground-based observatories with much-improved optical properties and data capabilities. For this reason, the open questions from Zdeněk's early work may soon find some answers, and it is timely to consider how these questions arose.

As a personal note about this paper: I learned a great deal from Zdeněk, and our research paths intersected frequently. Our names appear together on only 12 papers, however, notably the first of his series "Large-Scale Active Coronal Phenomena in *Yohkoh* SXT Images," with which he brought his *Skylab* experience to bear on data that I was closely familiar with. See Section 6 below for my views of how this developed.

2. Optical Spectroscopy of Flare Emissions

2.1. Spectroscopy Then

Zdeněk's research career began with optical spectroscopy of solar flares – the tools available at the time permitted high spectral resolution and therefore supported analyses of radiation physics in the flaring medium, but these tools were cumbersome. A good observation typically produced a well-defined spectrum with not-so-good seeing, limited metadata, and perhaps another one or two at different phases (and locations of the long slit) on the flaring region. Typically, the observer would place the slit at the brightest point of the flare and be thankful if good signal-to-noise properties developed. Zdeněk worked extensively in

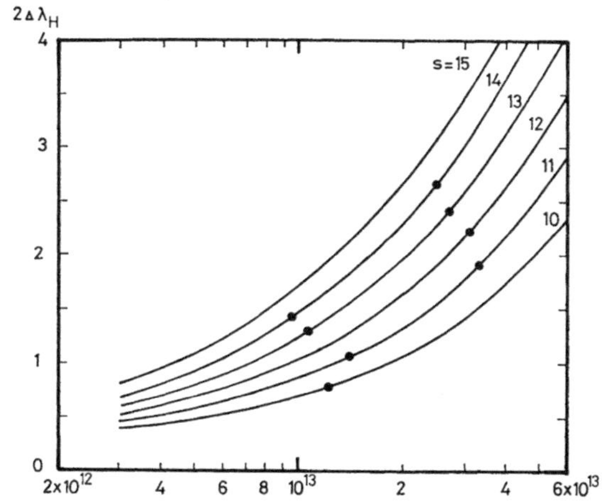

Figure 1 Balmer-series observations of two major flares (SOL1956-12-29 and SOL1960-08-07), showing results as two sets of black dots, showing the half-widths (Angstroms) of series lines 10 – 14 plotted against electron density, cm^{-3}. These correspond to mean densities $n_e = 1.2 \times 10^{13}$ and $2.9 \times 10^{13}\ \mathrm{cm}^{-3}$, respectively. Adapted from Švestka (1972).

this area and led the development of flare spectroscopy at Ondřejov observatory in the then Czechoslovakia. This work taught us a great deal about the behavior of the lower solar atmosphere during a flare (*e.g.* Berlicki, 2007), and also about the "loop prominence systems" (see below), which interestingly contained low-temperature gas embedded magnetically in the hot corona. It also led to review papers (Švestka, 1966, 1972) and to his monograph "Solar Flares" (Švestka, 1976). This book remains current, perhaps partly because the attention of solar-flare researchers turned away from optical ground-based observations and the complexities of the atmospheric structures they reveal, as the new spectacular observations from space developed – for example, with the *Skylab* observatory, which boosted Zdeněk into a new career phase, as described below.

High-resolution optical spectroscopy thus led to many insights in flare physics. Chief among Zdeněk's particular contributions was the work on Stark broadening in the profiles of the Balmer-series lines. These showed that high densities must prevail (Suemoto and Hiei, 1959; Švestka, 1972); estimates from that era put $n_e > 10^{12}\ \mathrm{cm}^{-3}$ for flares observed against the disk. We readily understand this as regards the flare ribbons, embedded in the dense chromosphere, and the direct implication is that the chromosphere has been substantially ionized by the flare process, when compared with models of the quiet chromosphere. Figure 1 shows representative observations of this effect.

For limb flares, we are looking at the mis-named (Švestka, 2007) "loop prominence systems," to be discussed in more detail in Section 4 below along with their mis-namings at other wavelengths. These structures may also have high densities, such that with high spatial resolution one can see emission in the core of Hα even when the coronal structures are projected against the solar disk. Hiei, Okamoto, and Tanaka (1983) found a coronal electron density $n_e \approx 10^{12.8}\ \mathrm{cm}^{-3}$ for the limb flare SOL1982-06-12.

2.2. Spectroscopy Now

How has the optical spectroscopy of solar flares developed since Zdeněk's time? Ground-based observations have been hit-or-miss; we could point to the multi-slit spectra (with film readout) of (*e.g.*, Martin *et al.*, 1974) or the initial CCD-based observations of Wülser (1987). Such observations struggled with the basic requirement for imaging spectroscopy at simultaneous (and high) spatial, temporal, and spectral resolution. We note also that observations of stellar flares, although without the advantage of spatial resolution, offer different views of probably the same physics. "Sun-as-a-star" observations of high-resolution optical spectra can also be achieved with instruments not originally designed for solar work (*e.g.* Johns-Krull *et al.*, 1997); such observations make use of cross-dispersed echelle spectrographs with broad spectral coverage.

It would be fair to say that observations of the lower solar atmosphere during flares, enabled by the advanced kinds of imaging spectroscopy using current technology, are only just returning to the mainstream of research into solar flares. In the meanwhile, the advance of technology has made stellar spectroscopy immeasurably better, and although this may seem embarrassing (why not study the nearest star most diligently?), the stellar observations provide a wonderful complement. The period of slower development in the meanwhile has an obvious explanation: access to space made it possible to do absolutely new things, and at the same time it was clear to all that the lower solar atmosphere presents huge problems for model development.

3. "Proton Flares"

3.1. Proton Flares Then

The phenomenon of the "proton flare" refers to the acceleration of high-energy particles (protons above 10 MeV) at or near the Sun, which then appear near the Earth as "solar cosmic rays." The presence of such particles definitively establishes the fundamental non-thermality of flare processes, since we typically observe energetic protons at 10 MeV, some 10^5 times the mean thermal energy in the corona. We now term these particles SEPs (Solar Energetic Particles) and recognize a huge variety of properties in terms of elemental abundances, ionic states, and even isotopic composition. In the 1950s it had already been well-established that there were often strong connections between sunspots (or flares) and geomagnetic storms (see, for example Chapman and Bartels, 1940).

In the 1950s the problem therefore was to establish observational signatures of the extreme particle acceleration that could occur in association with flares. This presumably was what led Zdeněk to creating an atlas of proton flares from 1938 to 1955, detected as the "Polar Cap Absorption" (PCA) signature in ionospheric soundings. From 1942, neutron monitors entered the scene, and of course the remarkable event SOL1956-02-23 (Meyer, Parker, and Simpson, 1956). This just escaped Zdeněk's catalog, but in several subsequent papers he explored the statistical aspects of the proton morphology in the context of the flare process. Eventually, he helped to organize a "Proton Flare Project" of dedicated observations, as reported by Švestka and Simon (1969). Although many technical conclusions appeared, there was no obvious success at achieving a basic physical understanding of how flares could accelerate all of these particles.

The relationship between the particles detected in interplanetary space and those required for flare development also remained murky, although at this early time it was by no means clear that particle acceleration played the central role in flare physics that we now recognize. Nevertheless, Zdeněk created the cartoon shown in Figure 2 (Švestka, 1970), showing that protons could in fact heat the lower solar atmosphere – this was not a new result, but the analysis was excellent. He also went on to provide estimates for the numbers of energetic neutrons directly arriving at Earth from a flare (Švestka, 1971), a phenomenon predicted by Biermann, Haxel, and Schlüter (1951), elaborated by Lingenfelter *et al.* (1965), and subsequently discovered by Chupp *et al.* (1982).

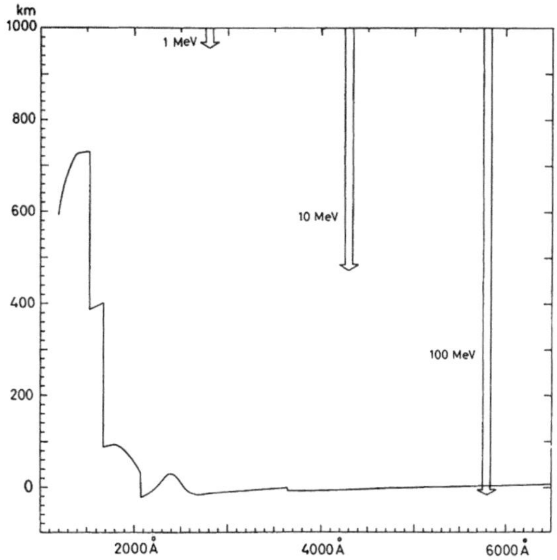

Figure 2 An early view of the thick-target model for solar flares, showing height in km *vs.* optical wavelength in the photosphere and low chromosphere. The solid line shows the location of the unit optical depth as a function of wavelength, and the arrows show the penetration of vertically incident protons accelerated in the corona. Protons require about 100 MeV to reach the visible photosphere (adapted from Švestka, 1970).

The impulsive phase of the flare, however, has a close identification with the bremsstrahlung emission of 10 – 100 keV electrons, not protons as suggested by the "proton flare" association and Figure 2. This change resulted in the "thick-target model" of an *electron beam* originating in the corona, rather than a *proton* beam (*e.g.* Kane and Donnelly, 1971; Brown, 1971; Hudson, 1972). It seems unlikely now, for various reasons, that Zdeněk's proton beam can play a role in the impulsive phase, or if it does, then the proton acceleration would likely be in the lower atmosphere directly because of the timescales involved.

3.2. Proton Flares Now

After the SOL1956-02-23, still ranking as the most powerful SEP event on the records, space research developed rapidly and flare-associated particle events have now become a major component of "space weather." Nevertheless, confusion still exists about the role of flares in SEP production, even though comprehensive evidence exists establishing that high-energy particles do indeed exist in flare plasmas. The acceleration of SEPs has a clear relationship with the interplanetary propagation of large-scale shock waves driven by CME expansion (*e.g.* Reames, 2013), and so these particular particles do not have any obvious association with those responsible for either the bremsstrahlung hard X-rays or the γ-ray emission by solar flares. For both of these components there is strong evidence for the non-thermal particle acceleration to take place on closed magnetic-field structures in the core of an active region, rather than in large-scale coronal structures involved with the CME (*e.g.* Fletcher *et al.*, 2011).

Recently, modern data have jeopardized the thick-target paradigm (powerful electron beams); Martínez Oliveros *et al.* (2012) successfully made absolute height determinations for the hard X-ray sources in the white-light flare SOL2011-02-24 and found them to be at the photosphere, out of the range of any plausible electron beam. At present we have no good working model for the process: neither a Švestka-style proton beam nor an electron beam seem to be appropriate.

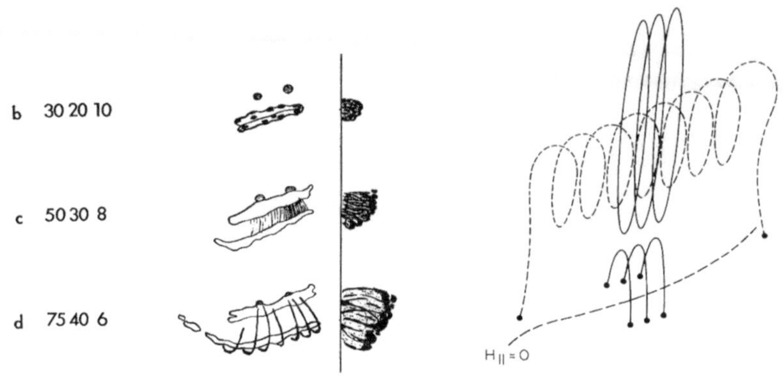

Figure 3 Left, adapted from Bruzek (1964), sketching Hα loop systems for three stages of flares as they would be seen on the disk and off the limb. The numbers in the legend show minutes after onset, height in Mm, and upward velocity in $\mathrm{km\,s^{-1}}$. Right, from Švestka (1983); this illustrates how 2D reconnection would simply create coronal toroids, whereas 3D reconnection could produce a flux rope.

4. Loop Prominence Systems

4.1. Loop Prominences Then

One of Zdeněk's earliest papers dealt with observations of solar flares in Hα, as a part of his work at Ondřejov in the group of Prof. F. Link (Švestka, 1949), and with a publication date 58 years later, he concluded his independent research activities with a comment on the nomenclature and physics of the coronal manifestations of solar flares in Švestka (2007). Both of these papers dealt with the coronal effects of solar flares.

The loop-prominence phenomenon played a major role in laying the framework of our current understanding of flare processes in the solar corona. Before Zdeněk's time, one had the clear knowledge that loop prominence systems had a strong association with the occurrence of major solar flares, and Bruzek (1964) provided the now-iconic view (*e.g.*, Figure 3, left panel) of the growth of large systems of coronal loops, now termed "arcades," which Hα observations revealed most clearly when they rose high enough to appear above the limb. The Bruzek paper also firmly established the existence of a relationship between the proton flares and the formation of a loop-prominence system. We can also nicely trace the ideas arising from these flaring loops, and their ribbon-like footpoints, to the popular flare model of large-scale magnetic reconnection now termed the "CSHKP" model (Carmichael, Sturrock, Hirayama, Kopp–Pneuman); Zdeněk was captivated by the description Kopp and Pneuman (1976) gave, but he recognized (Figure 3, right panel) the need for three-dimensionality.

The coronal development of an eruptive flare has many signatures, ranging actually across the entire electromagnetic spectrum that is accessible to us. Accordingly, many misleading neologisms have popped up. In the corona the optical emission lines (*e.g.*, the "green line" of Fe XIV at 5303 Å, but especially the "yellow line" of Ca XV at 3594 Å) spawned the terminology of the "sporadic coronal condensation," and the microwave observations created the terminology "post-burst increase". There has recently been some use of the term "post-eruption arcade" (PEA), even though an arcade can appear without any eruption (*e.g.* Gaizauskas *et al.*, 1998). Zdeněk felt that "post-flare loops," and presumably all of these others, might more efficiently just be called "flares" (Švestka, 2007).

4.2. Loop Prominences Now

The MHD modeling of "eruptive flares" – including those with CMEs, has become a major activity with the recognition of the importance of "space weather" and the need to anticipate or predict the occurrence of extreme events capable of major damage to human society (Baker *et al.*, 2013). The sketch of data and the ensuing cartoon description (Figure 3) have evolved into a major activity involving many research workers and participating agencies, including commercial users of space-weather data and models.

As the solar aftermath of an eruption (although not uniquely so, as noted above), a set of flare loops usually provides an excellent guide to the heliographic region underlying a CME event. Typically, powerful events have associations with active regions, but sometimes they do not – filament eruptions in quiet regions (*disparitions brusques*) may have effects strongly resembling those of active-region flares (*e.g.* Harvey, Sheeley, and Harvey, 1986). Such eruptions may involve coronal footprints much larger than those of active-region eruptive flares (*e.g.* Hudson, Haisch, and Strong, 1995), even if the ensuing CME may occupy a similar fraction of the heliosphere. These events have a shadowy research history since they do not show up readily in GOES soft X-rays because they never achieve high temperatures, and overlap in parameter space with the few cases of "stealth CMEs" that have virtually no detectable low-coronal counterpart (*e.g.* Robbrecht, Patsourakos, and Vourlidas, 2009).

The loop prominence systems, in modern data, follow the pattern that led to the development of the standard reconnection models. A large two-ribbon flare consists of an orchestrated array of individual loops; each (according to the accepted picture) forms via large-scale reconnection. The energy newly deposited in an elemental newly formed loop renders it under-dense and over-pressured in terms of hydrostatic equilibrium, so that plasma flows from the chromosphere below must try to restore this equilibrium as the loop cools. Eventually, the temperature drops to the point of Field's thermal instability (Goldsmith, 1971), condensation ensues, and "coronal rain" results when Hα and other chromospheric emissions suddenly appear out of nowhere and slide down along the field structure. Recently, Martínez Oliveros *et al.* (2014) discovered that the *Helioseismic and Magnetic Imager* on SDO can readily detect major arcade systems in the continuum near the magnetic line 6173 Å. These observations have high spectroscopic resolution (but only six points over a narrow band) and full polarization capability, exploited by Saint-Hilaire *et al.* (2014) to show that an appreciable component of the continuum emission at this wavelength is in fact due to electron scattering. The observations also reveal the presence of post-flare coronal rain, as expected. This serendipitous capability, plus the wealth of coronal information from AIA, should allow future research to understand the energetics of these arcade structures more precisely. Section 6 contains some related information.

5. The Flare Build-up Study

5.1. Flare Build-up Then

The Flare Build-up Study (FBS, organized by Zdeněk and Paul Simon) began with a workshop in 1975 and continued until 1986 (Gaizauskas and Švestka, 1987). Its premise was the need for a proof of the concept that flare energy could be stored magnetically in the solar atmosphere, and to find out how it could be released. This large-scale organization involved about a hundred scientists and dozens of participating observatories, with several workshops during the course of the activity; important programs involved included the *Solar Maximum*

Mission spacecraft among others, the *Very Large Array* and other radio facilities, the Kitt Peak Observatory among several ground-based observatories, and the initial coordinated observations focused on NOAA Active Region 5022 during western-hemisphere daylight hours in June, 1982.

Indeed, much data resulted and provided material avidly discussed during the workshops that ensued. Gaizauskas (1989) provided a general and still-current description of preflare activity based upon these results, for example. The topics addressed sound familiar now, three decades later: flux emergence, shear, magnetic complexity, flare homology, magnetic reconnection, large-scale photospheric magnetic patterns, not to mention flux cancellation at the small scales. At this time, the initial difficult observations of the vector magnetic field were just appearing (Hagyard *et al.*, 1982; see also Severny, 1964), and there were no systematic soft X-ray or EUV coronal images with which to define the coronal geometry in adequate detail.

5.2. Flare Build-up Now

In retrospect, the Flare Build-up Study did not provide decisive quantitative answers to the key questions: i) Does the coronal field uniquely store the flare energy before its release, ii) can we establish a "relaxation oscillator" (loading/unloading) behavior that directly links energy input and release, and iii) what are the basic instabilities of the structure that create the flare? The latter two of these questions (not exactly as asked by the FBS) still remain open to this day, but we can recognize clear signs of progress. As Zdeněk and the organizers of these early campaigns realized, the data of that day were incomplete – but were they inadequate? Perhaps not, because flares inevitably involve large scales, even if the unobservable microphysics does not. So in my view, the main problem was, and remains, theoretical understanding. Things have improved enormously, but still with no convergence on the answers to questions of this sort. In the following I comment on each of these items in turn:

i) We have seen vast improvement in our ability to characterize the coronal magnetic field based on steadily improving photospheric observations of the vector field, including much sharper ways of assessing the free energy available for release in a flare. Not only do these extrapolations reveal the basic requirement, namely the storage of sufficient energy for a major flare, they also in some sense localize it (De Rosa *et al.*, 2009; Wiegelmann and Sakurai, 2012, *e.g.*). In many cases now, these methods actually show a time-series decrease in the total coronal magnetic energy of an active region at the time of a major flare (*e.g.*, Sun *et al.*, 2012).

ii) In a relaxation oscillator, a steady input of energy into a system that can become unstable results in characteristic forms of time structure. If the trigger mechanism of the instability is itself well-determined, as in an electronic circuit, then the time structure is periodic. If not, then there is a proportionality between the energy release of the instability and the time interval before, or after (*e.g.* Hudson *et al.*, 1998), the trigger action. If the energy input is not steady, the pattern becomes correspondingly confused. Many natural systems show this kind of behavior (*e.g.* Lewin *et al.*, 1976; Middleditch *et al.*, 2006) – perhaps even the human heartbeat!

The Flare Build-up Study led to searches for this kind of effect (Gaizauskas, 1982), making use of sequences of homologous flares where one should have the best opportunity to detect a pattern, on the plausible assumptions that flux emergence has some persistence and that the homology of the resulting flares indicates a consistent trigger threshold. This study did not provide convincing evidence of the expected pattern, but the several assumptions required offer possible explanations. Most obviously, the lack of a correlation could reflect the irregularity of the input of free energy from the photosphere into the corona, closely related to possibly complicated pattern of emerging flux; how persistent is it? In addition, we still do not understand the physical mechanism of the flare instability, nor understand how it may be triggered. Nevertheless, the detection of a significant correlation would confirm the simplest picture of the process.

iii) The identification of which kind of instability can produce a flare remains contentious. We have seen one major development since 1986, in that the notion of "shear" has turned into a substantial effort to understand the description of *helicity* in the magnetized coronal plasma and its role in defining the accessible stored energy, and in linking flares with global field restructurings such as, filament eruptions, and CMEs (*e.g.* Rust, 1994; Low, 2001).

6. Giant Arches

6.1. Giant Arches Then

The loop prominence systems identified with CMEs had suggested the CSHKP model of magnetic reconnection, and the data from *Skylab* (1973 – 1974) and the *Solar Maximum Mission* (SMM; 1980 – 1989) really solidified these ideas. Zdeněk participated in data analyses from both of these missions, and his ideas about "giant arches" began with an analysis of a flare that was well observed by SMM and radio observatories (Švestka *et al.*, 1982): SOL1980-05-21 (X1), the same event from which Hoyng *et al.* (1981) observed the relationship between hard X-ray footpoint sources and flare ribbons. Both of these observational results derived from hard X-ray observations by the HXIS instrument onboard SMM, a remarkable hard X-ray telescope (van Beek *et al.*, 1980) based on innovative technology never again utilized. The hard X-ray imaging results from this instrument, both in the lower solar atmosphere and high up in the corona, anticipated many of the results now associated with the *Yohkoh* and RHESSI hard X-ray imaging.

The description "arch" obviously leaps to mind from Figure 4, from Švestka *et al.* (1982). The paper argued convincingly that this arch could be distinguished from the post-flare arcade and had a separate dynamical development in the outermost part of the active region. The structure seen in X-rays and in meter-wavelength images from Culgoora (right panel) remained static, rather than growing with time as expected for reconnection products. This structure could thus not readily be understood in terms of the Kopp and Pneuman (1976) picture. This paper attracted some interest, but it would be fair to say that much of the community remained unconvinced that anything beyond the Kopp and Pneuman model would be needed.

Yohkoh arrived, carrying the first CCD-based soft X-ray telescope with good resolution in space and good sampling in time. We immediately could get a better view of what SMM had glimpsed, and Zdeněk eagerly wrote a series of papers on the flare arcades and giant arches (Švestka *et al.*, 1995, 1996, 1997; Fárník *et al.*, 1996), specifically showing that the new and better soft X-ray imaging confirmed the existence of the somewhat shadowy "giant arches." In the final paper of this series we see the suggestion that large-scale reconnection was occurring on two scales. The event analyzed in the paper, SOL1992-10-28T10:21 (C3.1), illustrated this well. The discussion suggested that the reconnection leading to the re-formation of the coronal streamer could be identified with the giant arches, and that the process leading to the flare loops proceeded independently, even though both sets of phenomena appeared in large-scale two-ribbon flare events. Kopp and Poletto (1990) provided a

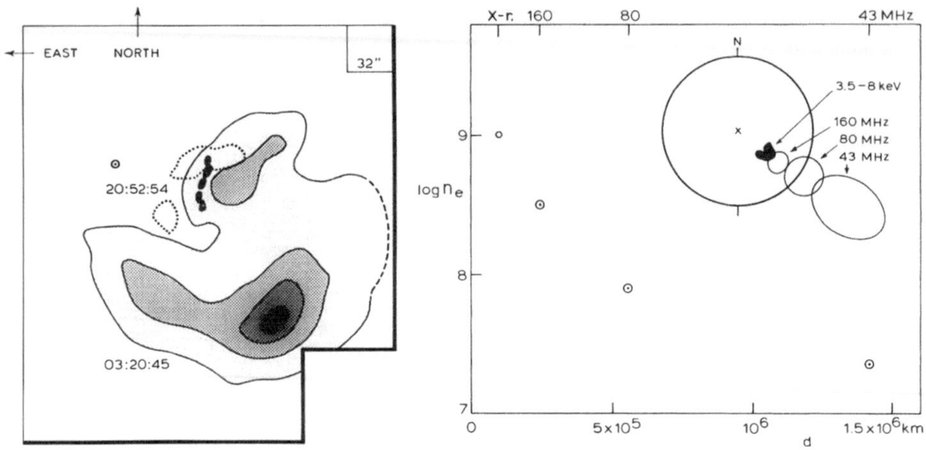

Figure 4 Left: the long-lasting coronal arch structure observed by HXIS as a late development of SOL1980-05-21 (energy 3.5 – 5.5 keV). The odd shape of the field of view is a part of the design of this instrument; the box at upper right shows its 32″ pixel size for this observation. Right: observations of the same event at several frequencies from the Culgoora radio telescope. The points (axis to left) represent density estimates for the sources seen at the three Culgoora wavelengths and in HXIS soft X-rays. Figure adapted from Švestka *et al.* (1982).

full discussion of this distinction and what it implied within the framework of the HXIS-era data (but see also West and Seaton, 2015).

A final paper in Zdeněk's series of papers entitled "Large-scale active coronal phenomena in *Yohkoh* SXT images" series dealt with the hot fan-like structure of the late phase of the limb event SOL1992-08-28 (C1.5), as illustrated in Figure 5 (Švestka *et al.*, 1998).

Events of this kind contain the dynamical phenomena we term "supra-arcade downflows", McKenzie and Hudson (1999). These could only have been discovered via *Yohkoh*/SXT observations by a revised pattern of exposures not implemented until 1998, but their discovery dramatically confirmed the presence of dynamics closely resembling that expected of a global reconnection model such as that of Kopp and Pneuman.

The soft X-ray images of the Sun also revealed large loop structures interconnecting active regions, including the interesting trans-equatorial loops. Because of the large scale of these structures and their association with flare or CME activity, they had interest in the context of the giant arches (Švestka *et al.*, 1995; Fárník, Karlický, and Švestka, 1999).

6.2. Giant Arches Now

Zdeněk's active period of research came to an end just a decade or so too soon; the voluminous high-resolution data from the *Solar Dynamics Observatory* has clarified much of what he inferred from much inferior data. We note that these data have wavelength and temperature sensitivities that are different from the soft X-ray observations of SMM/HXIS, *Yohkoh*/SXT, or *Hinode*/XRT, but they have wonderful coverage, angular resolution, and continuity. We note two important developments that he just missed: the "supra-thermal arcade downflows" (SADs; McKenzie and Hudson, 1999) and the "EUV late phase" with its associated morphology (Woods *et al.*, 2011). Both of these concern our interpretation of the giant arches.

Figure 5 SOL1992-08-28 (C1.5), a beautiful flare event at the limb that displays the hot fan-like structure studied in Švestka *et al.* (1998). The four panels show soft X-ray emission at various stages of the event, as described in the text. The hot, spiky fan structure in the late image (02:41 UT, some five hours after the initiation of the event) is remarkable (adapted from Švestka *et al.*, 1998).

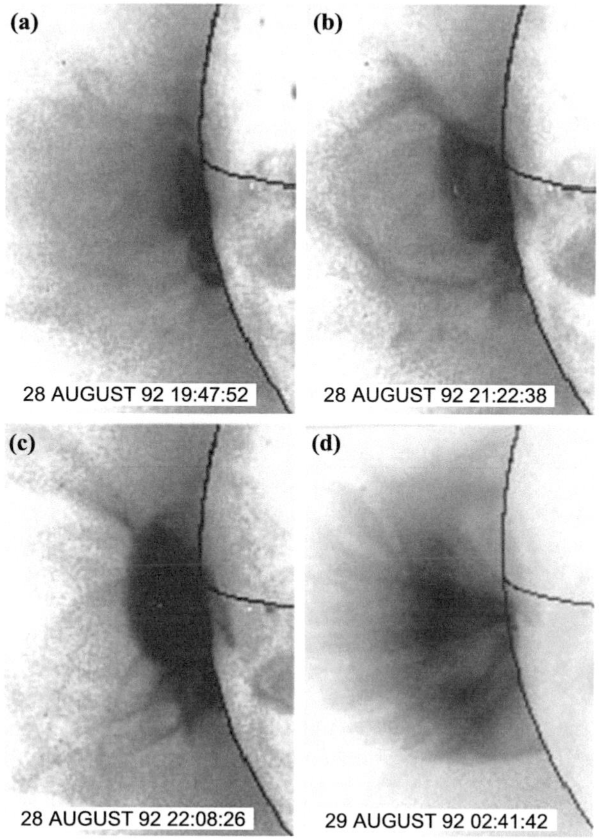

6.2.1. SADs

Exactly the spiky fan structures seen in soft X-rays (Figure 5) turned out to support a characteristic pattern of downflows in the form of dark blobs seen in soft X-ray images (McKenzie and Hudson, 1999). These structures occupy the region where a reconnection outflow jet would appear in the standard model, but have sub-Alfvénic speeds and a gradual deceleration, features not predicted either in the cartoon or in related MHD simulations of large-scale reconnection processes in solar flares (*e.g.* Shibata and Magara, 2011); such models also predict that the reconnection outflow would terminate in a fast-mode MHD shock, which is not observed. Thus the SADs and the spiky arcade tops described by Švestka *et al.* (1998) simultaneously offer striking evidence of large-scale magnetic reconnection upon very complicated separatrix structures, but also have not yet been convincingly modeled.

The SADs can originate in the corona well above where the original soft X-ray observations lost sight of them; Sheeley, Warren, and Wang (2004) could track the flows out into the domain of LASCO. Their presence in the high corona might make them a candidate for the modern identification of the phenomena that Zdeněk called "giant arches" in some sense, but current thinking associates them with the main flare eruption and the restructuring following it: the flare loops. Recent work on SADs suggests that the whole concept of coronal magnetic reconnection at low plasma beta may need revision (*e.g.* Shibasaki, 2001; McKenzie, 2013). This radical change in our thinking about the microphysics may fit comfortably enough within our well-developed observational understanding of the magnetic restructuring involved in a major eruptive flare; in other words, Zdeněk may not have been too surprised by this interesting possible revision.

6.2.2. EUV Late Phase

The "EUV late phase" phenomenon (Woods *et al.*, 2011) appeared first in the Sun-as-a-star spectroscopy from the EVE instrument on SDO. These data, although simple EUV spectra, are sensitive and precise. Furthermore, the AIA instrument on SDO allows us to identify the spectral variations with image features (Woods, 2014), which include eruptive signatures as detected directly in image motions as well as via coronal dimmings (*e.g.* Hudson and Cliver, 2001). These data showed clearly that in addition to the flare disturbance as seen by GOES at high temperatures (above Fe XX, say), sometimes other disturbances followed as a part of the late development of the overall event. These later disturbances typically appear most prominently at Fe XVI – elevated temperatures, but not as high as in the impulsive or gradual phases previously known. They involve large-scale loop structures not directly associated with the flare gradual phase, and Liu *et al.* (2013) noted that a CME may or may not accompany an EUV late-phase event. As a speculation, we suggest the identification of the EUV late-phase events with Svestka's giant arches.

7. Conclusion

I have reviewed some of Zdeněk Švestka's most interesting research activities, in some of which I was fortunate enough to help out. It is quite remarkable how much progress has ensued in the recent few years, and I have tried to point out where Zdeněk's discoveries and insights may fit into the new context. Of course, during his day the data were hugely less competent, and we could only get glimpses of phenomena now laid out for us in striking detail – if sometimes still perplexing. It is most unfortunate that Zdeněk did not see some of these developments, especially those which bore out some of his favorite ideas.

The optical spectroscopy of solar flares, especially white-light flares, remains as a lasting legacy of Ondřejov observatory, and Zdeněk in particular. In 1859 this phenomenon opened the door for many kinds of research, including the current major developments in "space weather." We should remember that this original flare predated both Heaviside and Röntgen, whose discoveries later in the 19th century eventually led to an understanding of direct solar flare effects. Zdeněk's research work began almost halfway in between then and now.

Acknowledgements This work was supported by NASA under grant NNX11AP05G for studies of white-light flares and by contract No. NAS 5-98033 for RHESSI at UC Berkeley. I would like to thank the University of Glasgow for hospitality during the preparation of this paper, and several friends for comments on the first draft of the paper: Franta Fárník, Vic Gaizauskas, and Giannina Poletto.

References

Baker, D.N., Li, X., Pulkkinen, A., Ngwira, C.M., Mays, M.L., Galvin, A.B., Simunac, K.D.C.: 2013, A major solar eruptive event in July 2012: Defining extreme space weather scenarios. *Space Weather* **11**, 585. DOI. ADS.

Berlicki, A.: 2007, Observations and modeling of line asymmetries in chromospheric flares. In: Heinzel, P., Dorotovič, I., Rutten, R.J. (eds.) *The Physics of Chromospheric Plasmas*, *Astronomical Society of the Pacific Conference Series* **368**, 387. ADS.

Biermann, L., Haxel, O., Schlüter, A.: 1951, Neutrale Ultrastrahlung von der Sonne. *Z. Naturforsch. Teil A* **6**, 47. ADS.

Brown, J.C.: 1971, The deduction of energy spectra of non-thermal electrons in flares from the observed dynamic spectra of hard X-ray bursts. *Solar Phys.* **18**, 489. DOI. ADS.

Bruzek, A.: 1964, On the association between loop prominences and flares. *Astrophys. J.* **140**, 746. DOI. ADS.

Carrington, R.C.: 1859, Description of a singular appearance seen in the Sun on September 1, 1859. *Mon. Not. Roy. Astron. Soc.* **20**, 13.

Chapman, S., Bartels, J.: 1940, *Geomagnetism*, Oxford University Press, London.

Chupp, E.L., Forrest, D.J., Ryan, J.M., Heslin, J., Reppin, C., Pinkau, K., Kanbach, G., Rieger, E., Share, G.H.: 1982, A direct observation of solar neutrons following the 0118 UT flare on 1980 June 21. *Astrophys. J. Lett.* **263**, L95. DOI. ADS.

De Rosa, M.L., Schrijver, C.J., Barnes, G., Leka, K.D., Lites, B.W., Aschwanden, M.J., Amari, T., Canou, A., McTiernan, J.M., Régnier, S., Thalmann, J.K., Valori, G., Wheatland, M.S., Wiegelmann, T., Cheung, M.C.M., Conlon, P.A., Fuhrmann, M., Inhester, B., Tadesse, T.: 2009, A critical assessment of nonlinear force-free field modeling of the solar corona for active region 10953. *Astrophys. J.* **696**, 1780. DOI. ADS.

Fárník, F., Karlický, M., Švestka, Z.: 1999, Long transequatorial interconnecting loops of the new solar cycle. *Solar Phys.* **187**, 33. DOI. ADS.

Fárník, F., Švestka, Z., Hudson, H.S., Uchida, Y.: 1996, Large-scale active coronal phenomena in Yohkoh SXT images. *Solar Phys.* **168**, 331. DOI. ADS.

Fletcher, L., Dennis, B.R., Hudson, H.S., Krucker, S., Phillips, K., Veronig, A., Battaglia, M., Bone, L., Caspi, A., Chen, Q., Gallagher, P., Grigis, P.T., Ji, H., Liu, W., Milligan, R.O., Temmer, M.: 2011, An observational overview of solar flares. *Space Sci. Rev.* **159**, 19. DOI. ADS.

Gaizauskas, V.: 1982, The relation of solar flares to the evolution and proper motions of magnetic fields. *Adv. Space Res.* **2**, 11. DOI. ADS.

Gaizauskas, V.: 1989, Preflare activity. *Solar Phys.* **121**, 135. DOI. ADS.

Gaizauskas, V., Švestka, Z.: 1987, Highlights of the flare build-up study. *Solar Phys.* **114**, 389. DOI. ADS.

Gaizauskas, V., Mandrini, C.H., Demoulin, P., Luoni, M.L., Rovira, M.G.: 1998, Interactions between nested sunspots. II. A confined X1 flare in a delta-type sunspot. *Astron. Astrophys.* **332**, 353. ADS.

Goldsmith, D.W.: 1971, Thermal effects in the formation of loop prominences. *Solar Phys.* **19**, 86. DOI. ADS.

Hagyard, M.J., Cumings, N.P., West, E.A., Smith, J.E.: 1982, The MSFC vector magnetograph. *Solar Phys.* **80**, 33. DOI. ADS.

Harvey, K., Sheeley, N. Jr., Harvey, J.: 1986, HE I 10830 a observations of two-ribbon flare-like events associated with filament disappearances. In: Simon, P.A., Heckman, G., Shea, M.A. (eds.) *Solar-Terrestrial Predictions*, 198. ADS.

Hiei, E., Okamoto, T., Tanaka, K.: 1983, Observation of the flare of 12 June 1982 by Norikura coronagraph and HINOTORI. *Solar Phys.* **86**, 185. DOI. ADS.

Hodgson, R.: 1859, On a curious appearance seen in the Sun. *Mon. Not. Roy. Astron. Soc.* **20**, 15.

Hoyng, P., Duijveman, A., Machado, M.E., Rust, D.M., Švestka, Z., Boelee, A., de Jager, C., Frost, K.T., Lafleur, H., Simnett, G.M., van Beek, H.F., Woodgate, B.E.: 1981, Origin and location of the hard X-ray emission in a two-ribbon flare. *Astrophys. J. Lett.* **246**, L155. DOI. ADS.

Hudson, H.S.: 1972, Thick-target processes and white-light flares. *Solar Phys.* **24**, 414. DOI. ADS.

Hudson, H.S.: 2007, Chromospheric flares. In: Heinzel, P., Dorotovič, I., Rutten, R.J. (eds.) *The Physics of Chromospheric Plasmas, Astronomical Society of the Pacific Conference Series* **368**, 365. ADS.

Hudson, H.S., Cliver, E.W.: 2001, Observing coronal mass ejections without coronagraphs. *J. Geophys. Res.* **106**, 25199. DOI. ADS.

Hudson, H., Haisch, B., Strong, K.T.: 1995, Comment on 'The solar flare myth' by J.T. Gosling. *J. Geophys. Res.* **100**, 3473. DOI. ADS.

Hudson, H.S., Labonte, B.J., Sterling, A.C., Watanabe, T.: 1998, NOAA 7978: The last best old-cycle region. In: Watanabe, T., Kosugi, T. (eds.) *Observational Plasma Astrophysics: Five Years of YOHKOH and Beyond, Astrophysics and Space Science Library* **229**, 237. ADS.

Johns-Krull, C.M., Hawley, S.L., Basri, G., Valenti, J.A.: 1997, Hamilton echelle spectroscopy of the 1993 March 6 solar flare. *Astrophys. J. Suppl.* **112**, 221. DOI. ADS.

Kane, S.R., Donnelly, R.F.: 1971, Impulsive hard X-ray and ultraviolet emission during solar flares. *Astrophys. J.* **164**, 151. DOI. ADS.

Kopp, R.A., Pneuman, G.W.: 1976, Magnetic reconnection in the corona and the loop prominence phenomenon. *Solar Phys.* **50**, 85. DOI. ADS.

Kopp, R.A., Poletto, G.: 1990, Formation and cooling of the giant HXIS arches of November 6–7, 1980. *Solar Phys.* **127**, 267. DOI. ADS.

Kosovichev, A.G., Zharkova, V.V.: 1998, X-Ray flare sparks quake inside Sun. *Nature* **393**, 317. DOI. ADS.

Lewin, W.H.G., Doty, J., Clark, G.W., Rappaport, S.A., Bradt, H.V.D., Doxsey, R., Hearn, D.R., Hoffman, J.A., Jernigan, J.G., Li, F.K., Mayer, W., McClintock, J., Primini, F., Richardson, J.: 1976, The discovery of rapidly repetitive X-ray bursts from a new source in Scorpius. *Astrophys. J. Lett.* **207**, L95. DOI. ADS.
Lingenfelter, R.E., Flamm, E.J., Canfield, E.H., Kellman, S.: 1965, High-energy solar neutrons, 1, production in flares. *J. Geophys. Res.* **70**, 4077. DOI. ADS.
Liu, K., Zhang, J., Wang, Y., Cheng, X.: 2013, On the origin of the extreme-ultraviolet late phase of solar flares. *Astrophys. J.* **768**, 150. DOI. ADS.
Low, B.C.: 2001, Coronal mass ejections, magnetic flux ropes, and solar magnetism. *J. Geophys. Res.* **106**, 25141. DOI. ADS.
Martin, S.F., Ramsey, H.E., Carroll, G.A., Martin, D.C.: 1974, Multi-slit spectrograph and H alpha Doppler system. *Solar Phys.* **37**, 343. DOI. ADS.
Martínez Oliveros, J.-C., Hudson, H.S., Hurford, G.J., Krucker, S., Lin, R.P., Lindsey, C., Couvidat, S., Schou, J., Thompson, W.T.: 2012, The height of a white-light flare and its hard X-ray sources. *Astrophys. J. Lett.* **753**, L26. DOI. ADS.
Martínez Oliveros, J.-C., Krucker, S., Hudson, H.S., Saint-Hilaire, P., Bain, H., Lindsey, C., Bogart, R., Couvidat, S., Scherrer, P., Schou, J.: 2014, Chromospheric and coronal observations of solar flares with the helioseismic and magnetic imager. *Astrophys. J. Lett.* **780**, L28. DOI. ADS.
McKenzie, D.E.: 2013, Turbulent dynamics in solar flare sheet structures measured with local correlation tracking. *Astrophys. J.* **766**, 39. DOI. ADS.
McKenzie, D.E., Hudson, H.S.: 1999, X-Ray observations of motions and structure above a solar flare arcade. *Astrophys. J. Lett.* **519**, L93. DOI. ADS.
Meyer, P., Parker, E.N., Simpson, J.A.: 1956, Solar cosmic rays of February, 1956 and their propagation through interplanetary space. *Phys. Rev.* **104**, 768. DOI. ADS.
Middleditch, J., Marshall, F.E., Wang, Q.D., Gotthelf, E.V., Zhang, W.: 2006, Predicting the starquakes in PSR J0537-6910. *Astrophys. J.* **652**, 1531. DOI. ADS.
Reames, D.V.: 2013, The two sources of solar energetic particles. *Space Sci. Rev.* **175**, 53. DOI. ADS.
Robbrecht, E., Patsourakos, S., Vourlidas, A.: 2009, No trace left behind: STEREO observation of a coronal mass ejection without low coronal signatures. *Astrophys. J.* **701**, 283. DOI. ADS.
Rust, D.M.: 1994, Spawning and shedding helical magnetic fields in the solar atmosphere. *Geophys. Res. Lett.* **21**, 241. DOI. ADS.
Saint-Hilaire, P., Schou, J., Martínez Oliveros, J.-C., Hudson, H.S., Krucker, S., Bain, H., Couvidat, S.: 2014, Observations of linear polarization in a solar coronal loop prominence system observed near 6173 Å. *Astrophys. J. Lett.* **786**, L19. DOI. ADS.
Severny, A.B.: 1964, Solar flares. *Annu. Rev. Astron. Astrophys.* **2**, 363. DOI. ADS.
Sheeley, N.R. Jr., Warren, H.P., Wang, Y.-M.: 2004, The origin of postflare loops. *Astrophys. J.* **616**, 1224. DOI. ADS.
Shibasaki, K.: 2001, High-beta disruption in the solar atmosphere. *Astrophys. J.* **557**, 326. DOI. ADS.
Shibata, K., Magara, T.: 2011, Solar flares: Magnetohydrodynamic processes. *Living Rev. Solar Phys.* **8**, 6. DOI. ADS.
Suemoto, Z., Hiei, E.: 1959, Balmer series lines of the flare and its structure. *Proc. Astr. Soc. Japan* **11**, 185. ADS.
Sun, X., Hoeksema, J.T., Liu, Y., Wiegelmann, T., Hayashi, K., Chen, Q., Thalmann, J.: 2012, Evolution of magnetic field and energy in a major eruptive active region based on SDO/HMI observation. *Astrophys. J.* **748**, 77. DOI. ADS.
Švestka, Z.: 1949, Observations of solar flares and high-speed prominences from July to December 1948. *Bull. Astron. Inst. Czechoslov.* **1**, 73. ADS.
Švestka, Z.: 1966, Optical observations of solar flares. *Space Sci. Rev.* **5**, 388. DOI. ADS.
Švestka, Z.: 1970, The phase of particle acceleration in the flare development. *Solar Phys.* **13**, 471. DOI. ADS.
Švestka, Z.: 1971, On the flux of neutrons from flares. *Solar Phys.* **19**, 202. DOI. ADS.
Švestka, Z.: 1972, Spectra of solar flares. *Annu. Rev. Astron. Astrophys.* **10**, 1. DOI. ADS.
Švestka, Z.: 1976, *Solar Flares*. ADS.
Švestka, Z.: 1983, Post-flare coronal arches. *Space Sci. Rev.* **35**, 259. DOI. ADS.
Švestka, Z.: 2007, The Misnomer of "Post-Flare Loops". *Solar Phys.* **246**, 393. DOI. ADS.
Švestka, Z.: 2010, Sixty years in solar physics. *Solar Phys.* **267**, 235. DOI. ADS.
Švestka, Z., Simon, P.: 1969, Proton flare project, 1966. Summary of the August/September particle events in the McMath region 8461 (Invited review paper). *Solar Phys.* **10**, 3. DOI. ADS.
Švestka, Z., Hoyng, P., van Tend, W., Boelee, A., de Jager, C., Stewart, R.T., Acton, L.W., Bruner, E.C., Gabriel, A.H., Rapley, C.G., de Jager, C., LaFleur, H., Nelson, G., Simnett, G.M., van Beek, H.F., Wagner, W.J.: 1982, Observations of a post-flare radio burst in X-rays. *Solar Phys.* **75**, 305. DOI. ADS.

Švestka, Z., Fárník, F., Hudson, H.S., Uchida, Y., Hick, P., Lemen, J.R.: 1995, Large-scale active coronal phenomena in YOHKOH SXT images, I. *Solar Phys.* **161**, 331. DOI. ADS.

Švestka, Z., Farnik, F., Hudson, H.S., Uchida, Y., Hick, P., Lemen, J.R.: 1996, Large-scale active coronal phenomena in YOHKOH SXT images. In: Uchida, Y., Kosugi, T., Hudson, H.S. (eds.) *IAU Colloq. 153: Magnetodynamic Phenomena in the Solar Atmosphere – Prototypes of Stellar Magnetic Activity*, 609. ADS.

Švestka, Z., Farnik, F., Hick, P., Hudson, H.S., Uchida, Y.: 1997, Large-scale active coronal phenomena in YOHKOH SXT images. III. Enhanced post-flare streamer. *Solar Phys.* **176**, 355. DOI. ADS.

Švestka, Z., Fárník, F., Hudson, H.S., Hick, P.: 1998, Large-scale active coronal phenomena in Yohkoh SXT images IV. Solar wind streams from flaring active regions. *Solar Phys.* **182**, 179. DOI. ADS.

van Beek, H.F., Hoyng, P., Lafleur, B., Simnett, G.M.: 1980, The hard X-ray imaging Spectrometer/HXIS/. *Solar Phys.* **65**, 39. DOI. ADS.

Webb, D.F., Cliver, E.W., Gopalswamy, N., Hudson, H.S., St. Cyr, O.C.: 1998, The solar origin of the January 1997 coronal mass ejection, magnetic cloud and geomagnetic storm. *Geophys. Res. Lett.* **25**, 2469. DOI. ADS.

West, M.J., Seaton, D.B.: 2015, SWAP observations of post-flare giant arches in the long-duration 14 October 2014 solar eruption. *Astrophys. J. Lett.* **801**, L6. DOI. ADS.

Wiegelmann, T., Sakurai, T.: 2012, Solar force-free magnetic fields. *Living Rev. Solar Phys.* **9**, 5. DOI. ADS.

Woods, T.N.: 2014, Extreme ultraviolet late-phase flares: Before and during the solar dynamics observatory mission. *Solar Phys.* **289**, 3391. DOI. ADS.

Woods, T.N., Hock, R., Eparvier, F., Jones, A.R., Chamberlin, P.C., Klimchuk, J.A., Didkovsky, L., Judge, D., Mariska, J., Warren, H., Schrijver, C.J., Webb, D.F., Bailey, S., Tobiska, W.K.: 2011, New solar extreme-ultraviolet irradiance observations during flares. *Astrophys. J.* **739**, 59. DOI. ADS.

Wülser, J.-P.: 1987, H-alpha line profile observations of solar flares with high temporal resolution. *Solar Phys.* **114**, 115. ADS.

Xu, Y., Cao, W., Liu, C., Yang, G., Qiu, J., Jing, J., Denker, C., Wang, H.: 2004, Near-infrared observations at 1.56 microns of the 2003 October 29 X10 white-light flare. *Astrophys. J. Lett.* **607**, L131. DOI. ADS.

Reprinted from *Solar Physics* Journal, DOI 10.1007/s11207-015-0748-2

Extreme Ultra-Violet Spectroscopy of the Lower Solar Atmosphere During Solar Flares (Invited Review)

Ryan O. Milligan[1,2,3]

Received: 31 January 2015 / Accepted: 23 July 2015 / Published online: 18 August 2015
© Springer Science+Business Media Dordrecht 2015

Abstract The extreme ultra-violet (EUV) portion of the solar spectrum contains a wealth of diagnostic tools for probing the lower solar atmosphere in response to an injection of energy, particularly during the impulsive phase of solar flares. These include temperature- and density-sensitive line ratios, Doppler-shifted emission lines, nonthermal broadening, abundance measurements, differential emission measure profiles, continuum temperatures and energetics, among others. In this article I review some of the recent advances that have been made using these techniques to infer physical properties of heated plasma at footpoint and ribbon locations during the initial stages of solar flares. I primarily focus on studies that have utilised spectroscopic EUV data from *Hinode/EUV Imaging Spectrometer* (EIS) and *Solar Dynamics Observatory/EUV Variability Experiment* (SDO/EVE), and I also provide some historical background and a summary of future spectroscopic instrumentation.

Keywords Chromosphere, active · Flares, dynamics · Flares, impulsive phase · Flares, spectrum · Heating, chromospheric · Heating, in flares · Spectral line, broadening · Spectral line, intensity and diagnostics · Spectrum, continuum · Spectrum, ultraviolet

1. Introduction

The spectroscopy of extreme ultra-violet (EUV) emission is a crucial diagnostic tool for determining the composition and dynamics of the flaring solar atmosphere in response to

Solar and Stellar Flares: Observations, Simulations, and Synergies
Guest Editors: Lyndsay Fletcher and Petr Heinzel

✉ R.O. Milligan
r.milligan@qub.ac.uk

1 Astrophysics Research Centre, School of Mathematics and Physics, Queen's University Belfast, University Road, Belfast, BT7 1NN, Northern Ireland, UK

2 Solar Physics Laboratory, Heliophysics Division, NASA Goddard Space Flight Center, Greenbelt, MD 20771, USA

3 Department of Physics, Catholic University of America, 620 Michigan Avenue, Northeast, Washington, DC 20064, USA

an injection of energy. It is well known that the chromosphere[1] is the dominant source of radiated energy during a solar flare (Neidig, 1989; Hudson *et al.*, 1992; Hudson, Wolfson, and Metcalf, 2006; Kretzschmar *et al.*, 2010; Kretzschmar, 2011; Woods *et al.*, 2004; Woods, Kopp, and Chamberlin, 2006), as well as being the origin of material that is ultimately observed in the corona through the process of chromospheric evaporation (Neupert, 1968). But how the energy initially stored in the coronal magnetic field is transported to the chromosphere, where it drives increased emission across a range of wavelengths, remains poorly understood. It is commonly assumed that the delivery mechanism is a 'beam' of accelerated (nonthermal) electrons since enhanced chromospheric (and occasionally photospheric) emission is temporally and/or spatially correlated with hard X-ray (HXR) emission due to thick-target bremsstrahlung. However, other processes such as thermal conduction fronts, ion beams, Alfvén waves, and return currents may also play a role. While imaging instruments provide important context information on the morphology and structure of solar features, the images themselves are usually broadband, comprising several different ion species that can bias the interpretation of the observations. Spectroscopy offers the advantage of providing quantifiable measurements of parameters such as temperature, density, and velocity that can then be compared with predictions from theoretical models.

The EUV portion of the solar spectrum spans the 100 – 1200 Å wavelength range (Tobiska and Bouwer, 2006), although some in the aeronomy discipline use 300 Å as the lower limit due to instrumental factors (Woods *et al.*, 2005), while others also include the prominent Lyman-alpha (Lyα) line at 1216 Å (Lilensten *et al.*, 2008), the strongest line in the solar spectrum (see Section 4 for a discussion on Lyα flare observations). The most prominent chromospheric features in the EUV range are the free–bound continua of H I and He I, and to a lesser extent, He II, with recombination edges at 912 Å, 504 Å, and 228 Å, respectively, as well as the resonance line of He II at 304 Å. During large flares free–free (thermal bremsstrahlung) continuum emission from the corona can also encroach into the EUV spectral range (Milligan *et al.*, 2012; Milligan and McElroy, 2013). However, the EUV range also contains myriad emission lines with intrinsic formation temperatures from 10^4 K, which are characteristically chromospheric, up to a few times 10^7 K that are only found during flares or in the cores of large active regions. It is when this inherently coronal emission of perhaps a few megaKelvin is observed to be spatially and/or temporally correlated with enhanced chromospheric emission that we are witnessing the direct manifestation of flare heating. Analysis of these lines and continua can therefore allow us to deduce the state of the heated plasma, although as they are predominantly optically thin, most of the deduced properties can be considered to be weighted averages. Energetically speaking, even the strongest chromospheric EUV features contain, at most, a few per cent of a flare's total energy (Emslie, Brown, and Donnelly, 1978; Milligan *et al.*, 2012, 2014), but the diagnostic information contained therein is invaluable for quantifying changes in the plasma parameters in response to injected energy.

The EUV component of radiated emission is also known to be a major energy input into the Earth's upper atmosphere and the geospace environment, heating the thermosphere and generating the ionosphere (Lean, Picone, and Emmert, 2009; Qian, Solomon, and Mlynczak, 2010). In particular, He II 304 Å photons are the most important energy contributor to overall heating in the thermosphere at all levels of solar activity, while chromospheric Lyα emission (along with soft X-rays; SXR) generates the D-layer of the ionosphere (80 – 100 km;

[1] I use the terms chromosphere, lower transition region, and lower solar atmosphere interchangeably, since we do not know how to adequately describe this region during a flare's impulsive phase. In essence I am referring to the footpoints or ribbons of a flaring loop, which are often spatially coincident with HXR emission. This may also include the photosphere in the case of white-light flares.

Tobiska *et al.*, 2000). The bulk of the Sun's EUV emission, however, is absorbed by the F-layer (> 150 km), while the E-layer (100 – 150 km) is affected by X-ray ultra-violet radiation (XUV; 1 – 100 Å). Changes in the Sun's output at EUV wavelengths, particularly during flares, can drive fluctuations in the ionospheric density, which increases drag on orbiting satellites, disrupts long-wave radio communication and compromises GPS accuracy. The EUV component of radiation emitted during the early stages of solar flares often originates in the chromosphere. The study of chromospheric EUV emission is therefore not only important for understanding flare heating, but also in determining the origins of a major component of space weather.

Despite the amazing diagnostic potential currently available to us through instruments such as the *EUV Imaging Spectrometer* (EIS; Culhane *et al.*, 2007) onboard *Hinode* (Kosugi *et al.*, 2007) and the *EUV Variability Experiment* (EVE; Woods *et al.*, 2005) on the *Solar Dynamics Observatory* (SDO; Pesnell, Thompson, and Chamberlin, 2012) – and previously from the *Coronal Diagnostic Spectrometer* (CDS; Harrison *et al.*, 1995) and *Solar Ultraviolet Measurements of Emitted Radiation* (SUMER; Wilhelm *et al.*, 1995) instruments onboard the *Solar and Heliospheric Observatory* (SOHO) – there are frustratingly few references in recent literature that focus on *bona fide* spectroscopic EUV observations of the lower solar atmosphere during flares. (See Table 1 for a list of the 18 flares successfully observed by EIS during the impulsive phase to date in the nine years since *Hinode* was launched, along with the associated publications in which they appear.) This is perhaps in part due to the difficulties associated with positioning the slit from a rastering spectrometer with a reduced field of view precisely over the footpoints or ribbons during the crucial few minutes of an impulsive phase, or detecting weak chromospheric enhancements against the strong EUV background in full-disc irradiance measurements. A reduced duty cycle is a factor for instruments such as *Thermosphere Ionosphere Mesosphere Energetics Dynamics/Solar EUV Experiment* (TIMED/SEE; Woods *et al.*, 2004) that only made measurements a few times per day, as well as spectrographs such as SERTS (*Solar EUV Rocket Telescope and Spectrograph*) and EUNIS (*EUV Normal Incidence Spectrographs*) that flew on sounding rockets and only observed the Sun for a few minutes at a time. The few references that are available, however, have gone a long way in increasing our understanding of the dynamic response of the lower solar atmosphere during flares. In doing so, these studies have also highlighted the gaps in our knowledge, and pointed to the questions that can yet be answered by spectroscopic EUV observations, particularly when used in conjunction with those from other instruments at different wavelengths, as well as theoretical modelling. This review paper aims to summarise many of the significant findings that have arisen from spectroscopic observations of flare footpoints and ribbons over the past decade or so in addition to raising awareness of the issues that remain to be addressed.

2. Historical Observations

Impulsive EUV bursts associated with a flare's impulsive phase were first discovered in the late 1960s and early 1970s using (photometric) data from the *Orbiting Solar Observatory* (OSO) III satellite (Hall and Hinteregger, 1969; Hall, 1971). These bursts were primarily observed in cool emission lines, such as He II, C II, O V, and Si III. Improved observations by the Harvard College Observatory (HCO) *EUV Spectroheliometer* on *Skylab* showed similar bursts at even higher temperatures (*e.g.* Mg X; Emslie and Noyes, 1978). Around the same time, increases in EUV emission were also being inferred from Sudden Frequency Deviations (a subset of Sudden Ionospheric Disturbances) observations by Donnelly (1969),

Table 1 Flares whose footpoints were observed by *Hinode*/EIS during the impulsive phase and the publications in which they appear.

Flare ID	GOES Class	Associated publications
SOL2007-01-16T02:22	C4.1	Watanabe *et al.* (2009), Li and Ding (2011)
SOL2007-05-22T23:26	B2.7	Del Zanna *et al.* (2011), Graham *et al.* (2013)
SOL2007-06-02T20:32	B2.7[a,b,c]	Del Zanna (2008), Warren, Feldman, and Brown (2008)
SOL2007-06-05T15:50	C6.6	Graham, Fletcher, and Hannah (2011)
SOL2007-06-06T16:55	C9.7	Watanabe *et al.* (2010)
SOL2007-06-07T00:37	B7.6[b]	Milligan (2008)
SOL2007-12-07T04:35	B1.4	Chen and Ding (2010)
SOL2007-12-14T15:22	B8.8	Graham *et al.* (2013)
SOL2007-12-14T15:54	B8.8	Graham *et al.* (2013)
SOL2007-12-14T01:39	B9.6	Graham *et al.* (2013), Graham, Fletcher, and Labrosse (2014)
SOL2007-12-14T14:16	C1.1	Milligan and Dennis (2009), Milligan (2011), Ning and Cao (2011), Graham *et al.* (2013), Graham, Fletcher, and Labrosse (2014)
SOL2007-12-16T06:23	B1.8	Graham *et al.* (2013)
SOL2011-02-13T13:44	C4.7	Li, Qiu, and Ding (2014)
SOL2011-02-16T01:32	M1.0	Li, Qiu, and Ding (2012)
SOL2011-02-16T07:35	M1.1	Young *et al.* (2013)
SOL2012-03-07T18:49	C1.4[c]	Brosius (2013a)
SOL2012-03-09T03:22	M6.3[a]	Doschek, Warren, and Young (2013)
SOL2012-11-21T18:20	C1.9[c]	Brosius (2013b)

[a]Both full CCDs were read out for these events.

[b]Footpoints were observed but the flare was not impulsive (*i.e.* there were no associated hard X-rays).

[c]Not included in the GOES Event List.

Kane and Donnelly (1971), and Donnelly and Kane (1978). While these observations were not spectroscopic *per se*, they were found to be temporally correlated with HXR emission. This suggested that the EUV emission was predominantly chromospheric in nature given that the ionosphere is sensitive to emission in the 10 – 1030 Å range, which is dominated by chromospheric free–bound and bound–bound transitions (Emslie, Brown, and Donnelly, 1978).

Observations of the Lyman continuum (LyC) of neutral hydrogen from the quiet Sun detected by OSO-IV (and later OSO-VI) were first reported by Noyes and Kalkofen (1970) and Vernazza and Noyes (1972). These data were used to determine the structure of the chromosphere, including temperature, density, and opacity, using the Eddington–Barbier relationship. Machado and Noyes (1978) later applied this same technique to a number of solar flares observed using HCO to reveal that the ground level of hydrogen is closer to local thermodynamic equilibrium during flares than in quiescent times, and that the layer at which LyC is formed during flares is lower in the atmosphere. This was verified by later studies (*e.g.* Machado *et al.*, 1980), including what appears to be the only reference in the literature of imaging a flare in the Lyman continuum (Hirayama *et al.*, 1985). This was carried out using the Japanese sounding rocket S520-5CN. These findings were also used to determine the mass of the material evaporated into the overlying loops. As LyC is formed in the upper chromosphere/lower transition region, it is more sensitive to heating from above compared to the Balmer and Paschen continua, which are formed in deeper layers. While this feature of LyC emission points to an important flare diagnostic, few other reports of continuum enhancements exist in the literature. Despite this lack of observations, Ding and Schleicher (1997) improved on the method of Machado and Noyes (1978) to include stimulation by nonthermal electrons in anticipation of future observations of these vital emissions. Similar measurements of the Lyman continua of He I (Zirin, 1975) and He II (Linsky *et al.*, 1976) were also carried out in the 1970s.

3. EUV Diagnostics

The introduction of imaging spectrometers such CDS, and its successor EIS, allowed spatially resolved line profiles to be obtained by stepping a slit across a chosen section of the solar atmosphere, building up a three-dimensional raster $(x(t), y, \lambda)$. Exposure (and read-out) times were initially quite long ($\gtrsim 30$ s), and so depending upon the chosen field-of-view, a single raster could often take as much as an hour or more to complete. EIS improved upon this with its greater sensitivity and broader temperature coverage, and rasters featuring a breadth of emission lines could be run in minutes, making it more ideal for flare studies. For example, on 14 December 2007, EIS observed a simple, isolated C1.1 flare while running a fast-raster (3.5 minute cadence) study that contained over 40 individual emission lines covering the 10^4-10^7 K temperature range, including several density-sensitive line pairs. Figure 1a shows the raster taken during the flare's impulsive phase, clearly showing the two footpoints (in Fe XV in this case) that coincided with the 20–25 keV HXR emission observed by the *Ramaty High Energy Solar Spectroscopic Imager* (RHESSI; Lin *et al.*, 2002 – orange contours). Figures 1b, c, and d show the corresponding Doppler velocity, nonthermal velocity, and density maps (from Fe XIV line ratios), respectively. The data obtained from this raster showed that material at the footpoints of this flare was hot, evaporating, turbulent, and dense, and EIS is able to do this for multiple emission lines simultaneously. However, by January 2008, *Hinode* had lost the use of its X-band transmitter, and so the available telemetry was greatly reduced. The findings that resulted from this and other, similar studies are discussed in Sections 3.1, 3.2, 3.3, and 3.4, while Section 3.5 deals with recombination continuum observations from EVE.

Many of the diagnostic methods discussed in the following subsections implicitly assume that the plasma under investigation is in ionisation equilibrium. This assumption may be valid at high densities when ionisation timescales are short (Bradshaw, 2009; Graham *et al.*, 2013; Caspi, McTiernan, and Warren, 2014), but at lower values departures from equilibrium can be significant, particularly for cooler emission lines (Doyle *et al.*, 2013). Such departures have been found during impulsive heating of cool, diffuse, coronal plasma (Bradshaw and Cargill, 2006) and in sunspot plumes (Doyle *et al.*, 1985). The consequence of non-equilibrium ionisation during flare heating would be that emission lines under investigation could be formed at higher or lower temperatures than otherwise assumed. The problem can be exacerbated for plasma in motion. For material moving upwards (or downwards) into regions of higher (or lower) temperatures, recombination rates can be out of synchronisation with ionisation. For example, Francis (1981) found the formation temperature of the C IV resonance lines dropped from $\log T = 5.0$ to $\log T = 4.65$ for a downflow of 6 $\mathrm{km\,s^{-1}}$. The integration times of the observations under investigation will also play a role. For long (> tens of seconds) exposure times, short-lived departures from equilibrium

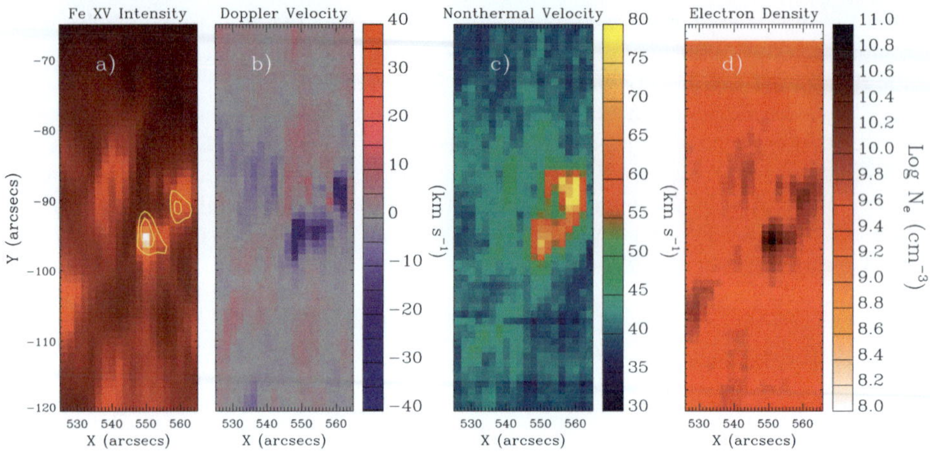

Figure 1 Derived plasma parameters from a single *Hinode*/EIS raster taken during the impulsive phase of a C1.1 flare that occurred on 14 December 2007. a) An image showing the spatial distribution of the Fe XV 284 Å(2 MK) line intensity. Two bright footpoints of an overlying flare loop are visible near the middle of the raster. Overlaid are the contours of the 20 – 25 keV sources as observed by RHESSI that denote the location of the HXR footpoints. b) The corresponding Doppler velocity map derived from shifts in the line centroid relative to a quiet-Sun value. Positive velocities (redshifts) indicate downflows, while negative velocities (blueshifts) indicate upflows. c) Map of the nonthermal velocity from the line widths over and above the thermal-plus-instrumental widths. d) Spatial distribution of electron density from the ratio of two Fe XIV lines that are formed at a similar temperature to that of Fe XV. This figure shows that plasma at the location of the HXR emission (believed to be in the chromosphere) was hot, upflowing, turbulent, and dense. From Milligan (2011). Reproduced by permission of the AAS.

may well be averaged out. However, as instrumentation becomes more advanced, and observational timescales become shorter, departures from equilibrium may become significant enough to bias the interpretation (see Section 6).

3.1. Emission Line Intensities

The intensity of emission lines formed at discrete temperatures can be used to reconstruct a differential emission measure (DEM; $\mathrm{d}EM(T)/\mathrm{d}T$) profile of the emitting plasma. The DEM is a powerful tool for investigating the temperature distribution of the emitting plasma, providing crucial clues to the underlying heating processes. It is a measure of the square of the electron density integrated over the column depth along the line of sight for any given temperature ($n_e^2\mathrm{d}z/\mathrm{d}T$). Despite decades of study, there is still no consensus on what the flare DEM profile looks like (*e.g.* McTiernan, Fisher, and Li, 1999), particularly during the impulsive phase. For example, the often-assumed flare DEM used in the CHIANTI atomic database is based on a single observation by Dere and Cook (1979) taken during the decay phase of an M-class flare. However, significant advances are currently being made using EVE data, with RHESSI data being used to constrain the highest temperature emission (Warren, Mariska, and Doschek, 2013; Caspi, McTiernan, and Warren, 2014). It is also possible to determine the DEM from a flaring footpoint using spatially resolved line profiles. The right-hand panel of Figure 2 shows a regularised DEM solution from Graham *et al.* (2013) of a spatially resolved footpoint from the impulsive phase of the 14 December 2007 flare using data from EIS using the methods of Hannah and Kontar (2012). The coloured lines are the EM loci curves (ratio of the data

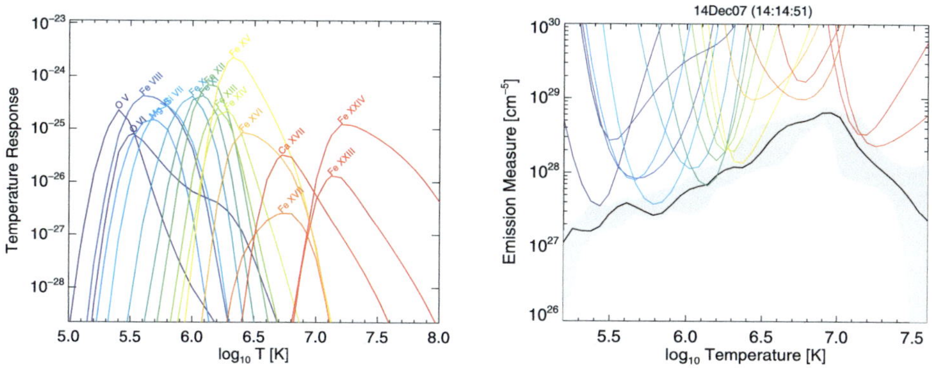

Figure 2 Left: Contribution functions for 16 lines observed by *Hinode*/EIS. Right: A DEM for a flare footpoint during the impulsive phase of the 14 December 2007 C-class flare. Grey shaded area denotes the corresponding error bars. From Graham *et al.* (2013). Reproduced by permission of the AAS.

and the temperature response) that define the isothermal (and hence maximum) emission possible at a given temperature. Graham *et al.* (2013) also used this technique on a number of B-class flares to establish the distribution of temperature at the footpoints. They found the slope of the distribution between $\log T = 5.5-7.0$ in each case to be ~ 1 ($EM(T) \propto T$), which they concluded was consistent with the layers beneath the electron deposition site being heated conductively.

By contrast, looking at larger (> M1) events observed by EVE, Kennedy *et al.* (2013) found a double-peaked DEM (top right-hand panel of Figure 3). Although EVE provides spatially integrated (Sun-as-a-star) observations, comparison between light curves binned by temperature (top left-hand panel of Figure 3; see Chamberlin, Milligan, and Woods, 2012) and associated AIA images revealed emission with temperatures approaching 8 MK to be both co-temporal and co-spatial with He II 304 Å ribbons during the impulsive phase (bottom panel of Figure 3). This implied that EVE spectra taken during this time, and therefore the resultant DEM, would be indicative of predominantly footpoint emission. The reason for the different DEM profiles obtained by Graham *et al.* (2013) and Kennedy *et al.* (2013) is unclear, although it may simply be attributed to the two studies focusing on events of very different magnitudes. The nature of the observations may also play a role. While EIS provides spatially resolved line profiles, they are often only for a single time interval when the slit of the spectrometer happened to raster over the footpoint. EVE, on the other hand, can provide continuous spectra throughout a flare, but with no spatial information. Simultaneous flare observations acquired with both instruments would greatly help in resolving this issue.

3.2. Emission Line Shifts

In the context of solar flares, SXR spectroscopy led to the first quantiative measurements of mass flows due to chromospheric evaporation through Doppler shifts of high-temperature line profiles (Doschek *et al.*, 1980; Antonucci and Dennis, 1983; Canfield *et al.*, 1987; Doschek *et al.*, 1992; Doschek, Mariska, and Sakao, 1996). However, these studies often focused on a single, high-temperature line and lacked any spatial information. Line profiles were found to be dominated by a strong stationary component with a blue-wing asymmetry indicative of upflows of several hundred $km\,s^{-1}$. It was initially thought that the stationary component emanated from the loop tops when flows had ceased or were

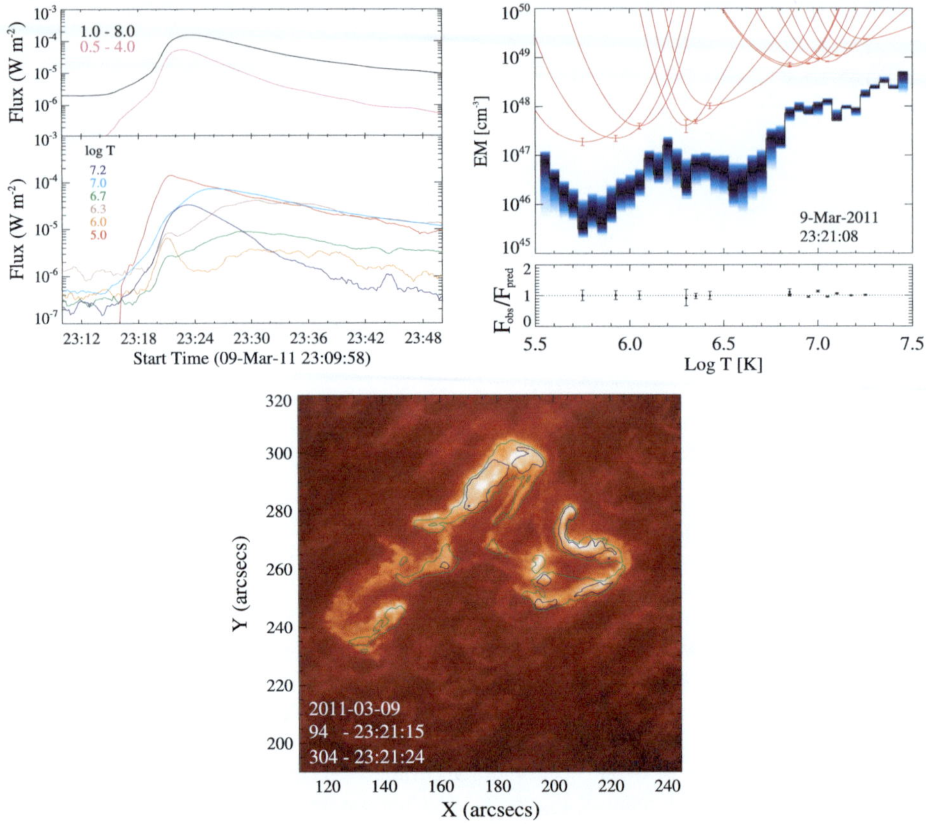

Figure 3 Top left panel: light curves of SXR emission (upper panel) and EUV emission binned by temperature (lower panel) for an M1.1 flare that occurred on 9 March 2011. Note that the log T = 6.0, 6.3, and 6.7 curves peak both during the impulsive phase as the chromosphere is heated – in sync with He II – and again during the decay phase as the flare loops cool. Top right panel: DEM derived from emission lines observed by EVE during the impulsive phase (23:21 UT). Bottom panel: SDO/AIA 304 Å image also taken during the impulsive phase. Overlaid are the contours of the associated 94 Å emission, which is dominated by Fe XVIII emission (log T = 6.9), and which also aligns with the He II ribbons. From Kennedy *et al.* (2013). Reproduced by permission of the AAS.

moving perpendicular to the line of sight, while the blue wing of the line was an indicator of evaporating material at the footpoints (see Doschek and Warren, 2005). The introduction of CDS on SOHO enabled spatially resolved line profiles to be measured across a range of temperatures, although the sampling was still quite sparse (Czaykowska *et al.*, 1999; Teriaca *et al.*, 2003; Milligan *et al.*, 2006a, 2006b; Del Zanna *et al.*, 2006; Raftery *et al.*, 2009). Despite the provision of spatially resolved line profiles from CDS, the lines still appeared to be dominated by a stationary component in contrast to the predictions of theory (*e.g.* Li, Emslie, and Mariska, 1989). The use of sit-and-stare observations further allowed flow velocities to be determined as a function of time, but at the cost of large uncertainties in the pointing aspect (Brosius, 2003; Brosius and Phillips, 2004; Brosius and Holman, 2007, 2009; Brosius, 2009).

EIS now allows spatially resolved line profiles to be measured using emission lines formed over a broad range of temperatures simultaneously (Milligan and Dennis, 2009;

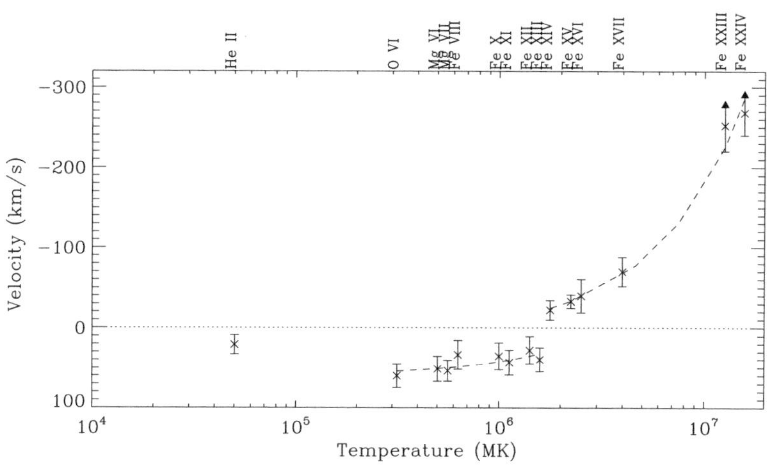

Figure 4 Derived Doppler velocities at a flaring footpoint for 15 emission lines formed between 0.05 – 16 MK. Negative velocities (blueshifts) represent upflows, while positive velocities (redshifted) denote downflows. From Milligan and Dennis (2009). Reproduced by permission of the AAS.

Del Zanna *et al.*, 2011; Graham, Fletcher, and Hannah, 2011). Figure 4 shows the derived Doppler velocities for 15 emission lines, formed between 0.05 – 16 MK, taken from one of the footpoints of the 14 December 2007 flare (Milligan and Dennis, 2009). This shows a linear relationship between the temperature of the heated plasma and the rate at which it rises ($v_{up} \propto T$). The material that was redshifted (chromospheric condensation, driven downward by the overpressure of the evaporated material required to conserve momentum; Canfield *et al.*, 1987, 1990) recoiled at approximately the same speed regardless of temperature, but was also observed at much higher temperatures than previously expected from chromospheric evaporation models (up to 1.5 MK). Fisher, Canfield, and McClymont (1985) believed that the temperature at which the Doppler velocity changed from redshifted to blueshifted (dubbed the 'Flow Reversal Point' by Brannon and Longcope, 2014) marked the layer of the solar atmosphere at which the electrons deposited the bulk of their energy. From Milligan and Dennis (2009), this implied that either the electrons lost their energy in the corona (unlikely, due to the low densities involved) or that the chromosphere was being heated as it recoiled. More recent simulations now show that hot redshifted material is possible if chromospheric heating is sustained ($\sim$60 s; *e.g.*, Liu, Petrosian, and Mariska, 2009), rather than by a single burst (5 s), as assumed by Fisher, Canfield, and McClymont (1985). Similar findings of hot redshifts have also been made by Milligan (2008), Chen and Ding (2010), and Graham, Fletcher, and Hannah (2011), although the precise 'crossover' temperature varied between 0.5 and 2 MK for different events. In one case presented by Li and Ding (2011), it was found to be as high as 5 MK.

Despite its modest spectral resolution, Hudson *et al.* (2011) were able to use EVE data to detect redshifted He II emission associated with chromospheric condensation during flares. By measuring Doppler velocities during the impulsive phase relative to those in the decay phase, they found downflows of 10 – 20 km s^{-1}, consistent with those from EIS observations. Diminishing blueshifts in Fe XXIV, as expected from chromospheric evaporation theory, were also detected throughout the main phase of the same events, although, as with the He II velocities, the authors correctly advised caution when attempting to derive the absolute magnitude of the flows. Many lines in the EVE data are likely to be significantly blended, and the lack of spatial information introduces line-of-sight effects due to flare location and loop geometry.

The issue of whether the stationary or blueshifted component should dominate the high-temperature ($\gtrsim$ 10 MK) line profiles is still an open question, with different studies promoting either case: Milligan and Dennis (2009) and Li and Ding (2011) both found stationary emission to dominate, while Watanabe *et al.* (2010) and Young *et al.* (2013) found lines to be completely blueshifted. In one particularly complex flare, Doschek, Warren, and Young (2013) found both scenarios to occur at different locations along the ribbons. It may be the case that current instrumentation does not have high enough spatial resolution to resolve the evaporation flows as they commence before they become 'diluted' with coronal material. New observations of Fe XXI emission (1354.1 Å) from the recently launched *Interface Region Imaging Spectrograph* (IRIS; De Pontieu *et al.*, 2014) may help address this issue (Young, Tian, and Jaeggli, 2015; Graham and Cauzzi, 2015).

3.3. Emission Line Widths

The width of spectral emission lines contains important information on the temperature and turbulence of the emitting plasma. Line width is generally made up of at least three components: the intrinsic instrumental resolution, the thermal Doppler width, and any excess (nonthermal) broadening, which may be an indicator of possible turbulence, or pressure or opacity broadening. Many studies have reported excess SXR line broadening over and above that expected from thermal emission during a flare's impulsive phase, indicating possible turbulent motion (Doschek *et al.*, 1980; Feldman *et al.*, 1980; Gabriel *et al.*, 1981; Antonucci *et al.*, 1982), although this emission was integrated over the entire solar disc; While turbulence is often cited as the most likely explanation of excess line broadening, there is also the possibility that a given line can comprise plasma with a range of temperatures. This could also result in a broader line width depending on the overall DEM profile and the contribution function of the line (Antonucci and Dodero, 1995).

Milligan (2011) presented a detailed investigation into the nature of spatially resolved line broadening of EUV emission lines during the impulsive phase of the 14 December 2007 flare using EIS thanks to its superior spectral and spatial resolution. Line profiles, co-spatial with HXR emission observed by RHESSI, were found to be broadened beyond their thermal widths (Figure 1c). Using techniques similar to those used to establish the nature of line broadening in quiescent active region spectra (Hara *et al.*, 2008; Doschek *et al.*, 2008; Bryans, Young, and Doschek, 2010; Peter, 2010), it was found that a strong correlation existed between Doppler velocity and nonthermal velocity for the Fe XV and Fe XVI lines at one of the footpoints. This suggested that the line broadening at these temperatures was a signature of unresolved plasma flows along the line of sight during the process of chromospheric evaporation. Analysis of Fe XIV lines, on the other hand, which showed no conclusive correlation between Doppler and nonthermal velocities, showed a stronger correlation between electron density and nonthermal width (Figure 5), suggesting that the excess line broadening at these temperatures may have been due to opacity or pressure broadening as emitted photons were scattered along the line of sight. However, estimating the opacity of the emission, τ_0, using

$$\tau_0 = 1.16 \times 10^{-14} \lambda f_{ij} \sqrt{\frac{M}{T}} \frac{n_{\text{ion}}}{n_{el}} \frac{n_{el}}{n_H} \frac{n_H}{N_e} N_e h, \tag{1}$$

where λ is the wavelength of the line, f_{ij} is the oscillator strength, M is the mass of the ion, n_{ion}/n_{el} is the ionisation fraction, n_{el}/n_H is the abundance of iron relative to hydrogen, N_e is

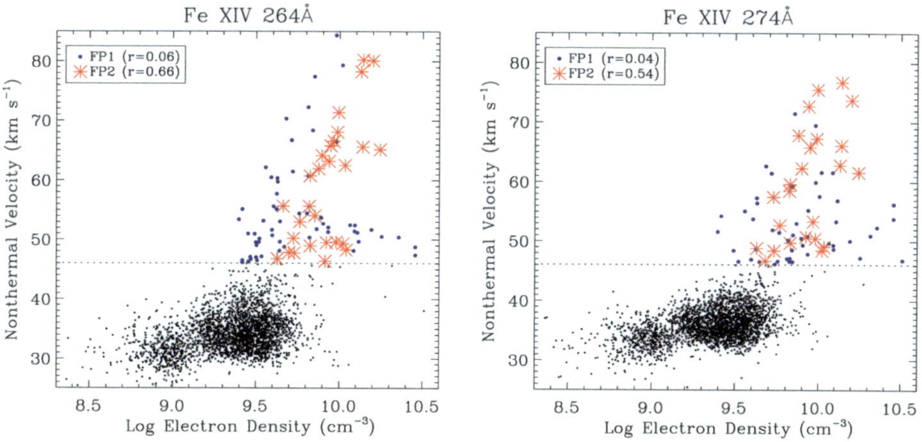

Figure 5 Plot of nonthermal velocity derived from the width of two Fe XIV lines (264 Å; left, 274 Å; right) against electron density derived from the ratio of two Fe XIV lines during the impulsive phase of the 14 December 2007 flare. Black dots denote the surrounding quiescent active region, while blue circles and red asterisks are data points from the two footpoints. A clear correlation (where r is the Pearson correlation coefficient) exists between the two parameters for at least one of the footpoints (FP2). From Milligan (2011). Reproduced by permission of the AAS.

the electron density, and h is the column depth, and the amount of pressure broadening, $\Delta\lambda$, using

$$\Delta\lambda = \frac{\lambda^2}{c}\frac{1}{\pi\Delta t_0} \approx \frac{\lambda^2}{c}\frac{N_e\sigma}{\pi}\sqrt{\frac{2k_BT}{M}}, \tag{2}$$

where Δt_0 is the timescale of the emitting photon and σ is the collisional excitation cross-sectional area (from Bloomfield *et al.*, 2002 and Christian *et al.*, 2006), suggested that the amount of excess broadening should have been negligible in each case. Perhaps the assumptions made in solving Equations (1) and (2) were incorrect (*e.g.* lines were not formed in ionisation equilibrium), or the broadening was due to a culmination of different effects, or perhaps it was due to a different mechanism altogether not considered (*e.g.* Stark broadening). Perhaps larger, more energetic events, or density diagnostics of higher (or lower) temperature plasmas, will show these effects to be more substantial. Brosius (2013b) found nonthermal velocities derived from the width of Mg VI, Si VII, Fe XIV, Fe XVI, and Fe XXIII emission lines increased by a factor of ~ 3 compared to quiet-Sun widths at the site of a flare ribbon using sit-and-stare observations. This was assumed to be an indicator of turbulent motion associated with chromospheric evaporation. Line broadening cannot only reveal important information with regard to the heating processes during flares, but can also be a crucial diagnostic of the fundamental atomic physics (abundances, ionisation fractions, *etc.*) and must therefore be a component of future flare modelling.

3.4. Density Diagnostics

Values of the electron density of a given plasma can be found by taking the ratio of the flux of two emission lines from the same ionisation stage when one of the lines is derived from a metastable transition (see Chapter 2.8.2 of Mariska, 1992 for a detailed derivation of this technique and Chapter 4.4 for early applications to solar observations). This approach is more robust than using broadband filter ratios (*e.g.* White, Thomas, and Schwartz, 2005) that require an accurate knowledge of the volume of the emitting plasma, which can be difficult for plane-of-sky observations. A filling factor of unity is also often assumed. Line ratio techniques have more recently been employed to infer electron densities in active regions (Gallagher *et al.*, 2001; Milligan *et al.*, 2005; Doschek *et al.*, 2007), jets (Chifor *et al.*, 2008) and flare loops (Milligan *et al.*, 2012).

Milligan (2011) used five pairs of density-sensitive line ratios formed at different temperatures (Mg VII, Si X, Fe XII, Fe XIII, and Fe XIV) to measure density enhancements at the footpoints during the 14 December 2007 flare. Peak values of $> 10^{11}$ cm^{-3} (above the high-density limit) were found using Fe XII and Fe XIII ratios, indicating enhancements of several orders of magnitude above quiescent active region levels. It was found that density enhancements were only visible in lines formed above ~ 1 MK as cooler lines are formed deeper in the solar atmosphere where densities are already high (Figure 1d and top left-hand panels in Figure 6). A recent follow-up study by Graham, Fletcher, and Labrosse (2014) found density values approaching 10^{12} cm^{-3} using O V lines for the same event. Graham, Fletcher, and Hannah (2011) also found comparable values from Fe XII, Fe XIII, and Fe XIV ratios at the ribbon location during a C6.6 flare. Sit-and-stare EIS observations from Brosius (2013b) also revealed a rapid increase in electron density of an order of magnitude using an Fe XIV line pair, but this increase appeared to be independent of the corresponding upflow velocity, as would be expected in a beam-heated chromosphere scenario; the density (and intensity) enhancements ceased while blueshifts persisted, which the author proposed as evidence for an energy-release process in the chromosphere itself as opposed to heating from above. However, care must be taken when interpreting observations from single-slit measurements as pointing accuracy and co-alignment with context images can be ambiguous. Similarly, the use of any line ratio technique to quantify electron densities depends greatly on the knowledge of the underlying atomic physics. Quite often these values are inaccurate, which can lead to large uncertainties in the theoretically predicted values (Young *et al.*, 2009).

Using the values derived for the electron densities, it is possible to compute the column depth, h, of the emitting material at the formation temperature of the lines. The intensity of a given emission line, I, can be expressed as

$$4\pi I = 0.83 \int G(T, N_e) N_e^2 \, \mathrm{d}h, \tag{3}$$

where $G(T, N_e)$ is the contribution function for a given line and N_e is the electron number density. By approximating the contribution function as a delta function at $T_{\max}$ and assuming that the density is constant across each pixel, Equation (3) can be written as

$$4\pi I = 0.83 G_0 N_e^2 h. \tag{4}$$

The EIS_DENSITY.PRO routine calculates G_0 for a given electron density, which allows the value of h to be derived for each pixel within a raster for which the density is known (see Young, 2011 for more details). The top right-hand panels of Figure 6 show the five column depth maps (in arcseconds) generated from the corresponding density maps from the 14 December 2007 flare (top left panels of Figure 6). Unsurprisingly, the spatial distribution of column depth closely resembles that of the density distributions, with footpoint emission exhibiting smaller column depths than the surrounding active region; less than 15″ in most cases, and as little as 0.01″ in some pixels. Similar column depths (~ 10 km) were found by Del Zanna *et al.* (2011) during a B-class flare from electron densities of 10^{10-11} cm s^{-1} using

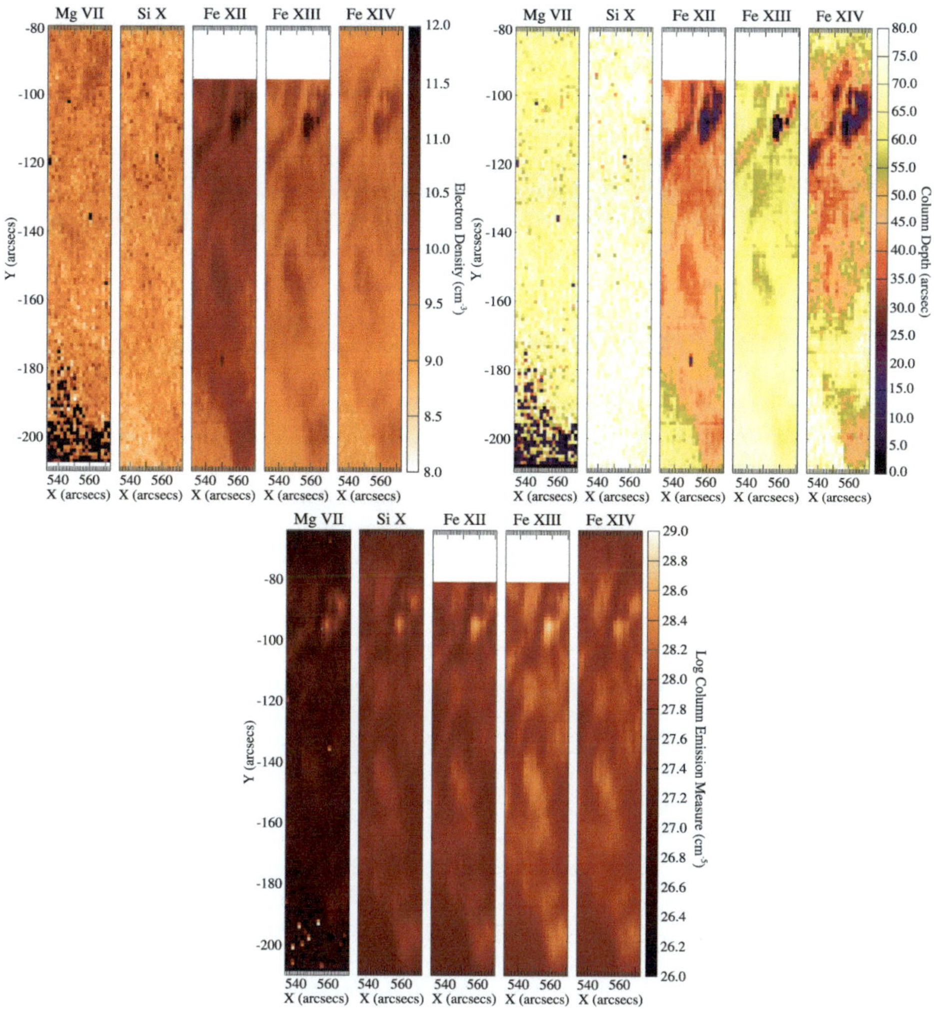

Figure 6 Five EIS rasters in the Mg VII, Si X, Fe XII, Fe XIII, and Fe XIV lines, taken during the impulsive phase of the 14 December 2007 flare. Top left: Electron density maps formed using line ratio techniques. Top right: Maps of column depth formed using Equation (4). Both figures are taken from Milligan (2011). Bottom: Column emission measure maps generated by combining data from the top two panels. From Graham *et al.* (2013). Reproduced by permission of the AAS.

Fe XIV ratios. This was a particularly well-observed event in which the ribbons were aligned in a N–S direction, parallel to the EIS slit, and were observed in 10 sequential rasters.

Following from Milligan (2011), Graham *et al.* (2013) combined the electron density and column depth measurements to estimate the column emission measure (EM_{col}) for a given line at a footpoint, given that $EM_{\mathrm{col}} = \int N_e^2 \,\mathrm{d}h$. The bottom five panels in Figure 6 show the column emission measure maps for the impulsive phase of the 14 December 2007 event. It can be seen in each of the five rasters that higher column emission measures were found at the footpoints compared to the surrounding active region, even in the Mg VII and Si X lines, which showed no discernible density enhancements. It is also worth noting that the value of the footpoint column emission measure at $1-2$ MK of $\sim 10^{28}$ cm^{-5} using this technique are comparable to those found using the regularised inversion method described by Hannah and Kontar (2012; right-hand panel of Figure 2).

3.5. Continuum Emission

Simulations by Allred *et al.* (2005) using the RADYN code of Carlsson and Stein (1994, 1995, 1997), which models the chromospheric response to both electron beam heating and backwarming from XEUV (X-ray and EUV) photons, suggested that chromospheric emission is energetically dominated by recombination (free–bound) continua, as opposed to line (bound–bound) emission. This includes the Lyman, Balmer, and Paschen continua of hydrogen, and the He I and He II continua. However, definitive observations of free–bound emission during solar flares have been scarce in recent years as many modern space-based instruments do not have the sensitivity, wavelength coverage, or duty cycle required to capture unambiguous continuum enhancements during flares. One exception was a serendipitous detection of increased Lyman continuum (~ 70 % at 910 Å) during an X-class flare via scattered light in the SUMER instrument onboard SOHO reported by Lemaire *et al.* (2004). There was also an analogous case in stellar flare spectra obtained from the *Extreme Ultra-Violet Explorer* (EUVE) mission by Christian *et al.* (2003). Tentative evidence for increased He I continuum emission was also found.

Temporally resolved measurements of the recombination continua of H and He have now been unambiguously observed during solar flares by Milligan *et al.* (2012, 2014). The MEGS-B (*Multiple EUV Grating Spectrograph*) component of the EVE instrument, which covers the 370 – 1050 Å range (top panel of Figure 7), includes the Lyman continuum of H I and He I with recombination edges at 912 Å and 504 Å, respectively. Using a variety of fitting techniques, the time profiles of these emissions during flares were found to be impulsive (bottom four panels of Figure 7), peaking in concert with the associated HXR emission. This suggested that the increase in continuum emission is due to recombination with free electrons that were liberated during chromospheric heating by nonthermal electrons. MEGS-A[2] also observed the He II continuum with a recombination edge at 228 Å, but it is inherently weak and only observed during the largest events. It also competes with the underlying free–free continuum, as well as numerous emission lines, adding to the complexity of fitting it and deriving physically meaningful parameters. By integrating under the light curves displayed in the bottom row of Figure 7, Milligan *et al.* (2014) found that the Lyman and He I continua contained ~ 1 % and ~ 0.1 %, respectively, of the total nonthermal electron energy as deduced from RHESSI observations during the X2.2 flare that occurred on 15 February 2011. By comparison, the Lyα and He II 304 Å lines radiated away significantly more energy (~ 6 % and ~ 2 %, respectively) in contrast to the predictions of Allred *et al.* (2005).

Although energetically very weak, these continua serve as powerful diagnostic tools for determining the conditions in the solar atmosphere at which they are formed. Following from Machado and Noyes (1978), Ding and Schleicher (1997) stated that the temperature of the continuum (relative to a blackbody) can be ascertained by taking the slope of the continuum (relative intensity at two distinct wavelengths, I_{λ_1} and I_{λ_2}) using

$$T = \frac{hc}{k}\left(\frac{1}{\lambda_1} - \frac{1}{\lambda_2}\right)\left[\ln\left(\frac{I_{\lambda_2}\lambda_2^5}{I_{\lambda_1}\lambda_1^5}\right)\right]^{-1}. \tag{5}$$

[2] Currently not functioning due to a power anomaly that occurred on 26 May 2014.

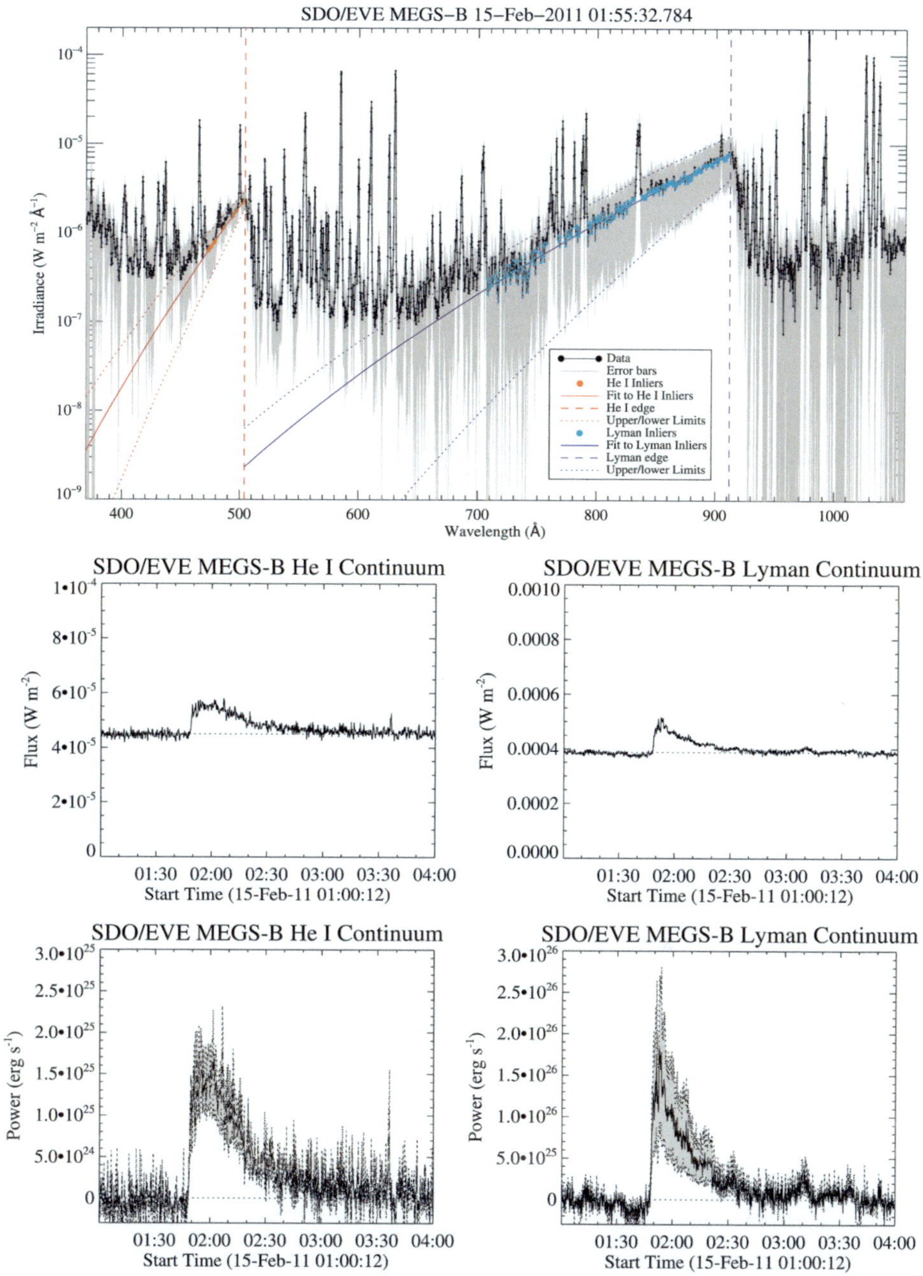

Figure 7 Top panel: EVE MEGS-B spectra taken near the peak of the X2.2 flare that occurred on 15 February 2011. Overlaid in orange and cyan are the fits to the He I and Lyman continua, respectively, using the RANSAC method (Fischler and Bolles, 1981). Middle row: full-Sun light curves for He I (left) and Lyman (right) continua from integrating under the fits at each 10 s time step. Bottom row: Light curves of He I (left) and Lyman (right) continua emission in units of energy after performing a pre-flare background subtraction. From Milligan *et al.* (2014). Reproduced by permission of the AAS.

The derived temperature can then be used to yield values of the coefficient of departure from blackbody emission, b_1, given by

$$b_1 = \frac{B_\lambda(T)}{I_\lambda} = \frac{2hc^2}{I_\lambda \lambda^5} \exp\left(\frac{-hc}{\lambda k T}\right), \quad (6)$$

where $B_\lambda(T)$ is the Planck function.

EVE measurements of flare-related increases in Lyman continuum emission carried out in conjunction with observations in the Balmer and Paschen continua as recently reported by Heinzel and Kleint (2014) and Kerr and Fletcher (2014), respectively, would greatly advance our knowledge of the distribution of energy throughout the lower solar atmosphere, as well as the dominant emission mechanism, particularly when compared to radiative hydrodynamic simulations such as those from Allred *et al.* (2005).

4. Lyα Flare Observations

Although the Lyα line of H I at 1216 Å straddles the boundary between the EUV and UV portions of the spectrum, its relevance to chromospheric flare heating and the current availability of photometric Lyα data from EVE MEGS-P warrants a brief summary of past observations and the potential for future diagnostics. True spectroscopic observations of Lyα have been few and far between lately, largely due to the technical challenges inherent in measuring the brightest line in the solar spectrum. Among the first spectrally resolved Lyα flare observations were those carried out using the S082B spectrograph onboard *Skylab*, although saturation was often an issue (Lites and Cook, 1979; Canfield and van Hoosier, 1980; Doyle and Cook, 1992). Before this, Lyα photometers had been flown on sounding rockets and early satellites, but no flare-related enhancements were detected (Chubb *et al.*, 1957; Kreplin, Chubb, and Friedmann, 1962). The *Solar Stellar Irradiance Comparison Experiment* (SOLSTICE) instrument onboard the *Solar Radiation and Climate Experiment* (SORCE) satellite typically takes spectral irradiance measurements across the EUV range once per orbit. However, during the Halloween flares of October 2003, it was fortuitously scanning through the Lyα line at a one-minute cadence. Woods *et al.* (2004) reported only a 20 % increase in the core of the line during the 28 October 2003 X28 flare, but the wings of the line were found to increase by a factor of two. This is significantly higher than the 6 % increase reported by Brekke *et al.* (1996) during the decay phase of an X3 flare on 27 February 1992 using the SOLSTICE instrument on UARS. Johnson *et al.* (2011) were able to detect enhancements in Lyα luminosity from scattered light using off-limb observations from SOHO/*Ultraviolet Coronagraph Spectrometer* (UVCS), while Rubio da Costa *et al.* (2009) estimated that Lyα emission accounts for < 10 % of a flare's radiated energy budget using broadband measurements from the *Transition Region And Coronal Explorer* (TRACE; Handy *et al.*, 1999). Flare-related enhancements in Lyα in EVE data have also been reported by Milligan *et al.* (2012) for the 15 February 2011 X2.2 flare, and a follow-up study also revealed Lyα emission to contribute 6 – 8 % of the total measured radiated losses from the chromosphere (Milligan *et al.*, 2014). In contrast, Kretzschmar, Dominique, and Dammasch (2013) found only a 0.6 % increase in Lyα emission during an M2 flare using the LYRA radiometer onboard PROBA2. More strikingly, however, was that its temporal behaviour appeared to mimic that of the GOES SXR emission, rather than of the impulsive HXR emission as one might expect for intrinsic chromospheric emission. They found that by taking the time derivative of the Lyα profile, the resulting peaks closely resembled those of the

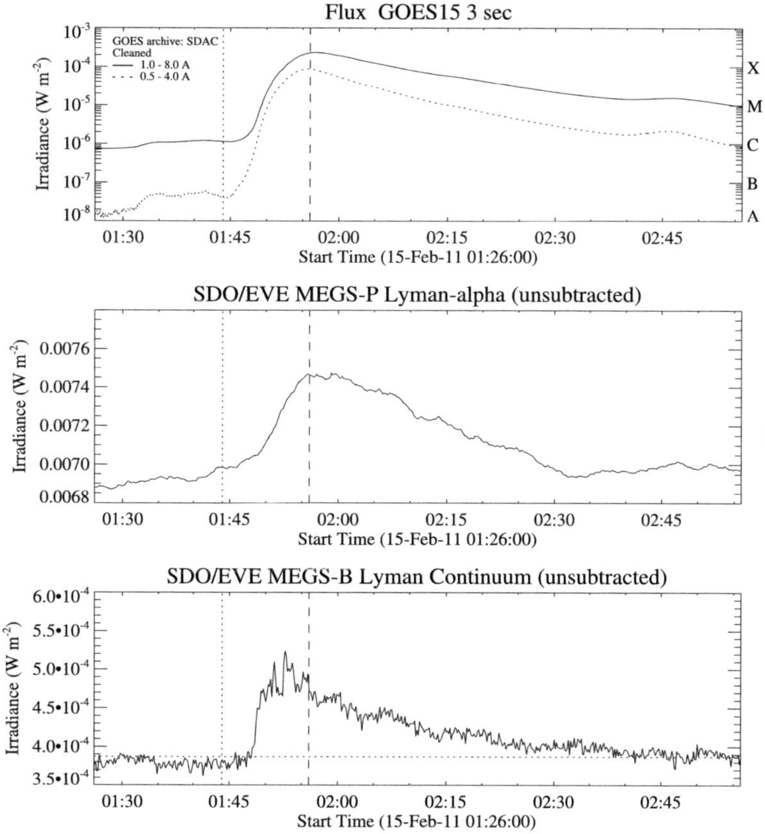

Figure 8 Top panel: GOES SXR light curves during the X2.2 flare that occurred on 15 February 2011 (SOL2011-02-15) in 1 – 8 Å (solid curve) and 0.5 – 4 Å (dotted curve). Middle panel: Lyα light curve taken using SDO/EVE MEGS-P for the same event. Bottom panel: LyC light curve from SDO/EVE MEGS-B data formed using the techniques described in Section 3.5. The vertical dotted and dashed lines in each panel denote the start and peak times of the flare, respectively, as stated in the GOES event list.

derivative of the SXR light curve. The authors cited this as evidence for pre-flare heating prior to the onset of accelerated particles. The possibility of gradual heating by thermal conduction cannot be ruled out, but it is more likely that instrumental factors were contributing to this anomalous behaviour (see below). Kretzschmar, Dominique, and Dammasch (2013) also provided a brief overview of many recent Lyα flare observations.

SDO/EVE now provides routine, broadband (100 Å) observations of flares in Lyα through its MEGS-P diode. Despite the lack of spectral and spatial information, these data are useful in terms of quantifying the energy budget of radiative losses in the chromosphere. However, similar to the temporal behaviour found by Kretzschmar, Dominique, and Dammasch (2013), the light curves of Lyα from EVE appear to show a more gradual, "GOES-like" profile. Figure 8 shows the SXR (top), Lyα (middle), and LyC (bottom) time profiles for the 15 February 2011 X-class flare. Although Lyα and LyC both result from electron transitions to the ground state of hydrogen, the LyC profile appears impulsive, as one might expect for a chromospheric plasma heated via Coulomb collisions, while Lyα displays a rise time of 10 – 20 minutes, similar to that seen in SXR. The reason for this disparity is unclear at present. The spectrally resolved Lyα profiles shown by Woods *et al.* (2004) for the 28 October 2003 flare using SORCE/SOLSTICE were synchronised with the derivative of the SXR, confirming an impulsively heated atmosphere. This suggests that the broadband nature of the EVE (and LYRA) measurements are in some way "smoothing out" this intrinsic, bursty nature. Perhaps other, higher temperature lines or continua are contributing to the passband, or a data processing algorithm in the EVE pipeline is responsible. This issue is discussed in more detail by Milligan and Chamberlin (2015). Once resolved, simultaneous Lyα and LyC observations (along with Lyβ, Lyγ, *etc.*, which are also observed by MEGS-B) will be a valuable tool for investigating heating of the lower solar atmosphere, particularly in conjunction with IRIS. It may also help in the interpretation of flare observations from the *EUV Monitor* onboard the *Mars Atmosphere and Volatile Evolution* (MAVEN) spacecraft as well as those from *Solar Orbiter*, which will also carry an *EUV Imager* (EUI) that will include a Lyα channel. Resolving the discrepancies with current Lyα observations is therefore paramount to ensure future measurements are interpreted correctly, particularly during explosive events.

5. Concluding Remarks

EUV spectroscopy has advanced considerably over the past 50 years or so. However, its application to flares, particularly at footpoints and ribbons during the impulsive phase, has been somewhat lacking given the diagnostic potential available. The studies referenced in this article highlight the capabilities (and limitations) of our current spectrometers and the scientific accomplishments that have been achieved. (For other reviews on chromospheric flares, see Hudson, 2007; Fletcher *et al.*, 2011; Fletcher, 2012, and Fletcher and Simoes, 2013.) However, many of these works have been carried out in isolation, or perhaps in conjunction with imaging data for context information. I believe that the greatest advances will be made by combining datasets from both EIS and EVE together with those from other instruments such as SDO/AIA, *Hinode*/SOT, RHESSI, IRIS, and ground-based facilities like those at the *Dunn Solar Telescope* (DST). On 29 March 2014, an X1.0 flare was observed by all of these instruments, making it one of the best observed flares ever, which will no doubt yield some interesting results as well as validating the advantage of coordinated observations. October 2014 also saw a hugely successful coordinated observing campaign facilitated by the Max Millennium Program for Solar Flare Research[3] in support of Service Mode Operations at the DST with coordinated observations from RHESSI, SDO/EVE MEGS-B, IRIS, and all three *Hinode* instruments. These Service Mode Operations are how flare observations are anticipated to be run at the *Daniel K. Inoue Solar Telescope* (DKIST; formerly the *Advanced Technologies Solar Telescope* – ATST) when it begins operations in 2018. Multi-wavelength campaigns such as these should be strongly encouraged.

As we enter the declining phase of Solar Cycle 24, the opportunities for coordinated campaigns may be somewhat limited. However, both EIS and EVE have several years of data waiting to be explored and exploited (see Watanabe, Masuda, and Segawa, 2012 for a complete list of flares observed by EIS to date and a similar list[4] has been compiled for all flares observed during EVE MEGS-B campaigns). This will help us move away from looking at individual events and begin investigating the collective properties of a statistically significant number of events. EVE data may also permit a comparison between flares in solar and

[3] http://solar.physics.montana.edu/max_millennium/.

[4] http://lasp.colorado.edu/eve/data_access/evewebdata/interactive/flare_campaign_observations.html.

stellar atmospheres by looking at analogous events observed by EUVE, for example, which covered a similar spectral range. But perhaps the most meaningful collaborations will be those that attempt to model the observed EUV features and their derived parameters through numerical simulations. Many current flare heating models are now capable of recreating atmospheric conditions such as temperature and density, as well as optically thin emission lines themselves, in response to an injection of energy (*e.g.* Bradshaw and Mason, 2003; Klimchuk, Patsourakos, and Cargill, 2008; Liu, Petrosian, and Mariska, 2009; Kašparová *et al.*, 2009). This injection can, and should, be based on nonthermal electron parameters derived from RHESSI (or *Fermi*) data with the model output being directly compared with solar observations. Any serious discrepancies between predictions and reality may mean refining the derived parameters of the electron spectrum (*e.g.* the low-energy cutoff; Ireland *et al.*, 2013), or considering other forms of energy transport, such as thermal conduction (Brannon and Longcope, 2014) or Alfvén waves (Russell and Fletcher, 2013). Modelling optically thick emission adds another layer of complexity in the form of solving the equations of radiative transfer (Allred *et al.*, 2005). Ultimately, a comparison between the outputs of chromospheric heating models and the corresponding EUV observations – based on a common energy input – is needed to gain a greater insight into the driving mechanism behind enhanced flare emission, at all wavelengths, in the lower solar atmosphere.

6. Future Prospects

The continued operation of EIS and EVE into the decay of Solar Cycle 24 promises that more interesting discoveries lie ahead. In the longer term, EUV spectroscopy remains a high priority for future missions. The current plan for *Solar-C* (the successor to *Hinode*) features an imaging spectrometer similar to EIS called the *EUV Spectroscopic Telescope* (EUVST). By performing spectral analysis of emission lines, EUVST will measure the velocity, temperature, and density of solar plasmas. It will have more than five times the spatial resolution and ten times the throughput of the currently operational spectrometers, and will be able to observe over a wide wavelength range (170 Å to 1280 Å), seamlessly covering the transition region from the chromosphere to the coronal and extend to flare temperatures. A high throughput performance will enable spectroscopic observations with short exposure times of around 1 s, so that it can reliably capture the rapid temporal fluctuations of magnetic structures and waves resulting from magnetic reconnection, jets, and photospheric motions.

Solar Orbiter, due to launch in 2018, will include the *Spectral Imaging of the Coronal Environment* (SPICE) instrument, an EUV imaging spectrograph designed to observe both on the solar disc and out in the corona to remotely characterise plasma properties at and near the Sun. Specific scientific topics to be addressed by SPICE include studies of solar wind origin by matching *in situ* composition signatures in solar wind streams to surface feature composition; studies of the physical processes that inject material from closed structures into solar wind streams; and studies of SEP source regions by imaging remotely the suprathermal ions thought to be seed populations of SEPs. The SPICE observing strategy is to produce 2D spectroheliograms of selected line profiles and line intensities. Like EUVST, the selected lines represent the full range of temperatures and heights in the solar atmosphere, from the chromosphere to the flaring corona.

With Solar Cycle 25 set to peak around 2025, the prospect of coordinated observations between *Solar-C* and *Solar Orbiter*, along with *Solar Probe Plus* and the DKIST, will mark an auspicious time to further our understanding of the most dynamic, complex, and perplexing region of the Sun's atmosphere, and that of other stars. Given the importance of the chromosphere and lower transition region to flare energetics, the acceleration of the solar wind, and the overarching field of space weather, diagnosing this crucial layer should be a primary focus of future missions.

Acknowledgements The author is grateful for financial support from NASA for LWS/TR&T grant NNX11AQ53G and LWS/SDO Data Analysis grant NNX14AE07G. He also thanks Brian Dennis, Mihalis Mathioudakis, Peter Young, Hugh Hudson, Helen Mason, David Graham, Lyndsay Fletcher (Guest Editor) and the anonymous referee for their useful comments and feedback on the manuscript, as well as the organisers of the meeting on Solar and Stellar Flares: Observations, Simulations, and Synergies in Prague, Czech Republic in June 2014 for the invitation to present this review.

References

Allred, J.C., Hawley, S.L., Abbett, W.P., Carlsson, M.: 2005, Radiative hydrodynamic models of the optical and ultraviolet emission from solar flares. *Astrophys. J.* **630**, 573. DOI.

Antonucci, E., Dennis, B.R.: 1983, Observation of chromospheric evaporation during the Solar Maximum Mission. *Solar Phys.* **86**, 67. DOI.

Antonucci, E., Dodero, M.A.: 1995, Temperature dependence of nonthermal motions in solar flare plasmas observed with the flat crystal spectrometer on SMM. *Astrophys. J.* **438**, 480. DOI.

Antonucci, E., Gabriel, A.H., Acton, L.W., Leibacher, J.W., Culhane, J.L., Rapley, C.G., Doyle, J.G., Machado, M.E., Orwig, L.E.: 1982, Impulsive phase of flares in soft X-ray emission. *Solar Phys.* **78**, 107. DOI.

Bloomfield, D.S., Mathioudakis, M., Christian, D.J., Keenan, F.P., Linsky, J.L.: 2002, Opacity in the upper atmosphere of AU Mic. *Astron. Astrophys.* **390**, 219. DOI.

Bradshaw, S.J.: 2009, A numerical tool for the calculation of non-equilibrium ionisation states in the solar corona and other astrophysical plasma environments. *Astron. Astrophys.* **502**, 409. DOI.

Bradshaw, S.J., Cargill, P.J.: 2006, Explosive heating of low-density coronal plasma. *Astron. Astrophys.* **458**, 987. DOI.

Bradshaw, S.J., Mason, H.E.: 2003, A self-consistent treatment of radiation in coronal loop modelling. *Astron. Astrophys.* **401**, 699. DOI.

Brannon, S., Longcope, D.: 2014, Modeling properties of chromospheric evaporation driven by thermal conduction fronts from reconnection shocks. *Astrophys. J.* **792**, 50. DOI.

Brekke, P., Rottman, G.J., Fontenla, J., Judge, P.G.: 1996, The ultraviolet spectrum of a 3B class flare observed with SOLSTICE. *Astrophys. J.* **468**, 418. DOI.

Brosius, J.W.: 2003, Chromospheric evaporation and warm rain during a solar flare observed in high time resolution with the coronal diagnostic spectrometer aboard the solar and heliospheric observatory. *Astrophys. J.* **586**, 1417. DOI.

Brosius, J.W.: 2009, Conversion from explosive to gentle chromospheric evaporation during a solar flare. *Astrophys. J.* **701**, 1209. DOI.

Brosius, J.W.: 2013a, Chromospheric evaporation in solar flare loop strands observed with the extreme-ultraviolet imaging spectrometer on board Hinode. *Astrophys. J.* **762**, 133. DOI.

Brosius, J.W.: 2013b, Rapid evolution of the solar atmosphere during the impulsive phase of a microflare observed with the extreme-ultraviolet imaging spectrometer aboard Hinode: hints of chromospheric magnetic reconnection. *Astrophys. J.* **777**, 135. DOI.

Brosius, J.W., Holman, G.D.: 2007, Chromospheric evaporation in a remote solar flare-like transient observed at high time resolution with SOHO's CDS and RHESSI. *Astrophys. J. Lett.* **659**, L73. DOI.

Brosius, J.W., Holman, G.D.: 2009, Observations of the thermal and dynamic evolution of a solar microflare. *Astrophys. J.* **692**, 492. DOI.

Brosius, J.W., Phillips, K.J.H.: 2004, Extreme-ultraviolet and X-ray spectroscopy of a solar flare loop observed at high time resolution: a case study in chromospheric evaporation. *Astrophys. J.* **613**, 580. DOI.

Bryans, P., Young, P.R., Doschek, G.A.: 2010, Multiple component outflows in an active region observed with the EUV imaging spectrometer on Hinode. *Astrophys. J.* **715**, 1012. DOI.

Canfield, R.C., Metcalf, T.R., Strong, K.T., Zarro, D.M.: 1987, A novel observational test of momentum balance in a solar flare. *Nature* **326**, 165. DOI.

Canfield, R.C., Metcalf, T.R., Zarro, D.M., Lemen, J.R.: 1990, Momentum balance in four solar flares. *Astrophys. J.* **348**, 333. DOI.

Canfield, R.C., van Hoosier, M.E.: 1980, Observed L-alpha profiles for two solar flares: 14:12 UT 15 June, 1973 and 23:16 UT 21 January, 1974. *Solar Phys.* **67**, 339. DOI.

Carlsson, M., Stein, R.F.: 1994, Radiation shock dynamics in the solar chromosphere – results of numerical simulations. In: Carlsson, M. (ed.) *Chromospheric Dynamics, Proc. Workshop, Univ. Oslo*, 47.

Carlsson, M., Stein, R.F.: 1995, Does a nonmagnetic solar chromosphere exist? *Astrophys. J. Lett.* **440**, L29. DOI.

Carlsson, M., Stein, R.F.: 1997, Formation of solar calcium H and K bright grains. *Astrophys. J.* **481**, 500. DOI.

Caspi, A., McTiernan, J.M., Warren, H.P.: 2014, Constraining solar flare differential emission measures with EVE and RHESSI. *Astrophys. J. Lett.* **788**, L31. DOI.

Chamberlin, P.C., Milligan, R.O., Woods, T.N.: 2012, Thermal evolution and radiative output of solar flares observed by the EUV variability experiment (EVE). *Solar Phys.* **279**, 23. DOI.

Chen, F., Ding, M.D.: 2010, Evidence of explosive evaporation in a microflare observed by Hinode/EIS. *Astrophys. J.* **724**, 640. DOI.

Chifor, C., Young, P.R., Isobe, H., Mason, H.E., Tripathi, D., Hara, H., Yokoyama, T.: 2008, An active region jet observed with Hinode. *Astron. Astrophys.* **481**, L57. DOI.

Christian, D.J., Mathioudakis, M., Bloomfield, D.S., Dupuis, J., Keenan, F.P., Pollacco, D.L., Malina, R.F.: 2006, Opacity in the upper atmospheres of active stars. II. AD Leonis. *Astron. Astrophys.* **454**, 889. DOI.

Christian, D.J., Mathioudakis, M., Jevremović, D., Dupuis, J., Vennes, S., Kawka, A.: 2003, The extreme-ultraviolet continuum of a strong stellar flare. *Astrophys. J. Lett.* **593**, L105. DOI.

Chubb, T.A., Friedman, H., Kreplin, R.W., Kupperian, J.E. Jr.: 1957, Lyman alpha and X-ray emissions during a small solar flare. *J. Geophys. Res.* **62**, 389. DOI.

Culhane, J.L., Harra, L.K., James, A.M., Al-Janabi, K., Bradley, L.J., Chaudry, R.A., Rees, K., Tandy, J.A., Thomas, P., Whillock, M.C.R., Winter, B., Doschek, G.A., Korendyke, C.M., Brown, C.M., Myers, S., Mariska, J., Seely, J., Lang, J., Kent, B.J., Shaughnessy, B.M., Young, P.R., Simnett, G.M., Castelli, C.M., Mahmoud, S., Mapson-Menard, H., Probyn, B.J., Thomas, R.J., Davila, J., Dere, K., Windt, D., Shea, J., Hagood, R., Moye, R., Hara, H., Watanabe, T., Matsuzaki, K., Kosugi, T., Hansteen, V., Wikstol, Ø.: 2007, The EUV imaging spectrometer for Hinode. *Solar Phys.* **243**, 19. DOI.

Czaykowska, A., de Pontieu, B., Alexander, D., Rank, G.: 1999, Evidence for chromospheric evaporation in the late gradual flare phase from SOHO/CDS observations. *Astrophys. J. Lett.* **521**, L75. DOI.

De Pontieu, B., Title, A.M., Lemen, J.R., Kushner, G.D., Akin, D.J., Allard, B., Berger, T., Boerner, P., Cheung, M., Chou, C., Drake, J.F., Duncan, D.W., Freeland, S., Heyman, G.F., Hoffman, C., Hurlburt, N.E., Lindgren, R.W., Mathur, D., Rehse, R., Sabolish, D., Seguin, R., Schrijver, C.J., Tarbell, T.D., Wülser, J.P., Wolfson, C.J., Yanari, C., Mudge, J., Nguyen-Phuc, N., Timmons, R., van Bezooijen, R., Weingrod, I., Brookner, R., Butcher, G., Dougherty, B., Eder, J., Knagenhjelm, V., Larsen, S., Mansir, D., Phan, L., Boyle, P., Cheimets, P.N., DeLuca, E.E., Golub, L., Gates, R., Hertz, E., McKillop, S., Park, S., Perry, T., Podgorski, W.A., Reeves, K., Saar, S., Testa, P., Tian, H., Weber, M., Dunn, C., Eccles, S., Jaeggli, S.A., Kankelborg, C.C., Mashburn, K., Pust, N., Springer, L., Carvalho, R., Kleint, L., Marmie, J., Mazmanian, E., Pereira, T.M.D., Sawyer, S., Strong, J., Worden, S.P., Carlsson, M., Hansteen, V.H., Leenaarts, J., Wiesmann, M., Aloise, J., Chu, K.C., Bush, R.I., Scherrer, P.H., Brekke, P., Martinez-Sykora, J., Lites, B.W., McIntosh, S.W., Uitenbroek, H., Okamoto, T.J., Gummin, M.A., Auker, G., Jerram, P., Pool, P., Waltham, N: 2014, The Interface Region Imaging Spectrograph (IRIS). *Solar Phys.* **289**, 2733. DOI.

Del Zanna, G.: 2008, Flare lines in Hinode EIS spectra. *Astron. Astrophys.* **481**, L69. DOI.

Del Zanna, G., Berlicki, A., Schmieder, B., Mason, H.E.: 2006, A multi-wavelength study of the compact M1 flare on October 22, 2002. *Solar Phys.* **234**, 95. DOI.

Del Zanna, G., Mitra-Kraev, U., Bradshaw, S.J., Mason, H.E., Asai, A.: 2011, The 22 May 2007 B-class flare: new insights from Hinode observations. *Astron. Astrophys.* **526**, A1. DOI.

Dere, K.P., Cook, J.W.: 1979, The decay of the 1973 August 9 flare. *Astrophys. J.* **229**, 772. DOI.

Ding, M.D., Schleicher, H.: 1997, Lyman continuum as a diagnostic for nonthermal processes in solar flares. *Astron. Astrophys.* **322**, 674.

Donnelly, R.F.: 1969, Energetic X-ray and extreme-ultraviolet flashes of solar flares. *Astrophys. J. Lett.* **158**, L165. DOI.

Donnelly, R.F., Kane, S.R.: 1978, Impulsive extreme-ultraviolet and hard X-ray emission during solar flares. *Astrophys. J.* **222**, 1043. DOI.

Doschek, G.A., Feldman, U., Kreplin, R.W., Cohen, L.: 1980, High-resolution X-ray spectra of solar flares. III – General spectral properties of X1-X5 type flares. *Astrophys. J.* **239**, 725. DOI.

Doschek, G.A., Mariska, J.T., Sakao, T.: 1996, Soft X-ray flare dynamics. *Astrophys. J.* **459**, 823. DOI.

Doschek, G.A., Mariska, J.T., Warren, H.P., Culhane, L., Watanabe, T., Young, P.R., Mason, H.E., Dere, K.P.: 2007, The temperature and density structure of an active region observed with the extreme-ultraviolet imaging spectrometer on Hinode. *Publ. Astron. Soc. Japan* **59**, 707.

Doschek, G.A., Mariska, J.T., Watanabe, T., Hiei, E., Lang, J., Culhane, J.L., Bentley, R.D., Brown, C.M., Feldman, U., Phillips, A.T., Phillips, K.J.H., Sterling, A.C.: 1992, YOHKOH Bragg crystal spectrometer

observations of the dynamics and temperature behavior of a soft X-ray flare. *Publ. Astron. Soc. Japan* **44**, L95.
Doschek, G.A., Warren, H.P.: 2005, Chromospheric evaporation in solar flares revisited. *Astrophys. J.* **629**, 1150. DOI.
Doschek, G.A., Warren, H.P., Mariska, J.T., Muglach, K., Culhane, J.L., Hara, H., Watanabe, T.: 2008, Flows and nonthermal velocities in solar active regions observed with the EUV imaging spectrometer on Hinode: a tracer of active region sources of heliospheric magnetic fields? *Astrophys. J.* **686**, 1362. DOI.
Doschek, G.A., Warren, H.P., Young, P.R.: 2013, Chromospheric evaporation in an M1.8 flare observed by the extreme-ultraviolet imaging spectrometer on Hinode. *Astrophys. J.* **767**, 55. DOI.
Doyle, J.G., Cook, J.W.: 1992, The sun as a star – high spectral resolution solar data degraded to low-dispersion IUE resolution. *Astrophys. J.* **391**, 393. DOI.
Doyle, J.G., Giunta, A., Madjarska, M.S., Summers, H., O'Mullane, M., Singh, A.: 2013, Diagnosing transient ionization in dynamic events. *Astron. Astrophys.* **557**, L9. DOI.
Doyle, J.G., Raymond, J.C., Noyes, R.W., Kingston, A.E.: 1985, The extreme ultraviolet spectrum of sunspot plumes – part two – spectral diagnostics and implications for cooling. *Astrophys. J.* **297**, 816. DOI.
Emslie, A.G., Brown, J.C., Donnelly, R.F.: 1978, The inter-relationship of hard X-ray and EUV bursts during solar flares. *Solar Phys.* **57**, 175. DOI.
Emslie, A.G., Noyes, R.W.: 1978, The characteristics of impulsive solar EUV bursts. *Solar Phys.* **57**, 373.
Feldman, U., Doschek, G.A., Kreplin, R.W., Mariska, J.T.: 1980, High-resolution X-ray spectra of solar flares. IV – General spectral properties of M type flares. *Astrophys. J.* **241**, 1175. DOI.
Fischler, M.A., Bolles, R.C.: 1981, Random sample consensus: a paradigm for model fitting with applications to image analysis and automated cartography. *Commun. ACM* **24**(6), 381. DOI.
Fisher, G.H., Canfield, R.C., McClymont, A.N.: 1985, Flare loop radiative hydrodynamics – Part Six – Chromospheric evaporation due to heating by nonthermal electrons. *Astrophys. J.* **289**, 425. DOI.
Fletcher, L.: 2012, Solar chromospheric flares: energy release, transport and radiation. In: Golub, L., De Moortel, I., Shimizu, T. (eds.) *Fifth Hinode Science Meeting*, *Astron. Soc. Pac. Conf. Ser.* **456**, 183.
Fletcher, L., Dennis, B.R., Hudson, H.S., Krucker, S., Phillips, K., Veronig, A., Battaglia, M., Bone, L., Caspi, A., Chen, Q., Gallagher, P., Grigis, P.T., Ji, H., Liu, W., Milligan, R.O., Temmer, M.: 2011, An observational overview of solar flares. *Space Sci. Rev.* **159**, 19. DOI.
Fletcher, L., Simoes, P.J.A.: 2013, The impulsive phase in solar flares: recent multi-wavelength results and their implications for microwave modeling and observations. arXiv.
Francis, M.H.: 1981, Non-equilibrium ionization in the transition region network. *Solar Phys.* **69**, 239. DOI.
Gabriel, A.H., Phillips, K.J.H., Acton, L.W., Wolfson, C.J., Culhane, J.L., Rapley, C.G., Bentley, R.D., Kayat, M.A., Jordan, C., Antonucci, E.: 1981, Observations of the limb solar flare on 1980 April 30 with the SMM X-ray polychromator. *Astrophys. J. Lett.* **244**, L147. DOI.
Gallagher, P.T., Phillips, K.J.H., Lee, J., Keenan, F.P., Pinfield, D.J.: 2001, The extreme-ultraviolet structure and properties of a newly emerged active region. *Astrophys. J.* **558**, 411. DOI.
Graham, D.R., Cauzzi, G.: 2015, Temporal evolution of multiple evaporating ribbon sources in a solar flare. *Astrophys. J.* **807**, L22. DOI.
Graham, D.R., Fletcher, L., Hannah, I.G.: 2011, Hinode/EIS plasma diagnostics in the flaring solar chromosphere. *Astron. Astrophys.* **532**, A27. DOI.
Graham, D.R., Fletcher, L., Labrosse, N.: 2014, Determining energy balance in the flaring chromosphere from oxygen V line ratios. arXiv.
Graham, D.R., Hannah, I.G., Fletcher, L., Milligan, R.O.: 2013, The emission measure distribution of impulsive phase flare footpoints. *Astrophys. J.* **767**, 83. DOI.
Hall, L.A.: 1971, Solar flares in the extreme ultraviolet. *Solar Phys.* **21**, 167. DOI.
Hall, L.A., Hinteregger, H.E.: 1969, Solar EUV enhancements associated with flares. In: de Jager, C., Svestka, Z. (eds.) *Solar Flares and Space Research*, North-Holland, Amsterdam, 81.
Handy, B.N., Acton, L.W., Kankelborg, C.C., Wolfson, C.J., Akin, D.J., Bruner, M.E., Caravalho, R., Catura, R.C., Chevalier, R., Duncan, D.W., Edwards, C.G., Feinstein, C.N., Freeland, S.L., Friedlaender, F.M., Hoffmann, C.H., Hurlburt, N.E., Jurcevich, B.K., Katz, N.L., Kelly, G.A., Lemen, J.R., Levay, M., Lindgren, R.W., Mathur, D.P., Meyer, S.B., Morrison, S.J., Morrison, M.D., Nightingale, R.W., Pope, T.P., Rehse, R.A., Schrijver, C.J., Shine, R.A., Shing, L., Strong, K.T., Tarbell, T.D., Title, A.M., Torgerson, D.D., Golub, L., Bookbinder, J.A., Caldwell, D., Cheimets, P.N., Davis, W.N., Deluca, E.E., McMullen, R.A., Warren, H.P., Amato, D., Fisher, R., Maldonado, H., Parkinson, C.: 1999, The transition region and coronal explorer. *Solar Phys.* **187**, 229. DOI.
Hannah, I.G., Kontar, E.P.: 2012, Differential emission measures from the regularized inversion of Hinode and SDO data. *Astron. Astrophys.* **539**, A146. DOI.
Hara, H., Watanabe, T., Harra, L.K., Culhane, J.L., Young, P.R., Mariska, J.T., Doschek, G.A.: 2008, Coronal plasma motions near footpoints of active region loops revealed from spectroscopic observations with Hinode EIS. *Astrophys. J. Lett.* **678**, L67. DOI.

Harrison, R.A., Sawyer, E.C., Carter, M.K., Cruise, A.M., Cutler, R.M., Fludra, A., Hayes, R.W., Kent, B.J., Lang, J., Parker, D.J., Payne, J., Pike, C.D., Peskett, S.C., Richards, A.G., Gulhane, J.L., Norman, K., Breeveld, A.A., Breeveld, E.R., Al Janabi, K.F., McCalden, A.J., Parkinson, J.H., Self, D.G., Thomas, P.D., Poland, A.I., Thomas, R.J., Thompson, W.T., Kjeldseth-Moe, O., Brekke, P., Karud, J., Maltby, P., Aschenbach, B., Bräuninger, H., Kühne, M., Hollandt, J., Siegmund, O.H.W., Huber, M.C.E., Gabriel, A.H., Mason, H.E., Bromage, B.J.I.: 1995, The coronal diagnostic spectrometer for the solar and heliospheric observatory. *Solar Phys.* **162**, 233. DOI.

Heinzel, P., Kleint, L.: 2014, Hydrogen Balmer continuum in solar flares detected by the Interface Region Imaging Spectrograph (IRIS). *Astrophys. J. Lett.* **794**, L23. DOI.

Hirayama, T., Tanaka, K., Watanabe, T., Akita, K., Sakurai, T., Nishi, K.: 1985, Rocket observation of the EUV images of a solar flare and active regions. *Solar Phys.* **95**, 281. DOI.

Hudson, H.S.: 2007, Chromospheric flares. In: Heinzel, P., Dorotovič, I., Rutten, R.J. (eds.) *The Physics of Chromospheric Plasmas*, *Astron. Soc. Pac. Conf. Ser.* **368**, 365.

Hudson, H.S., Acton, L.W., Hirayama, T., Uchida, Y.: 1992, White-light flares observed by YOHKOH. *Publ. Astron. Soc. Japan* **44**, L77.

Hudson, H.S., Wolfson, C.J., Metcalf, T.R.: 2006, White-light flares: a TRACE/RHESSI overview. *Solar Phys.* **234**, 79. DOI.

Hudson, H.S., Woods, T.N., Chamberlin, P.C., Fletcher, L., Del Zanna, G., Didkovsky, L., Labrosse, N., Graham, D.: 2011, The EVE Doppler sensitivity and flare observations. *Solar Phys.* **273**, 69. DOI.

Ireland, J., Tolbert, A.K., Schwartz, R.A., Holman, G.D., Dennis, B.R.: 2013, Estimating the properties of hard X-ray solar flares by constraining model parameters. *Astrophys. J.* **769**, 89. DOI.

Johnson, H., Raymond, J.C., Murphy, N.A., Giordano, S., Ko, Y.K., Ciaravella, A., Suleiman, R.: 2011, Transition region emission from solar flares during the impulsive phase. *Astrophys. J.* **735**, 70. DOI.

Kane, S.R., Donnelly, R.F.: 1971, Impulsive hard X-ray and ultraviolet emission during solar flares. *Astrophys. J.* **164**, 151. DOI.

Kašparová, J., Varady, M., Heinzel, P., Karlický, M., Moravec, Z.: 2009, Response of optical hydrogen lines to beam heating. I. Electron beams. *Astron. Astrophys.* **499**, 923. DOI.

Kennedy, M.B., Milligan, R.O., Mathioudakis, M., Keenan, F.P.: 2013, Solar flare impulsive phase emission observed with SDO/EVE. *Astrophys. J.* **779**, 84. DOI.

Kerr, G.S., Fletcher, L.: 2014, Physical properties of white-light sources in the 2011 February 15 solar flare. *Astrophys. J.* **783**, 98. DOI.

Klimchuk, J.A., Patsourakos, S., Cargill, P.J.: 2008, Highly efficient modeling of dynamic coronal loops. *Astrophys. J.* **682**, 1351. DOI.

Kosugi, T., Matsuzaki, K., Sakao, T., Shimizu, T., Sone, Y., Tachikawa, S., Hashimoto, T., Minesugi, K., Ohnishi, A., Yamada, T., Tsuneta, S., Hara, H., Ichimoto, K., Suematsu, Y., Shimojo, M., Watanabe, T., Shimada, S., Davis, J.M., Hill, L.D., Owens, J.K., Title, A.M., Culhane, J.L., Harra, L.K., Doschek, G.A., Golub, L.: 2007, The Hinode (Solar-B) mission: an overview. *Solar Phys.* **243**, 3. DOI.

Kreplin, R.W., Chubb, T.A., Friedmann, H.: 1962, X-ray and Lyman-alpha emission from the Sun as measured from the NRL SR-1 satellite. *J. Geophys. Res.* **67**, 2231. DOI.

Kretzschmar, M.: 2011, The Sun as a star: observations of white-light flares. *Astron. Astrophys.* **530**, A84. DOI.

Kretzschmar, M., de Wit, T.D., Schmutz, W., Mekaoui, S., Hochedez, J.F., Dewitte, S.: 2010, The effect of flares on total solar irradiance. *Nat. Phys.* **6**, 690. DOI.

Kretzschmar, M., Dominique, M., Dammasch, I.E.: 2013, Sun-as-a-star observation of flares in Lyman α by the PROBA2/LYRA radiometer. *Solar Phys.* **286**, 221. DOI.

Lean, J.L., Picone, J.M., Emmert, J.T.: 2009, Quantitative forecasting of near-term solar activity and upper atmospheric density. *J. Geophys. Res.* **114**, 7301. DOI.

Lemaire, P., Gouttebroze, P., Vial, J.C., Curdt, W., Schühle, U., Wilhelm, K.: 2004, Flare observation of the Sun as a star by SUMER/SOHO in the hydrogen Lyman continuum. *Astron. Astrophys.* **418**, 737. DOI.

Li, P., Emslie, A.G., Mariska, J.T.: 1989, Soft X-ray diagnostics of electron-heated solar flare atmospheres. *Astrophys. J.* **341**, 1075. DOI.

Li, Y., Ding, M.D.: 2011, Different patterns of chromospheric evaporation in a flaring region observed with Hinode/EIS. *Astrophys. J.* **727**, 98. DOI.

Li, Y., Qiu, J., Ding, M.D.: 2012, Analysis and modeling of two flare loops observed by AIA and EIS. *Astrophys. J.* **758**, 52. DOI.

Li, Y., Qiu, J., Ding, M.D.: 2014, Heating and dynamics of two flare loop systems observed by AIA and EIS. *Astrophys. J.* **781**, 120. DOI.

Lilensten, J., Dudok de Wit, T., Kretzschmar, M., Amblard, P.O., Moussaoui, S., Aboudarham, J., Auchère, F.: 2008, Review on the solar spectral variability in the EUV for space weather purposes. *Ann. Geophys.* **26**, 269. DOI.

Lin, R.P., Dennis, B.R., Hurford, G.J., Smith, D.M., Zehnder, A., Harvey, P.R., Curtis, D.W., Pankow, D., Turin, P., Bester, M., Csillaghy, A., Lewis, M., Madden, N., van Beek, H.F., Appleby, M., Raudorf, T., McTiernan, J., Ramaty, R., Schmahl, E., Schwartz, R., Krucker, S., Abiad, R., Quinn, T., Berg, P., Hashii, M., Sterling, R., Jackson, R., Pratt, R., Campbell, R.D., Malone, D., Landis, D., Barrington-Leigh, C.P., Slassi-Sennou, S., Cork, C., Clark, D., Amato, D., Orwig, L., Boyle, R., Banks, I.S., Shirey, K., Tolbert, A.K., Zarro, D., Snow, F., Thomsen, K., Henneck, R., McHedlishvili, A., Ming, P., Fivian, M., Jordan, J., Wanner, R., Crubb, J., Preble, J., Matranga, M., Benz, A., Hudson, H., Canfield, R.C., Holman, G.D., Crannell, C., Kosugi, T., Emslie, A.G., Vilmer, N., Brown, J.C., Johns-Krull, C., Aschwanden, M., Metcalf, T., Conway, A.: 2002, The reuven Ramaty High-Energy Solar Spectroscopic Imager (RHESSI). *Solar Phys.* **210**, 3. DOI.

Linsky, J.L., Glackin, D.L., Chapman, R.D., Neupert, W.M., Thomas, R.J.: 1976, The solar XUV spectrum of He II. *Astrophys. J.* **203**, 509. DOI.

Lites, B.W., Cook, J.W.: 1979, A semiempirical model of the upper flare chromosphere. *Astrophys. J.* **228**, 598. DOI.

Liu, W., Petrosian, V., Mariska, J.T.: 2009, Combined modeling of acceleration, transport, and hydrodynamic response in solar flares. I. The numerical model. *Astrophys. J.* **702**, 1553. DOI.

Machado, M.E., Avrett, E.H., Vernazza, J.E., Noyes, R.W.: 1980, Semiempirical models of chromospheric flare regions. *Astrophys. J.* **242**, 336. DOI.

Machado, M.E., Noyes, R.W.: 1978, Lyman continuum observations of solar flares. *Solar Phys.* **59**, 129. DOI.

Mariska, J.T.: 1992, *The Solar Transition Region*, *Cambridge Astrophys. Ser.*, Cambridge University Press, New York.

McTiernan, J.M., Fisher, G.H., Li, P.: 1999, The solar flare soft X-ray differential emission measure and the Neupert effect at different temperatures. *Astrophys. J.* **514**, 472. DOI.

Milligan, R.O.: 2008, A hot microflare observed with RHESSI and Hinode. *Astrophys. J. Lett.* **680**, L157. DOI.

Milligan, R.O.: 2011, Spatially resolved nonthermal line broadening during the impulsive phase of a solar flare. *Astrophys. J.* **740**, 70. DOI.

Milligan, R.O., Chamberlin, P.C.: 2015, The temporal behaviour of Lyman-alpha during solar flares using SDO/EVE. arXiv.

Milligan, R.O., Chamberlin, P.C., Hudson, H.S., Woods, T.N., Mathioudakis, M., Fletcher, L., Kowalski, A.F., Keenan, F.P.: 2012, Observations of enhanced extreme ultraviolet continua during an X-class solar flare using SDO/EVE. *Astrophys. J. Lett.* **748**, L14. DOI.

Milligan, R.O., Dennis, B.R.: 2009, Velocity characteristics of evaporated plasma using Hinode/EUV imaging spectrometer. *Astrophys. J.* **699**, 968. DOI.

Milligan, R.O., Gallagher, P.T., Mathioudakis, M., Bloomfield, D.S., Keenan, F.P., Schwartz, R.A.: 2006a, RHESSI and SOHO CDS observations of explosive chromospheric evaporation. *Astrophys. J. Lett.* **638**, L117. DOI.

Milligan, R.O., Gallagher, P.T., Mathioudakis, M., Keenan, F.P.: 2006b, Observational evidence of gentle chromospheric evaporation during the impulsive phase of a solar flare. *Astrophys. J. Lett.* **642**, L169. DOI.

Milligan, R.O., Gallagher, P.T., Mathioudakis, M., Keenan, F.P., Bloomfield, D.S.: 2005, Plasma diagnostics of active-region evolution and implications for coronal heating. *Mon. Not. Roy. Astron. Soc.* **363**, 259. DOI.

Milligan, R.O., Kennedy, M.B., Mathioudakis, M., Keenan, F.P.: 2012, Time-dependent density diagnostics of solar flare plasmas using SDO/EVE. *Astrophys. J. Lett.* **755**, L16. DOI.

Milligan, R.O., Kerr, G.S., Dennis, B.R., Hudson, H.S., Fletcher, L., Allred, J.C., Chamberlin, P.C., Ireland, J., Mathioudakis, M., Keenan, F.P.: 2014, The radiated energy budget of chromospheric plasma in a major solar flare deduced from multi-wavelength observations. *Astrophys. J.* **793**, 70. DOI.

Milligan, R.O., McElroy, S.A.: 2013, Continuum contributions to the SDO/AIA passbands during solar flares. *Astrophys. J.* **777**, 12. DOI.

Neidig, D.F.: 1989, The importance of solar white-light flares. *Solar Phys.* **121**, 261. DOI.

Neupert, W.M.: 1968, Comparison of solar X-ray line emission with microwave emission during flares. *Astrophys. J. Lett.* **153**, L59. DOI.

Ning, Z., Cao, W.: 2011, Hard X-ray source distributions on EUV bright Kernels in a solar flare. *Solar Phys.* **269**, 283. DOI.

Noyes, R.W., Kalkofen, W.: 1970, The solar Lyman continuum and the structure of the solar chromosphere. *Solar Phys.* **15**, 120. DOI.

Pesnell, W.D., Thompson, B.J., Chamberlin, P.C.: 2012, The Solar Dynamics Observatory (SDO). *Solar Phys.* **275**, 3. DOI.

Peter, H.: 2010, Asymmetries of solar coronal extreme ultraviolet emission lines. *Astron. Astrophys.* **521**, A51. DOI.

Qian, L., Solomon, S.C., Mlynczak, M.G.: 2010, Model simulation of thermospheric response to recurrent geomagnetic forcing. *J. Geophys. Res.* **115**, 10301. DOI.

Raftery, C.L., Gallagher, P.T., Milligan, R.O., Klimchuk, J.A.: 2009, Multi-wavelength observations and modelling of a canonical solar flare. *Astron. Astrophys.* **494**, 1127. DOI.

Rubio da Costa, F., Fletcher, L., Labrosse, N., Zuccarello, F.: 2009, Observations of a solar flare and filament eruption in Lyman α and X-rays. *Astron. Astrophys.* **507**, 1005. DOI.

Russell, A.J.B., Fletcher, L.: 2013, Propagation of Alfvénic waves from corona to chromosphere and consequences for solar flares. *Astrophys. J.* **765**, 81. DOI.

Teriaca, L., Falchi, A., Cauzzi, G., Falciani, R., Smaldone, L.A., Andretta, V.: 2003, Solar and heliospheric observatory/coronal diagnostic spectrograph and ground-based observations of a two-ribbon flare: spatially resolved signatures of chromospheric evaporation. *Astrophys. J.* **588**, 596. DOI.

Tobiska, W.K., Bouwer, S.D.: 2006, New developments in SOLAR2000 for space research and operations. *Adv. Space Res.* **37**, 347. DOI.

Tobiska, W.K., Woods, T., Eparvier, F., Viereck, R., Floyd, L., Bouwer, D., Rottman, G., White, O.R.: 2000, The SOLAR2000 empirical solar irradiance model and forecast tool. *J. Atmos. Solar-Terr. Phys.* **62**, 1233. DOI.

Vernazza, J.E., Noyes, R.W.: 1972, Inhomogeneous structure of the solar chromosphere from Lyman-continuum data. *Solar Phys.* **22**, 358. DOI.

Warren, H.P., Feldman, U., Brown, C.M.: 2008, Solar observations of high-temperature emission with the extreme-ultraviolet imaging spectrometer on Hinode. *Astrophys. J.* **685**, 1277. DOI.

Warren, H.P., Mariska, J.T., Doschek, G.A.: 2013, Observations of thermal flare plasma with the EUV variability experiment. *Astrophys. J.* **770**, 116. DOI.

Watanabe, K., Masuda, S., Segawa, T.: 2012, Hinode flare catalogue. *Solar Phys.* **279**, 317. DOI.

Watanabe, T., Hara, H., Sterling, A.C., Harra, L.K.: 2010, Production of high-temperature plasmas during the early phases of a C9.7 flare. *Astrophys. J.* **719**, 213. DOI.

Watanabe, T., Hara, H., Yamamoto, N., Kato, D., Sakaue, H.A., Murakami, I., Kato, T., Nakamura, N., Young, P.R.: 2009, Fe XIII density diagnostics in the EIS observing wavelengths. *Astrophys. J.* **692**, 1294. DOI.

White, S.M., Thomas, R.J., Schwartz, R.A.: 2005, Updated expressions for determining temperatures and emission measures from goes soft X-ray measurements. *Solar Phys.* **227**, 231. DOI.

Wilhelm, K., Curdt, W., Marsch, E., Schühle, U., Lemaire, P., Gabriel, A., Vial, J.C., Grewing, M., Huber, M.C.E., Jordan, S.D., Poland, A.I., Thomas, R.J., Kühne, M., Timothy, J.G., Hassler, D.M., Siegmund, O.H.W.: 1995, SUMER – Solar Ultraviolet Measurements of Emitted Radiation. *Solar Phys.* **162**, 189. DOI.

Woods, T.N., Eparvier, F.G., Bailey, S.M., Chamberlin, P.C., Lean, J., Rottman, G.J., Solomon, S.C., Tobiska, W.K., Woodraska, D.L.: 2005, Solar EUV experiment (SEE): mission overview and first results. *J. Geophys. Res.* **110**, 1312. DOI.

Woods, T.N., Eparvier, F.G., Fontenla, J., Harder, J., Kopp, G., McClintock, W.E., Rottman, G., Smiley, B., Snow, M.: 2004, Solar irradiance variability during the October 2003 solar storm period. *Geophys. Res. Lett.* **31**, 10802. DOI.

Woods, T.N., Kopp, G., Chamberlin, P.C.: 2006, Contributions of the solar ultraviolet irradiance to the total solar irradiance during large flares. *J. Geophys. Res.* **111**, 10. DOI.

Young, P.R.: 2011, Deriving densities, column depths and filling factors from Hinode/EIS data, Version 2.2. EIS Software Note 15. http://hesperia.gsfc.nasa.gov/ssw/hinode/eis/doc/eis_notes/15_DENSITY/eis_swnote_15.pdf.

Young, P.R., Doschek, G.A., Warren, H.P., Hara, H.: 2013, Properties of a solar flare Kernel observed by Hinode and SDO. *Astrophys. J.* **766**, 127. DOI.

Young, P.R., Tian, H., Jaeggli, S.: 2015, The 2014 March 29 X-flare: subarcsecond resolution observations of Fe XXI λ1354.1. *Astrophys. J.* **799**, 218. DOI.

Young, P.R., Watanabe, T., Hara, H., Mariska, J.T.: 2009, High-precision density measurements in the solar corona. I. Analysis methods and results for Fe XII and Fe XIII. *Astron. Astrophys.* **495**, 587. DOI.

Zirin, H.: 1975, The helium chromosphere, coronal holes, and stellar X-rays. *Astrophys. J. Lett.* **199**, L63. DOI.

Reprinted from *Solar Physics* Journal, DOI 10.1007/s11207-015-0710-3

From Coronal Observations to MHD Simulations, the Building Blocks for 3D Models of Solar Flares (Invited Review)

M. Janvier[1] · G. Aulanier[2] · P. Démoulin[2]

Received: 2 March 2015 / Accepted: 18 May 2015 / Published online: 10 June 2015
© Springer Science+Business Media Dordrecht 2015

Abstract Solar flares are energetic events taking place in the Sun's atmosphere, and their effects can greatly impact the environment of the surrounding planets. In particular, eruptive flares, as opposed to confined flares, launch coronal mass ejections into the interplanetary medium, and as such, are one of the main drivers of space weather. After briefly reviewing the main characteristics of solar flares, we summarise the processes that can account for the build-up and release of energy during their evolution. In particular, we focus on the development of recent 3D numerical simulations that explain many of the observed flare features. These simulations can also provide predictions of the dynamical evolution of coronal and photospheric magnetic field. Here we present a few observational examples that, together with numerical modelling, point to the underlying physical mechanisms of the eruptions.

Keywords Flares, dynamics · Flares, relation to magnetic field · Magnetic fields, models · Coronal mass ejections · Magnetohydrodynamics

1. Introduction

Solar flares are sudden brightenings occurring in the atmosphere of our Sun that can be monitored in a large range of the electromagnetic spectrum. They are associated with intense radiations, the release of energetic particles, as well as in some cases the ejection of solar plasma and magnetic field in the form of coronal mass ejections (hereafter, CMEs). As such, solar flares are major drivers of space weather (Gosling *et al.*, 1991) as they can impact planetary environments (Prangé *et al.*, 2004) and in particular human activities (Schrijver *et al.*, 2014; Lugaz, 2015). Further away from our solar system, flares are also seen

Solar and Stellar Flares: Observations, Simulations, and Synergies
Guest Editors: Lyndsay Fletcher and Petr Heinzel

✉ M. Janvier
mjanvier@maths.dundee.ac.uk

1 Division of Mathematics, University of Dundee, Dundee DD1 4HN, Scotland, UK

2 LESIA, UMR 8109 (CNRS), Observatoire de Paris, 92195 Meudon Principal Cedex, France

to occur on other stars (Linsky, 2000), as reported by the new findings from the NASA-Kepler mission (Walkowicz *et al.*, 2011; Wichmann *et al.*, 2014). Therefore, understanding the mechanisms of solar flares is important, not only to better predict the changes occurring in the Earth's environment, but also, to a large extent, to understand star–planet interactions. This will ultimately help to assess planetary habitability conditions (Poppenhaeger, 2014; Strugarek *et al.*, 2014).

The first solar flare ever recorded was also one of the most powerful (although its energy content has been greatly debated; see McCracken *et al.*, 2001; Tsurutani *et al.*, 2003; Wolff *et al.*, 2012; Cliver and Dietrich, 2013). It occurred on 1 September 1859, as reported by Carrington and Hodgson (Carrington, 1859; Hodgson, 1859). To date, the flare of 4 November 2003 is the most energetic flare recorded with recent instrumentation. Its soft X-ray emission was above the X28 range. The existence of super-flares on other stars (Maehara *et al.*, 2012) therefore requires investigating whether flares more intense than that observed so far can occur on our aging Sun (Schrijver *et al.*, 2012; Aulanier *et al.*, 2013). This question, linked with the dynamo evolution in the Sun's interior, also requires a further understanding on how the energy of flares is stored, and what mechanisms can account for its release. Note that flares are seen to occur over a large range of the energy spectrum: small-scale flares, associated with rather low energy events (10^{16} to 10^{20} J), are monitored in the quiet regions of the Sun. These so-called nanoflares or subflares will not be the focus of the present article.

With ever-increasing spatial and temporal resolutions, instruments onboard more recent spacecraft now monitor the Sun in different wavelengths, providing detailed observations of the evolution of the photospheric, chromospheric and coronal plasma and magnetic field. They therefore provide strong constraints on models of solar flares. As such, understanding their mechanisms encompasses a variety of approaches, from deciphering the observational features to developing numerical models that capture their underlying physics. Dedicated reviews of solar flares are numerous: while some review the relevant physics to describe the magnetic conversion process during flares (see *e.g.* Priest and Forbes, 2002; Shibata and Magara, 2011), others report on the variety of their observational features (see *e.g.* Krucker *et al.*, 2008; Fletcher *et al.*, 2011). Although reviews on the models and simulations developed to explain flares are also available (*e.g.* Chen, 2011), they generally focus on a 2D or 2.5D description of the magnetic field in the solar atmosphere. To fill this gap, we present in the following sections a review of the recent modelling developments that incorporate the intrinsic, and therefore essentially 3D nature of solar flares. We will see the wealth of methods now developed to extrapolate the coronal magnetic field, to characterise 3D topological features, and to reproduce the dynamics of the field in 3D. The present review aims at explaining how these techniques and observations provide the building blocks to construct 3D models of solar flares. In particular, we propose in the final section how different features of eruptive flares can be combined to construct a comprehensive, yet rather complete, standard 3D model for eruptive flares.

Section 2 briefly reviews the observational characteristics of solar flares, with a particular focus on the differences between confined and eruptive flares. Section 3 reviews the current understanding of energy storage and introduces the concept of topology to understand where flare energy release is initiated. Section 4 discusses the evolution of flares, from their onset through their decline, and in Section 5 we focus on eruptive flares leading to the ejection of flux ropes and the characteristics of the 3D magnetic reconnection process. Finally, conclusions to this article are presented in Section 6.

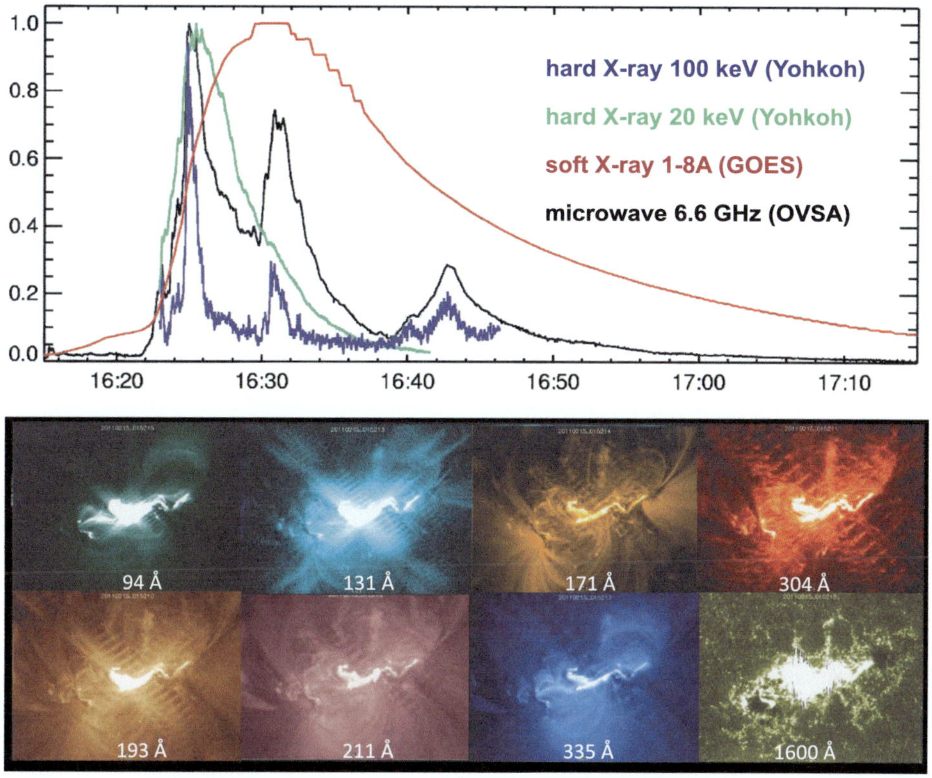

Figure 1 Top: Emissions of the 19 October 2001 X1.6 flare (SOL2001-10-19T16:30) observed with different instruments. The hard X-ray observations were made with the HXT/*Yohkoh* (Kosugi *et al.*, 1991), the soft X-ray observations with the GOES spacecraft, and finally, the microwave observations come from the *Owens Valley Solar Array* (OVSA, Gary and Hurford, 1990). The figure is adapted from Qiu, Gary, and Fleishman (2009). Bottom: the 15 February 2011 X2.2 flare (SOL2011-02-15T01:56) in AR 11198 (at 01:45 UT) as observed with the different filters of the AIA/SDO instrument. The seven EUV channels (from 94 Å to 335 Å) are separated into the ionised iron-dominated coronal channels (first three columns) and the He II-dominated chromosphere 304 Å, while the 1600 Å in the EUV shows the emission in the upper photosphere. The ribbons are most clearly visible at 335 Å while the flare loops are seen at a variety of wavelengths (*e.g.* 94 Å and 131 Å) emitting at different temperatures (adapted from Schrijver *et al.*, 2011).

2. Observational Characteristics of Confined and Eruptive Flares

2.1. Flare Emissions

Flares are defined as impulsive releases of radiative energy over a large range of the electromagnetic spectrum (see Figure 1). The classification of flares is made with soft X-rays, using the 1 – 8 Å band of the GOES-5 satellite (*Geostationary Orbiting Environmental Satellites*). This classification ranks flares from the A class to the X class (A, B, C, M, X), with the X class associated with the most energetic flares (with a GOES flux exceeding 10^{-4} Wm^{-2} at Earth, and the other classes decreasing by one order of magnitude). Weak events occurring at an energy level about 10^{-9} times lower than large flares have been named "nanoflares", and they are being investigated as a possible mechanism for the heating of the corona (see Klimchuk and Bradshaw, 2014, and references therein), following the seminal idea of Parker

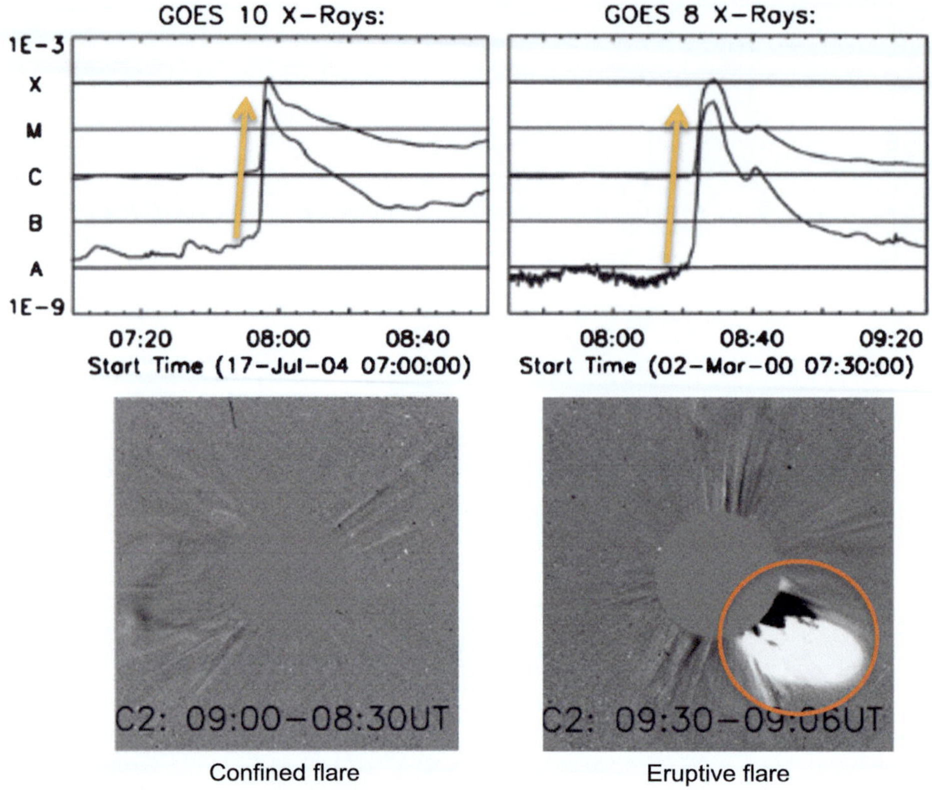

Figure 2 An example of two flares showing a similar GOES Soft X-ray evolution (top) with a similar impulsive and decay phase time scale, while the running difference coronagraph images (bottom) show the large-scale effect of an eruptive flare with the launch of a CME only for one of them (adapted from Wang and Zhang, 2007).

(1988). In between, microflares, with energies of order 10^{-6} times those of large flares, have been statistically investigated by Hannah *et al.* (2011). The authors, using the *Reuven Ramaty High Energy Solar Spectroscopic Imager* (RHESSI, Lin *et al.*, 2002), showed that the physics that underly larger flare events can be extended down to smaller events. This is a remarkable outcome of the survey of flares and their emission within a large range of the electromagnetic spectrum. Indeed, the distribution of flare parameters, such as their estimated radiated energy, can be represented by a power law (see *e.g.* Figure 3 of Schrijver *et al.*, 2012 and references in the article).

A distinction can be made between confined and eruptive flaring events, a classification that was proposed already back in the 1980s by Švestka (1986). Confined flares are generally impulsive in time and compact in space, and the flare loops (see Section 2.2) contain most of the energy. They are accompanied by a population of accelerated particles and intense plasma radiation indicating that strong heating occurs. Eruptive flares, on the contrary, generally extend to a large volume of the corona and lead to the ejection of solar material in the form of coronal mass ejections. Although a distinction cannot be made between those two flare categories by only looking at the light curves, a coronagraph is generally a straightforward way to define them, as shown in Figure 2. Note that if eruptive flares are defined by the presence of a CME, the inverse is not true: CMEs have been seen leaving the solar atmosphere without an associated flare emission. These cases, referred to as stealth CMEs (see the review of Howard and Harrison, 2013, and references therein), can be associated with coronal phenomena invisible to present instruments, and therefore should be considered as a space weather prediction pitfall, as they are less well monitored.

Yashiro *et al.* (2005) investigated the link between CMEs and flares by associating events observed over a period of 7 years, and found that the CME association rate with the intensity of the flare increases from 20 % for C-class flares (between C3 and C9 levels) to 100 % for large flares (above the X3 level). However, X-class flares are not necessarily all associated with CMEs as the recent series of X-class flares of October 2014 in AR 12192 has shown (Thalmann *et al.*, 2015; see also the studies by Feynman and Hundhausen, 1994 and Green *et al.*, 2002). This therefore leaves the question of the relation between intense flares and eruptiveness still open.

Both eruptive and confined events share some similarities in the evolution of the flare emission, as they are both characterised by an increase in radiation and are associated with emissions due to a population of non-thermal electrons, which is generally not detected in the non-flaring corona. However, eruptive flares are generally associated with long duration events, while confined flares typically have a much shorter time scale, as was shown in the study of Yashiro *et al.* (2006). The flare emissions for both types of events in different wavelengths do not peak at the same time (see also Fletcher *et al.*, 2011, and references therein): typically, a brightening in soft X-rays of a few keV is first detected, associated with the chromospheric plasma heated to coronal temperatures. Also, radio emissions in the metric range, such as type III radio bursts caused by escaping electron beams, give an indication that the pre-flare phase is not entirely thermal. Most of the flare energy is then released during a short-span phase referred to as the impulsive phase, which is highly non-thermal. It is generally associated with hard X-ray emissions due to energetic electrons, as well as gamma-ray line emission caused by energetic ions. It is believed that most of the free coronal energy is released during this phase under the form of non-thermal particles, with electrons reaching energies up to 300 MeV, and ions up to 10 GeV. As the flare progresses during this phase, the soft X-ray emission increases in size and total flux, while the lower chromosphere also radiates more, as indicated by a growing Hα emission. After the impulsive phase the decay phase starts, where most of the emissions due to non-thermal particles have disappeared. The hot plasma decreases in temperature to the point that it also completely disappears, and flare loops, filled with evaporated dense plasma, emit in Hα.

2.2. Flare Loops, Ribbons and Kernels

One of the most remarkable features of flares is the appearance of very hot coronal loops ($T \geq 10$ MK), referred to as flare loops, which remain above the flaring region (contrary to the CME during eruptive flares). Since they emit in different temperature ranges, it is possible to see their evolution in time with different filters. Looking at this evolution in just one filter is, however, misleading, as these loops would be seen growing in size (Figure 3(a)). However, following the same loop, it typically appears first in hot filters and progressively shifts to colder filters while it shrinks in height (Forbes and Acton, 1996). Moreover, newly formed loops are forming above already existing loops (*e.g.* Raftery *et al.*, 2009; Sun *et al.*, 2013).

Another characteristics of these flare loops is the evolution of their shear (Figure 3(b)). The shear angle is defined between the polarity inversion line (PIL) and the line that joins their footpoints, and as such, requires flares with a well-defined two-ribbon configuration and that are mostly observed near center-disk (so as to avoid any projection effects). This evolution was studied in various papers (Asai *et al.*, 2003; Su *et al.*, 2006; Aulanier, Janvier,

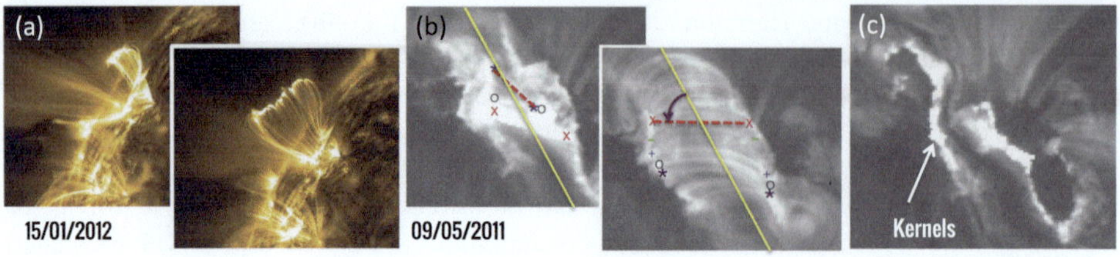

Figure 3 An example of the typical evolution of flare loops. (a) Low-to-high-altitude evolution of flare loops as seen at 171 Å in the 15 January 2012 flare (SOL2012-01-15T23:59) with the AIA/SDO (Lemen *et al.*, 2012). (b) Strong-to-weak shear evolution of flare loops as seen at 195 Å in the 11 May 2011 flare observed with the EUVI/STEREO A (Wuelser *et al.*, 2004). The shear angle is defined between the photospheric inversion line (PIL, yellow line) of the photospheric magnetic field and the line joining the two footpoints of one loop (red line). (c) The two ribbons appearing in the early phase of the 09 May 2011 (SOL2011-05-09T20:59) flare show clumps of brightenings, or kernels (adapted from Aulanier, Janvier, and Schmieder, 2012).

and Schmieder, 2012), where conjugate footpoints at first indicated strongly sheared field lines, and later reveal less sheared coronal connections (Figure 3(b)). Su *et al.* (2007) also studied 50 X- and M-class flares observed with TRACE, associated with an arcade of flare loops, and found that 86 % of these flares showed a general decrease in the shear angle.

The flare ribbons, discussed above, are intense emissions forming at the footpoints of flare loops, and they are generally well seen before the appearance of the flare loops, mostly in Hα, but also in other EUV filters, as shown in Figure 3(b), (c). Flare ribbons appear in their fully developed state as coherent and elongated structures. They are frequently accompanied by spatially localised and strongly brightened clumps, referred to as kernels (Figure 3(c)). Because of their strong emission in chromospheric lines, it is believed that kernels and ribbons form as a result of strong heating of the chromospheric plasma, most likely by collisions with non-thermal electrons. The evaporation of this strongly heated plasma can be seen as Doppler shifts in the diagnostics of spectral lines, for example with the EIS/HINODE instrument, as investigated by del Zanna *et al.* (2006), Milligan and Dennis (2009) and Young *et al.* (2013). Some parts of flare ribbons are also associated with the sites of HXR emissions (see Fletcher and Hudson, 2001; Asai *et al.*, 2002; Reid *et al.*, 2012).

Eruptive flares generally display two ribbons with a large portion generally parallel to each other, and which also present a hook shape at their ends (Figure 3(c); see also Chandra *et al.*, 2009). It was shown, *e.g.* in Janvier *et al.* (2014), that this hook surrounds a dimming region generally associated with the footpoints of the ejected CME. For confined flares, flare ribbons can be more numerous (see for example Wang *et al.*, 2014) or have different shapes (see for example a circular ribbon configuration in Masson *et al.*, 2009). As such, the presence of two ribbons with a J-shape structure is interpreted as evidence of an eruptive flare.

2.3. Filaments/Prominences and CMEs

Prominences, which appear on the limb as bright structures, or filaments, dark filamentary objects on disk, are potential progenitors of eruptive flares (Figure 4(a)). These structures, which are observationally complex to understand (see the recent reviews of Schmieder and Aulanier, 2012; Engvold, 2015 and references therein), are indeed often associated with CMEs (Figure 4(b) and (c)). This association has been thoroughly investigated in a series of papers (Subramanian and Dere, 2001; see also the reviews of Schmieder, Malherbe, and Wu, 2014 and Webb, 2015). In particular, filament/prominence eruptions and CMEs are

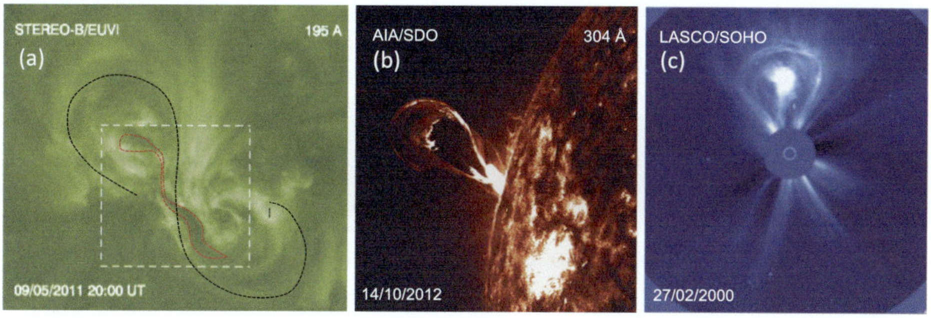

Figure 4 (a) Flare precursors can be seen prior to eruptions in some active regions. Here, the AR 11193, which led to a C-class flare on 09 May 2011, is shown one hour before the event. It features a filament (indicated with the red dashed lines) and a sigmoidal structure (black dashed lines). (b) The early phase of a CME eruption during the 14 October 2012 flare, accompanying an intense emission lower in the atmosphere, as seen with AIA/SDO. (c) A CME seen by the coronagraph LASCO onboard SOHO.

generally well associated, from 56 % to 92 % of the former associated with CMEs (Hori and Culhane, 2002; Gopalswamy *et al.*, 2003; Jing *et al.*, 2004).

Note that some flares, which would be confined by definition (no CME observed), clearly show the beginning of a prominence or filament eruption, although instead of lifting off to greater altitude, the structure remains in or returns to the region. Such cases are generally called failed eruptions (see for example in Ji *et al.*, 2003; Gilbert, Alexander, and Liu, 2007 and Joshi *et al.*, 2015).

In active regions, CME precursors are often seen as bright structure made of S- or J-shaped coronal loops, forming what is called a sigmoid (Rust and Kumar, 1996; Canfield, Hudson, and McKenzie, 1999; Green and Kliem, 2009). They are present in the vicinity of the filament (see the example in Figure 4(a)), and they are generally well seen in the X-ray range (see the catalogue of sigmoids in Savcheva *et al.*, 2014).

3. Energy Build-up and Release

The energy of a solar flare is typically estimated to be between 10^{28} to a few 10^{32} erg (see Schrijver *et al.*, 2012). With a plasma $\beta < 1$, the magnetic field drives the coronal activity, and its associated energy is the only component that can account for flare power. Looking at the evolution of light curves (for example those in Figure 2), several questions arise: How to explain the long duration energy storage phase, lasting from a few days to a few weeks? What are the mechanisms for the sudden energy release? What triggers flares, and how is the energy converted into other types of energy (*i.e.*, kinetic energy, radiation)? In the following, we review recent theoretical and numerical developments that try to answer these questions.

3.1. Mechanisms for Energy Build-up

For a given distribution of the normal component of a photospheric field, the potential field (current free) has the lowest magnetic energy. Any excess is called "free energy" and it is stored as currents in non-potential field configurations. Indeed, there is indirect observational evidence of strong electric currents stored in magnetic structures prior to an eruption onset.

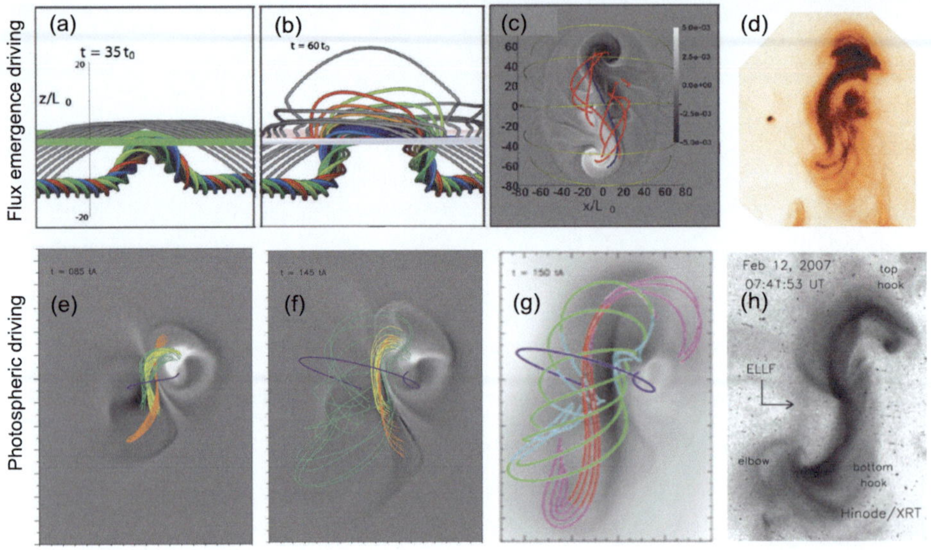

Figure 5 Two examples of 3D MHD simulations forming twisted flux tubes in the corona. (a, b) Leake, Linton, and Török (2013) proposed a flux-rope emergence simulation. The twisting motion of the polarities resulting from the emergence leads to current-carrying coronal loops. These loops are represented with the red lines in panel (c), while the background represents the normalised vertical current density J_z. The shapes of the red lines, especially the elongated one that runs almost parallel to the inversion line, are quite similar to the J/S-shape coronal loops found in the sigmoid of 12 February 2007 (panel (d) McKenzie and Canfield, 2008). (e, f) The simulation of Aulanier *et al.* (2010) forms a flux-rope structure via photospheric motions (twisting) and diffusion at the photosphere (panels (e) and (f)). This creates a set of field lines running above the inversion line (in green and orange) and a sigmoidal distribution of currents (background grey levels). The sigmoidal field lines and the integrated current (panel (g)) also reproduces well the shape of the 12 February 2007 sigmoid (panel (h)).

This is particularly the case with the X-ray emitting sigmoids, as investigated by *e.g.* Green *et al.* (2007), McKenzie and Canfield (2008), Savcheva *et al.* (2012a), and in the presence of twisted filaments/prominences (*e.g.* Williams *et al.*, 2005; Koleva *et al.*, 2012). How can such magnetic fields that store their energy in aligned currents be created?

Two mechanisms can account for the energy storage. One is the emergence of sub-photospheric current-carrying flux tubes from the convection zone (Figure 5, top row). Several authors have reported observational evidence that the Sun's magnetic field is often twisted on emergence (Leka *et al.*, 1996; van Driel-Gesztelyi, 1997; Luoni *et al.*, 2011).

Numerical simulations are heavily used to study the photospheric emergence since the physical problem is complex (see the reviews by Fan, 2009; Hood, Archontis, and MacTaggart, 2012; Cheung and Isobe, 2014). This complexity is in particular linked to the presence of a large stratification of the plasma and magnetic field (since the scale height of the photosphere is only about 150 km). This implies that a flux rope coming from the convective zone is no longer buoyant below the photosphere. There, its magnetic field is blocked forming a flat structure until enough magnetic flux has accumulated. This leads to the development of the so-called Parker instability, allowing the magnetic flux to emerge in pieces into the chromosphere and corona. As a result, the magnetic field is completely changed from its initial state during the emergence process, and finally it is a new flux rope that reforms in the corona (Hood *et al.*, 2009; Toriumi and Yokoyama, 2012). In some simulations, the coronal flux rope is eventually ejected (*e.g.* Archontis and Hood, 2012;

Kusano *et al.*, 2012). However, the key parameters leading to such an ejection are still not understood (see Section 4).

Another mechanism that can account for the formation of a current-carrying magnetic field is slow photospheric motions, for example by twisting the polarities, or by inducing shearing motions parallel to the inversion line. Moreover, flux ropes can form by reconnection of low field lines, such as in the model of van Ballegooijen and Martens (1989). This reconnection at the PIL occurs as the magnetic field diffuses gradually at the photosphere as found in numerical simulations (*e.g.* Amari *et al.*, 2003b; Aulanier *et al.*, 2010; Amari *et al.*, 2011; Amari, Canou, and Aly, 2014). There, the presence of current density layers at bald patches (regions where the magnetic field is tangent to the photosphere) leads to the reconnection of magnetic field lines that create a newly formed structure made of twisted flux bundles, embedded in an overlying, almost potential arcade. High electric currents confined in this structure are similar to the J- or S-shaped sigmoids often seen in observations (see Figure 5, bottom row).

3.2. Reconnection: A Mechanism for Flares

As discussed in the previous section, magnetic reconnection is thought to be the key mechanism that permits the evolution of the field in the storage phase and the eruptive phase of flares. Already recognising its importance back in the 1950s, Sweet (1958a) and Parker (1957) developed a theory, based on a work by Dungey (1953), that could explain how the magnetic energy contained in the corona could be released efficiently while producing the intense radiation recorded during flares. Such a mechanism is related to the fact that in an evolving magnetic configuration with a null point, a current sheet typically forms. This thin sheet introduces small scales where dissipative effects become important. In such a current sheet, the magnetic energy is dissipated into heat, kinetic energy and energetic particles. By applying this theory to solar flares, they defined a flaring configuration with four domains of connectivity. The reconnection of field lines between these regions converts magnetic energy into other forms of energy, while transforming the magnetic field configuration.

Since these pioneering works, there has been a tremendous number of studies on magnetic reconnection, especially in relation to flares, and the reader is referred to previous reviews (Forbes, 2010; Shibata and Magara, 2011; Priest, 2014, Chapter 12). Here, we only note that in 2D, *i.e.* in the presence of separatrices, which are special field lines separating different domains of connectivity, the definition of reconnection is clear. It can be defined both mathematically by a change of connectivity domain for some field lines, and physically by the dissipation of the magnetic energy within thin layers. A definition that is both mathematical and physical is an important distinction to make, as in 3D, mathematically well-defined separations of connectivity domains do not necessarily exist (see Section 3.4), while the reconnection, defined physically as the dissipation of magnetic energy in thin electric current layers, remains a valid concept.

There is much evidence that the reconnection process takes place in the corona. One aspect of it is the formation of a population of non-thermal particles, which is also observed in plasma experiments and in *in-situ* data of the magnetosphere (see Chapter V.B in the review of Yamada, Kulsrud, and Ji, 2010). For solar flares, several authors have reported on the existence of hard X-ray sources above the flare loop tops (*e.g.* Masuda *et al.*, 1994; Hudson *et al.*, 2001; Sui and Holman, 2003), indicative of a source forming high-energy particles (see Figure 6(c)). In particular, Masuda *et al.* (1994) proposed a heating process due to high speed jets produced by reconnection, colliding with the top of the reconnected flare loops. Furthermore, flare loops are ordered in temperature with the outermost loops

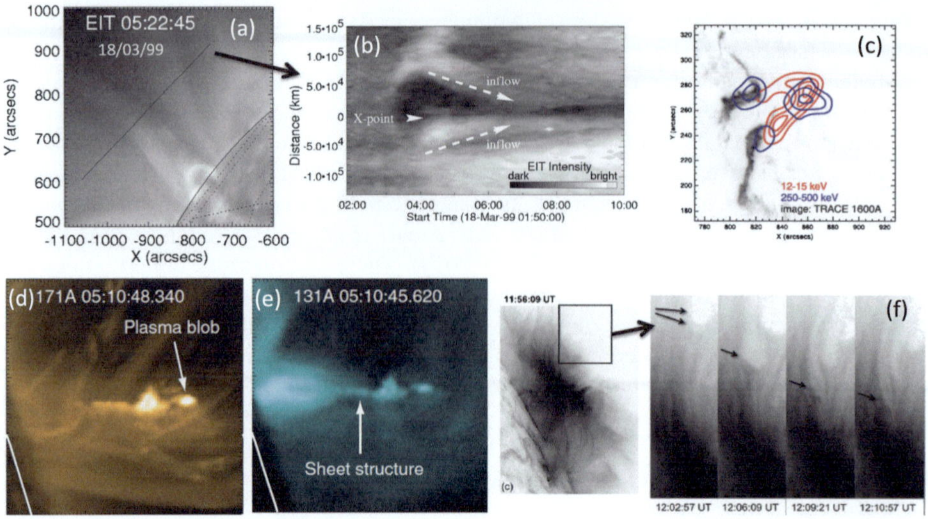

Figure 6 Examples of reconnection signatures in the solar corona. (a, b) Converging motions (inflows) toward a thin sheet-like region, above flare loops during the 18 March 1999 (SOL1999-03-18T04:04) event as observed by EIT/SOHO (Yokoyama *et al.*, 2001). Panel (b) is the time evolution of the emission in 195 Å along the slit shown in (a). (c) HXR emissions from RHESSI during the 20 January 2005 event (SOL2005-01-20T06:30, Krucker *et al.*, 2008), showing the presence of a strong emission source (> 250 keV, blue) above the top of the loop (red contours). (d, e) Observation of a plasmoid ejection during the 18 August 2010 flare (SOL2010-08-18T05:48) and its associated reconnection region, seen within two filters of AIA/SDO instrument (Takasao *et al.*, 2012). (f) Observation of a supra-arcade downflow seen as a void propagating sunward in the region above the flare loops, during the 22 October 2011 event (SOL2011-10-22T10:00) in the AIA 131 Å filter (Savage, McKenzie, and Reeves, 2012).

being the hottest, as expected for a reconnection process that forms them sequentially at larger height (see *e.g.* Forbes and Acton, 1996; Tsuneta, 1996).

Another strong evidence that magnetic reconnection is taking place is given by the presence of inflows and outflows from and to the reconnection site. For example, as reported by Yokoyama *et al.* (2001), the authors used EUV and soft X-ray observations of a flare with the EIT/SOHO (Delaboudinière *et al.*, 1995) and SXT/Yohkoh (Tsuneta *et al.*, 1991) instruments, and pointed out the pulling process of coronal loops in a region underneath a plasmoid ejection and above the top of the flare loops (Figure 6(a) and (b)). Using diagnostics from the soft X-ray emissions, they were able to deduce an inflow speed. Outflows, on the other hand, are often seen as void structures in EUV and SXR emission ranges above flare loops. Sunward-flowing voids are referred to as supra-arcade downflows (SADs; see McKenzie, 2011 and Figure 6(f)), propagating with a speed ranging from 50 to 500 $\mathrm{km\,s^{-1}}$ (sub-Alfvénic). They first have been interpreted as cross-sections of evacuated loops retracting from the reconnection region, and then re-interpreted as low-density trailing regions or wakes of much thinner retracting flux tubes (Savage, McKenzie, and Reeves, 2012). Inversely, plasmoids ejected away from the Sun have also been reported (Nishizuka *et al.*, 2010) in increasing numbers with the temporal and spatial capacity improvements of AIA onboard SDO (Takasao *et al.*, 2012 and Figure 6(d) and (e)). Using radio emissions, more refined analysis of the dynamics of these plasmoids provides more insights into the reconnection mechanism at the origin of their ejection (*e.g.* Nishizuka *et al.*, 2015). Finally, inflows and outflows are often seen together, as reported in the coronal observations of Yokoyama *et al.* (2001), Li and Zhang (2009), Savage *et al.* (2012), Takasao *et al.* (2012), Su *et al.* (2013).

When magnetic energy is released in the corona, a significant part of this energy is also transported along field lines by thermal conduction fronts, and/or energetic particles, and/or Poynting flux (*e.g.* Fletcher and Hudson, 2008; Birn *et al.*, 2009; Longcope and Tarr, 2012) toward the lower atmosphere. This generally makes the first energy release of flares (see Krucker, Fivian, and Lin, 2005; Saint-Hilaire and Benz, 2005, and references therein). Then the chromospheric plasma becomes hot (up to tens of millions of degrees, as seen in soft X-rays), as non-thermal particles collide with the particles of this atmospheric layer. Intense radiation can be seen as hard X-ray sources at the footpoints of bright flare loops, which used to be referred to as post-flare loops, (*e.g.* kernels; see Section 2.2), but also at the footpoints of twisted loops, as reported by Liu and Alexander (2009) and Guo *et al.* (2012). Note that for both cases, HXR sources were found at the endpoints of the associated filaments, appearing during a failed eruption. Eruptive flares do not usually show such intense emissions, although it is possible to find such cases (Musset, Vilmer, and Bommier, 2015). As the chromosphere expands, evaporation of hot plasma ensues, filling up coronal loops that are then seen as flare loops (see Section 2.2). Much of the energy coming from the non-thermal population is eventually converted into thermal energy and radiation, seen dominantly as the UV and optical components of the flare (see *e.g.* Milligan *et al.*, 2014). Therefore, measuring the radiative losses of flares is almost equivalent to the flare's energy release. Almost, since one would also need to measure the bulk kinetic energy from the ejecta and the escaping particle populations to obtain an accurate description of the total energy of a flare.

3.3. Using Topology to Find Reconnection Regions

In order to localise where reconnection could occur, one needs to search for the regions where the evolving magnetic field can form current layers, in other words, where ideal MHD breaks down. These regions are found by analysing the magnetic topology of the field.

The first models of reconnection were 2D. In 2D, the magnetic field can be separated in different domains of connectivity by special field lines that are called separatrices. Two separatrices cross at a null point, where the magnetic field vanishes (Sweet, 1958b). From the 1990s, a great deal of work has been done to extend those mathematically defined features from 2D to 3D. Separatrices in 3D are surfaces, which delimit different domains of connectivity. The intersection of two separatrices defines a separator, which is a privileged location for reconnection to occur. It is a special field line joining two null points. The complete topology description of a magnetic field is given by the skeleton formed by null points, separators and separatrices. Reconnection taking place at each of these topological features in 3D has been at the core of several works which explored the complexity of possible magnetic topologies (*e.g.* Longcope and Cowley, 1996; Parnell *et al.*, 1996; Priest and Titov, 1996; Pontin, Hornig, and Priest, 2004; Pontin, Hornig, and Priest, 2005; Pontin, Bhattacharjee, and Galsgaard, 2007; Priest and Pontin, 2009; Parnell, Haynes, and Galsgaard, 2010, and references therein).

The application of the above theory to solar flares started to be developed around the 1980s by placing a few magnetic sources below the photosphere, so as to reproduce the observed magnetograms. The topology is then computed and compared to flare observations (Baum and Bratenahl, 1980; Gorbachev and Somov, 1988; Lau, 1993). The complexity of the model was later increased to include many more sources with positions and strengths being computed by a least square fit of the model to the magnetogram of a specific observation. This procedure defines the coronal magnetic field and its topology (via magnetic connectivity between magnetic sources). Several flare studies found that the observed flare ribbons are located in the vicinity of the computed separatrices. This is evidence that the energy is released by magnetic reconnection (*e.g.* Mandrini *et al.*, 1993; Démoulin *et al.*, 1994).

This approach was further developed in the so-called Magnetic Charge Topology (MCT) model (*e.g.* Barnes, Longcope, and Leka, 2005; Longcope, 2005), in particular by estimating the amount of electric current flowing along the separator before reconnection and the amount of flux reconnected during the observed flare. Although successful in describing reconnection signatures for a variety of solar active regions, the MCT model is limited since it requires one to first split the observed magnetogram into a series of separated polarities, which is ambiguous for unipolar regions. The topology is computed as if every polarity, modelled by a charge, is surrounded by flux-free regions. The above approach introduces an ensemble of null points and separatrices that are not all present in the coronal field computed from the photospheric magnetogram using force-free extrapolation methods.

3.4. Quasi-separatrix Layers

Flares with comparable characteristics are observed with and without magnetic null points and their associated separatrices (Démoulin, Hénoux, and Mandrini, 1994). Therefore, extending the concept of separatrices to regions where the magnetic field connectivity remains continuous but changes with a strong gradient, Démoulin *et al.* (1996) introduced the concept of quasi-separatrix layers (QSLs). These are mathematically defined with the Jacobi matrix of connectivity transformation from a photospheric positive polarity domain to a negative polarity domain, or simply put, by the field line mapping of the magnetic field. Regions where the magnetic field is strongly distorted, or "squashed", are defined by a strong value of the squashing degree Q (Titov, Hornig, and Démoulin, 2002; Pariat and Démoulin, 2012). As such, computing QSLs allows the magnetic topology of an active region to be defined when the coronal configuration is computed by any means (*e.g.* magnetic extrapolation or MHD simulation).

QSLs are a generalisation in 3D of the concept of separatrices. In the limit where photospheric unipolar regions are separated by flux-free regions, or when a magnetic null enters in the coronal domain, or when a pair of nulls appears after a bifurcation, the central parts of QSLs become separatrices (Démoulin *et al.*, 1996; Restante, Aulanier, and Parnell, 2009). QSLs with large Q values still remain on both sides of a separatrix (where Q becomes infinite, Masson *et al.*, 2009). Finally, because of the strong distortion of the magnetic field, strong current density build-up typically occurs at QSLs during the magnetic field evolution (as shown in Aulanier, Pariat, and Démoulin, 2005; Büchner, 2006; Effenberger *et al.*, 2011; Janvier *et al.*, 2013). As such, QSLs are preferential locations for reconnection to take place. This was evidenced observationally by the correspondence between flare ribbons and QSL locations and by the presence of concentrated electric currents located at the border of the QSLs (Démoulin *et al.*, 1997; Mandrini *et al.*, 1997).

Hesse and Schindler (1988) and Schindler, Hesse, and Birn (1988) developed a general framework for 3D reconnection in the absence of null points and separatrices. Their formulation of the magnetic field with Euler potentials allows the description of the evolution of field lines and to pinpoint how and where they are cut, therefore accessing the physics of 3D reconnection. It occurs where localised non-ideal regions are set. QSLs are the locations where these non-idealness can occur as strong current density typically builds up there during an evolution. Therefore, the two approaches are complementary, with the QSLs providing the large-scale topology, localising where reconnection occurs, and the formalism of Hesse and Schindler describing the local reconnection process within the MHD framework (Démoulin, Priest, and Lonie, 1996; Richardson and Finn, 2012). These results were recently extended to kinetic numerical simulations (Wendel *et al.*, 2013; Finn *et al.*, 2014).

The QSL theory has been tested in several simple configurations, for example in cases where two interacting bipoles in a solar active region could easily be identified (Démoulin *et al.*, 1997). The theory has also been extensively used to understand more complex cases. A sample of the many other studies comparing successfully the location of reconnection signatures, typically flare ribbons with the locations of QSLs, can be found in Schmieder *et al.* (1997), Bagalá *et al.* (2000), Wang *et al.* (2000), Masson *et al.* (2009), Chandra *et al.* (2011), Savcheva *et al.* (2012b), Zhao *et al.* (2014), Dudík *et al.* (2014), Dalmasse *et al.* (2015). This close correspondence between flare ribbons and separatrices/QSLs found in various magnetic configurations, and derived from the photospheric magnetograms, is strong evidence that it is magnetic reconnection which is releasing the free magnetic energy in flares.

In the above flare applications, computing QSLs in a magnetic field volume is more demanding of computer resources than locating the null points in the MCT model, due to the large amount of field lines to process. However, the advantage of the QSL method is that it can be applied to any magnetic extrapolation technique, and any theoretical model (for example see Section 5.1). Furthermore, the calculation of QSLs can be made faster with the use of a refinement algorithm using an adaptive grid that progressively computes field lines only where the thin QSLs are located (Démoulin *et al.*, 1996; Pariat and Démoulin, 2012).

4. Magnetic Environment of Flares

4.1. Confined Flares: Emerging Flux

A model for emerging flux and energy dissipation was proposed by Heyvaerts, Priest, and Rust (1977) then Forbes and Priest (1984) for confined flares: as new flux emerges from the photosphere, it presses against overlying field structures, leading to the formation of a current density layer. As this process goes on, the current in the sheet grows, until a critical threshold is reached and a micro-instability is triggered. The occurrence of such an instability can lead to an increase in the local resistivity of the plasma, which in turn can rapidly dissipate the current (see Section 3.2). Then flux emergence is not only associated with the transfer of current-carrying magnetic field structures (see Section 3.1) from the solar convection zone to the atmosphere, it can also account for triggering energy releases by forcing the transformation of the magnetic field configuration.

On the other hand, large-scale flux emergence can also occur with the appearance of new polarities emerging in older regions, without being necessarily associated with an already formed flux rope. Such conditions then lead to the appearance of several magnetic field connectivity domains. For example, quite a few examples (*e.g.* Mandrini *et al.*, 1993; Gaizauskas *et al.*, 1998) demonstrated that almost oppositely oriented bipole that emerged between the two main polarities of an AR created topological conditions for current layers to form, and as such for flaring activity to occur. In particular, in the AR 2372 studied by Mandrini *et al.* (1993), some of the Hα flaring kernels were directly associated with flux emergence processes. Therefore, the appearance of several domains of connectivity, as well as their associated photospheric motions, are both favourable conditions for magnetic reconnection to take place. Since flux emergence is ubiquitous and continuously takes place in the

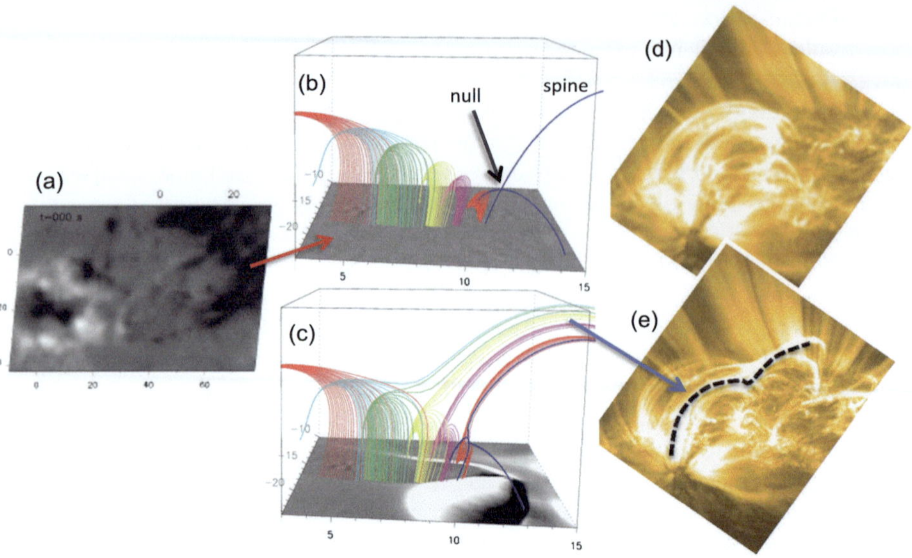

Figure 7 Example of a simulation of a confined flare. (a) Photospheric magnetogram on 16 November 2002 (SOL2002-11-16T13:58) with positive/negative polarity in light/dark grey. (b, c) Evolution of field lines in the MHD simulation before and after reconnection at the null point (OHM code, adapted from Masson *et al.*, 2009). The dark-blue field lines are separatrix field lines passing through the null point. The other coloured field lines are plotted with fixed footpoints in a positive flux area. The bottom boundary shows the distribution of the vertical electric current density (grey scale). (d, e) Comparison of the simulation to another confined flare: the flare loops before and after the occurrence of the flare on 11 March 2014 (SOL2014-03-11T03:50) recorded by AIA/SDO.

solar corona (Schrijver, 2009), this mechanism has been put forward as possibly an important process occurring in the majority of flares, and is in particular a rather straightforward process to trigger confined flares.

A test of flux emergence as a possible mechanism for confined flares has been proposed as follows. Masson *et al.* (2009) studied an important emergence occurring in the middle of AR 10191 and which led to a confined flare on 16/11/2002 (SOL2002-11-16T13:58). The photospheric motions are first modelled to reproduce the evolution of photospheric magnetograms from MDI/SOHO (Scherrer *et al.*, 1995), and a potential magnetic field extrapolation is used as an initial condition. The MHD evolution shows the build-up of localised electric currents, then the release of magnetic energy by reconnection. This magnetic reconnection takes place at the null point and the surrounding QSLs, and leads to the formation of newly reconnected field lines (Figure 7(b) and (c)) in good correspondence with coronal observations (Masson *et al.*, 2009). Note that although the simulation was defined as a case study for AR 10191, similar configurations of reconnected field lines can be found in other confined flares and now in other data-driven simulations of null-point reconnection. Figure 7(d) and (e) represent heated coronal loops before and after the 11 March 2014 (SOL2014-03-11T03:50) flare event obtained with the AIA instrument onboard SDO, and it is presented as an illustration.

4.2. Eruptive Flares: Unstable Flux Ropes

Since eruptive flares are defined as CME-accompanying flares, there is a necessity to understand what triggers the magnetic field configuration to erupt in the first place. Since this structure can remain stable for a period from days to weeks above the solar surface (see prominences/filaments in Section 2.3), a CME ejection can be seen as a sudden loss of balance between two acting forces: the magnetic tension of coronal arcades above the flux rope that pushes it downward, and the magnetic pressure of the flux rope that leads to further expansion. Then a search for a trigger mechanism must account for either the magnetic tension to decrease or for the magnetic pressure to increase. In the review of Aulanier (2014), the author discussed the necessary ingredients for a twisted structure to be expelled into the outer corona, which we briefly recall in the following.

Looking at the evolution of active regions or regions associated with prominences and filaments can help understand what the ingredients of triggering scenarios to be accounted for are. For example, during the decaying phase of active regions the dispersion of magnetic flux by convective motions implies a converging motion of opposite polarities along the PIL as well as flux cancellation taking place there (*e.g.* Martin, Livi, and Wang, 1985; Schmieder *et al.*, 2008; Green, Kliem, and Wallace, 2011). The coronal response to such photospheric evolution has been studied in many numerical models, such as the 3D simulations of Amari *et al.* (2003a), Mackay and van Ballegooijen (2006), Yeates and Mackay (2009), and Amari *et al.* (2011, 2014). While flux cancellation at the PIL leads to magnetic flux decrease in the photosphere, it transforms the magnetic arcades overlying the flux rope – in other words, the overlapping field lines, to field lines wrapping around this flux rope, so further building it. Then the photospheric evolution leads to a decrease in the coronal magnetic tension, a favourable condition to weaken the downward force acting on the flux rope.

Although several other processes can be identified as participating in the triggering processes (such as small- and large-scale reconnection inducing tension weakening, flare reconnection from below, shearing motions building up pressure below the overlying arcades, flux cancellation decreasing the arcade flux – see references in Aulanier, 2014), a question remains about the actual intrinsic mechanism that triggers flares. Such a model would take into account the different phenomena observed prior to flares, explaining the trigger of CME eruptions for a majority of the cases. Then several authors have studied flux-rope ejections with numerical simulations by bringing a flux rope to an unstable state (Figure 8): Amari *et al.* (2000), Lin, Forbes, and Isenberg (2001), Forbes *et al.* (2006), Fan and Gibson (2007), Aulanier *et al.* (2010), Olmedo and Zhang (2010), Zuccarello, Meliani, and Poedts (2012). The characterisation of this instability was proposed by Aulanier *et al.* (2010): from the formation of the flux rope (see Section 3.1), they studied its stability by turning down the driving forces (*i.e.* the photospheric boundary motions) at different times in the simulation. By doing so, they found that there is a threshold provided by the configuration of the flux rope, from which it becomes unstable. This instability is the torus instability (see also Kliem and Török, 2006).

Another scenario has been proposed for the magnetic tension reduction by Antiochos, DeVore, and Klimchuk (1999) and DeVore and Antiochos (2008). In their so-called breakout model, a current sheet is formed in a quadrupolar configuration at the null point located above a sheared magnetic field configuration (which could also host a flux rope, but not necessarily). At some point of the evolution a fast reconnection is triggered there. It rapidly removes part of the magnetic tension over the sheared arcade, inducing its eruption. Although this configuration can be observed in the solar atmosphere (see *e.g.* Aulanier *et al.*, 2000), it becomes problematic if too much flux accumulates at higher altitude, as then the magnetic tension reduction via reconnection becomes too inefficient to account for the sudden ejection of CMEs. Since this scenario involves mostly sheared arcades rather than a flux-rope structure, a main ingredient in the ejection of CMEs, we will focus in the following on simulations of torus-unstable flux ropes.

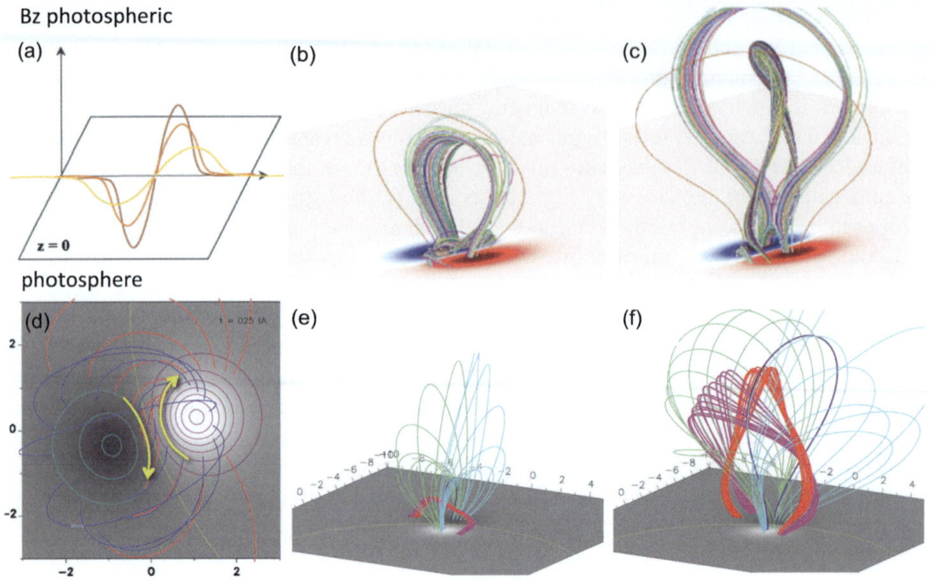

Figure 8 Examples of flux-rope formation and destabilisation in 3D numerical simulations. (a)–(c): magnetic flux dispersal (panel (a)) leading to a flux-rope formation then ejection (panels (b) and (c)), adapted from Amari *et al.* (2003b). (d)–(f): flux-rope formation at the bald patches following flux diffusion and shearing motion at the photosphere (yellow arrows panel (d)). Panels (e) and (f) represent two time steps before/after the flux rope has reached the torus-instability threshold, adapted from Aulanier *et al.* (2010).

5. Advances in Flare QSL Reconnection

5.1. QSLs in the Presence of Twisted Flux Tubes

It is necessary to understand the 3D topological features associated with flux ropes to explain the underlying processes linked to their ejection during an eruptive flare. In the analytical work of Démoulin, Priest, and Lonie (1996), it was shown that QSLs form a thin volume wrapping around flux tubes. Their mapping, down at the photospheric boundary, is representative of the twisted shape of the flux rope, with the hook part swirling more in a more twisted configuration (see Figure 9(a) and (b)). The evolution of QSLs during the formation of a flux rope by kinematic emergence through the photospheric boundary was investigated in Titov and Démoulin (1999). In particular, it was shown that as the flux rope fully emerges, the photospheric footprints of the QSLs split into two distinct traces, reminiscent of the Hα ribbons during flares (Titov, 2007).

This analytical flux rope (see Section 2 in Titov and Démoulin, 1999) was reproduced numerically by Savcheva *et al.* (2012b) for comparisons with MHD simulations and observations. The volume associated with the highest values of the squashing degree Q (see Section 3.4), defined as the hyperbolic flux tube (HFT, Titov, Hornig, and Démoulin, 2002), is located below the flux rope (see Figure 9(c)). This HFT volume maps all the way down to the photosphere to the beginning of the hook-shaped structure (see the red area in the photospheric footprints of the QSLs in Figure 9(d)). In the close vicinity of the HFT, different types of connectivities are present. Above are field lines from the flux rope, below are arcade-like field lines, and on the sides are large-scale arcade-like field lines surrounding the flux rope. This variety of connectivities in a small region implies that the QSLs, and

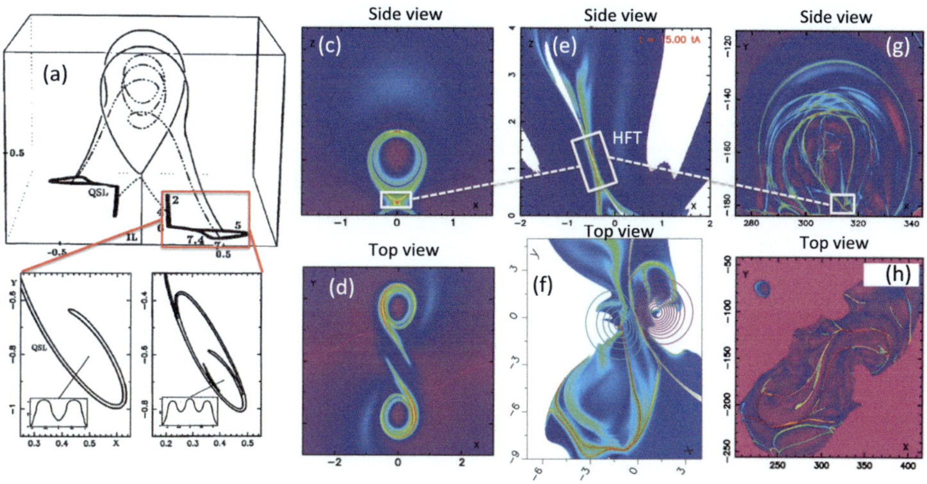

Figure 9 (a) QSLs footprints and field lines for an analytical model of a flux rope (Démoulin, Priest, and Lonie, 1996), (b) with different hook shapes depending on the degree of twist of the flux-rope structure. The footpoints of the flux rope are located inside the hooks of the photospheric QSLs. (c, d) Numerical computation from Pariat and Démoulin (2012) of the QSLs from the analytical Titov & Démoulin flux-rope model. The central cut (c) shows the tear-drop shape of the QSL volume surrounding the flux rope, and underneath it the HFT, where the squashing factor becomes the highest (in red). The top view (d) shows the photospheric signature with the hooks closing onto themselves. (e) QSLs seen from a 2D cut in the 3D flux-rope ejection numerical model (see Janvier *et al.*, 2013) showing the tear-drop shape above the HFT, as well as the cusp structure underneath it. (f) The photospheric footprints of the QSLs show an asymmetry, due to the magnetic polarity asymmetry. (g, h) QSLs computed from the flux-rope insertion extrapolation model of the 12 February 2007 event. The QSL computation renders a much more complex structure, although the tear-drop shape, cusp and HFT are discerned in the side view, as well as the hook-shaped photospheric QSLs (from Savcheva *et al.*, 2012b).

especially their core, the HFT, are privileged regions both for current layer build-up and for the occurrence of magnetic reconnection.

For configurations before eruptions, seen as on-disk sigmoid structures, it is interesting to compute the QSLs and this first requires one to compute the coronal magnetic field. One of the most successful methods to reproduce the presence of a twisted coronal magnetic structure is that of van Ballegooijen (2004), who proposed inserting a flux rope in the magnetic field of a sigmoid, extrapolated with a potential field, and leaving the system to relax using a magneto-frictional code. This method was successfully applied to several active regions containing sigmoids as evidence of the presence of a flux rope (Savcheva and van Ballegooijen, 2009; Savcheva *et al.*, 2012b; Su and van Ballegooijen, 2012). In particular, in Savcheva *et al.* (2012b), the authors calculated the QSLs using their extrapolation method. As shown in Figure 9(g) and (h), the structure of the QSLs, although more complicated due to the presence of more complex surrounding coronal loops, also show an HFT, tear-drop and cusp shapes above and beneath it, as well as hook-shaped photospheric signatures. As such, the presence of an HFT is an essential part of the sigmoidal structures seen in the corona, and as will be shown below, it holds the key to the underlying reconnection processes forming flaring structures.

From the results of the pre-eruptive magnetic field, one can predict that the flare reconnection will occur at the HFT. This analytical result was indeed verified in a dynamical MHD model reproducing the ejection of an unstable flux rope (Janvier *et al.*, 2013), starting from the original simulation set-up of Aulanier *et al.* (2010) but with no photospheric boundary driving applied. Computing the squashing degree in the 3D volume, the QSLs are seen to wrap around the modelled flux rope (Figure 9(e)) like in the analytical flux-rope model. With time, the tear-drop shape becomes bigger as the flux rope expands, and the HFT underneath it moves upward, following the eruption of the flux rope, while the cusp region becomes wider. Since this cusp structure indicates the delimitation between the regions where field lines have not reconnected (coronal field lines) and regions of reconnected field lines (the flux rope above the cusp, and the flare loops inside the cusp region), the cusp becomes wider as more flare loops are fed into the region.

The photospheric signatures of the QSLs, as seen in Figure 9(f), also present a hook-shaped structure, and the straight parts near the PIL move away from each other as the eruption evolves, while the hooks become rounder. Note that the hooks do not close onto themselves, as found in the analytical Titov–Démoulin model shown in Figure 9(c) and (d). This is because the flux-rope structure is not as twisted as that in the analytical model. A comparison with the morphology of the flare ribbons (see Section 5.3) indicates that flux ropes in the corona should not have twist much larger than one turn (Janvier *et al.*, 2014), contrary to what is suggested in kink-unstable flux-rope models.

5.2. Reconnection in 3D: Slipping Motion

Quasi-separatrix layers indicate regions of strong magnetic field distortion. When field lines are reconnecting with their neighbouring field lines in a QSL, their motion is seen as an apparent *flipping*, or *slipping* motion of field lines. This was analytically predicted by several authors (Priest and Forbes, 1992; Priest and Démoulin, 1995; Démoulin *et al.*, 1996).

Since it is impossible to follow elementary flux tubes in the corona, numerical simulations present an invaluable resource to study the dynamics of field lines during the reconnection process. Indeed, by fixing a field line starting point, one can evaluate its successive changes of connectivity by tracing the temporal evolution of this field line. Note that if the change of connectivity occurs at super-Alfvénic time scales, the apparent slipping motion is said to be "slip-running" (Aulanier *et al.*, 2006). This term refers to the fact that within the MHD paradigm information only travels with less than (or up to) the Alfvén speed in a low-β plasma. The fixed footpoint of the magnetic field line at one end does not actually "see" the change of connectivity if its opposite footpoint moves at super-Alfvénic speeds. Then this opposite footpoint is defined as the photospheric endpoint location of the same field line computed from the fixed footpoint location. At the scale of the system, for super-Alfvénic or slip-running motion, field lines behave as if they had reconnected in the presence of separatrices. Such motion has been reported in different simulation setups (Török *et al.*, 2009; Masson *et al.*, 2009; Masson *et al.*, 2012; Janvier *et al.*, 2013).

The characteristics of the slipping motion of field lines during an eruptive flare were studied by Janvier *et al.* (2013), and in particular the speed of the motion was quantified, for the flux-rope ejection simulation shown in the bottom row of Figure 8. An example of a slipping field line is presented in Figure 10 (panels (a) and (b)), where the same field line, defined by its footpoint anchored in the negative polarity, is traced at different times in the simulation. Since the location of the other footpoint is moving as the consequence of the field line reconnecting with its neighbouring field lines, it is possible to report for each time step the distance travelled (Figure 10(c), where the photospheric QSL footprints are also shown), and hence the slipping motion speed, which is shown in Figure 10(e). This slipping motion speed is not constant in time: rather, it has spikes, and its values can be sub- or super-Alfvénic.

As the location of the non-fixed footpoint moves along the QSL in the positive polarity, the other fixed point remains at the same location in the negative polarity. It is anchored

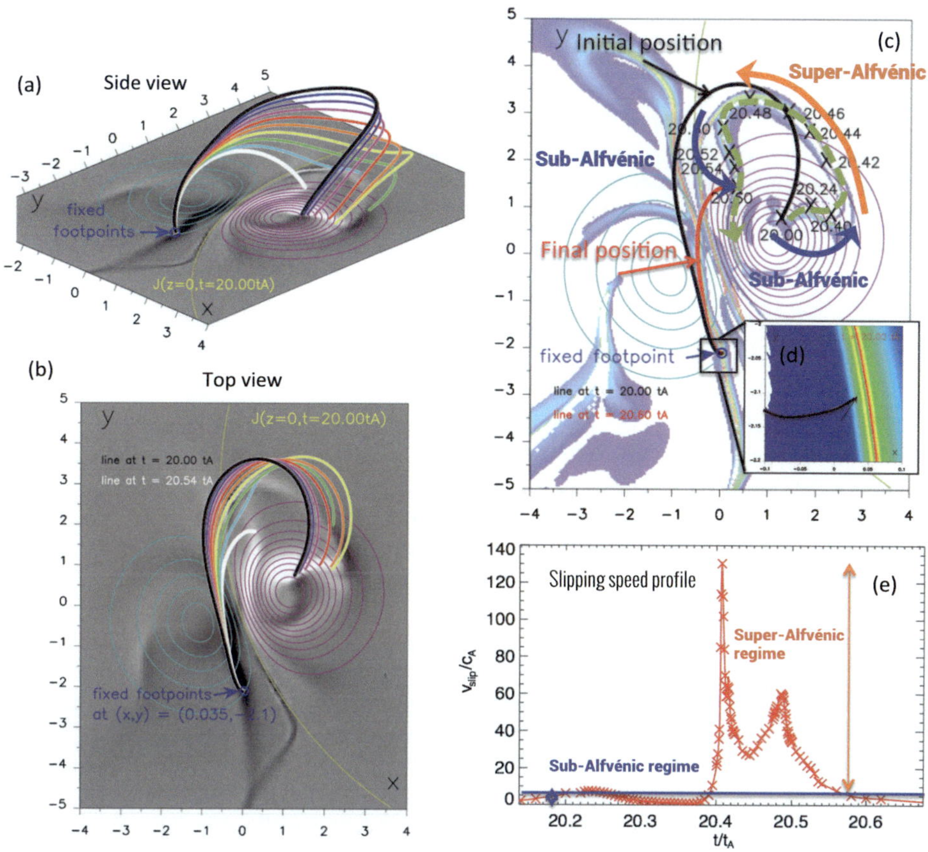

Figure 10 Deducing the speed of the slipping motion of a field line from a simulation (see Janvier *et al.*, 2013 for the simulation setup). (a)–(b): Side and top views, respectively, of the same field line plotted at different times as it reconnects with neighbouring field lines. The thick black line represents the initial position, while the change of colours from black to white represents different times in the simulation. The contourplots indicate the strength of the magnetic field for both the negative (cyan) and the positive (purple) polarities, while the yellow line represents the PIL. The background plot in grey scale indicates the strength of the photospheric electric current densities, where the black (resp. white) contours indicate the strongest negative (resp. positive) J_z. The positions of the moving footpoint as well as their corresponding time in the simulation are reported on a photospheric map of the QSLs (panel (c)), along with the initial and final position of the loop (in black and red). The regimes where the slipping speed is sub-Alfvénic or super-Alfvénic are reported as well. Sub-panel (d) shows a zoom of the photospheric footprint of the QSL surrounding the fixed footpoint region. (e) The speed of the slipping motion is reported at each timestep of the simulation, showing its non-constant evolution.

outside of the QSL before the reconnection takes place, as indicated in the zoomed region in Figure 10(d). However, as time evolves, this QSL moves away from the PIL, and hence "sweeps" the field line footpoint location. Indeed, as the flux rope is ejected away, the trailing reconnecting region moves upward, so that field lines passing through it reconnect with each other and the photospheric footprints of the QSLs in each polarity are then seen to move away from each other (similar to the CSHKP model, Carmichael, 1964; Sturrock, 1966; Hirayama, 1974; Kopp and Pneuman, 1976). As such, while the motion of the footpoint location in the positive polarity is *along* the QSL, the footpoint location in the negative polarity has a relative motion with respect to the QSL *perpendicular* to it. Then the speed of the moving footpoint location is super-Alfvénic when the fixed footpoint is in the core of the QSL, and sub-Alfvénic on both sides. Note that taking a field line that would be defined by a fixed footpoint location in the positive polarity would give the inverse result: the fixed footpoint would move perpendicularly with respect to the QSL footprint in the positive polarity, while its corresponding displaced footpoint location in the negative polarity would move along the QSL footprint. Then the slipping speed of the moving footpoint location is directly related to the distortion of the magnetic field, given by the field line mapping (N; see Equation (4) in Démoulin *et al.*, 1996) and the speed at which the QSL is displaced is: $v_{\text{slipping}} = N \times v_{\text{QSL}}$ (see Janvier *et al.*, 2013).

From its first prediction in the 1990s (Priest and Forbes, 1992; Priest and Démoulin, 1995; Démoulin *et al.*, 1996) and its numerical modelling (Aulanier, Pariat, and Démoulin, 2005; Török *et al.*, 2009; Masson *et al.*, 2009; Janvier *et al.*, 2013), the slipping motion of field lines has been observed for the first time in coronal loops (Aulanier *et al.*, 2007) and during eruptive flares (Sun *et al.*, 2013; Dudík *et al.*, 2014; Li and Zhang, 2014) very recently. These observations, at high temporal cadence, can only provide evidence of sub-Alfvénic motions, since it is the only regime for which the response of the plasma to the change of connectivity can be observed. The slipping field lines of the coronal loops (Aulanier *et al.*, 2007) was monitored with the XRT instrument onboard *Hinode*, and were seen to exchange their footpoint positions with a motion speed estimated between 30 to 150 $\text{km}\,\text{s}^{-1}$. Dudík *et al.* (2014) reported the slipping motion of field lines in the X-class eruptive 1 July 2012 (SOL2012-07-12T16:50) event observed the AIA/SDO instrument, and also showed the propagation of kernel brightenings along the flare ribbons, reminiscent of the motion of footpoints obtained in the MHD simulation of a similar-looking active region (Janvier *et al.*, 2013). The authors reported on apparent velocities ranging from 4 $\text{km}\,\text{s}^{-1}$ to 140 $\text{km}\,\text{s}^{-1}$, and the successive kernel brightenings were associated with the footpoints of the field lines forming the envelope of the flux rope that eventually gets ejected during the flare (a CME was indeed recorded for this event). Note that since this process should be in principle intrinsic to any flare (as they all have a 3D structure), it is not related to the particular zipping brightening motion reported in some flares (Tripathi, Isobe, and Mason, 2006). The latter can be explained by the propagation of the reconnection site in one direction due to an asymmetric filament eruption, in which case both slipping motion and zipping motion can take place, although not necessarily on the same timescales.

5.3. Electric Current Evolution

The formation of electric current layers is an essential component of flaring activity. These thin layers are predicted to form in the regions where the magnetic field is strongly distorted, *i.e.* at similar locations as the Quasi-Separatrix Layers (see Section 3.4). On the other hand, as shown in Figure 9, QSLs form a coronal volume that maps all the way down to the photospheric boundary, with a photospheric signature presenting a J-shape. Such a structure can be described in a 3D model of eruptive flares, where the core of the CME (*i.e.*, the flux rope) fills a coronal volume embedded in the QSLs (see Figure 9) mapping into the hooks of the QSL footprints, flare loops are embedded in the cusp region, and field lines are slipping within the QSL. The representation of such a model is given in Figure 11, panel (a), where these main characteristics have been put together. Then, do current layers have a similar morphology as that of the QSLs?

3D numerical modelling of eruptive flares has been able to answer this question by mapping the regions of high current densities. For example, in Kliem *et al.* (2013), the authors

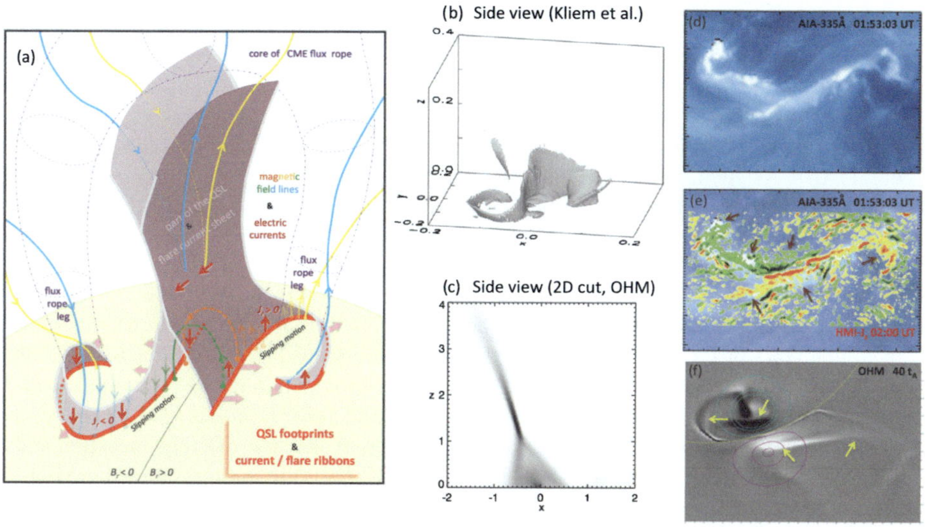

Figure 11 (a) The 3D standard model for eruptive flares (Janvier *et al.*, 2014) derived from the interpretation of eruptive flares with simulations of a torus-unstable flux rope. The grey surface indicates the QSL volume that wraps around the flux-rope structure (purple dashed lines), as well as the current density volume that maps to a similar location (although it does not extend above the flux rope; see panel (c)). The coloured field lines represent coronal loops that have slip-reconnected within the QSL, as indicated by the series of footpoints representing the slipping motion. The blue and yellow loops form the outer envelope of the flux rope, while the green and orange loops represent the newly formed flare loops. The footpoints of the flux rope are located within the hook of the QSL/current volume photospheric footprints, while the footpoints of the flare loops are located in the region delimited by the straight portion of these footprints. (b) Isosurface of (10 %) the maximum value of the current density, in the simulation of data-driven unstable flux rope (Kliem *et al.*, 2013), showing its extension within the simulated volume and its mapping all the way down to the photospheric boundary, and showing the hook structure wrapping around the flux rope. (c) 2D cross section of the current density volume, taken at the central part of the flux rope, in a torus-unstable flux-rope ejection model (Aulanier, Janvier, and Schmieder, 2012), showing the thin intense current portion (at a similar, though not exact location as the HFT, see Janvier *et al.*, 2013), as well as the cusp shape similar to that found for QSLs (see Figure 9). (d) Flare ribbons seen in the 335 Å filter of the AIA/SDO instrument during the 15 February 2011 X2.2 flare (SOL2011-02-15T01:56), showing a similar J-shaped structure as the photospheric current ribbons and QSLs. The intense emission locations have been compared with the location of strong electric currents derived from the HMI magnetograms, as indicated with the brown arrows in panel (e). (f) Current ribbons in the OHM simulation of an eruptive flux rope, showing similar changes (intense patches of current density, outward motion and hooks becoming rounder) as those found for current and flare ribbons from the observations (panels (d) and (e)).

used an extrapolated field of the AR 11060 as an initial condition to run their numerical code. The associated 3D current layer for the present model, obtained as a current density itself derived from the components of magnetic field obtained in the simulation, is shown in Figure 11(b). It presents a similar morphology as for the QSLs, although the high current density layer does not wrap around the flux rope as the QSLs do. This 3D view also shows the rounded hooks that surround the core region of the flux rope. In Janvier *et al.* (2013), the authors studied the temporal evolution of the current layer: a 2D cut in the central part of the flux-rope 3D volume clearly shows the cusp region as well as the current layer above it where reconnection takes place. Note, however, that the region of highest current density, found at the cusp in the 2D cut, is slightly offset compared with the location of the highest squashing factor location, namely the HFT (see Figure 2 of Janvier *et al.*, 2013 for comparisons). As the flux rope expands, this current layer moves upward and becomes thinner, while the hooks of the photospheric current layers (shown for example in Figure 11(f)) become rounder and the straight parts of each current "ribbon" separate from each other, away from the PIL, as more flux is fed into the flux-rope structure. As such, the 3D view contains the evolution of the field as represented by the 2D model. These modelled photospheric changes in electric current resemble those observed in bright flare ribbons during eruptive flares (*e.g.* Figure 11(d)). As such, the model for eruptive flares shown in Figure 11(a) incorporates the various ingredients of the evolution mentioned above. It represents a development from the 2D view of the standard model of eruptive flares in 2D (or CSHKP model) to its more complete and more accurate 3D version (see Janvier *et al.*, 2014).

A question that remains to be answered from the description above is how the flare ribbons and simulated photospheric current ribbons are related. Using the measurements of the three components of the magnetic field at the solar surface means that it is possible to observe the evolution of the photospheric current density, provided that there is a good spatial and temporal resolution. The HMI instrument onboard the *Solar Dynamics Observatory* has an unprecedented temporal resolution, providing maps of the line-of-sight measurements of the magnetic field every 12 mn. In Janvier *et al.* (2014), the authors used the Milne–Eddington inversion code UNNOFIT (see Bommier *et al.*, 2007) applied to the HMI magnetograms covering the active region NOAA 11158 during 4 hours of observation, including the X2.2 flare that took place on 15 February 2011 (SOL2011-02-15T01:56) and peaking at 01:56 UT. From this inversion process, maps of the vertical component of the current density vector J_z were deduced. Note that although the active region presents a quadrupolar configuration, the X-class flare of February 2011 was mainly confined within the two central bipoles, and as such had a configuration (including the asymmetry in the strength of the magnetic polarities) similar to that studied previously in the eruptive flare model of Aulanier, Janvier, and Schmieder (2012).

Investigating the evolution of J_z at different times before and after the peak of the flare, the authors reported an increase in the direct electric current component along both the current ribbons in the negative and in the positive polarities (in regions indicated with arrows in Figure 11(e)). These current ribbons delimitate the region around the PIL where the magnetic field becomes more and more horizontal, showing that these regions do correspond to the formation of new post-flare loops (Sun *et al.*, 2012; Wang *et al.*, 2012; Petrie, 2013).

These results seem *a priori* contradictory to the core idea that flares typically dissipate the stored free energy as the magnetic configuration evolves toward a more potential, and hence less current-carrying, one. While a strong dissipation of magnetic energy is expected in the thin current layers, at the same time the electric current density increases due to the collapse of the thin layers. Such an evolution is due to the upward ejective motion of the torus-unstable magnetic flux rope: The field lines underneath the flux rope get squashed in the trailing current region (Figure 11(c)), which in turn thins resulting in a current density increase there. Since this current volume maps all the way down to the photosphere (panel (b)), changes are then also expected to be seen at the photospheric level, as predicted in the numerical simulation (panel (f)). So although this mechanism competes with the strong diffusion taking place, the total current density is large at the beginning, during the flare impulsive phase, before it starts decreasing later on.

The location and the evolution of enhanced electric currents within flare ribbons, predicted with numerical simulations, have now been confirmed thanks to the enhanced temporal and spatial resolution of recent instruments. They are not only correlated with the evolution of flare ribbons, as summarised above, but also with a variety of other phenomena: they map to similar locations as strong twist of the magnetic field, as shown in a data-driven simulation of the same X2.2 flare of 15 February 2011 (SOL2011-02-15T01:56) by Inoue *et al.*

(2014), and regions of current changes have also been associated with HXR sources observed with RHESSI (Musset, Vilmer, and Bommier, 2015) as well as sunquakes (Zharkov *et al.*, 2011).

6. Conclusions

The present paper summarises the MHD building blocks of solar flare modelling, from the topology to the trigger mechanisms, linking observations to 3D numerical simulations. While a flare is typically classified by its strength depending on its soft X-ray emission, it can also be differentiated in two categories, either confined or eruptive (see Section 2). While confined flares do not have a large influence on the global corona, eruptive flares are typically associated with coronal mass ejections (CMEs; see Section 2.3), and can therefore be potentially disruptive for surrounding planetary environments. They share similar features, such as the existence of ribbons and flare loops (Section 2.2).

Observations of solar flares inevitably lead to the question of the storage and the release of the magnetic energy that fuels them. Several mechanisms have been proposed (Section 3): one scenario involves current-carrying magnetic structures that form underneath the photospheric boundary, in the convection zone, and that are transported all the way to the corona. 3D simulations that have been developed recently are able to reproduce this emergence process, sometimes all the way to the ejection of current-carrying structures such as flux ropes (see Section 3.1). Other scenarios involve the formation of non-potential fields by photospheric motions such as shearing or flux cancellation, and have been tested against observations of the evolution of active regions and coronal loops.

Magnetic reconnection has been proposed as a mechanism to dissipate the free energy (Section 3.2). This phenomenon involves the creation of thin layers of high current density where the plasma becomes non-ideal, and where the magnetic energy can be converted into heat and kinetic energy, while forming a population of non-thermal particles. Although the spatial scales at which reconnection is expected to occur are too small for the present-day instruments to detect, its consequences have been witnessed in numerous cases, with an ever-increasing resolution (both temporal and spatial, Figure 6) over the years, and support the present theory. The regions of non-idealness can be found by searching for the topological characteristics of the magnetic field, such as the presence of null points, separatrices, separators, and also quasi-separatrix layers (see Section 3.3). Current density layers, where the magnetic energy is dissipated, preferentially form at QSLs. Searching for them in coronal magnetic fields has become a powerful and successful tool to understand the signatures of flare evolution.

Using observations as well as numerical simulations, several major mechanisms can account for confined and eruptive flares. Emerging flux (Section 4.1) may possibly account for a great number of confined flares, although further understanding of the process driving them is required. Data-driven MHD numerical simulations can indeed explain such flares. Other observations of active regions have shown that several features appear prior to the onset of an eruptive flare, such as flux cancellation and flux dispersal, as well as shearing motions of the polarities. These mechanisms can provide the conditions for a flux rope to be formed. As discussed in the review by Aulanier (2014), those conditions can ultimately lead to a flux-rope ejection via the torus-unstable flux-rope model (Section 4.2). Next, we saw how the QSLs fill a thin coronal volume that maps all the way to the photosphere in the presence of a flux rope (Section 5.1). In their presence, the magnetic field lines change their connectivity successively, they are said to be slipping as time goes by (Section 5.2). This motion, first proposed as an analytical by-product of the presence of QSLs, has been investigated in detail in the numerical simulation of a flux-rope ejection (Janvier *et al.*, 2013). The slipping motion of field lines is intrinsic to the reconnection in 3D, and therefore, to all flares. Apparent motions of field lines and their associated kernels indeed have been observed with high temporal resolution imaging.

Magnetic reconnection dissipates the magnetic energy in highly localised regions of high current density, at the QSLs. Using numerical simulations of flux-rope ejection, different studies show a similar behaviour and geometry: the coronal current layer where reconnection takes place forms a volume that maps to the photosphere, following the QSL mapping. The impulsive phase of an eruptive flare is then dictated by the sudden thinning of the layer, where the electric current density increases, allowing reconnection to take place on a short time scale. This change of regimes has been confirmed by direct measurements of the photospheric vertical currents, with a high temporal-cadence instrument such as the HMI onboard SDO. This underlines the importance of using numerical models as they can also pinpoint consequences of such mechanisms that remain difficult to observe. They therefore provide the physical understanding to interpret observations, as well as the ideas for where and how to look at them. Taking into account the latest developments of numerical flare modelling, we report in Figure 11, panel (a), on the extension of a 3D flare model from the previous CSHKP cartoon that remains valid only in 2D.

As discussed in the paper, MHD simulations of flares have been able to reproduce observational features, help to interpret the underlying mechanisms, and make predictions. Nonetheless, several issues still remain to be addressed, with further developments needed in the future. Indeed, within the MHD framework, the kinetic physics of the magnetic energy dissipation within the 3D current layer cannot be addressed. The length scales that account for this dissipation (the ion or electron length scales) are several orders of magnitude smaller than the typical lengthscale of coronal loops. As such, approximating the physics that takes place within the current layer (such as turbulence processes, plasmoid formation, Hall effect, ion heating, kinetic effects) is one solution, given the computer power available as of today, along with the development of hybrid codes. This also means that the MHD framework needs to be extended, so as to explain how a population of non-thermal particles is created. Recent developments in laboratory plasma physics, and in particular of the partition of energy between the different species (Yamada *et al.*, 2014) may shed light on the processes happening in the corona. Furthermore, knowing the energy partition given to non-thermal particles is of high importance to understand how the energy is deposited from the corona to the photosphere (Milligan *et al.*, 2014).

In conclusion, the correspondences between the 3D MHD models developed in the recent years, and the observations obtained with an increasing time and spatial resolution, provide strong evidence that several physical ingredients important to flares are included. Such models are of great importance not only to understand the flaring activity of our host star, but also to interpret data of other flaring stars, for which observations remain limited.

Acknowledgements M.J. acknowledges partial funding from the SOC of the "Solar and stellar flares: observations, simulations and synergies" conference, held from 23 – 27 June 2014, where this work was presented. The authors would like to thank the anonymous referee and Lyndsay Fletcher for their useful comments that helped improve the article.

References

Amari, T., Canou, A., Aly, J.-J.: 2014, Characterizing and predicting the magnetic environment leading to solar eruptions. *Nature* **514**, 465. DOI. ADS.

Amari, T., Luciani, J.F., Mikić, Z., Linker, J.: 2000, A twisted flux rope model for coronal mass ejections and two-ribbon flares. *Astrophys. J. Lett.* **529**, L49. DOI. ADS.

Amari, T., Luciani, J.F., Aly, J.J., Mikić, Z., Linker, J.: 2003a, Coronal mass ejection: initiation, magnetic helicity, and flux ropes. I. Boundary motion-driven evolution. *Astrophys. J.* **585**, 1073. DOI. ADS.

Amari, T., Luciani, J.F., Aly, J.J., Mikić, Z., Linker, J.: 2003b, Coronal mass ejection: initiation, magnetic helicity, and flux ropes. II. Turbulent diffusion-driven evolution. *Astrophys. J.* **595**, 1231. DOI. ADS.

Amari, T., Aly, J.-J., Luciani, J.-F., Mikić, Z., Linker, J.: 2011, Coronal mass ejection initiation by converging photospheric flows: toward a realistic model. *Astrophys. J. Lett.* **742**, L27. DOI. ADS.

Antiochos, S.K., DeVore, C.R., Klimchuk, J.A.: 1999, A model for solar coronal mass ejections. *Astrophys. J.* **510**, 485. DOI. ADS.

Archontis, V., Hood, A.W.: 2012, Magnetic flux emergence: a precursor of solar plasma expulsion. *Astron. Astrophys.* **537**, A62. DOI. ADS.

Asai, A., Masuda, S., Yokoyama, T., Shimojo, M., Isobe, H., Kurokawa, H., Shibata, K.: 2002, Difference between spatial distributions of the Hα kernels and hard X-ray sources in a solar flare. *Astrophys. J. Lett.* **578**, L91. DOI. ADS.

Asai, A., Ishii, T.T., Kurokawa, H., Yokoyama, T., Shimojo, M.: 2003, Evolution of conjugate footpoints inside flare ribbons during a great two-ribbon flare on 2001 April 10. *Astrophys. J.* **586**, 624. DOI. ADS.

Aulanier, G.: 2014, The physical mechanisms that initiate and drive solar eruptions. In: Schmieder, B., Malherbe, J.-M., Wu, S.T. (eds.) *Nature of Prominences and Their Role in Space Weather*, *IAU Symposium* **300**, 184. DOI. ADS.

Aulanier, G., Janvier, M., Schmieder, B.: 2012, The standard flare model in three dimensions. I. Strong-to-weak shear transition in post-flare loops. *Astron. Astrophys.* **543**, A110. DOI. ADS.

Aulanier, G., Pariat, E., Démoulin, P.: 2005, Current sheet formation in quasi-separatrix layers and hyperbolic flux tubes. *Astron. Astrophys.* **444**, 961. DOI. ADS.

Aulanier, G., DeLuca, E.E., Antiochos, S.K., McMullen, R.A., Golub, L.: 2000, The topology and evolution of the Bastille Day flare. *Astrophys. J.* **540**, 1126. DOI. ADS.

Aulanier, G., Pariat, E., Démoulin, P., DeVore, C.R.: 2006, Slip-running reconnection in quasi-separatrix layers. *Solar Phys.* **238**, 347. DOI. ADS.

Aulanier, G., Golub, L., DeLuca, E.E., Cirtain, J.W., Kano, R., Lundquist, L.L., Narukage, N., Sakao, T., Weber, M.A.: 2007, Slipping magnetic reconnection in coronal loops. *Science* **318**, 1588. DOI. ADS.

Aulanier, G., Török, T., Démoulin, P., DeLuca, E.E.: 2010, Formation of torus-unstable flux ropes and electric currents in erupting sigmoids. *Astrophys. J.* **708**, 314. DOI. ADS.

Aulanier, G., Démoulin, P., Schrijver, C.J., Janvier, M., Pariat, E., Schmieder, B.: 2013, The standard flare model in three dimensions. II. Upper limit on solar flare energy. *Astron. Astrophys.* **549**, A66. DOI. ADS.

Bagalá, L.G., Mandrini, C.H., Rovira, M.G., Démoulin, P.: 2000, Magnetic reconnection: a common origin for flares and AR interconnecting arcs. *Astron. Astrophys.* **363**, 779. ADS.

Barnes, G., Longcope, D.W., Leka, K.D.: 2005, Implementing a magnetic charge topology model for solar active regions. *Astrophys. J.* **629**, 561. DOI. ADS.

Baum, P.J., Bratenahl, A.: 1980, Flux linkages of bipolar sunspot groups – a computer study. *Solar Phys.* **67**, 245. DOI. ADS.

Birn, J., Fletcher, L., Hesse, M., Neukirch, T.: 2009, Energy release and transfer in solar flares: simulations of three-dimensional reconnection. *Astrophys. J.* **695**, 1151. DOI. ADS.

Bommier, V., Landi Degl'Innocenti, E., Landolfi, M., Molodij, G.: 2007, UNNOFIT inversion of spectro-polarimetric maps observed with THEMIS. *Astron. Astrophys.* **464**, 323. DOI. ADS.

Büchner, J.: 2006, Locating current sheets in the solar corona. *Space Sci. Rev.* **122**, 149. DOI. ADS.

Canfield, R.C., Hudson, H.S., McKenzie, D.E.: 1999, Sigmoidal morphology and eruptive solar activity. *Geophys. Res. Lett.* **26**, 627. DOI. ADS.

Carmichael, H.: 1964, A process for flares. *NASA Spec. Publ.* **50**, 451. ADS.

Carrington, R.C.: 1859, Description of a singular appearance seen in the Sun on September 1, 1859. *Mon. Not. Roy. Astron. Soc.* **20**, 13. ADS.

Chandra, R., Schmieder, B., Aulanier, G., Malherbe, J.M.: 2009, Evidence of magnetic helicity in emerging flux and associated flare. *Solar Phys.* **258**, 53. DOI. ADS.

Chandra, R., Schmieder, B., Mandrini, C.H., Démoulin, P., Pariat, E., Török, T., Uddin, W.: 2011, Homologous flares and magnetic field topology in active region NOAA 10501 on 20 November 2003. *Solar Phys.* **269**, 83. DOI. ADS.

Chen, P.F.: 2011, Coronal mass ejections: models and their observational basis. *Living Rev. Solar Phys.* **8**, 1. DOI. ADS.

Cheung, M.C.M., Isobe, H.: 2014, Flux emergence (theory). *Living Rev. Solar Phys.* **11**, 3. DOI. ADS.

Cliver, E.W., Dietrich, W.F.: 2013, The 1859 space weather event revisited: limits of extreme activity. *J. Space Weather Space Clim.* **3**(27), A31. DOI. ADS.

Dalmasse, K., Chandra, R., Schmieder, B., Aulanier, G.: 2015, Can we explain atypical solar flares? *Astron. Astrophys.* **574**, A37. DOI. ADS.
del Zanna, G., Schmieder, B., Mason, H., Berlicki, A., Bradshaw, S.: 2006, The gradual phase of the X17 flare on October 28, 2003. *Solar Phys.* **239**, 173. DOI. ADS.
Delaboudinière, J.-P., Artzner, G.E., Brunaud, J., Gabriel, A.H., Hochedez, J.F., Millier, F., Song, X.Y., Au, B., Dere, K.P., Howard, R.A., Kreplin, R., Michels, D.J., Moses, J.D., Defise, J.M., Jamar, C., Rochus, P., Chauvineau, J.P., Marioge, J.P., Catura, R.C., Lemen, J.R., Shing, L., Stern, R.A., Gurman, J.B., Neupert, W.M., Maucherat, A., Clette, F., Cugnon, P., van Dessel, E.L.: 1995, EIT: Extreme-ultraviolet Imaging Telescope for the SOHO mission. *Solar Phys.* **162**, 291. DOI. ADS.
Démoulin, P., Hénoux, J.C., Mandrini, C.H.: 1994, Are magnetic null points important in solar flares ? *Astron. Astrophys.* **285**, 1023. ADS.
Démoulin, P., Priest, E.R., Lonie, D.P.: 1996, Three-dimensional magnetic reconnection without null points 2. Application to twisted flux tubes. *J. Geophys. Res.* **101**, 7631. DOI. ADS.
Démoulin, P., Mandrini, C.H., Rovira, M.G., Hénoux, J.C., Machado, M.E.: 1994, Interpretation of multi-wavelength observations of November 5, 1980 solar flares by the magnetic topology of AR 2766. *Solar Phys.* **150**, 221. DOI. ADS.
Démoulin, P., Hénoux, J.C., Priest, E.R., Mandrini, C.H.: 1996, Quasi-separatrix layers in solar flares. I. Method. *Astron. Astrophys.* **308**, 643. ADS.
Démoulin, P., Bagala, L.G., Mandrini, C.H., Hénoux, J.C., Rovira, M.G.: 1997, Quasi-separatrix layers in solar flares. II. Observed magnetic configurations. *Astron. Astrophys.* **325**, 305. ADS.
DeVore, C.R., Antiochos, S.K.: 2008, Homologous confined filament eruptions via magnetic breakout. *Astrophys. J.* **680**, 740. DOI. ADS.
Dudík, J., Janvier, M., Aulanier, G., Del Zanna, G., Karlický, M., Mason, H.E., Schmieder, B.: 2014, Slipping magnetic reconnection during an X-class solar flare observed by SDO/AIA. *Astrophys. J.* **784**, 144. DOI. ADS.
Dungey, J.W.: 1953, The motion of magnetic fields. *Mon. Not. Roy. Astron. Soc.* **113**, 679. ADS.
Effenberger, F., Thust, K., Arnold, L., Grauer, R., Dreher, J.: 2011, Numerical simulation of current sheet formation in a quasiseparatrix layer using adaptive mesh refinement. *Phys. Plasmas* **18**(3), 032902. DOI. ADS.
Engvold, O.: 2015, Description and classification of prominences. In: Vial, J.-C., Engvold, O. (eds.) *Solar Prominences, Astrophysics and Space Science Library* **415**, 31. DOI. ADS.
Fan, Y.: 2009, Magnetic fields in the solar convection zone. *Living Rev. Solar Phys.* **6**, 4. DOI. ADS.
Fan, Y., Gibson, S.E.: 2007, Onset of coronal mass ejections due to loss of confinement of coronal flux ropes. *Astrophys. J.* **668**, 1232. DOI. ADS.
Feynman, J., Hundhausen, A.J.: 1994, Coronal mass ejections and major solar flares: the great active center of March 1989. *J. Geophys. Res.* **99**, 8451. DOI. ADS.
Finn, J.M., Billey, Z., Daughton, W., Zweibel, E.: 2014, Quasi-separatrix layer reconnection for nonlinear line-tied collisionless tearing modes. *Plasma Phys. Control. Fusion* **56**(6), 064013. DOI. ADS.
Fletcher, L., Hudson, H.: 2001, The magnetic structure and generation of EUV flare ribbons. *Solar Phys.* **204**, 69. DOI. ADS.
Fletcher, L., Hudson, H.S.: 2008, Impulsive phase flare energy transport by large-scale Alfvén waves and the electron acceleration problem. *Astrophys. J.* **675**, 1645. DOI. ADS.
Fletcher, L., Dennis, B.R., Hudson, H.S., Krucker, S., Phillips, K., Veronig, A., Battaglia, M., Bone, L., Caspi, A., Chen, Q., Gallagher, P., Grigis, P.T., Ji, H., Liu, W., Milligan, R.O., Temmer, M.: 2011, An observational overview of solar flares. *Space Sci. Rev.* **159**, 19. DOI. ADS.
Forbes, T.: 2010, Models of coronal mass ejections and flares. In: Schrijver, C.J., Siscoe, G.L. (eds.) *Heliophysics: Space Storms and Radiation: Causes and Effects*, Cambridge University Press, Cambridge, 159. ADS.
Forbes, T.G., Priest, E.R.: 1984, Numerical simulation of reconnection in an emerging magnetic flux region. *Solar Phys.* **94**, 315. DOI. ADS.
Forbes, T.G., Acton, L.W.: 1996, Reconnection and field line shrinkage in solar flares. *Astrophys. J.* **459**, 330. DOI. ADS.
Forbes, T.G., Linker, J.A., Chen, J., Cid, C., Kóta, J., Lee, M.A., Mann, G., Mikić, Z., Potgieter, M.S., Schmidt, J.M., Siscoe, G.L., Vainio, R., Antiochos, S.K., Riley, P.: 2006, CME theory and models. *Space Sci. Rev.* **123**, 251. DOI. ADS.
Gaizauskas, V., Mandrini, C.H., Démoulin, P., Luoni, M.L., Rovira, M.G.: 1998, Interactions between nested sunspots. II. A confined X1 flare in a delta-type sunspot. *Astron. Astrophys.* **332**, 353. ADS.
Gary, D.E., Hurford, G.J.: 1990, Multifrequency observations of a solar microwave burst with two-dimensional spatial resolution. *Astrophys. J.* **361**, 290. DOI. ADS.
Gilbert, H.R., Alexander, D., Liu, R.: 2007, Filament kinking and its implications for eruption and re-formation. *Solar Phys.* **245**, 287. DOI. ADS.

Gopalswamy, N., Shimojo, M., Lu, W., Yashiro, S., Shibasaki, K., Howard, R.A.: 2003, Prominence eruptions and coronal mass ejection: a statistical study using microwave observations. *Astrophys. J.* **586**, 562. DOI. ADS.

Gorbachev, V.S., Somov, B.V.: 1988, Photospheric vortex flows as a cause for two-ribbon flares – a topological model. *Solar Phys.* **117**, 77. DOI. ADS.

Gosling, J.T., McComas, D.J., Phillips, J.L., Bame, S.J.: 1991, Geomagnetic activity associated with earth passage of interplanetary shock disturbances and coronal mass ejections. *J. Geophys. Res.* **96**, 7831. DOI.

Green, L.M., Kliem, B.: 2009, Flux rope formation preceding coronal mass ejection onset. *Astrophys. J. Lett.* **700**, L83. DOI. ADS.

Green, L.M., Kliem, B., Wallace, A.J.: 2011, Photospheric flux cancellation and associated flux rope formation and eruption. *Astron. Astrophys.* **526**, A2. DOI. ADS.

Green, L.M., Matthews, S.A., van Driel-Gesztelyi, L., Harra, L.K., Culhane, J.L.: 2002, Multi-wavelength observations of an X-class flare without a coronal mass ejection. *Solar Phys.* **205**, 325. DOI. ADS.

Green, L.M., Kliem, B., Török, T., van Driel-Gesztelyi, L., Attrill, G.D.R.: 2007, Transient coronal sigmoids and rotating erupting flux ropes. *Solar Phys.* **246**, 365. DOI. ADS.

Guo, Y., Ding, M.D., Schmieder, B., Démoulin, P., Li, H.: 2012, Evolution of hard X-ray sources and ultraviolet solar flare ribbons for a confined eruption of a magnetic flux rope. *Astrophys. J.* **746**, 17. DOI. ADS.

Hannah, I.G., Hudson, H.S., Battaglia, M., Christe, S., Kašparová, J., Krucker, S., Kundu, M.R., Veronig, A.: 2011, Microflares and the statistics of X-ray flares. *Space Sci. Rev.* **159**, 263. DOI. ADS.

Hesse, M., Schindler, K.: 1988, A theoretical foundation of general magnetic reconnection. *J. Geophys. Res.* **93**, 5559. DOI. ADS.

Heyvaerts, J., Priest, E.R., Rust, D.M.: 1977, An emerging flux model for the solar flare phenomenon. *Astrophys. J.* **216**, 123. DOI. ADS.

Hirayama, T.: 1974, Theoretical model of flares and prominences. I: evaporating flare model. *Solar Phys.* **34**, 323. DOI. ADS.

Hodgson, R.: 1859, On a curious appearance seen in the Sun. *Mon. Not. Roy. Astron. Soc.* **20**, 15. ADS.

Hood, A.W., Archontis, V., MacTaggart, D.: 2012, 3D MHD flux emergence experiments: idealised models and coronal interactions. *Solar Phys.* **278**, 3. DOI. ADS.

Hood, A.W., Archontis, V., Galsgaard, K., Moreno-Insertis, F.: 2009, The emergence of toroidal flux tubes from beneath the solar photosphere. *Astron. Astrophys.* **503**, 999. DOI. ADS.

Hori, K., Culhane, J.L.: 2002, Trajectories of microwave prominence eruptions. *Astron. Astrophys.* **382**, 666. DOI. ADS.

Howard, T.A., Harrison, R.A.: 2013, Stealth coronal mass ejections: a perspective. *Solar Phys.* **285**, 269. DOI. ADS.

Hudson, H.S., Kosugi, T., Nitta, N.V., Shimojo, M.: 2001, Hard X-radiation from a fast coronal ejection. *Astrophys. J. Lett.* **561**, L211. DOI. ADS.

Inoue, S., Hayashi, K., Magara, T., Choe, G.S., Park, Y.D.: 2014, Magnetohydrodynamic simulation of the X2.2 solar flare on 2011 February 15. I. Comparison with the observations. *Astrophys. J.* **788**, 182. DOI. ADS.

Janvier, M., Aulanier, G., Pariat, E., Démoulin, P.: 2013, The standard flare model in three dimensions. III. Slip-running reconnection properties. *Astron. Astrophys.* **555**, A77. DOI. ADS.

Janvier, M., Aulanier, G., Bommier, V., Schmieder, B., Démoulin, P., Pariat, E.: 2014, Electric currents in flare ribbons: observations and three-dimensional standard model. *Astrophys. J.* **788**, 60. DOI. ADS.

Ji, H., Wang, H., Schmahl, E.J., Moon, Y.-J., Jiang, Y.: 2003, Observations of the failed eruption of a filament. *Astrophys. J. Lett.* **595**, L135. DOI. ADS.

Jing, J., Yurchyshyn, V.B., Yang, G., Xu, Y., Wang, H.: 2004, On the relation between filament eruptions, flares, and coronal mass ejections. *Astrophys. J.* **614**, 1054. DOI. ADS.

Joshi, A.D., Forbes, T.G., Park, S.-H., Cho, K.-S.: 2015, A trio of confined flares in AR 11087. *Astrophys. J.* **798**, 97. DOI. ADS.

Kliem, B., Török, T.: 2006, Torus instability. *Phys. Rev. Lett.* **96**(25), 255002. DOI. ADS.

Kliem, B., Su, Y.N., van Ballegooijen, A.A., DeLuca, E.E.: 2013, Magnetohydrodynamic modeling of the solar eruption on 2010 April 8. *Astrophys. J.* **779**, 129. DOI. ADS.

Klimchuk, J.A., Bradshaw, S.J.: 2014, Are chromospheric nanoflares a primary source of coronal plasma? *Astrophys. J.* **791**, 60. DOI. ADS.

Koleva, K., Madjarska, M.S., Duchlev, P., Schrijver, C.J., Vial, J.-C., Buchlin, E., Dechev, M.: 2012, Kinematics and helicity evolution of a loop-like eruptive prominence. *Astron. Astrophys.* **540**, A127. DOI. ADS.

Kopp, R.A., Pneuman, G.W.: 1976, Magnetic reconnection in the corona and the loop prominence phenomenon. *Solar Phys.* **50**, 85. DOI. ADS.

Kosugi, T., Makishima, K., Murakami, T., Sakao, T., Dotani, T., Inda, M., Kai, K., Masuda, S., Nakajima, H., Ogawara, Y., Sawa, M., Shibasaki, K.: 1991, The Hard X-ray Telescope (HXT) for the SOLAR-A mission. *Solar Phys.* **136**, 17. DOI. ADS.
Krucker, S., Fivian, M.D., Lin, R.P.: 2005, Hard X-ray footpoint motions in solar flares: comparing magnetic reconnection models with observations. *Adv. Space Res.* **35**, 1707. DOI. ADS.
Krucker, S., Battaglia, M., Cargill, P.J., Fletcher, L., Hudson, H.S., MacKinnon, A.L., Masuda, S., Sui, L., Tomczak, M., Veronig, A.L., Vlahos, L., White, S.M.: 2008, Hard X-ray emission from the solar corona. *Astron. Astrophys. Rev.* **16**, 155. DOI. ADS.
Kusano, K., Bamba, Y., Yamamoto, T.T., Iida, Y., Toriumi, S., Asai, A.: 2012, Magnetic field structures triggering solar flares and coronal mass ejections. *Astrophys. J.* **760**, 31. DOI. ADS.
Lau, Y.-T.: 1993, Magnetic nulls and topology in a class of solar flare models. *Solar Phys.* **148**, 301. DOI. ADS.
Leake, J.E., Linton, M.G., Török, T.: 2013, Simulations of emerging magnetic flux. I. The formation of stable coronal flux ropes. *Astrophys. J.* **778**, 99. DOI. ADS.
Leka, K.D., Canfield, R.C., McClymont, A.N., van Driel-Gesztelyi, L.: 1996, Evidence for current-carrying emerging flux. *Astrophys. J.* **462**, 547. DOI. ADS.
Lemen, J.R., Title, A.M., Akin, D.J., Boerner, P.F., Chou, C., Drake, J.F., Duncan, D.W., Edwards, C.G., Friedlaender, F.M., Heyman, G.F., Hurlburt, N.E., Katz, N.L., Kushner, G.D., Levay, M., Lindgren, R.W., Mathur, D.P., McFeaters, E.L., Mitchell, S., Rehse, R.A., Schrijver, C.J., Springer, L.A., Stern, R.A., Tarbell, T.D., Wuelser, J.-P., Wolfson, C.J., Yanari, C., Bookbinder, J.A., Cheimets, P.N., Caldwell, D., Deluca, E.E., Gates, R., Golub, L., Park, S., Podgorski, W.A., Bush, R.I., Scherrer, P.H., Gummin, M.A., Smith, P., Auker, G., Jerram, P., Pool, P., Soufli, R., Windt, D.L., Beardsley, S., Clapp, M., Lang, J., Waltham, N.: 2012, The Atmospheric Imaging Assembly (AIA) on the Solar Dynamics Observatory (SDO). *Solar Phys.* **275**, 17. DOI. ADS.
Li, L., Zhang, J.: 2009, Observations of the magnetic reconnection signature of an M2 flare on 2000 March 23. *Astrophys. J.* **703**, 877. DOI. ADS.
Li, T., Zhang, J.: 2014, Slipping magnetic reconnection triggering a solar eruption of a triangle-shaped flag flux rope. *Astrophys. J. Lett.* **791**, L13. DOI. ADS.
Lin, J., Forbes, T.G., Isenberg, P.A.: 2001, Prominence eruptions and coronal mass ejections triggered by newly emerging flux. *J. Geophys. Res.* **106**, 25053. DOI. ADS.
Lin, R.P., Dennis, B.R., Hurford, G.J., Smith, D.M., Zehnder, A., Harvey, P.R., Curtis, D.W., Pankow, D., Turin, P., Bester, M., Csillaghy, A., Lewis, M., Madden, N., van Beek, H.F., Appleby, M., Raudorf, T., McTiernan, J., Ramaty, R., Schmahl, E., Schwartz, R., Krucker, S., Abiad, R., Quinn, T., Berg, P., Hashii, M., Sterling, R., Jackson, R., Pratt, R., Campbell, R.D., Malone, D., Landis, D., Barrington-Leigh, C.P., Slassi-Sennou, S., Cork, C., Clark, D., Amato, D., Orwig, L., Boyle, R., Banks, I.S., Shirey, K., Tolbert, A.K., Zarro, D., Snow, F., Thomsen, K., Henneck, R., McHedlishvili, A., Ming, P., Fivian, M., Jordan, J., Wanner, R., Crubb, J., Preble, J., Matranga, M., Benz, A., Hudson, H., Canfield, R.C., Holman, G.D., Crannell, C., Kosugi, T., Emslie, A.G., Vilmer, N., Brown, J.C., Johns-Krull, C., Aschwanden, M., Metcalf, T., Conway, A.: 2002, The Reuven Ramaty High-Energy Solar Spectroscopic Imager (RHESSI). *Solar Phys.* **210**, 3. DOI. ADS.
Linsky, J.L.: 2000, Stellar flares: how common? How Important? In: Holt, S.S., Zhang, W.W. (eds.) *COSMIC EXPLOSIONS: Tenth Astrophysics Conference, American Institute of Physics Conference Series* **522**, 389. DOI. ADS.
Liu, R., Alexander, D.: 2009, Hard X-ray emission in kinking filaments. *Astrophys. J.* **697**, 999. DOI. ADS.
Longcope, D.W.: 2005, Topological methods for the analysis of solar magnetic fields. *Living Rev. Solar Phys.* **2**, 7. DOI. ADS.
Longcope, D.W., Cowley, S.C.: 1996, Current sheet formation along three-dimensional magnetic separators. *Phys. Plasmas* **3**, 2885. DOI. ADS.
Longcope, D.W., Tarr, L.: 2012, The role of fast magnetosonic waves in the release and conversion via reconnection of energy stored by a current sheet. *Astrophys. J.* **756**, 192. DOI. ADS.
Lugaz, N.: 2015, Eruptive prominences and their impact on the Earth and our life. In: Vial, J.-C., Engvold, O. (eds.) *Solar Prominences, Astrophysics and Space Science Library* **415**, 433. DOI. ADS.
Luoni, M.L., Démoulin, P., Mandrini, C.H., van Driel-Gesztelyi, L.: 2011, Twisted flux tube emergence evidenced in longitudinal magnetograms: magnetic tongues. *Solar Phys.* **270**, 45. DOI. ADS.
Mackay, D.H., van Ballegooijen, A.A.: 2006, Models of the large-scale corona. I. Formation, evolution, and liftoff of magnetic flux ropes. *Astrophys. J.* **641**, 577. DOI. ADS.
Maehara, H., Shibayama, T., Notsu, S., Notsu, Y., Nagao, T., Kusaba, S., Honda, S., Nogami, D., Shibata, K.: 2012, Superflares on solar-type stars. *Nature* **485**, 478. DOI. ADS.
Mandrini, C.H., Rovira, M.G., Démoulin, P., Hénoux, J.C., Machado, M.E., Wilkinson, L.K.: 1993, Evidence for magnetic reconnection in large-scale magnetic structures in solar flares. *Astron. Astrophys.* **272**, 609. ADS.

Mandrini, C.H., Démoulin, P., Bagala, L.G., van Driel-Gesztelyi, L., Hénoux, J.C., Schmieder, B., Rovira, M.G.: 1997, Evidence of magnetic reconnection from Hα, soft X-ray and photospheric magnetic field observations. *Solar Phys.* **174**, 229. DOI. ADS.

Martin, S.F., Livi, S.H.B., Wang, J.: 1985, The cancellation of magnetic flux. II – in a decaying active region. *Aust. J. Phys.* **38**, 929. ADS.

Masson, S., Pariat, E., Aulanier, G., Schrijver, C.J.: 2009, The nature of flare ribbons in coronal null-point topology. *Astrophys. J.* **700**, 559. DOI. ADS.

Masson, S., Aulanier, G., Pariat, E., Klein, K.-L.: 2012, Interchange slip-running reconnection and sweeping SEP beams. *Solar Phys.* **276**, 199. DOI. ADS.

Masuda, S., Kosugi, T., Hara, H., Tsuneta, S., Ogawara, Y.: 1994, A loop-top hard X-ray source in a compact solar flare as evidence for magnetic reconnection. *Nature* **371**, 495. DOI. ADS.

McCracken, K.G., Dreschhoff, G.A.M., Zeller, E.J., Smart, D.F., Shea, M.A.: 2001, Solar cosmic ray events for the period 1561 – 1994: 1. Identification in polar ice, 1561 – 1950. *J. Geophys. Res.* **106**, 21585. DOI. ADS.

McKenzie, D.E.: 2011, Observational studies of reconnection in the solar corona. *Phys. Plasmas* **18**(11), 111205. DOI. ADS.

McKenzie, D.E., Canfield, R.C.: 2008, Hinode XRT observations of a long-lasting coronal sigmoid. *Astron. Astrophys.* **481**, L65. DOI. ADS.

Milligan, R.O., Dennis, B.R.: 2009, Velocity characteristics of evaporated plasma using Hinode/EUV imaging spectrometer. *Astrophys. J.* **699**, 968. DOI. ADS.

Milligan, R.O., Kerr, G.S., Dennis, B.R., Hudson, H.S., Fletcher, L., Allred, J.C., Chamberlin, P.C., Ireland, J., Mathioudakis, M., Keenan, F.P.: 2014, The radiated energy budget of chromospheric plasma in a major solar flare deduced from multi-wavelength observations. *Astrophys. J.* **793**, 70. DOI. ADS.

Musset, S., Vilmer, N., Bommier, V.: 2015, Relation between electric current densities and X-ray emissions from particles accelerated during solar flares. *Astrophys. J.*, submitted.

Nishizuka, N., Takasaki, H., Asai, A., Shibata, K.: 2010, Multiple plasmoid ejections and associated hard X-ray bursts in the 2000 November 24 flare. *Astrophys. J.* **711**, 1062. DOI. ADS.

Nishizuka, N., Karlický, M., Janvier, M., Bárta, M.: 2015, Particle acceleration in plasmoid ejections derived from radio drifting pulsating structures. *Astrophys. J.* **799**, 126. DOI. ADS.

Olmedo, O., Zhang, J.: 2010, Partial torus instability. *Astrophys. J.* **718**, 433. DOI. ADS.

Pariat, E., Démoulin, P.: 2012, Estimation of the squashing degree within a three-dimensional domain. *Astron. Astrophys.* **541**, A78. DOI. ADS.

Parker, E.N.: 1957, Sweet's mechanism for merging magnetic fields in conducting fluids. *J. Geophys. Res.* **62**, 509. DOI. ADS.

Parker, E.N.: 1988, Nanoflares and the solar X-ray corona. *Astrophys. J.* **330**, 474. DOI. ADS.

Parnell, C.E., Haynes, A.L., Galsgaard, K.: 2010, Structure of magnetic separators and separator reconnection. *J. Geophys. Res.* **115**, 2102. DOI. ADS.

Parnell, C.E., Smith, J.M., Neukirch, T., Priest, E.R.: 1996, The structure of three-dimensional magnetic neutral points. *Phys. Plasmas* **3**, 759. DOI. ADS.

Petrie, G.J.D.: 2013, A spatio-temporal description of the abrupt changes in the photospheric magnetic and Lorentz-force vectors during the 15 February 2011 X2.2 flare. *Solar Phys.* **287**, 415. DOI. ADS.

Pontin, D.I., Hornig, G., Priest, E.R.: 2004, Kinematic reconnection at a magnetic null point: spine-aligned current. *Geophys. Astrophys. Fluid Dyn.* **98**, 407. DOI. ADS.

Pontin, D.I., Hornig, G., Priest, E.R.: 2005, Kinematic reconnection at a magnetic null point: fan-aligned current. *Geophys. Astrophys. Fluid Dyn.* **99**, 77. DOI. ADS.

Pontin, D.I., Bhattacharjee, A., Galsgaard, K.: 2007, Current sheet formation and nonideal behavior at three-dimensional magnetic null points. *Phys. Plasmas* **14**(5), 052106. DOI. ADS.

Poppenhaeger, K.: 2014, Stellar magnetic activity and star–planet Interactions (invited review). arXiv.

Prangé, R., Pallier, L., Hansen, K.C., Howard, R., Vourlidas, A., Courtin, R., Parkinson, C.: 2004, An interplanetary shock traced by planetary auroral storms from the Sun to Saturn. *Nature* **432**, 78. DOI. ADS.

Priest, E.: 2014, *Magnetohydrodynamics of the Sun*. ADS.

Priest, E.R., Forbes, T.G.: 1992, Magnetic flipping – reconnection in three dimensions without null points. *J. Geophys. Res.* **97**, 1521. DOI. ADS.

Priest, E.R., Démoulin, P.: 1995, Three-dimensional magnetic reconnection without null points. 1. Basic theory of magnetic flipping. *J. Geophys. Res.* **100**, 23443. DOI. ADS.

Priest, E.R., Titov, V.S.: 1996, Magnetic reconnection at three-dimensional null points. *Phil. Trans. Roy. Soc., Math. Phys. Eng. Sci.* **354**, 2951. DOI. ADS.

Priest, E.R., Forbes, T.G.: 2002, The magnetic nature of solar flares. *Astron. Astrophys. Rev.* **10**, 313. DOI. ADS.

Priest, E.R., Pontin, D.I.: 2009, Three-dimensional null point reconnection regimes. *Phys. Plasmas* **16**(12), 122101. DOI. ADS.
Qiu, J., Gary, D.E., Fleishman, G.D.: 2009, Evaluating mean magnetic field in flare loops. In: *AAS/Solar Physics Division Meeting* **40**, 19.06. ADS.
Raftery, C.L., Gallagher, P.T., Milligan, R.O., Klimchuk, J.A.: 2009, Multi-wavelength observations and modelling of a canonical solar flare. *Astron. Astrophys.* **494**, 1127. DOI. ADS.
Reid, H.A.S., Vilmer, N., Aulanier, G., Pariat, E.: 2012, X-ray and ultraviolet investigation into the magnetic connectivity of a solar flare. *Astron. Astrophys.* **547**, A52. DOI. ADS.
Restante, A.L., Aulanier, G., Parnell, C.E.: 2009, How skeletons turn into quasi-separatrix layers in source models. *Astron. Astrophys.* **508**, 433. DOI. ADS.
Richardson, A.S., Finn, J.M.: 2012, Quasi-separatrix layers and three-dimensional reconnection diagnostics for line-tied tearing modes. *Commun. Nonlinear Sci. Numer. Simul.* **17**, 2132. DOI. ADS.
Rust, D.M., Kumar, A.: 1996, Evidence for helically kinked magnetic flux ropes in solar eruptions. *Astrophys. J. Lett.* **464**, L199. DOI. ADS.
Saint-Hilaire, P., Benz, A.O.: 2005, Thermal and non-thermal energies of solar flares. *Astron. Astrophys.* **435**, 743. DOI. ADS.
Savage, S.L., Holman, G., Reeves, K.K., Seaton, D.B., McKenzie, D.E., Su, Y.: 2012, Low-altitude reconnection inflow–outflow observations during a 2010 November 3 solar eruption. *Astrophys. J.* **754**, 13. DOI. ADS.
Savage, S.L., McKenzie, D.E., Reeves, K.K.: 2012, Re-interpretation of supra-arcade downflows in solar flares. *Astrophys. J. Lett.* **747**, L40. DOI. ADS.
Savcheva, A., van Ballegooijen, A.: 2009, Nonlinear force-free modeling of a long-lasting coronal sigmoid. *Astrophys. J.* **703**, 1766. DOI. ADS.
Savcheva, A.S., Green, L.M., van Ballegooijen, A.A., DeLuca, E.E.: 2012a, Photospheric flux cancellation and the build-up of sigmoidal flux ropes on the Sun. *Astrophys. J.* **759**, 105. DOI. ADS.
Savcheva, A., Pariat, E., van Ballegooijen, A., Aulanier, G., DeLuca, E.: 2012b, Sigmoidal active region on the Sun: comparison of a magnetohydrodynamical simulation and a nonlinear force-free field model. *Astrophys. J.* **750**, 15. DOI. ADS.
Savcheva, A.S., McKillop, S.C., McCauley, P.I., Hanson, E.M., DeLuca, E.E.: 2014, A new sigmoid catalog from Hinode and the solar dynamics observatory: statistical properties and evolutionary histories. *Solar Phys.* **289**, 3297. DOI. ADS.
Scherrer, P.H., Bogart, R.S., Bush, R.I., Hoeksema, J.T., Kosovichev, A.G., Schou, J., Rosenberg, W., Springer, L., Tarbell, T.D., Title, A., Wolfson, C.J., Zayer, I., MDI Engineering Team: 1995, The solar oscillations investigation – Michelson Doppler Imager. *Solar Phys.* **162**, 129. DOI. ADS.
Schindler, K., Hesse, M., Birn, J.: 1988, General magnetic reconnection, parallel electric fields, and helicity. *J. Geophys. Res.* **93**, 5547. DOI. ADS.
Schmieder, B., Aulanier, G.: 2012, New perspectives on solar prominences. In: Faurobert, M., Fang, C., Corbard, T. (eds.) *EAS Publications Series* **55**, 149. DOI. ADS.
Schmieder, B., Malherbe, J.-M., Wu, S.T. (eds.): 2014, *Nature of Prominences and Their Role in Space Weather, IAU Symposium* **300**. ADS.
Schmieder, B., Aulanier, G., Demoulin, P., van Driel-Gesztelyi, L., Roudier, T., Nitta, N., Cauzzi, G.: 1997, Magnetic reconnection driven by emergence of sheared magnetic field. *Astron. Astrophys.* **325**, 1213. ADS.
Schmieder, B., Bommier, V., Kitai, R., Matsumoto, T., Ishii, T.T., Hagino, M., Li, H., Golub, L.: 2008, Magnetic causes of the eruption of a quiescent filament. *Solar Phys.* **247**, 321. DOI. ADS.
Schrijver, C.J.: 2009, Driving major solar flares and eruptions: a review. *Adv. Space Res.* **43**, 739. DOI. ADS.
Schrijver, C.J., Aulanier, G., Title, A.M., Pariat, E., Delannée, C.: 2011, The 2011 February 15 X2 flare, ribbons, coronal front, and mass ejection: interpreting the three-dimensional views from the solar dynamics observatory and STEREO guided by magnetohydrodynamic flux-rope modeling. *Astrophys. J.* **738**, 167. DOI. ADS.
Schrijver, C.J., Beer, J., Baltensperger, U., Cliver, E.W., Güdel, M., Hudson, H.S., McCracken, K.G., Osten, R.A., Peter, T., Soderblom, D.R., Usoskin, I.G., Wolff, E.W.: 2012, Estimating the frequency of extremely energetic solar events, based on solar, stellar, lunar, and terrestrial records. *J. Geophys. Res.* **117**, 8103. DOI. ADS.
Schrijver, C.J., Dobbins, R., Murtagh, W., Petrinec, S.M.: 2014, Assessing the impact of space weather on the electric power grid based on insurance claims for industrial electrical equipment. *Space Weather* **12**, 487. DOI. ADS.
Shibata, K., Magara, T.: 2011, Solar flares: magnetohydrodynamic processes. *Living Rev. Solar Phys.* **8**, 6. DOI. ADS.
Strugarek, A., Brun, A.S., Matt, S.P., Réville, V.: 2014, On the diversity of magnetic interactions in close-in star–planet systems. *Astrophys. J.* **795**, 86. DOI. ADS.

Sturrock, P.A.: 1966, Model of the high-energy phase of solar flares. *Nature* **211**, 695. DOI. ADS.
Su, Y.N., Golub, L., van Ballegooijen, A.A., Gros, M.: 2006, Analysis of magnetic shear in an X17 solar flare on October 28, 2003. *Solar Phys.* **236**, 325. DOI. ADS.
Su, Y., Golub, L., van Ballegooijen, A., Deluca, E.E., Reeves, K.K., Sakao, T., Kano, R., Narukage, N., Shibasaki, K.: 2007, Evolution of the sheared magnetic fields of two X-class flares observed by Hinode/XRT. *Publ. Astron. Soc. Japan* **59**, 785. DOI. ADS.
Su, Y., van Ballegooijen, A.: 2012, Observations and magnetic field modeling of a solar polar crown prominence. *Astrophys. J.* **757**, 168. DOI. ADS.
Su, Y., Veronig, A.M., Holman, G.D., Dennis, B.R., Wang, T., Temmer, M., Gan, W.: 2013, Imaging coronal magnetic-field reconnection in a solar flare. *Nat. Phys.* **9**, 489. DOI. ADS.
Subramanian, P., Dere, K.P.: 2001, Source regions of coronal mass ejections. *Astrophys. J.* **561**, 372. DOI. ADS.
Sui, L., Holman, G.D.: 2003, Evidence for the formation of a large-scale current sheet in a solar flare. *Astrophys. J. Lett.* **596**, L251. DOI. ADS.
Sun, X., Hoeksema, J.T., Liu, Y., Wiegelmann, T., Hayashi, K., Chen, Q., Thalmann, J.: 2012, Evolution of magnetic field and energy in a major eruptive active region based on SDO/HMI observation. *Astrophys. J.* **748**, 77. DOI. ADS.
Sun, X., Hoeksema, J.T., Liu, Y., Aulanier, G., Su, Y., Hannah, I.G., Hock, R.A.: 2013, Hot spine loops and the nature of a late-phase solar flare. *Astrophys. J.* **778**, 139. DOI. ADS.
Švestka, Z.: 1986, On the varieties of solar flares. In: Neidig, D.F. (ed.) *The Lower Atmosphere of Solar Flares, Proceedings of the Solar Maximum Mission Symposium, Sunspot, NM, Aug. 20–24, 1985 (A87-26201 10-92), Sunspot, NM, National Solar Observatory, 1986*, 332. ADS.
Sweet, P.A.: 1958a, The neutral point theory of solar flares. In: Lehnert, B. (ed.) *Electromagnetic Phenomena in Cosmical Physics, IAU Symposium* **6**, 123. ADS.
Sweet, P.A.: 1958b, The topology of force-free magnetic fields. *Observatory* **78**, 30. ADS.
Takasao, S., Asai, A., Isobe, H., Shibata, K.: 2012, Simultaneous observation of reconnection inflow and outflow associated with the 2010 August 18 solar flare. *Astrophys. J. Lett.* **745**, L6. DOI. ADS.
Thalmann, J.K., Su, Y., Temmer, M., Veronig, A.M.: 2015, The confined x-class flares of solar active region 2192. *Astrophys. J. Lett.* **801**, L23. DOI. ADS.
Titov, V.S.: 2007, Generalized squashing factors for covariant description of magnetic connectivity in the solar corona. *Astrophys. J.* **660**, 863. DOI. ADS.
Titov, V.S., Démoulin, P.: 1999, Basic topology of twisted magnetic configurations in solar flares. *Astron. Astrophys.* **351**, 707. ADS.
Titov, V.S., Hornig, G., Démoulin, P.: 2002, Theory of magnetic connectivity in the solar corona. *J. Geophys. Res.* **107**, 1164. DOI. ADS.
Toriumi, S., Yokoyama, T.: 2012, Large-scale 3D MHD simulation on the solar flux emergence and the small-scale dynamic features in an active region. *Astron. Astrophys.* **539**, A22. DOI. ADS.
Török, T., Aulanier, G., Schmieder, B., Reeves, K.K., Golub, L.: 2009, Fan-Spine topology formation through two-step reconnection driven by twisted flux emergence. *Astrophys. J.* **704**, 485. DOI. ADS.
Tripathi, D., Isobe, H., Mason, H.E.: 2006, On the propagation of brightening after filament/prominence eruptions, as seen by SoHO-EIT. *Astron. Astrophys.* **453**, 1111. DOI. ADS.
Tsuneta, S.: 1996, Structure and dynamics of magnetic reconnection in a solar flare. *Astrophys. J.* **456**, 840. DOI. ADS.
Tsuneta, S., Acton, L., Bruner, M., Lemen, J., Brown, W., Caravalho, R., Catura, R., Freeland, S., Jurcevich, B., Morrison, M., Ogawara, Y., Hirayama, T., Owens, J.: 1991, The soft X-ray telescope for the SOLAR-A mission. *Solar Phys.* **136**, 37. DOI. ADS.
Tsurutani, B.T., Gonzalez, W.D., Lakhina, G.S., Alex, S.: 2003, The extreme magnetic storm of 1–2 September 1859. *J. Geophys. Res.* **108**, 1268. DOI. ADS.
van Ballegooijen, A.A.: 2004, Observations and modeling of a filament on the Sun. *Astrophys. J.* **612**, 519. DOI. ADS.
van Ballegooijen, A.A., Martens, P.C.H.: 1989, Formation and eruption of solar prominences. *Astrophys. J.* **343**, 971. DOI. ADS.
van Driel-Gesztelyi, L.: 1997, Emerging flux tube geometry and sunspot proper motions. In: Schmieder, B., del Toro Iniesta, J.C., Vazquez, M. (eds.) *1st Advances in Solar Physics Euroconference. Advances in Physics of Sunspots, Astronomical Society of the Pacific Conference Series* **118**, 81. ADS.
Walkowicz, L.M., Basri, G., Batalha, N., Gilliland, R.L., Jenkins, J., Borucki, W.J., Koch, D., Caldwell, D., Dupree, A.K., Latham, D.W., Meibom, S., Howell, S., Brown, T.M., Bryson, S.: 2011, White-light flares on cool stars in the Kepler quarter 1 data. *Astron. J.* **141**, 50. DOI. ADS.
Wang, H., Yan, Y., Sakurai, T., Zhang, M.: 2000, Topology of magnetic field and coronal heating in solar active regions – II. The role of quasi-separatrix layers. *Solar Phys.* **197**, 263. DOI. ADS.

Wang, Y., Zhang, J.: 2007, A comparative study between eruptive X-class flares associated with coronal mass ejections and confined X-class flares. *Astrophys. J.* **665**, 1428. DOI. ADS.

Wang, S., Liu, C., Liu, R., Deng, N., Liu, Y., Wang, H.: 2012, Response of the photospheric magnetic field to the X2.2 flare on 2011 February 15. *Astrophys. J. Lett.* **745**, L17. DOI. ADS.

Wang, H., Liu, C., Deng, N., Zeng, Z., Xu, Y., Jing, J., Cao, W.: 2014, Study of two successive three-ribbon solar flares on 2012 July 6. *Astrophys. J. Lett.* **781**, L23. DOI. ADS.

Webb, D.F.: 2015, Eruptive prominences and their association with coronal mass ejections. In: Vial, J.-C., Engvold, O. (eds.) *Solar Prominences, Astrophysics and Space Science Library* **415**, 411. DOI. ADS.

Wendel, D.E., Olson, D.K., Hesse, M., Aunai, N., Kuznetsova, M., Karimabadi, H., Daughton, W., Adrian, M.L.: 2013, The relation between reconnected flux, the parallel electric field, and the reconnection rate in a three-dimensional kinetic simulation of magnetic reconnection. *Phys. Plasmas* **20**(12), 122105. DOI. ADS.

Wichmann, R., Fuhrmeister, B., Wolter, U., Nagel, E.: 2014, Kepler super-flare stars: what are they? *Astron. Astrophys.* **567**, A36. DOI. ADS.

Williams, D.R., Török, T., Démoulin, P., van Driel-Gesztelyi, L., Kliem, B.: 2005, Eruption of a kink-unstable filament in NOAA active region 10696. *Astrophys. J. Lett.* **628**, L163. DOI. ADS.

Wolff, E.W., Bigler, M., Curran, M.A.J., Dibb, J.E., Frey, M.M., Legrand, M., McConnell, J.R.: 2012, The Carrington event not observed in most ice core nitrate records. *Geophys. Res. Lett.* **39**, 8503. DOI. ADS.

Wuelser, J.-P., Lemen, J.R., Tarbell, T.D., Wolfson, C.J., Cannon, J.C., Carpenter, B.A., Duncan, D.W., Gradwohl, G.S., Meyer, S.B., Moore, A.S., Navarro, R.L., Pearson, J.D., Rossi, G.R., Springer, L.A., Howard, R.A., Moses, J.D., Newmark, J.S., Delaboudinière, J.-P., Artzner, G.E., Auchere, F., Bougnet, M., Bouyries, P., Bridou, F., Clotaire, J.-Y., Colas, G., Delmotte, F., Jerome, A., Lamare, M., Mercier, R., Mullot, M., Ravet, M.-F., Song, X., Bothmer, V., Deutsch, W.: 2004, EUVI: the STEREO-SECCHI extreme ultraviolet imager. In: Fineschi, S., Gummin, M.A. (eds.) *Telescopes and Instrumentation for Solar Astrophysics, Society of Photo-Optical Instrumentation Engineers (SPIE) Conference Series* **5171**, 111. DOI. ADS.

Yamada, M., Kulsrud, R., Ji, H.: 2010, Magnetic reconnection. *Rev. Mod. Phys.* **82**, 603. DOI. ADS.

Yamada, M., Yoo, J., Jara-Almonte, J., Ji, H., Kulsrud, R.M., Myers, C.E.: 2014, Conversion of magnetic energy in the magnetic reconnection layer of a laboratory plasma. *Nat. Commun.* **5**, 4774. DOI. ADS.

Yashiro, S., Gopalswamy, N., Akiyama, S., Michalek, G., Howard, R.A.: 2005, Visibility of coronal mass ejections as a function of flare location and intensity. *J. Geophys. Res.* **110**(A9), 12. DOI. ADS.

Yashiro, S., Akiyama, S., Gopalswamy, N., Howard, R.A.: 2006, Different power-law indices in the frequency distributions of flares with and without coronal mass ejections. *Astrophys. J. Lett.* **650**, L143. DOI. ADS.

Yeates, A.R., Mackay, D.H.: 2009, Initiation of coronal mass ejections in a global evolution model. *Astrophys. J.* **699**, 1024. DOI. ADS.

Yokoyama, T., Akita, K., Morimoto, T., Inoue, K., Newmark, J.: 2001, Clear evidence of reconnection inflow of a solar flare. *Astrophys. J. Lett.* **546**, L69. DOI. ADS.

Young, P.R., Doschek, G.A., Warren, H.P., Hara, H.: 2013, Properties of a solar flare kernel observed by Hinode and SDO. *Astrophys. J.* **766**, 127. DOI. ADS.

Zhao, J., Li, H., Pariat, E., Schmieder, B., Guo, Y., Wiegelmann, T.: 2014, Temporal evolution of the magnetic topology of the NOAA active region 11158. *Astrophys. J.* **787**, 88. DOI. ADS.

Zharkov, S., Green, L.M., Matthews, S.A., Zharkova, V.V.: 2011, 2011 February 15: sunquakes produced by flux rope eruption. *Astrophys. J. Lett.* **741**, L35. DOI. ADS.

Zuccarello, F.P., Meliani, Z., Poedts, S.: 2012, Numerical modeling of the initiation of coronal mass ejections in active region NOAA 9415. *Astrophys. J.* **758**, 117. DOI. ADS.

Reprinted from *Solar Physics* Journal, DOI 10.1007/s11207-015-0712-1

Flare-CME Models: An Observational Perspective (Invited Review)

B. Schmieder[1] · **G. Aulanier**[1] · **B. Vršnak**[2]

Received: 18 November 2014 / Accepted: 23 May 2015 / Published online: 10 June 2015
© Springer Science+Business Media Dordrecht 2015

Abstract Eruptions, flares, and coronal mass ejection (CMEs) are due to physical phenomena mainly driven by an initially force-free current-carrying magnetic field. We review some key observations relevant to the current theoretical trigger mechanisms of the eruption and to the energy release *via* reconnection. Sigmoids observed in X-rays and UV, as well as the pattern (double J-shaped) of electric currents in the photosphere show clear evidence of the existence of currents parallel to the magnetic field and can be the signature of a flux rope that is detectable in CMEs. The magnetic helicity of filaments and active regions is an interesting indirectly measurable parameter because it can quantify the twist of the flux rope. On the other hand, the magnetic helicity of the solar structures allows us to associate solar eruptions and magnetic clouds in the heliosphere. The magnetic topology analysis based on the 3D magnetic field extrapolated from vector magnetograms is a good tool for identifying the reconnection locations (null points and/or the 3D large volumes – hyperbolic flux tube, HFT). Flares are associated more with quasi-separatrix layers (QSLs) and HFTs than with a single null point, which is a relatively rare case. We review various mechanisms that have been proposed to trigger CMEs and their observable signatures: by "breaking" the field lines overlying the flux rope or by reconnection below the flux rope to reduce the magnetic tension, or by letting the flux rope to expand until it reaches a minimum threshold height (loss of equilibrium or torus instability). Additional mechanisms are commonly operating in the solar atmosphere. Examples of observations are presented throughout the article and are discussed in this framework.

Keywords Sun: flare · Sun: coronal mass ejection

Solar and Stellar Flares: Observations, Simulations, and Synergies
Guest Editors: Lyndsay Fletcher and Petr Heinzel

✉ B. Schmieder
Brigitte.Schmieder@obspm.fr

B. Vršnak
bvrsnak@geof.hr

1 Observatoire de Paris, LESIA, 92190 Meudon, France

2 Hvar Observatory, Faculty of Geodesy, University of Zagreb, Kačićeva 26, 10000 Zagreb, Croatia

1. Introduction

Coronal mass ejections (CMEs) are bright features in the solar corona observed in white light with coronagraphs. CMEs propagate in the interplanetary space as interplanetary coronal mass ejections (ICMEs). Some of them take the form of the so-called magnetic clouds, which are characterized by a smooth rotation of the magnetic field, a decreasing front-to-rear speed profile, a depression of the proton temperature, and low plasma beta (Lepping, Burlaga, and Jones, 1990; Liu *et al.*, 2011). ICMEs drive geomagnetic storms that affect the space environment around Earth (Schwenn, 2006; Pulkkinen, 2007; Chandra *et al.*, 2010).

An armada of spacecraft (in the past *e.g.*, the *Solar and Heliospheric Observatory* (SOHO), *Yohkoh* and the *Transition Region And Coronal Explorer* (TRACE), and presently *Hinode*, the *Solar TErrestrial RElations Observatory* (STEREO), and the *Solar Dynamics Observatory* (SDO)) provided flare observations from space over Solar Cycles 22, 23, and 24. In a first approach, the relationship between flares, filament eruptions, and CMEs is given by statistics (Harrison, 1995; Zhang *et al.*, 2001; Subramanian and Dere, 2001; Vršnak, Sudar, and Ruždjak, 2005; Maričić *et al.*, 2007; Bein *et al.*, 2012). According to the most recent article (Bein *et al.*, 2012), up to 25 % of CMEs are only associated with filament eruptions, without a flare detected by GOES. All the others are related to flares, with or without a filament eruption. It is nevertheless frequently observed that a CME associated with a flare is also associated with a filament eruption (70 %). Flares and CMEs belong to the same magnetically driven event (Webb and Howard, 2012), *i.e.*, they are two different manifestations of the same magnetic process. However, apparently, they are not in a cause–effect relationship, as demonstrated by Zhang *et al.* (2001), who clearly rejected the idea of CMEs being driven by flare thermal pressure-pulse.

CMEs are associated with so-called eruptive flares, as opposed to so-called confined flares, which represent flares that are not associated with an observable ejection of mass (*cf.*, Priest, 1981; see also, *e.g.*, Schmieder *et al.*, 1997; Guo *et al.*, 2012; Dalmasse *et al.*, 2014). There are also flares that are associated with "failed eruptions" (Török and Kliem, 2005), *i.e.*, the event starts like a normal eruptive flare, but the ejection is halted after reaching a certain height and fails to leave the Sun. McCauley *et al.* (2015) found that 72 % of filament eruptions have associated CMEs, implying that 28 % of filament eruptions fail to escape. The eruption either ends at an upper equilibrium height (Vršnak *et al.*, 1990), or the elevated structure falls back (Wang and Sheeley, 2002). It has been argued that flares without CMEs on average have a higher temperature, presumably because all the magnetic energy released is concentrated in the heating of the plasma, whereas in CME-associated flares a significant part is spent in accelerating the ejecta, *i.e.*, on its kinetic energy (Yashiro *et al.*, 2006).

In the past 50 years, flare and CME models have significantly progressed from the standard two-dimensional (2D) CSHKP model (Carmichael, 1964; Sturrock, 1966; Hirayama, 1974; Kopp and Pneuman, 1976). Subsequently, these basically phenomenological models, based on the concept of flux-rope eruption, were formulated in a quantitative way (*e.g.*, Anzer and Pneuman, 1982; Chen, 1989; Vršnak, 1990). On the other hand, Shibata (1996), Forbes and Malherbe (1986), and Forbes (1986) combined ("unified") the models, showing three necessary conditions for defining flare–CMEs events: the existence of a magnetic flux rope, reconnection below the flux rope, and reconnected loops or post-flare loops. However, it should be noted that many problems are not addressed in this "standard" model. For example, the flux rope could be formed from a part of the magnetic arcade during the destabilization process (Mikic and Linker, 1994). Furthermore, for some models the existence of a flux rope is apparently not necessary. The large overlying arcade is pushed from below. This type of models could be relevant for the understanding of CMEs that are associated with large-scale loops connecting different active regions, like trans-equatorial loops (Delannée and Aulanier, 1999; Sun *et al.*, 2014). Currently, popular CME-triggering mechanisms are the tether-cutting that considers field line reconnection below the flux rope (Moore *et al.*, 2001), the break out through the overlying field that reconnects above the sheared arcade or the flux rope (Antiochos, DeVore, and Klimchuk, 1999; Jacobs *et al.*, 2009), and the torus instability (Kliem and Török, 2006; Kliem *et al.*, 2010; Török and Kliem, 2005).

Recent reviews approach these topics by focusing on various aspects: historical, observational, and theoretical. They are concentrated on the CME morphology, the solar sources, or their propagation through the heliosphere (Schwenn, 2006; Pulkkinen, 2007; Chen, 2011; Webb and Howard, 2012). Some of the reviews are focused on the CME triggering mechanisms (Forbes *et al.*, 2006; Shibata and Magara, 2011; Chen, 2011; Kliem *et al.*, 2014a). 3D MHD models have mainly been developed and discussed in this framework (see the review by Aulanier (2014) and references herein).

Numerical simulations are very helpful for testing the role of the different parameters involved in CME triggers. However, the analysis of observations is essential to constrain and test theoretical models. Therefore, in this article we revisit a set of observations from a critical point of view to answer some open questions concerning the CME triggering mechanisms that are still not answered by the theoretical models alone. We perform a meta-analysis that is not exhaustive, however. Our intention is to highlight specific points that are important in current CMEs models. For example: What do we learn from the new generation of magnetographs operating with high polarimetric sensitivity, high spatial resolution, and high cadence? Can we detect the flux rope formation in the photosphere or in the corona by using the observations and/or the new magnetic field extrapolation techniques? Can we follow the flux rope evolution until its eruption as CME? Can we detect where the reconnection site will be by analyzing the topology of active regions? Is the reconnection region below or above the flux rope?

We here present, as a first step, the basic characteristics of CMEs, including the CME morphology, their velocity, and acceleration. Then, we perform a meta-analysis of flare–CMEs events, from an observational perspective. We focus on four specific points: the formation of the magnetic flux rope, the twist, and the helicity of the magnetic flux rope, the topology of CME sources and the break-out mechanism, the CME acceleration, and the torus instability. Finally, we show that even though the observations are complex and models are oversimplified, some conclusions can still be drawn.

2. Characteristics of CMEs

2.1. Morphology and Statistics

CMEs often have three parts: a bright core surrounded by a dark cavity with a bright compression front (Illing and Hundhausen, 1986; Forbes, 2000; Schwenn, 2006) (Figures 1a, b). However, only 30 % to 40 % of CMEs show this standard three-part structure (Webb and Hundhausen, 1987; Vourlidas *et al.*, 2013). In most of the cases only bright loops or diffuse structures are visible in CME images. It strongly depends on the sensitivity of the coronagraph. We emphasize that often the coronagraphic CME images are analyzed by subtracting an image of the corona at rest, or by subtracting two successive images, to enhance the visibility of the propagation of the CME. Therefore, it strongly depends on the cadence of the instrument. The bright core is associated with a filament, the cavity is assumed to be the flux rope. Unfortunately, these observations do not give any information about the formation of a flux rope.

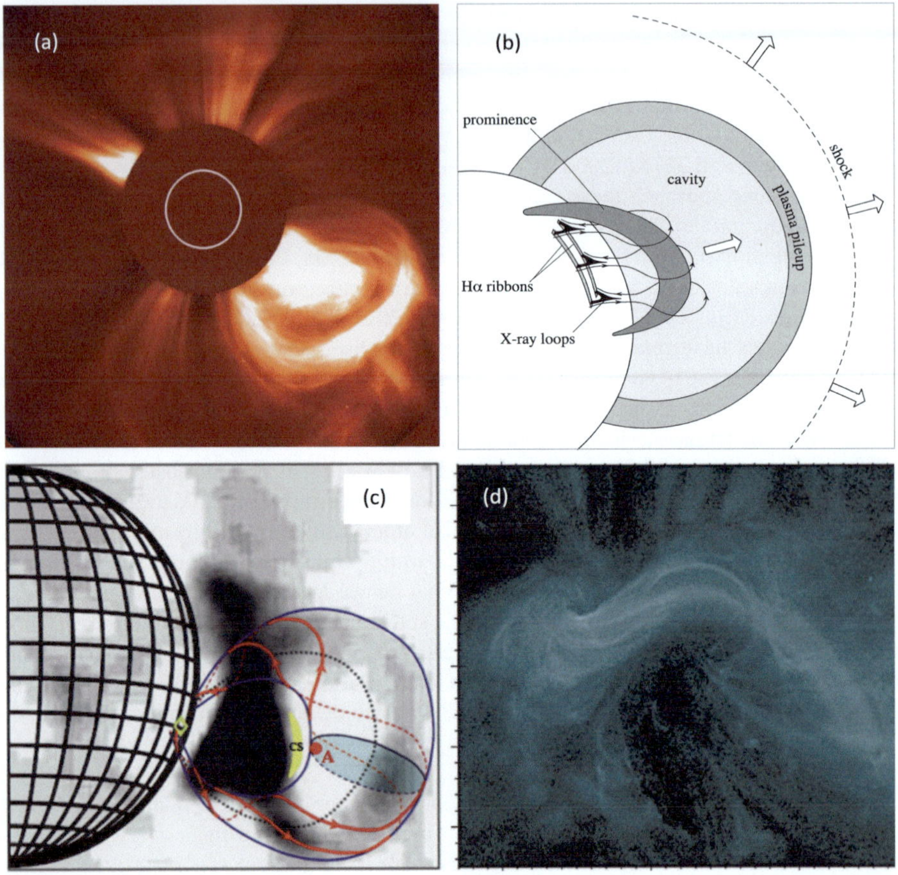

Figure 1 (a) White-light CME observed by SOHO/LASCO. (b) Three-part CME cartoon (Forbes, 2000). (c) CME observed in radio by the *Nançay Radioheliograph* (adapted from Démoulin *et al.*, 2012). (d) Sigmoid observed with SDO/AIA in EUV at 94 Å (from Dudík *et al.*, 2014).

In coronagraph images, the white-light emission comes from the Thomson-scattered photospheric radiation by free electrons in the corona. Enhanced brightness in white light means that the coronal density along the line of sight is increased. Since the emission is optically thin, this means that usually several structures are integrated along the line of sight due to projection effects, which further complicates the interpretation of the observed features. We note that the increase in density can also be due to a CME-driven compression of the surrounding coronal plasma, forming the so-called bow shock (Chen, 2011). Delannée *et al.* (2008) suggested that the CME is surrounded by a current shell that might be the location of compression and heating. This provides a way to explain the EIT waves by compressing and heating the surrounding plasma as the CME expands. This idea was checked by using different filters of the *Atmospheric Imaging Assembly* (AIA) of SDO, which show a hotter front of the CME when it is observed in projection on the disk as a propagating circular front (Schrijver *et al.*, 2011).

Early stages of CMEs are frequently observed in various spectral ranges: X-rays (Rust, 1983; Maričić *et al.*, 2004; Vršnak *et al.*, 2004), EUV (Dere *et al.*, 1997; Chen, 2009), and radio (Gopalswamy and Kundu, 1992; Maia *et al.*, 1999; Démoulin *et al.*, 2012; Car-

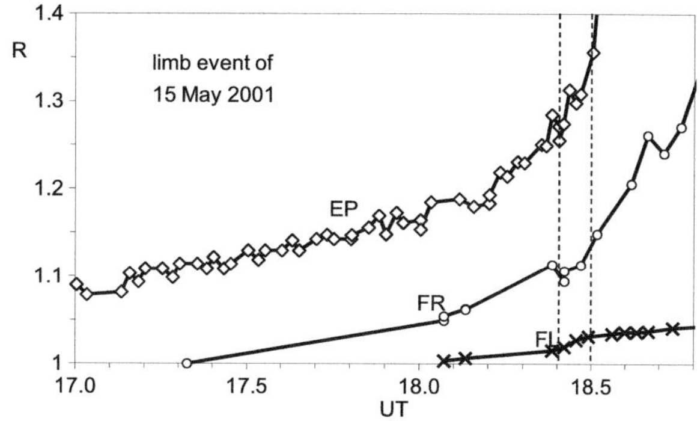

Figure 2 Kinematics of the eruptive prominence (EP) and the bottom of the embedding flux rope (FR), compared with the growth of the flare loop system (FL) in the behind-the-limb event of 15 May 2001. Radial distances, *R*, are expressed in solar radii. For details see Vršnak *et al.* (2004).

ley *et al.*, 2013). Démoulin *et al.* (2012) proposed that the radio CME illuminated by synchrotron emission from relativistic electrons corresponds to the white-light flux-rope cavity surrounded the bright CME front (Figure 1c).

A CME is called a halo CME when its brightness completely encircles the solar disk because of its propagation towards the observer. Vourlidas *et al.* (2013) showed different features visible in the halo CME of 15 February 2011 that was observed from three different angles by the LASCO and STEREO coronagraphs. The morphology of the LASCO halo CME was different in STEREO images, having a narrower angular width. Halo CMEs are commonly associated with energetic flares and are frequently geoeffective. However, we note that halo CMEs launched from regions close to the solar limb during Solar Cycle 23 were never very geoeffective (Cid *et al.*, 2012). To be strongly geoeffective, the CME source has to be close to the central meridian.

The characteristic mass of CMEs is $0.1-4 \times 10^{13}$ kg (Jackson *et al.*, 1985; Gopalswamy and Kundu, 1992; Chen, 2011). Only a few have a much lower mass, around 10^{11} kg (Vourlidas *et al.*, 2002). The occurrence rate of CMEs was found to be 0.2/3.5 *per* day before SOHO and 0.5/6 *per* day after SOHO during solar minimum/maximum activity, respectively. The velocity range is between 20 to 3000 $\mathrm{km\,s^{-1}}$. Given the indirect mass estimates, the kinetic energy is estimated to lie within between 10^{28} to 10^{32} ergs, depending on the event (Yashiro *et al.*, 2004). The CME speeds do not have a unique value. Measurements of, *e.g.*, the eruptive prominence, the CME leading edge, or the front edge of the flux-rope cavity give different values (Zhang *et al.*, 2001; Cheng *et al.*, 2014b; Vršnak *et al.*, 2004) (Figure 2). We note that the profile of the height–time curve characterizes the trigger mechanism (Schrijver *et al.*, 2008).

2.2. CME Acceleration

The main acceleration phase of CMEs and the impulsive phase of the associated flares are tightly related phenomena. For example, various statistical studies have shown that characteristics of CMEs, such as the velocity or kinetic energy, are correlated with the properties of the associated flares, *e.g.*, peak flux or integrated flux of the soft X-ray (SXR) burst (Moon *et al.*, 2002, 2003; Burkepile *et al.*, 2004; Vršnak, Sudar, and Ruždjak, 2005; Bein *et al.*, 2012).

In particular, Vršnak, Sudar, and Ruždjak (2005) and Bein *et al.* (2012) have shown that, generally, CME events ordered according to their acceleration are related to the importance of their associated flares: CMEs launched from spotless regions that are not associated with any significant flare signature show the weakest accelerations, but are still similar to that of active-region CMEs associated with weak flares of SXR-class A, B and C, whereas CMEs associated with flares of SXR-class M and X show considerably stronger acceleration. But there are no two *distinct* classes of CMEs (flare/non-flare), as proposed by Gosling *et al.* (1976) and MacQueen and Fisher (1983), which became a widely accepted concept that was frequently employed to interpret various aspects of CMEs (see, *e.g.*, Sheeley *et al.*, 1999; Andrews and Howard, 2001; Low and Zhang, 2002; Moon *et al.*, 2002). Instead, there is a continuum of events from weak to strong ones, the CME acceleration increasing with increasing strength of the associated flares. It is also important to note that although the CMEs associated with filament eruptions on average have significantly lower accelerations, their highest average speed is only marginally lower than that of flare-associated CMEs. This shows that the presence of a filament (*i.e.*, a dense/heavy prominence) only increases the inertia of the system, making the eruption more gradual in the take-off phase, but not preventing it from reaching high velocities (Vršnak, Sudar, and Ruždjak, 2005) if the acceleration lasts long enough.

Another important issue is the synchronization of the rising phase of the SXR burst and the main acceleration phase of the CME, reported by, *e.g.*, Kahler *et al.* (1988), Neupert *et al.* (2001), Zhang *et al.* (2001, 2004), Shanmugaraju *et al.* (2003), Maričić *et al.* (2004), Vršnak *et al.* (2004), Zhang and Dere (2006). In the context of these case-study results, Maričić *et al.* (2007) and Bein *et al.* (2012) performed a statistical analysis that unambiguously confirmed that such a synchronization exists in the majority of events. However, we stress that Maričić *et al.* (2007) noted that there are events that do not show such a behavior and, furthermore, even in the majority of synchronized events, the CME acceleration starts *before* the SXR burst and ends after it reaches maximum.

This relationship indicates that the CME main acceleration phase is related to the flare impulsive phase because the SXR burst growth is usually considered to be a consequence of the nonthermal energy release (the so-called Neupert effect, Neupert, 1968; see also Dennis and Zarro, 1993; Veronig *et al.*, 2002 and references therein). Indeed, Temmer *et al.* (2008) and Temmer *et al.* (2010) have recently demonstrated that the CME acceleration is roughly contemporaneous with the hard X-ray burst.

Since the nonthermal energy release rate in flares is correlated with the magnetic reconnection rate (for a direct evidence see, *e.g.*, Miklenic *et al.*, 2007; Miklenic, Veronig, and Vršnak, 2009), it can be hypothesized that in the majority of events there is a feedback relationship between the CME dynamics and the flare-related reconnection process in the wake of the CME – the increasing reconnection rate enhances the CME acceleration and *vice versa*, enhanced acceleration provides more efficient reconnection (for details see Vršnak, 2008; Temmer *et al.*, 2010). This concept is strongly supported by the results presented by Qiu and Yurchyshyn (2005) and Miklenic, Veronig, and Vršnak (2009), who found that the CME velocities and the strength of SXR flares are correlated with the total reconnected flux. However, the most direct evidence comes from the case study by Qiu *et al.* (2004), who revealed a synchronization of the eruptive prominence acceleration and the reconnection rate ("flux change rate") in the associated flare.

3. Formation of a Flux Rope

3.1. Definition of Flux Rope

A magnetic flux rope is a plasma structure in which the magnetic field is wrapped around its central axis. These structures are difficult to detect observationally. It has been proposed

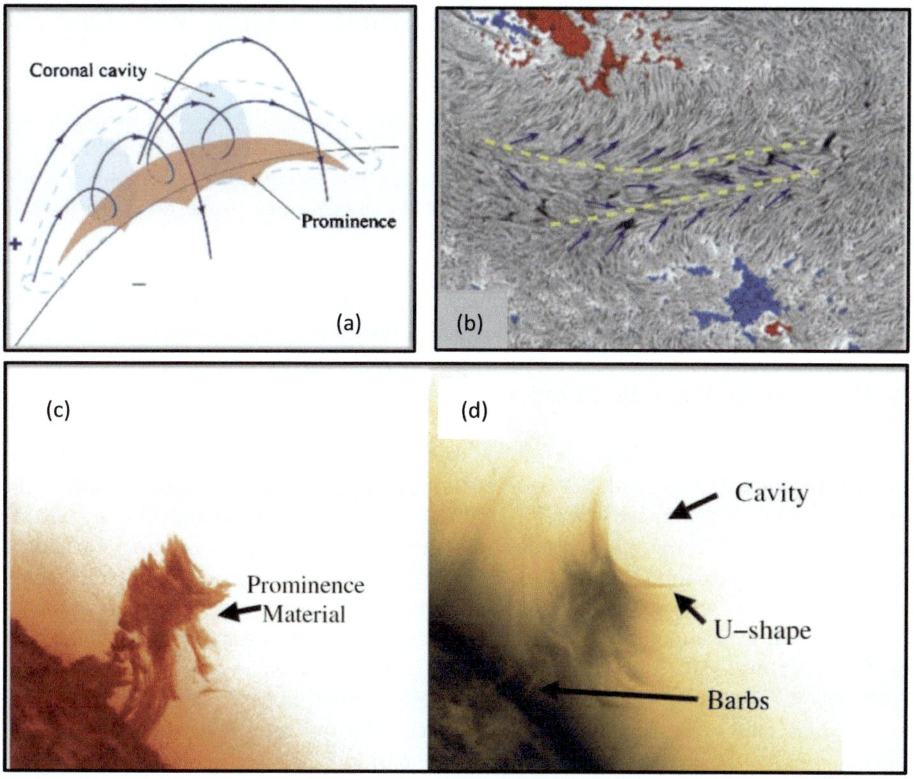

Figure 3 Relationship between cavity and prominence. (a) Cartoon of Filippov *et al.* (2015), (b) Big Bear Solar Observatory (BBSO) Hα filament overlaid by arrows indicating the orientation of the horizontal magnetic field component. The yellow lines represent the separatrices, the blue/red the positive/negative polarity (adapted from Filippov *et al.*, 2015). (c) Prominence observed in AIA 304 Å. (d) The corresponding cavity in AIA 171 Å (adapted from Régnier, Walsh, and Alexander, 2011).

that sigmoids with S or reverse-S-shaped bright loops observed in active regions in X-rays with the *Yohkoh Soft X-ray Telescope* (SXT) and the *Hinode X-Ray Telescope* (XRT) could represent a kinked flux rope (Acton *et al.*, 1992; Rust and Kumar, 1996; Aurass *et al.*, 1999) or its envelope (Green *et al.*, 2007; Green, Kliem, and Wallace, 2011; Savcheva *et al.*, 2012, 2014; Dudík *et al.*, 2014) (Figure 1d). Such sigmoid or bright loops can be the manifestation of a magnetic flux rope and can be observed before flares (Cheng *et al.*, 2014a; Dudík *et al.*, 2014). On the limb, bright twisted stretched loops observed by SDO/AIA in different filters at the CME onset may represent a flux rope (Cheng *et al.*, 2011, 2014b). Nevertheless, it remains a challenge to be able to prove that the sigmoids or the bright loops belong to a flux rope.

Recent observations at the limb give new insight into the relationship between cavity and prominence (Régnier, Walsh, and Alexander, 2011; Gibson *et al.*, 2006). The scheme proposed by Filippov *et al.* (2015) summarizes the AIA observations of the cavity and the prominence (Figures 3a, c, and d). The three-dimensional shape of the cavity is approximately a circular cylinder. The field lines of a flux rope have the form of helices. The prominence material is embedded at the bottom of the flux rope, *e.g.*, in the field-line dips. When observed on the disk, the cavity corresponds to the filament channel observed as a dark elongated region in EUV (Heinzel *et al.*, 2008; Schmieder *et al.*, 2014; Cheng *et al.*, 2014b) and in radio wavelengths (Marqué *et al.*, 2001).

One strong observational support for the existence of flux ropes above filament channels comes from the magnetic field measurements: they revealed that the filament magnetic field has a so-called inverse polarity as compared to that of the surface magnetic field on both sides of the filament channel (Bommier and Leroy, 1998; Aulanier and Démoulin, 2003). In the chromosphere, the horizontal magnetic field changes sense at the boundaries of the filament channel, parallel to the photospheric inversion line (PIL) (Filippov *et al.*, 2015). This is clearly seen in the orientation of the chromospheric fibrils (Figure 3b). Dips or bald patches have been detected in filament channels by using vector magnetograms, which also suggests the existence of a flux rope (Lites, 2005; López Ariste *et al.*, 2006).

3.2. Photospheric Magnetic Field Evolution

Most of MHD numerical CME models involve a flux rope in an active region that commonly is represented by a bipolar field. Some of the models start with an already existing flux rope (Wu *et al.*, 2004; Török and Kliem, 2005; Shiota *et al.*, 2010; Lugaz and Roussev, 2011). This type of models does not bring any information on flux rope formation. In another type of models, a flux rope is created in an idealized way by using some mechanisms that have been inferred from various observations. "Idealized" means that these models are not data driven, but they are still more or less realistic. These MHD models propose the flux-rope formation either after the eruption has already started by shearing motions (Mikic and Linker, 1994; Amari *et al.*, 2000; Jacobs *et al.*, 2009), or before the onset of the eruption by flux cancellation at sheared PILs (van Ballegooijen and Martens, 1989; Amari *et al.*, 2003; Aulanier *et al.*, 2010; Yeates, 2014; Xia, Keppens, and Guo, 2014) or by flux emergence (see the next section). In both the cancellation and the emergence cases, converging flows initiate reconnection near the PIL, and sheared fields are converted into a twisted field. The magnetic flux rope is created through flux cancellation in the photosphere or *via* tether-cutting in the corona. This formation of long twisted field lines by reconnection due to canceling flux was observed in filaments (Schmieder *et al.*, 2004).

Observations provide examples where one or two of these mechanisms could be efficient for the formation of sigmoids and subsequent observed CMEs. Many studies concern the evolution of the magnetic field by using data from the SOHO *Michelson Doppler Imager* (MDI) or the high-cadence SDO *Helioseismic and Magnetic Imager* (HMI). Flux cancellation is frequently observed along the PIL of active regions. A statistical analysis shows that two-thirds of the 72 analyzed long-lasting sigmoids observed before flares were created by flux cancellation (Savcheva *et al.*, 2014). The flux cancellation model is commonly based on the slow evolution of active regions. The photospheric field diffuses by dispersion or because of magnetic cancellation along the polarity inversion line. This occurs during the decaying phase of active regions (Démoulin *et al.*, 2002; van Driel-Gesztelyi *et al.*, 2003; Schmieder *et al.*, 2008). A simulation (Gibb *et al.*, 2014) driven by a sequence of seven days of line-of-sight component magnetograms from SOHO/MDI showed the evolution of the coronal magnetic field through a continuous series of nonlinear force-free states. This simulation describes the formation of a flux rope across the polarity inversion line during flux cancellation. The flux rope forms at the same location as an an X-ray sigmoid observed by *Yohkoh*.

Shear motions may also lead to the formation of a flux rope. A good example is the shear created between two emerging active regions in February 2011 (Figure 4a). The sigmoid, a signature of the flare onset, was in the region squeezed between the negative polarity of

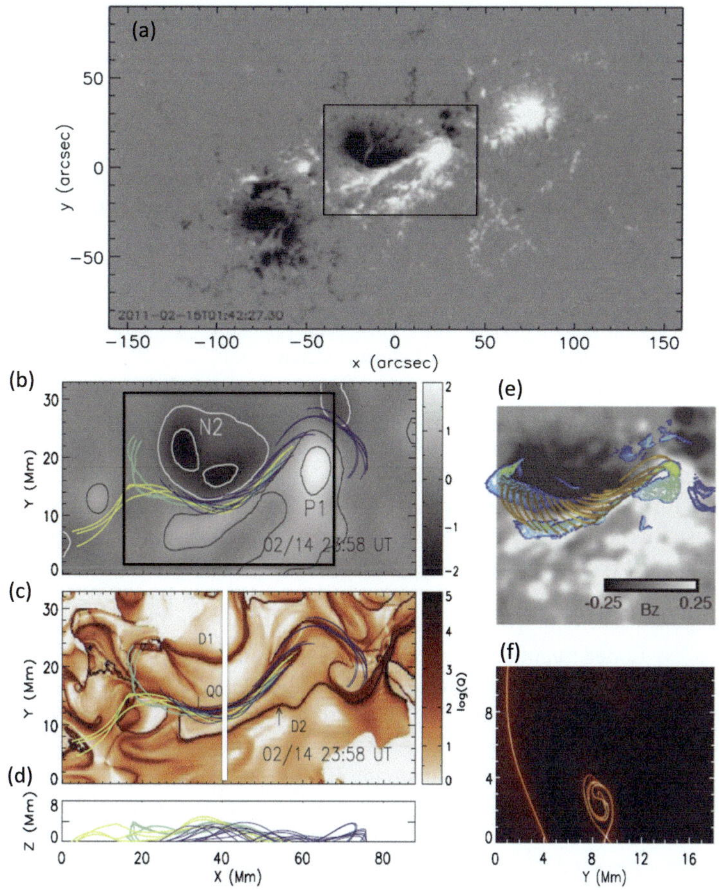

Figure 4 (a) AR 11158 on 15 February 2011, HMI longitudinal magnetogram (Li *et al.*, 2007; Schrijver *et al.*, 2011). The box shows the field of view of panels b and c. (b) Magnetic field lines and a flux rope over the central part of AR 11158 obtained from a nonlinear force-free extrapolation (Zhao *et al.*, 2014). (c) and (d) The flux rope viewed over the central quasi-separatrix-layer footprint in the photosphere and its side view. The vertical white line represents the cut in f, the box is the field of view of panel e. (e) Magnetic field lines over the magnetogram and the flare ribbons (in light blue) (Inoue *et al.*, 2014). (f) Zoom of the small flux rope shown in d and presented in Figure 7 (right panel) (from Zhao *et al.*, 2014).

the western active region and the positive polarity of the eastern active region, where a large amount of flux cancellation took place. This well-studied region was the site of many flares (Schrijver *et al.*, 2011; Toriumi *et al.*, 2013). The formation of the flux rope in this region has been proposed to be due to tether-cutting reconnection (Liu *et al.*, 2013; Inoue *et al.*, 2014) (see the next sections). Thus, it is not clear whether a unique mechanism alone is responsible for flux-rope formation.

3.3. Emergence Through the Photosphere

Some CME models start with a flux rope that emerges from the photosphere by buoyancy (Fan and Gibson, 2003; Manchester *et al.*, 2008). However, theoretical models including the photosphere and the subsurface convection zone as a stratified atmosphere show that it is difficult for a flux rope to emerge as a whole from below the solar surface. A high twist is necessary for a flux rope to survive its rise through the convection zone to the photosphere (Emonet and Moreno-Insertis, 1998; Jouve, Brun, and Aulanier, 2013). The full emergence is difficult because dense material is trapped in the dips of the field lines (Fan, 2001; Manchester *et al.*, 2004; Archontis, Hood, and Tsinganos, 2014; Leake, Linton, and Antiochos, 2014). Theoretical models reveal that it is impossible for a flux rope to fully emerge. They show that only the central axis of the flux rope can be above the solar surface, and a secondary flux rope could be formed in the emerged structure by reconnection resembling tether-cutting (Hood *et al.*, 2009; MacTaggart and Hood, 2010; Leake, Linton, and Török, 2013; Leake, Linton, and Antiochos, 2014).

From the observational point of view, it is difficult to see the emergence of a twisted flux rope. Transverse magnetic-field orientation change was interpreted by the passage of a magnetic flux rope through the photosphere (Okamoto *et al.*, 2008), but this observation has been revisited by Vargas Domínguez *et al.* (2012). Using an emergence model as a tool, these authors demonstrated that the observed signatures were not due to emergence, but rather due to canceling flux along the PIL.

During the formation of an active region, a flux rope tentatively rises through the photosphere. We usually observe highly fragmented polarities at the solar surface. The opposite polarities of the newly emerged bipole expand to form the sunspots of a new active region (see the review of flux emergence by Schmieder, Archontis, and Pariat (2014)). The polarities have an elongated shape (Luoni *et al.*, 2011; Chandra *et al.*, 2009) that is called "magnetic tongues" by López Fuentes *et al.* (2000). This signature is typical for the emergence of the twisted magnetic flux rope since it traces the azimuthal field of the flux rope. Magnetic tongues are also typically found in MHD simulations of flux rope emergence (Hood *et al.*, 2009, and references therein). Figures 5a and b show an example of an emerging new active region in September 2005 with two elongated polarities along the PIL (Li *et al.*, 2007; Canou *et al.*, 2009). Using a nonlinear force-free extrapolation, Canou *et al.* (2009) revealed a short, low-lying coronal flux rope over a long PIL (see Figure 5c). Similar studies, using different data, reported a flux rope along the PIL (Zhao *et al.*, 2014; Guo *et al.*, 2010a) (Figures 4d and f). In the case reported by Guo *et al.* (2010a), a long, low-altitude flux rope coexists with a sheared magnetic arcade along the PIL. Moreover, some eruptions are explained by small-scale emergence in the vicinity of a flux rope. These emerging bipoles can reconnect with the arcades that surround the flux rope and thus destabilize the system (Chen and Shibata, 2000; Török *et al.*, 2011; Toriumi *et al.*, 2013; Masson, Antiochos, and DeVore, 2013).

In conclusion, we note that the statistical analysis of Savcheva *et al.* (2014) indicates that the flux-rope formation correlates better with flux cancellation than with flux emergence. Simple emergence is also not as effective in initiating an unstable state.

4. Twist and Helicity of the Flux Rope

The coronal magnetic field must be highly non-potential before the eruption onset, *i.e.*, it must contain strong electric currents. Flux ropes (cavities or sigmoids) are the observational signatures of such electric currents. The bright twisted features mimic twisted field lines along which the currents could be aligned. This reveals the existence of free energy and magnetic helicity in the corona, which is necessary for the eruption and CME. It is nevertheless impossible to directly measure the electric current in the corona. Some attempts have been made to measure the vertical component of the electric current in the photosphere using high-resolution vector magnetograms, as we discuss in the next section.

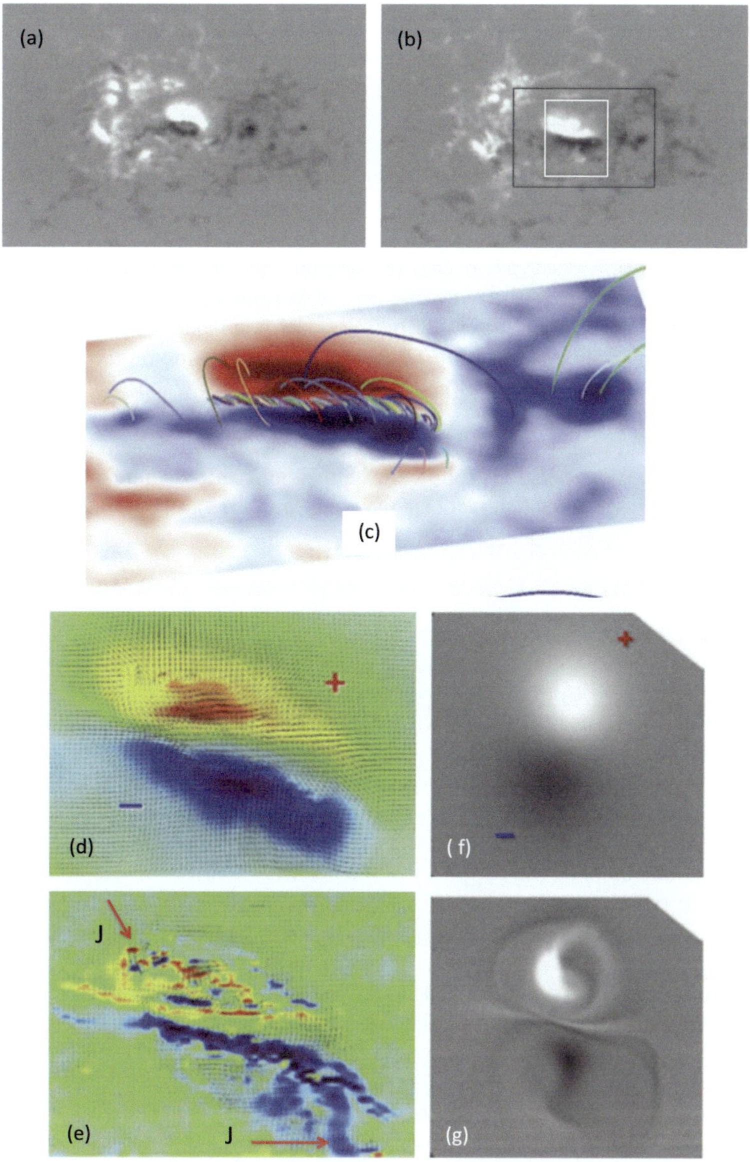

Figure 5 Emerging active region displaying a tongue pattern: (a) on 12 September 2005, and (b) 14 September 2005. The black box represents the field of view of panel c. (c) Evidence of a flux rope in a nonlinear force-free extrapolation of the THEMIS vector magnetogram (from Canou *et al.*, 2009. The white box in b represents the field of view of the lower panels d and e. The THEMIS observation on 14 September 2005 is shown in (d) with the vertical field component, B_z, as background and the transverse component represented by vectors. (e) The electric current density with a J shape over the two polarities (Schmieder and Aulanier, 2012). Note the return currents in the positive polarity. (f) MHD simulation of a dipole with two rotating polarities as the initial condition, and (g) the developing currents (adapted from Aulanier, Janvier, and Schmieder, 2012). Note the similar shape of the currents in e and g.

4.1. Morphological Determination of Magnetic Helicity

Magnetic helicity is an invariant in the solar environment. The question is how we can determine the magnetic helicity from the observations. Morphologically, the presence of electric currents may be inferred from the twist or shear of the magnetic structures embedded in the corona (Pevtsov, 2002). We note that electric currents are related to the magnetic helicity and the stored energy.

From the morphology of the different magnetic structures in the corona, it is possible to determine their helicity sign. The chirality of filaments is easy to determine from the barb orientations or the surrounding fibrils (Foukal, 1971; Martin, 1998; López Ariste *et al.*, 2006). In active regions observed in Hα, the bending of the fibrils or whirls around the sunspots (counter-clockwise – CCW/clockwise – CW) indicates a negative/positive magnetic helicity sign. Active regions commonly follow the hemispheric rule (positive/negative in the southern/northern hemisphere) by 55 % to 87 %, depending on the analyzed feature (see Table 1 in Wang, 2013). Active regions keep the same magnetic helicity sign during their life, even if some injection of opposite helicity occurs (Luoni *et al.*, 2004). The most recent statistical studies on the magnetic helicity sign of active regions in both hemispheres show that 75 % of the active regions follow the hemispheric rule, a value that is just between the previously obtained values (Liu, Hoeksema, and Sun, 2014).

4.2. Example of an Emerging Twisted Flux Tube

We give an example that shows that we can detect the helicity sign of an active region by its morphology and the direction of the magnetic field vector. For this purpose, we again focus on the example of the emerging active region in September 2005 (Figure 5). We computed the magnetic vector field from the *Télescope Héliographique pour l'Etude du Magnétisme et des Instabilités Solaires* (THEMIS) polarimetric observations. Figure 5d represents the vertical component of the field as the background, overlaid by the transverse component (arrows). The shift between the two opposite-sign magnetic polarities (tongues) along the PIL indicates a positive magnetic helicity. The horizontal field has a globally similar organization in both sunspots when mirrored from the central point of the bipole, and reversing the sign of the magnetic field. At the periphery of the positive polarity, the magnetic field vector is close to being radial, pointing away from the strong vertical, B_z, values, except close to the PIL, where strong magnetic shear is present. Indeed, the field in the vicinity of the PIL has emerged recently, so it still contains a large amount of free magnetic energy. Moreover, at the east side of the positive sunspot and at the west side of the negative one, the horizontal projection of the field lines (whirls) has the same curved shape (CW) as was found in chromospheric fibrils around some sunspots (Hale, 1927; Nakagawa *et al.*, 1971). The bending of these curved lines is an indication of a positive magnetic helicity, in agreement with the magnetic shear present all along the PIL between the spots. This is coherent with the above conclusion regarding the "magnetic tongues" (Section 3.3).

In this example it is found that the magnetic helicity sign is positive in the spot because of the right-handed twist for the corresponding emerging flux tube, and the clockwise rotational pattern for the horizontal magnetic fields around the spots. This is consistent with the hemispherical rule that favors a positive helicity for this active region because it is located in the southern hemisphere.

4.3. Electric Currents and Comparison with Simulations

Magnetic helicity has to be computed in a volume filled with connected magnetic field lines, otherwise it gives incorrect results (Pariat, Démoulin, and Nindos, 2007; Dalmasse *et al.*, 2013). In the example of September 2005 (see Section 4.2), we have shown that the two polarities belong to the same emerging flux tube. The magnetograms were taken five hours before a flare. The sign of the global magnetic helicity (H_B) in an active region is given by the dominant sign of the scalar product $\mathbf{j} \cdot \mathbf{B}$. This quantity indicates whether the field-aligned component of the so-called direct currents are parallel or antiparallel to the magnetic field. In Figure 5e, the current density is presented and is exactly what is observed in the vector field map. This analysis was possible thanks to the high sensitivity of THEMIS and can also be performed with the high-resolution vector magnetograms of *Hinode*/SOT (Georgoulis, 2012).

Inside the positive-spot umbra, the so-called return currents (negative, in blue) can be seen, interleaved with direct currents (positive, in yellow). The observed pattern of the vertical currents in the photosphere (Figures 5d and e) nicely matches the predictions based on various MHD models of twisted flux tubes anchored in the photosphere at both ends, *e.g.*, the model of a rotating bipole of Aulanier, Janvier, and Schmieder (2012) (Figures 5f, g). At the east side of the positive sunspot, two bands of positive vertical current can be seen, one aligned more east-west and the other more north-south, drawing a J-shaped curve. The symmetrical negative pair can be seen at the west side of the negative spot. These regions are denoted by J in the figure. This broad J-shaped pattern of the photospheric currents, together with the existence of "direct" and "return" currents inside each magnetic polarity, as named and described by Melrose (1991), agrees with the theoretical predictions of several MHD models of stable coronal flux ropes, either formed by flux emergence (Fan and Gibson, 2003, 2004), by simple and symmetric sunspot rotational motions (Aulanier, Pariat, and Démoulin, 2005), or by shearing motions combined with photospheric flux dispersal and cancellation (Aulanier *et al.*, 2010).

The observations of the emerging active region presented in Figures 5a and b that are zoomed in Figures 5d and e are consistent with the presence of a right-handed twisted flux tube in the corona that extends throughout the PIL between the two straight parts of the J-shaped current pattern, and which is rooted inside the hooks of the J. This is similar to what was obtained by Canou *et al.* (2009), who performed a nonlinear force free field (NLFFF) extrapolation and showed the existence of a right-handed twisted, but relatively short flux rope above the PIL. The NLFFF extrapolation confirms what we have found using the observations of the tongues and the magnetic vectors (see Section 3.3).

Current density vectors before and during an X-class flare have been computed for the complex active region of 15 February 2011 (Janvier *et al.*, 2014). As a result of the high cadence of HMI (12 minutes) vector magnetograms, the analysis of the evolution of the currents has allowed us to detect increasing electric currents in thin ribbons overlying the EUV flare ribbons during the flare, corresponding to the downward extension of the current sheet at the coronal reconnection site. Some hook-shaped currents have been interpreted as the footprints of the outer envelope of the flux rope that are already present one hour before the flare. This is exactly as the models predicted (Aulanier *et al.*, 2010; Aulanier, Janvier, and Schmieder, 2012; Janvier *et al.*, 2013). This erupting flux rope was seen as a CME by the different coronagraphs (*Large Angle and Spectrometric Coronagraph Experiment* (LASCO), STEREO/COR1, COR2) (Vourlidas *et al.*, 2013; Schrijver *et al.*, 2011).

After the flare, the currents correspond to the energy in the hook-shaped endings of the flare ribbons. The energy is released during the reconnection (Janvier *et al.*, 2014). HXR emissions can be seen within these hooks with the *Reuven Ramaty High Energy Solar Spectroscopic Imager* (RHESSI) (Musset, Vilmer, and Bommier, 2015). In the event described by Guo *et al.* (2012), different *Yohkoh*/HXR sources *versus* time were detected with RHESSI. The two outer sources were the strongest and the two inner sources were absent at the peak time. This implies that the outer HXR sources appeared at the footpoints of the large erupted flux rope.

5. Magnetic Topology of the CME Solar Sources and Reconnection

5.1. Solar CME Source, Site of Reconnection

Another interesting issue is understanding where the energy stored along the field lines will be released through reconnection. It is a challenge and should become a tool for flare predictions. In the standard 2D CSHKP model, the reconnection operates between two legs of an arcade at an X-type null point. In a 3D configuration, null points can be found in the corona using a linear force-free or nonlinear force-free extrapolation of the photospheric magnetic field and can be responsible for flares and jets (Guo *et al.*, 2013; Sun *et al.*, 2013; Mandrini *et al.*, 2014; see also Figures 6a, b). Nevertheless, flares may also occur without null points (Démoulin, Hénoux, and Mandrini, 1994; Démoulin, Priest, and Lonie, 1996; Li *et al.*, 2006). Thus, null points are just a particular case.

It is important to analyze the magnetic configuration of active regions using extrapolations of the photospheric magnetic field to map the different magnetic domains of connectivity, *i.e.* the separatrices. Separatrices, that is, surfaces separating two domains of connectivity, are the possible reconnection sites. A generalization of separatrices are quasi-separatrix layers (QSLs), volumes where reconnection is possible. In QSLs the gradient of field-line connectivity is sharp, but finite. In contrast, it is infinite in true separatrices, across which the connectivity is discontinuous. This means that separatrices can be regarded as the limit of QSLs when their thickness tends to zero. The thickness is characterized by a parameter

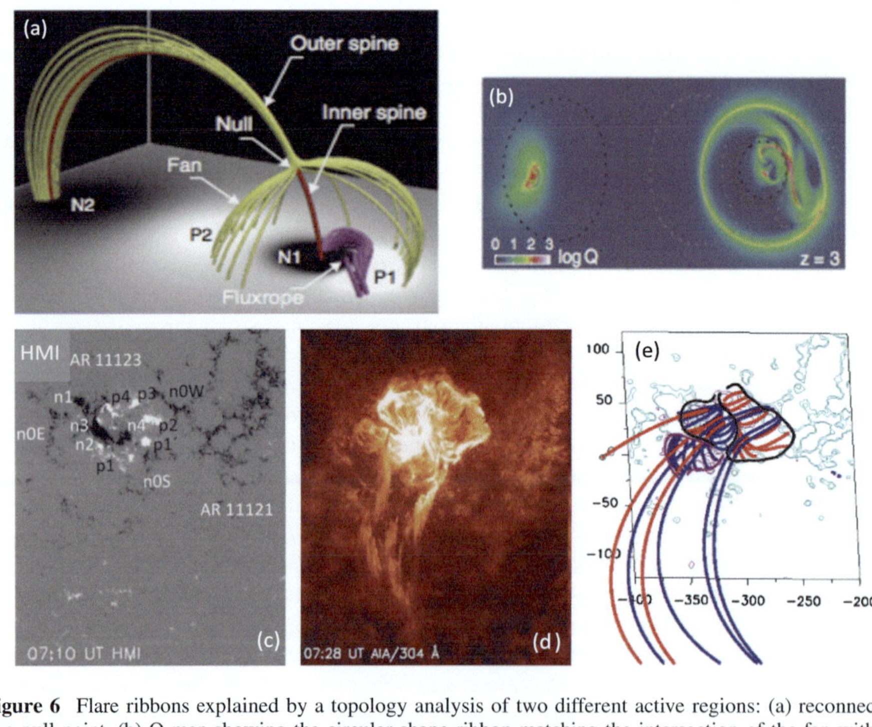

Figure 6 Flare ribbons explained by a topology analysis of two different active regions: (a) reconnection at a null point, (b) Q-map showing the circular shape ribbon matching the intersection of the fan with the chromosphere, (c) emerging flux in November 2010 forming AR 11123. (d) AIA 304 Å image of the flare ribbons. (e) The QSLs (black contours) match the flare ribbons shown in d. Panels a and b are adapted from the paper of Sun *et al.* (2013), panels c, d, and e from Mandrini *et al.* (2014).

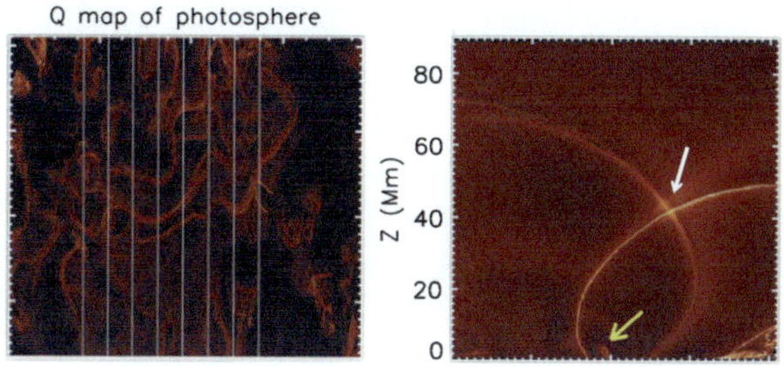

Figure 7 AR 11158 on 15 February 2011 (left panel): Q-map showing the quasi-separatrix layer footprints in the photosphere overlaid by white lines indicating the locations of vertical cuts. This map is also shown in Figure 4c with a different color table. (Right panel): Vertical plane over the central cut showing the two possible reconnection sites that correspond to the high-altitude HFT (white arrow) and to the low flux rope (yellow arrow). The low flux tube is zoomed in Figure 4f (adapted from Zhao *et al.*, 2014).

called the squashing degree Q (Démoulin *et al.*, 1997; Schmieder *et al.*, 1997; Titov, Démoulin, and Hornig, 2003; Janvier *et al.*, 2013), which is higher for smaller thicknesses. In the case described by Mandrini *et al.* (2014), where a new active region is emerging (Figure 6c), the flare ribbons have a complicated shape and could not be explained only by the presence of two nulls. It was nevertheless possible to reproduce the shape of the ribbons by computing the quasi-separatrices using a linear force-free extrapolation (Figure 6e). This demonstrates the robustness of the topology. The bright ribbons are produced by particles that are accelerated at the reconnection site and stream along the field lines downwards to the chromosphere (Jiang *et al.*, 2013).

The core of quasi-separatrix layers is contained in a volume called the hyperbolic flux tube (HFT) (Titov, Démoulin, and Hornig, 2003). The HFT is generally the region where reconnection can occur, and it therefore is a probable site of the energy release.

An interesting example is the active region of 15 February 2011, which has been well documented in many articles; we already referred to it above as the site of the large X1.2 flare (Janvier *et al.*, 2014), and CME (Vourlidas *et al.*, 2013; Schrijver *et al.*, 2011). The magnetic field configuration is relatively well established. The evolution of the topology of this region shows that it is formed from two active regions. The negative following polarity of the first active region was squeezed by the positive leading polarity of the second active region, leading to increasing shear, as we have reported earlier regarding the presence of a strong shear before the flare. The region looks like a quadrupolar region (Figure 4a) with a large HFT whose neighborhood can be assimilated to that around a null point (Zhao *et al.*, 2014). Figure 4c, zoom of Figure 7 (left panel), shows the Q map that represents the footprints of the quasi-separatrix layers in the photosphere. Figure 7 (right panel) shows one cut of the active-region volume perpendicular to the solar surface and the QSL direction. It suggests the existence of a large hyperbolic flux tube with an X-point indicated by the white arrow and a small flux rope near the solar surface indicated by the yellow arrow that is enlarged in Figure 4f. Below the high-altitude HFT, in the region of high shear (bipolar region), a second local HFT exists below the small flux rope; this is indicated by the yellow arrow in the right panel of Figure 7. These two different HFTs are regions favorable for reconnection, one at high altitude related to large-scale structures, and one very low (Figure 4d and f). The two brightest flare ribbons overlie the small flux rope in the central part of the region (Figure 4e), whereas the other ribbons are on the intersections of the secondary large-scale QSLs with the solar surface.

We discuss the two possible reconnection sites in the next section and give some arguments for reconnection at the low HFT leading to the X-flare of February 2011 (Zhao *et al.*,

2014). To conclude, it is interesting to study the magnetic topology of an active region before the flare, looking for the existence of null points or HFTs in order to understand where the flare could start.

5.2. Break-out and Tether-Cutting

Some CME initiation models can be tested by applying a topological analysis to observations since models require the presence and activation of some specific topological features.

The reconnection of the field lines above the flux rope reduces the tension of the overlying field (Antiochos, DeVore, and Klimchuk, 1999; van der Holst, Jacobs, and Poedts, 2007; Lynch *et al.*, 2008). Such a break-out process was modeled in quadrupolar configurations and was directly related to the onset of large flares (Aulanier *et al.*, 2000). However, this reconnection does not necessarily lead to a successful eruption if the overlying flux, although partially "cut", is still strong enough to confine the stressed core (Guo *et al.*, 2012). Observations of brightness or secondary ribbons far from the center of the active region indicate that the break-out is effective. Large flares can be accompanied by these pre-events. Mandrini *et al.* (2006) provided evidence of the break-out process in the X-class flare of 28 October 2003 by identifying EUV brightening at the periphery of the active region that corresponded to the secondary ribbons associated to the quasi-separatrices before the flare.

We continue the discussion started in the previous section by using the example of 15 February 2011. This example is interesting because the region is quadrupolar with a large-scale HFT, as shown in Figure 7 (Zhao *et al.*, 2014), and one might expect *a priori* a break-out eruption. Numerous quasi-separatrices were present before the flare, but no precursor enhancements were observed in SDO/AIA EUV images. The authors show that the brightness enhancement of the secondary ribbons appeared *after* the main central flare ribbon and not before, as we could expect in the case of the break-out (as proposed by Aulanier *et al.*, 2000 for another case). Thus, the break-out could not trigger the flare, it was a consequence of the eruption. The authors favor the torus instability after some efficient tether-cutting mechanism instead. Very low down in the middle of the sheared bipole, they detected low field lines forming a flux rope (Figures 4d and f). The small reconnection site (low HFT) is below the flux rope (Figure 4f). Inoue *et al.* (2014) modeled the field lines in this tube for the same event and concluded that they could reconnect in the low corona, leading to the formation of longer magnetic field lines belonging to a newly formed flux rope. They proposed that the tether-cutting mechanism led to the formation of a longer flux rope, which then became unstable and led to a CME (Zhao *et al.*, 2014; Inoue *et al.*, 2014, 2015). This example shows the constraints given by the observations. Nevertheless, it is always difficult to be sure about the trigger mechanism.

Thus, theoretical models suggest reconnection of field lines either above the flux rope (the break-out mechanism; Antiochos, DeVore, and Klimchuk, 1999) or below the flux rope (tether-cutting mechanism; Moore *et al.*, 2001). In the latter option, the field lines are progressively reconnecting below a flux tube or between legs of arcades, which reduces the tension. A detailed analysis of one MHD simulation and its comparison with electric-wire models showed that "photospheric flux-cancellation and tether-cutting coronal reconnection do not trigger CMEs in bipolar magnetic fields, but are key pre-eruptive mechanisms for flux ropes to build up and rise to a critical height above the photosphere at which the torus instability causes the eruption" (Aulanier *et al.*, 2010). Canceling flux along the PIL, emerging flux, or both, inflows in the corona, and shearing are responsible of this pre-eruptive phase. In any case, there is a long preparatory phase in the active region to store free energy and to reduce the tension that maintains the pre-eruptive flux rope.

6. CME Acceleration and Torus Instability

6.1. Forces Playing an Important Role

The CME/flare process is preceded either by a slow evolution of the coronal magnetic structure through a series of quasi-equilibrium states in a metastable system (driven by processes mentioned in the previous sections: flux cancellation, shearing, *etc.*) or by abrupt magnetic field restructuring (a confined flare before the eruption; see, *e.g.*, Aurass *et al.*, 1999; Priest, 1981). In the former case, a metastable structure evolves until it loses equilibrium and/or an MHD instability sets in, both options leading to the eruption. In the latter case, the abrupt restructuring creates a new magnetic configuration that is unstable and immediately erupts.

The eruption itself is driven by the Lorentz force, and the energy released in the CME/flare process is converted from the free energy stored in the non-potential magnetic field. Both aspects imply that there are electric currents circulating through the eruptive structure. In the crudest approximation, the otherwise very complex coronal current system can be represented by an arc-shaped line-current that is anchored with both ends in the conductive photosphere (Figure 8a). Under such circumstances, in the absence of reconnection the magnetic flux encircled by the line-current and the photosphere has to be conserved. This flux can be expressed as $\Phi = LI$, where L is the inductance of the system and I is the electric current (Batygin and Toptygin, 1962). Under this paradigm, some constraints can be placed on the evolution of the magnitude of the currents during the eruption, hence on the time variation of the CME acceleration. Since the inductance is proportional to the size of the line current (see, *e.g.*, Jackson, 1998), the electric current must decrease in the course of the eruption, *i.e.*, $I \propto h^{-1}$, where h is the height of the structure (at large heights one can take approximately $I \propto R^{-1}$, where R is the heliocentric distance). Consequently, the free energy of the structure, $W = LI^2$, decreases as $W \propto R^{-1}$. This energy is converted into the kinetic and potential energy, as well as into the work done against the aerodynamic drag (Vršnak, 2001). The Lorentz force decreases even more rapidly because in addition to the electric current, the magnetic field also decreases with height. Thus, the acceleration decreases rapidly, meaning that the maximum CME speed is strongly limited. Similar qualitative behaviors are obtained if in the MHD approximation, the conservation of magnetic flux during the eruption, or the conservation of the end-to-end twist during the expansion of the flux rope is applied (van Tend and Kuperus, 1978; Forbes, 1990; Forbes and Priest, 1995; Kliem and Török, 2006; Démoulin and Aulanier, 2010).

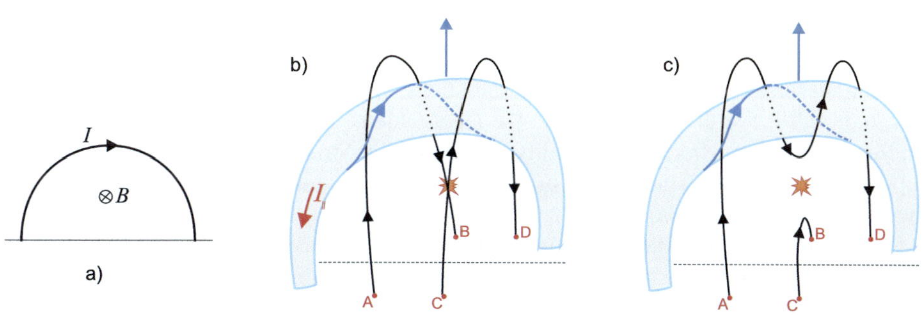

Figure 8 (a) Conceptual sketch of the coronal current and the conductive photosphere. (b) and (c) Flux-rope concept with reconnection (marked by red star) supplying the rope with additional poloidal flux. The thin dashed line represents the photospheric magnetic inversion line. Flare kernels occur at points C and B, which are connected by a flaring loop.

Thus, a somewhat better approximation is to replace the line current with the flux rope whose footpoints are anchored in the photosphere, so that the longitudinal and poloidal flux are conserved due to the frozen-in conditions (see, *e.g.*, Gold and Hoyle, 1960; Sakurai, 1976; Anzer, 1978; Mouschovias and Poland, 1978; van Tend, 1979; Pneuman, 1980; Anzer and Pneuman, 1982; Chen, 1989; Vršnak, 1990; Titov and Démoulin, 1999; Török and Kliem, 2003), as sketched in Figures 8b and c. In that case, the inductance and the Lorentz force can be specified in a form that enables quantitative analysis of the eruption dynamics. Now, the total inductance is the sum of an external and an internal contribution, which can be expressed as $L_e = \mu_0\lambda[\ln(8R/r) - 2]$, and $L_i = k\mu_0\lambda$, respectively, where μ_0 is the permeability, λ is the length of the flux rope, R/r is the ratio of the major and minor radius of the torus, and k is a constant that depends on the radial profile of the electric current density (for details see Chen, 1989; Žic, Vršnak, and Skender, 2007, and references therein). The Lorentz force can be expressed in terms of the hoop force, which basically consists of the upward-directed magnetic pressure gradient of the poloidal field component and the downward-directed magnetic tension of the axial component (*cf.* Chen, 1989; Vršnak, 1990 and references therein).

6.2. CME Model and Acceleration Profile

One of direct successes of the model discussed in the previous section is that it predicts that the eruption achieves the highest acceleration at a height comparable with the footpoint half-distance (Vršnak, 1990; Chen and Krall, 2003), which is fully consistent with observations (Vršnak, Ruždjak, and Rompolt, 1991; Chen *et al.*, 2006). However, to successively reproduce the CME eruptions, the described model requires a significant supply of poloidal magnetic flux in the course of the acceleration phase (for example, $\Phi \approx 10^{22}$ Mx), or otherwise the acceleration is too weak. Such a flux can be supplied by the flare-related reconnection (Figures 8b and c), since the reconnected flux is considerably higher than required by the model (Forbes *et al.*, 2006; Miklenic *et al.*, 2007; Miklenic, Veronig, and Vršnak, 2009).

Contemporaneously with Chen (1989), Vršnak (1990) proposed a similar model, which included the diamagnetic effect of the conductive photospheric surface (Kuperus and Raadu, 1974). Inclusion of this effect provided explanations for several observational phenomena. For example, the model explains how the change of some of the model parameters (*e.g.*, twist, density, longitudinal current) can cause a transition from a completely stable state to a metastable state (see Figure 5a of Vršnak, 2008). In this new situation, the structure is in a metastable equilibrium at a given height, and there is an unstable-equilibrium point above it. A small displacement from the metastable equilibrium will result in oscillations that can explain transversal prominence/filament oscillations. However, if the perturbation is strong enough, it will move the structure from the metastable-equilibrium height to the unstable-equilibrium position where the instability sets in, and the structure erupts. This can explain eruptions initiated by disturbances coming from a distant source (*e.g.*, a large-amplitude wave from another eruption).

Furthermore, for a certain domain in the parameter space, there is another stable-equilibrium above both the metastable and the unstable equilibrium position (see Figure 5b of Vršnak, 2008). This explains oscillations that are sometimes observed after failed eruptions (*e.g.*, Vršnak *et al.*, 1990). Finally, a persistent gradual change of model parameters (*e.g.*, increasing the twist and/or current, or decreasing the mass, *etc.*) eventually leads to a loss of equilibrium, approximately at a height comparable to the flux-rope footpoint half-distance.

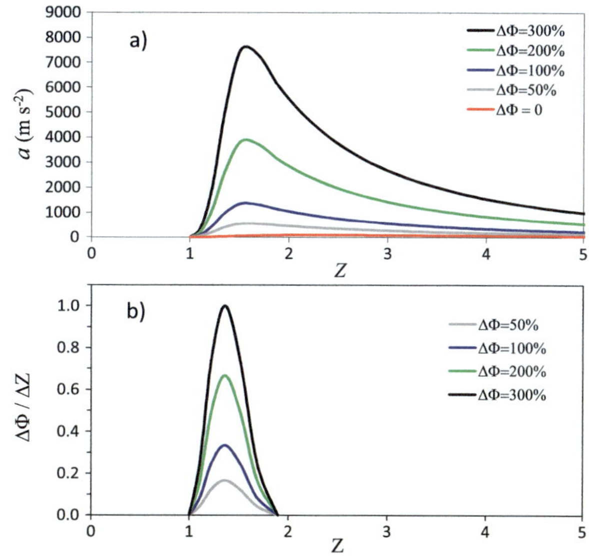

Figure 9 (a) Flux rope acceleration, a, presented as a function of the height expressed in units of the footpoint half-separation, Z. (b) The reconnection rate, expressed in the form $\Delta\Phi/\Delta Z$, where $\Delta\Phi$ represents the flux reconnected while the rope height increases with ΔZ. For details see Vršnak (2008).

However, in this model the accelerations reach values on the order of 100 m s^{-2} at most, which is too low compared to the observed values, which can be as high as several km s^{-2}. This is a direct consequence of the poloidal magnetic flux conservation and the fact that the inductance increases with the size of the structure. Thus, the only way to obtain strong accelerations is to break the requirement of the poloidal flux conservation, *i.e.*, to include reconnection (Figures 8b and c; see also Anzer and Pneuman, 1982; Steele and Priest, 1989). This is illustrated in Figure 9a, where the flux-rope acceleration is presented as a function of the height for different amounts of the reconnection-supplied poloidal flux (expressed in percentages of the initial flux). The rate at which the flux is supplied to the rope is presented in Figure 9b. In summary, Figure 9 shows that by supplying enough of the poloidal flux, the highest accelerations observed can be achieved by the model.

6.3. Instabilities

Catastrophe (Forbes and Isenberg, 1991; Forbes and Priest, 1995; Lin and Forbes, 2000) *versus* instability (Baty, 2001; Fan and Gibson, 2004; Kliem and Török, 2006; Isenberg and Forbes, 2007) for the eruption of a toroidal solar magnetic flux rope has been deeply studied, and it was shown that both are equivalent when one considers that the flux rope forms gradually (Démoulin and Aulanier, 2010; Kliem *et al.*, 2014a).

A flux rope can undergo two main types of instabilities: the kink instability (Sakurai, 1976; Rust and Kumar, 1996) and/or the torus instability (Kliem and Török, 2006). The kink instability of a helical flux rope may occur when the helical turn between the two legs of the flux rope exceeds 3π (1.5 turns). When the two legs approach each other, a current sheet is created. Simulations show that with a complex reconnection, the top part of the flux rope is pushed upward with possible ejections of plasmoids, while a low flux rope is formed and pushed downward (Kliem *et al.*, 2010). During the reconnection, particles are accelerated. Some recent observations of filaments seem to show a strong twist before the eruption and a rise as a kinking structure (Williams *et al.*, 2005; see also Kumar and Cho, 2014). However,

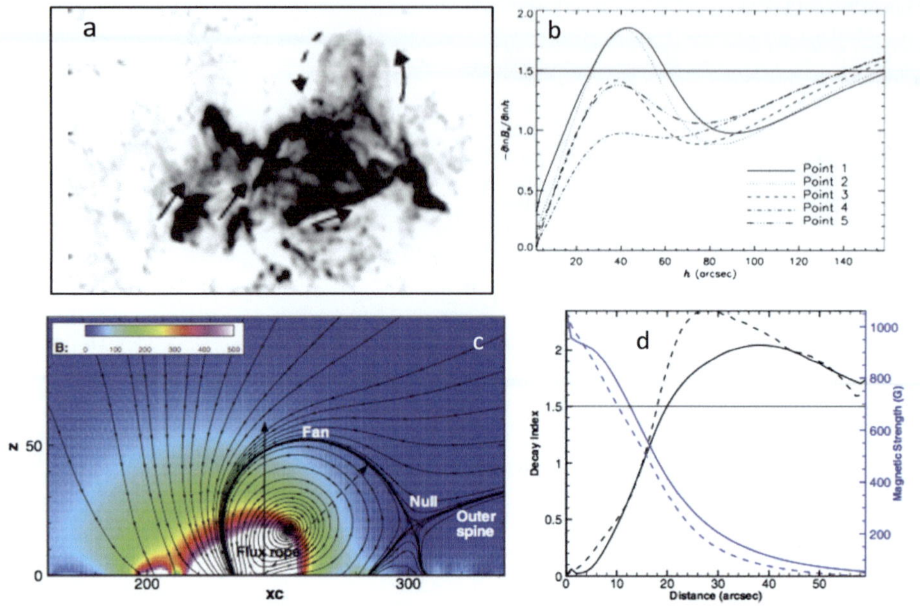

Figure 10 (a) TRACE observations of an erupting filament. The arrows (solid and dashed) around the loop show the contour of the failed eruption. (b) For the eruption shown in a, the decay index is below the critical value (represented by the horizontal line at 1.5) for initiating the torus instability of the loop (Guo *et al.*, 2010b). (c) Cut of a nonlinear force-free field extrapolation over the AR 11283 showing a null point, the fan, and the flux rope. (d) The decay index is higher than the threshold of the torus instability (Jiang *et al.*, 2013).

we argue that these are not direct observations of the kink instability. Estimating the twist in the filament in the way reported by, *e.g.,* Yan *et al.* (2014) is also questionable. Therefore, no straight conclusion from the observations can be drawn for the kink instability as a CME driver.

The torus instability may occur when the flux rope has ascended to a height where the background field decreases fast enough. The flux rope expands and the torus instability probably triggers and drives its impulsive acceleration (Kliem and Török, 2006; Kliem *et al.*, 2010; Török and Kliem, 2005, 2007; Dudík *et al.*, 2014). It is interesting to study the evolution *versus* time of the decay speed of the enveloping flux, quantified by a decay index ($n = \mathrm{d}\ln B/\mathrm{d}\ln z$), to see whether a torus instability really exists in the observed flux rope and can be the trigger of the CME. In the example studied by Guo *et al.* (2010b), the decay index value remains below 1.4, and it was considered as a failed eruption with no CME (Figures 10a, b).

AR 11283 produced a CME on 6 September 2011, and Jiang *et al.* (2013) showed a flux rope in this event using a nonlinear force-free extrapolation of the magnetic field. A cross-section of the field line configuration shows a null point and the flux rope (Figure 10c). The authors computed the decay indexes along two different directions from near the polarity inversion line (Figure 10d). In both directions, the flux rope reached the domain of the torus instability with values higher than the critical threshold of 1.5 (Török and Kliem, 2007; Aulanier *et al.*, 2010). The torus instability has now been detected in several modeled and/or observed eruptions (Kliem *et al.*, 2013; Nindos, Patsourakos, and Wiegelmann, 2012; Zuccarello *et al.*, 2014; Amari, Canou, and Aly, 2014; Inoue *et al.*, 2015).

7. Conclusions

Flares and CMEs belong to the same magnetically driven event, 75 % of the CMEs are strongly related to flares detected by GOES (Bein *et al.*, 2012). Their acceleration phase is related to the impulsive phase of flares (Temmer *et al.*, 2010).

In this article, we have tried to perform a meta-analysis of the CME-flare process from an observational point of view. Our study is not exhaustive and mainly focuses on a few specific points. The immediate outcome of this compilation is that the observations are always difficult to interpret in light of the models. Several mechanisms can explain the onset of CMEs. The observations also teach us that various mechanisms proposed by the theoretical models can exist or co-exist. Our main conclusions are the following.

1. Canceling flux is a quite plausible mechanism for flux-rope formation, since it is always present in active regions (van Driel-Gesztelyi *et al.*, 2003; Schmieder *et al.*, 2004; Green, Kliem, and Wallace, 2011). Conversely, direct evidence of the emergence of large flux ropes through the photosphere is not unambiguously observed (*e.g.* Canou *et al.* (2009), Schmieder, Archontis, and Pariat (2014) and references therein). Models that include the sub-surface layers do not succeed either in simulating the emergence of the full body of the rope (Fan, 2009; Leake, Linton, and Antiochos, 2014).

2. The existence of twist in flux tubes can be detected by a morphological pattern, *i.e.*, a tongue-shape of emerging polarities in longitudinal magnetograms (López Fuentes *et al.*, 2000). Vertical electric currents can be measured using vector magnetograms provided by high-sensitivity magnetographs such as THEMIS, *Hinode*/SOT, or SDO/HMI. We showed that the flux rope can sometimes exist before the flare, but this is not always clear. Enhancement of the current density has been detected during reconnection (Janvier *et al.*, 2014). A twisted flux rope can sometimes be reproduced by nonlinear force-free extrapolations (Canou *et al.*, 2009; Guo *et al.*, 2010a).

3. For the CME trigger mechanisms, it seems that a topological analysis of the active region is an interesting tool for predicting where the reconnection can occur (Démoulin *et al.*, 1997). The presence of a high-altitude hyperbolic flux tube (HFT) might be a good indication of a future break-out (Zhao *et al.*, 2014). Enhancement of brightenings before the eruption onset in the related quasi-separatrices needs to be detected to test the possibility of a real break-out through the overlying field lines (Aulanier *et al.*, 2000). A low HFT with associated double J-shape QSLs indicates the presence of a pre-eruptive flux rope and the site of the flare (Janvier *et al.*, 2013). We summarize our conclusions in Figure 11, where we show the observables (currents, flare ribbons, bright quasi-separatrix footprints,

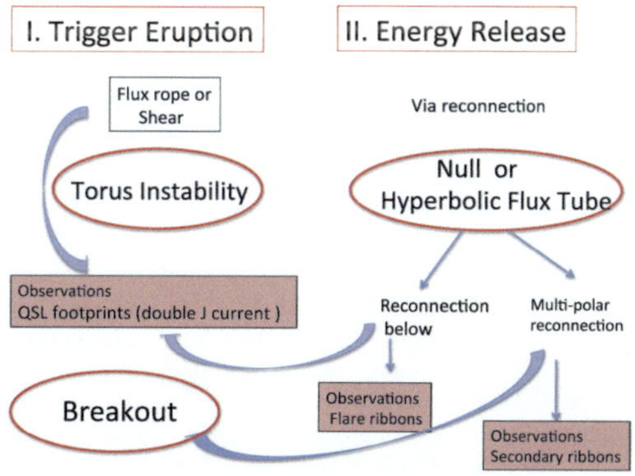

Figure 11 Scheme of flare–CME event triggering mechanisms with the observables (pink boxes): quasi-separatrix layer (QSL) footprints, main flare ribbons, and secondary ribbons.

and secondary ribbons) for the different trigger mechanisms that we have discussed in this article (break-out and torus instability).

4. The CME acceleration phase is closely associated with the impulsive phase of the associated flare, as evidenced by the synchronization of the CME acceleration peak and the peak of the hard X-ray burst or derivative of the soft X-ray flux (*e.g.* Bein *et al.* (2012) and references therein). This indicates a possible feed-back relationship between the CME acceleration and flare-related reconnection (Lin and Forbes, 2000; Vršnak, 2008). However, this relationship requires further investigation since i) the mentioned synchronization is violated in certain parts of events (Maričić *et al.*, 2007); ii) apparently, the ejection starts accelerating before the flare onset in most events (Bein *et al.*, 2012).

5. From the observational point of view, the best-understood possible drivers of the eruption are the break-out (Aulanier *et al.*, 2000; Ugarte-Urra, Warren, and Winebarger, 2007) and the torus instability (Zuccarello *et al.*, 2014; Nindos, Patsourakos, and Wiegelmann, 2012; Kliem *et al.*, 2014b). All the other mechanisms proposed by theoretical models are not really efficient in initiating the observed CMEs (*e.g.* Aulanier (2014) and references therein). The tether-cutting reconnection favors the formation of a flux rope, low or high in the corona, but it does not seem to initiate the onset of a CME (Aulanier *et al.*, 2010). The kink instability could prepare the flux rope to reach the torus instability (Török and Kliem, 2005). To resolve the ambiguity concerning the trigger, including the loss-of-equilibrium or break out effect, meticulous theory-led observational studies need to be made, like those reported in this article.

To conclude, we must always keep in mind that the theoretical models are highly simplified and are set up to highlight a specific problem, *e.g.*, the flux-rope formation by emergence, or shear or diffusion. Some models start with an already unstable rope and merely calculate its development together with its environment. With such models we cannot identify the true trigger mechanism.

Acknowledgements B. Vršnak acknowledges financial support by Croatian Science Foundation under the project 6212 SOLSTEL and the European Commissions Seventh Framework Programme (FP7/2007-2013) under the grant agreement No. 284461 (eHEROES; http://soteria-space.eu/eheroes/html/).

References

Acton, L., Tsuneta, S., Ogawara, Y., Bentley, R., Bruner, M., Canfield, R., Culhane, L., Doschek, G., Hiei, E., Hirayama, T.: 1992, The Yohkoh mission for high-energy solar physics. *Science* **258**, 618. ADS. DOI.

Amari, T., Canou, A., Aly, J.-J.: 2014, Characterizing and predicting the magnetic environment leading to solar eruptions. *Nature* **514**, 465. ADS. DOI.

Amari, T., Luciani, J.F., Mikic, Z., Linker, J.: 2000, A twisted flux rope model for coronal mass ejections and two-ribbon flares. *Astrophys. J. Lett.* **529**, L49. ADS. DOI.

Amari, T., Luciani, J.F., Aly, J.J., Mikic, Z., Linker, J.: 2003, Coronal mass ejection: Initiation, magnetic helicity, and flux ropes. II. Turbulent diffusion-driven evolution. *Astrophys. J.* **595**, 1231. ADS. DOI.

Andrews, M.D., Howard, R.A.: 2001, A two-type classification of LASCO coronal mass ejection. *Space Sci. Rev.* **95**, 147. ADS.

Antiochos, S.K., DeVore, C.R., Klimchuk, J.A.: 1999, A model for solar coronal mass ejections. *Astrophys. J.* **510**, 485. ADS. DOI.

Anzer, U.: 1978, Can coronal loop transients be driven magnetically. *Solar Phys.* **57**, 111. ADS. DOI.

Anzer, U., Pneuman, G.W.: 1982, Magnetic reconnection and coronal transients. *Solar Phys.* **79**, 129. ADS.

Archontis, V., Hood, A.W., Tsinganos, K.: 2014, Recurrent explosive eruptions and the "Sigmoid-to-arcade" transformation in the Sun driven by dynamical magnetic flux emergence. *Astrophys. J. Lett.* **786**, L21. ADS. DOI.

Aulanier, G.: 2014, The physical mechanisms that initiate and drive solar eruptions. In: Schmieder, B., Malherbe, J.-M., Wu, S.T. (eds.) *IAU Symposium* **300**, 184. ADS. DOI.

Aulanier, G., Démoulin, P.: 2003, Amplitude and orientation of prominence magnetic fields from constant-alpha magnetohydrostatic models. *Astron. Astrophys.* **402**, 769. ADS. DOI.

Aulanier, G., Janvier, M., Schmieder, B.: 2012, The standard flare model in three dimensions. I. Strong-to-weak shear transition in post-flare loops. *Astron. Astrophys.* **543**, A110. ADS. DOI.

Aulanier, G., Pariat, E., Démoulin, P.: 2005, Current sheet formation in quasi-separatrix layers and hyperbolic flux tubes. *Astron. Astrophys.* **444**, 961. ADS. DOI.

Aulanier, G., DeLuca, E.E., Antiochos, S.K., McMullen, R.A., Golub, L.: 2000, The topology and evolution of the Bastille day flare. *Astrophys. J.* **540**, 1126. ADS. DOI.

Aulanier, G., Török, T., Démoulin, P., DeLuca, E.E.: 2010, Formation of torus-unstable flux ropes and electric currents in erupting sigmoids. *Astrophys. J.* **708**, 314. ADS. DOI.

Aurass, H., Vršnak, B., Hofmann, A., Rudžjak, V.: 1999, Flares in sigmoidal coronal structures a case study. *Solar Phys.* **190**, 267. ADS. DOI.

Baty, H.: 2001, On the MHD stability of the $\mathbf{m} = 1$ kink mode in solar coronal loops. *Astron. Astrophys.* **367**, 321. ADS. DOI.

Batygin, V.V., Toptygin, I.N.: 1962, *Problems in Electrodynamics*, Academic Press, New York.

Bein, B.M., Berkebile-Stoiser, S., Veronig, A.M., Temmer, M., Vršnak, B.: 2012, Impulsive acceleration of coronal mass ejections. II. Relation to soft X-ray flares and filament eruptions. *Astrophys. J.* **755**, 44. ADS. DOI.

Bommier, V., Leroy, J.L.: 1998, Global pattern of the magnetic field vectors above neutral lines from 1974 to 1982: Pic-du-Midi observations of prominences. In: Webb, D.F., Schmieder, B., Rust, D.M. (eds.) *IAU Colloq. 167: New Perspectives on Solar Prominences*, *Astron. Soc. Pac. CS* **150**, 434. ADS.

Burkepile, J.T., Hundhausen, A.J., Stanger, A.L., Cyr, O.C.St., Seiden, J.A.: 2004, Role of projection effects on solar coronal mass ejection properties: 1. A study of CMEs associated with limb activity. *J. Geophys. Res.* **109**, A03103. ADS. DOI.

Canou, A., Amari, T., Bommier, V., Schmieder, B., Aulanier, G., Li, H.: 2009, Evidence for a pre-eruptive twisted flux rope using the THEMIS vector magnetograph. *Astrophys. J. Lett.* **693**, L27. DOI.

Carley, E.P., Long, D.M., Byrne, J.P., Zucca, P., Bloomfield, D.S., McCauley, J., Gallagher, P.T.: 2013, Quasiperiodic acceleration of electrons by a plasmoid-driven shock in the solar atmosphere. *Nat. Phys.* **9**, 811. ADS. DOI.

Carmichael, H.: 1964, *A Process for Flares*, *Series NASA SP* **50**, NASA, Washington, 451. ADS.

Chandra, R., Schmieder, B., Aulanier, G., Malherbe, J.M.: 2009, Evidence of magnetic helicity in emerging flux and associated flare. *Solar Phys.* **258**, 53. ADS. DOI.

Chandra, R., Pariat, E., Schmieder, B., Mandrini, C.H., Uddin, W.: 2010, How can a negative magnetic helicity active region generate a positive helicity magnetic cloud? *Solar Phys.* **261**, 127. ADS. DOI.

Chen, J.: 1989, Effects of toroidal forces in current loops embedded in a background plasma. *Astrophys. J.* **338**, 453. DOI.

Chen, J., Krall, J.: 2003, Acceleration of coronal mass ejections. *J. Geophys. Res.* **108**, 1410. DOI.

Chen, J., Marqué, C., Vourlidas, A., Krall, J., Schuck, P.W.: 2006, The flux-rope scaling of the acceleration of coronal mass ejections and eruptive prominences. *Astrophys. J.* **649**, 452. ADS. DOI.

Chen, P.F.: 2009, The relation between EIT waves and coronal mass ejections. *Astrophys. J. Lett.* **698**, L112. ADS. DOI.

Chen, P.F.: 2011, Coronal mass ejections: Models and their observational basis. *Living Rev. Solar Phys.* **8**, 1. ADS. DOI.

Chen, P.F., Shibata, K.: 2000, An emerging flux trigger mechanism for coronal mass ejections. *Astrophys. J.* **545**, 524. ADS. DOI.

Cheng, X., Zhang, J., Ding, M.D., Guo, Y., Su, J.T.: 2011, A comparative study of confined and eruptive flares in NOAA AR 10720. *Astrophys. J.* **732**, 87. DOI.

Cheng, X., Ding, M.D., Zhang, J., Sun, X.D., Guo, Y., Wang, Y.M., Kliem, B., Deng, Y.Y.: 2014a, Formation of a double-Decker magnetic flux rope in the sigmoidal solar active region 11520. *Astrophys. J.* **789**, 93. ADS. DOI.

Cheng, X., Ding, M.D., Guo, Y., Zhang, J., Vourlidas, A., Liu, Y.D., Olmedo, O., Sun, J.Q., Li, C.: 2014b, Tracking the evolution of a coherent magnetic flux rope continuously from the inner to the outer corona. *Astrophys. J.* **780**, 28. ADS. DOI.

Cid, C., Cremades, H., Aran, A., Mandrini, C., Sanahuja, B., Schmieder, B., Menvielle, M., Rodriguez, L., Saiz, E., Cerrato, Y., Dasso, S., Jacobs, C., Lathuillere, C., Zhukov, A.: 2012, Can a halo CME from the limb be geoeffective? *J. Geophys. Res.* **117**, 11102. ADS. DOI.

Dalmasse, K., Pariat, E., Valori, G., Démoulin, P., Green, L.M.: 2013, First observational application of a connectivity-based helicity flux density. *Astron. Astrophys.* **555**, L6. ADS. DOI.

Dalmasse, K., Pariat, E., Démoulin, P., Aulanier, G.: 2014, Photospheric injection of magnetic helicity: Connectivity-based flux density method. *Solar Phys.* **289**, 107. ADS. DOI.

Delannée, C., Aulanier, G.: 1999, CME associated with transequatorial loops and a bald patch flare. *Solar Phys.* **190**, 107. ADS. DOI.

Delannée, C., Török, T., Aulanier, G., Hochedez, J.-F.: 2008, A new model for propagating parts of EIT waves: A current shell in a CME. *Solar Phys.* **247**, 123. ADS. DOI.

Démoulin, P., Aulanier, G.: 2010, Criteria for flux rope eruption: Non-equilibrium versus torus instability. *Astrophys. J.* **718**, 1388. ADS. DOI.

Démoulin, P., Hénoux, J.C., Mandrini, C.H.: 1994, Are magnetic null points important in solar flares? *Astron. Astrophys.* **285**, 1023. ADS.

Démoulin, P., Priest, E.R., Lonie, D.P.: 1996, Three-dimensional magnetic reconnection without null points 2. Application to twisted flux tubes. *J. Geophys. Res.* **101**, 7631. ADS. DOI.

Démoulin, P., Bagala, L.G., Mandrini, C.H., Hénoux, J.C., Rovira, M.G.: 1997, Quasi-separatrix layers in solar flares. II. Observed magnetic configurations. *Astron. Astrophys.* **325**, 305. ADS.

Démoulin, P., Mandrini, C.H., van Driel-Gesztelyi, L., Thompson, B.J., Plunkett, S., Kovári, Z., Aulanier, G., Young, A.: 2002, What is the source of the magnetic helicity shed by CMEs? The long-term helicity budget of AR 7978. *Astron. Astrophys.* **382**, 650. DOI.

Démoulin, P., Vourlidas, A., Pick, M., Bouteille, A.: 2012, Initiation and development of the white-light and radio coronal mass ejection on 2001 April 15. *Astrophys. J.* **750**, 147. ADS. DOI.

Dennis, B.R., Zarro, D.M.: 1993, The Neupert effect – What can it tell us about the impulsive and gradual phases of solar flares? *Solar Phys.* **146**, 177. ADS. DOI.

Dere, K.P., Brueckner, G.E., Howard, R.A., Koomen, M.J., Korendyke, C.M., Kreplin, R.W., Michels, D.J., Moses, J.D., Moulton, N.E., Socker, D.G., Cyr, O.C.St., Delaboudinière, J.P., Artzner, G.E., Brunaud, J., Gabriel, A.H., Hochedez, J.F., Millier, F., Song, X.Y., Chauvineau, J.P., Marioge, J.P., Defise, J.M., Jamar, C., Rochus, P., Catura, R.C., Lemen, J.R., Gurman, J.B., Neupert, W., Clette, F., Cugnon, P., van Dessel, E.L., Lamy, P.L., Llebaria, A., Schwenn, R., Simnett, G.M.: 1997, EIT and LASCO observations of the initiation of a coronal mass ejection. *Solar Phys.* **175**, 601. ADS. DOI.

Dudík, J., Janvier, M., Aulanier, G., Del Zanna, G., Karlický, M., Mason, H.E., Schmieder, B.: 2014, Slipping magnetic reconnection during an X-class solar flare observed by SDO/AIA. *Astrophys. J.* **784**, 144. ADS. DOI.

Emonet, T., Moreno-Insertis, F.: 1998, The physics of twisted magnetic tubes rising in a stratified medium: Two-dimensional results. *Astrophys. J.* **492**, 804. ADS. DOI.

Fan, Y.: 2001, The emergence of a twisted Ω-tube into the solar atmosphere. *Astrophys. J. Lett.* **554**, L111. ADS. DOI.

Fan, Y.: 2009, The emergence of a twisted flux tube into the solar atmosphere: Sunspot rotations and the formation of a coronal flux rope. *Astrophys. J.* **697**, 1529. DOI.

Fan, Y., Gibson, S.E.: 2003, The emergence of a twisted magnetic flux tube into a preexisting coronal arcade. *Astrophys. J. Lett.* **589**, L105. ADS. DOI.

Fan, Y., Gibson, S.E.: 2004, Numerical simulations of three-dimensional coronal magnetic fields resulting from the emergence of twisted magnetic flux tubes. *Astrophys. J.* **609**, 1123. ADS. DOI.

Filippov, B., Martsenyuk, O., Srivastava, A.K., Uddin, W.: 2015, Solar magnetic flux ropes. *J. Astron. Astrophys.* ADS. arXiv. DOI

Forbes, T.G.: 1986, Fast-shock formation in line-tied magnetic reconnection models of solar flares. *Astrophys. J.* **305**, 553. ADS. DOI.

Forbes, T.G.: 1990, Numerical simulation of a catastrophe model for coronal mass ejections. *J. Geophys. Res.* **95**, 11919. ADS. DOI.

Forbes, T.G.: 2000, A review on the genesis of coronal mass ejections. *J. Geophys. Res.* **105**, 23153. DOI.

Forbes, T.G., Isenberg, P.A.: 1991, A catastrophe mechanism for coronal mass ejections. *Astrophys. J.* **373**, 294. DOI.

Forbes, T.G., Malherbe, J.M.: 1986, A shock condensation mechanism for loop prominences. *Astrophys. J. Lett.* **302**, L67. ADS. DOI.

Forbes, T.G., Priest, E.R.: 1995, Photospheric magnetic field evolution and eruptive flares. *Astrophys. J.* **446**, 377. ADS. DOI.

Forbes, T.G., Linker, J.A., Chen, J., Cid, C., Kóta, J., Lee, M.A., Mann, G., Mikić, Z., Potgieter, M.S., Schmidt, J.M., Siscoe, G.L., Vainio, R., Antiochos, S.K., Riley, P.: 2006, CME theory and models. *Space Sci. Rev.* **123**, 251. DOI.

Foukal, P.: 1971, Morphological relationships in the chromospheric Hα fine structure. *Solar Phys.* **19**, 59. ADS. DOI.

Georgoulis, M.K.: 2012, Are solar active regions with major flares more fractal, multifractal, or turbulent than others? *Solar Phys.* **276**, 161. ADS. DOI.

Gibb, G.P.S., Mackay, D.H., Green, L.M., Meyer, K.A.: 2014, Simulating the formation of a sigmoidal flux rope in AR10977 from SOHO/MDI magnetograms. *Astrophys. J.* **782**, 71. ADS. DOI.

Gibson, S.E., Foster, D., Burkepile, J., de Toma, G., Stanger, A.: 2006, The calm before the storm: The link between quiescent cavities and coronal mass ejections. *Astrophys. J.* **641**, 590. ADS. DOI.
Gold, T., Hoyle, F.: 1960, On the origin of solar flares. *Mon. Not. Roy. Astron. Soc.* **120**, 89. ADS.
Gopalswamy, N., Kundu, M.R.: 1992, Estimation of the mass of a coronal mass ejection from radio observations. *Astrophys. J. Lett.* **390**, L37. ADS. DOI.
Gosling, J.T., Hildner, E., MacQueen, R.M., Munro, R.H., Poland, A.I., Ross, C.L.: 1976, The speeds of coronal mass ejection events. *Solar Phys.* **48**, 389. ADS. DOI.
Green, L.M., Kliem, B., Wallace, A.J.: 2011, Photospheric flux cancellation and associated flux rope formation and eruption. *Astron. Astrophys.* **526**, A2. DOI.
Green, L.M., Kliem, B., Török, T., van Driel-Gesztelyi, L., Attrill, G.D.R.: 2007, Transient coronal sigmoids and rotating erupting flux ropes. *Solar Phys.* **246**, 365. DOI.
Guo, Y., Schmieder, B., Démoulin, P., Wiegelmann, T., Aulanier, G., Török, T., Bommier, V.: 2010a, Coexisting flux rope and dipped arcade sections along one solar filament. *Astrophys. J.* **714**, 343. ADS. DOI.
Guo, Y., Ding, M.D., Schmieder, B., Li, H., Török, T., Wiegelmann, T.: 2010b, Driving mechanism and onset condition of a confined eruption. *Astrophys. J. Lett.* **725**, L38. ADS. DOI.
Guo, Y., Ding, M.D., Schmieder, B., Démoulin, P., Li, H.: 2012, Evolution of hard x-ray sources and ultraviolet solar flare ribbons for a confined eruption of a magnetic flux rope. *Astrophys. J.* **746**, 17. ADS. DOI.
Guo, Y., Démoulin, P., Schmieder, B., Ding, M.D., Vargas Domínguez, S., Liu, Y.: 2013, Recurrent coronal jets induced by repetitively accumulated electric currents. *Astron. Astrophys.* **555**, A19. ADS. DOI.
Hale, G.E.: 1927, The fields of force in the atmosphere of the Sun. *Nature* **119**, 708. ADS. DOI.
Harrison, R.A.: 1995, The nature of solar flares associated with coronal mass ejection. *Astron. Astrophys.* **304**, 585. ADS.
Heinzel, P., Schmieder, B., Fárník, F., Schwartz, P., Labrosse, N., Kotrč, P., Anzer, U., Molodij, G., Berlicki, A., DeLuca, E.E., Golub, L., Watanabe, T., Berger, T.: 2008, Hinode, TRACE, SOHO, and ground-based observations of a quiescent prominence. *Astrophys. J.* **686**, 1383. ADS. DOI.
Hirayama, T.: 1974, Theoretical model of flares and prominences. I: Evaporating flare model. *Solar Phys.* **34**, 323. ADS. DOI.
Hood, A.W., Archontis, V., Galsgaard, K., Moreno-Insertis, F.: 2009, The emergence of toroidal flux tubes from beneath the solar photosphere. *Astron. Astrophys.* **503**, 999. DOI.
Illing, R.M.E., Hundhausen, A.J.: 1986, Disruption of a coronal streamer by an eruptive prominence and coronal mass ejection. *J. Geophys. Res.* **91**, 10951. ADS. DOI.
Inoue, S., Hayashi, K., Magara, T., Choe, G.S., Park, Y.D.: 2014, Magnetohydrodynamic simulation of the X2.2 solar flare on 2011 February 15. I. Comparison with the observations. *Astrophys. J.* **788**, 182. ADS. DOI.
Inoue, S., Hayashi, K., Magara, T., Choe, G.S., Park, Y.D.: 2015, Magnetohydrodynamic simulation of the X2.2 solar flare on 2011 February 15. II. Dynamics connecting the solar flare and the coronal mass ejection. *Astrophys. J.* **803**, 73. ADS. DOI.
Isenberg, P.A., Forbes, T.G.: 2007, A three-dimensional line-tied magnetic field model for solar eruptions. *Astrophys. J.* **670**, 1453. ADS. DOI.
Jackson, B.V., Howard, R.A., Koomen, M.J., Michels, D.J., Sheeley, N.R. Jr.: 1985, The mass distribution of coronal mass ejections. In: *Bull. Am. Astron. Soc.* **17**, 636. ADS.
Jackson, J.D.: 1998, *Classical Electrodynamics*, 3rd edn. Wiley, New York, 218.
Jacobs, C., Roussev, I.I., Lugaz, N., Poedts, S.: 2009, The internal structure of coronal mass ejections: Are all regular magnetic clouds flux ropes? *Astrophys. J. Lett.* **695**, L171. ADS. DOI.
Janvier, M., Aulanier, G., Pariat, E., Démoulin, P.: 2013, The standard flare model in three dimensions. III. Slip-running reconnection properties. *Astron. Astrophys.* **555**, A77. ADS. DOI.
Janvier, M., Aulanier, G., Bommier, V., Schmieder, B., Démoulin, P., Pariat, E.: 2014, Electric currents in flare ribbons: Observations and three-dimensional standard model. *Astrophys. J.* **788**, 60. ADS. DOI.
Jiang, C., Feng, X., Wu, S.T., Hu, Q.: 2013, Magnetohydrodynamic simulation of a sigmoid eruption of active region 11283. *Astrophys. J. Lett.* **771**, L30. ADS. DOI.
Jouve, L., Brun, A.S., Aulanier, G.: 2013, Global dynamics of subsurface solar active regions. *Astrophys. J.* **762**, 4. ADS. DOI.
Kahler, S.W., Moore, R.L., Kane, S.R., Zirin, H.: 1988, Filament eruptions and the impulsive phase of solar flares. *Astrophys. J.* **328**, 824. DOI.
Kliem, B., Török, T.: 2006, Torus instability. *Phys. Rev. Lett.* **96**(25), 255002. DOI.
Kliem, B., Linton, M.G., Török, T., Karlický, M.: 2010, Reconnection of a kinking flux rope triggering the ejection of a microwave and hard X-ray source II. Numerical modeling. *Solar Phys.* **266**, 91. DOI.
Kliem, B., Su, Y.N., van Ballegooijen, A.A., DeLuca, E.E.: 2013, Magnetohydrodynamic modeling of the solar eruption on 2010 April 8. *Astrophys. J.* **779**, 129. ADS. DOI.

Kliem, B., Lin, J., Forbes, T.G., Priest, E.R., Török, T.: 2014a, Catastrophe versus instability for the eruption of a toroidal solar magnetic flux rope. *Astrophys. J.* **789**, 46. ADS. DOI.
Kliem, B., Török, T., Titov, V.S., Lionello, R., Linker, J.A., Liu, R., Liu, C., Wang, H.: 2014b, Slow rise and partial eruption of a double-decker filament. II. A double flux rope model. *Astrophys. J.* **792**, 107. ADS. DOI.
Kopp, R.A., Pneuman, G.W.: 1976, Magnetic reconnection in the corona and the loop prominence phenomenon. *Solar Phys.* **50**, 85. ADS. DOI.
Kumar, P., Cho, K.-S.: 2014, Multiwavelength observation of a large-scale flux rope eruption above a kinked small filament. *Astron. Astrophys.* **572**, A83. ADS. DOI.
Kuperus, M., Raadu, M.A.: 1974, The support of prominences formed in neutral sheets. *Astron. Astrophys.* **31**, 189. http://esoads.eso.org/abs/1974A%26A....31..189K.
Leake, J.E., Linton, M.G., Antiochos, S.K.: 2014, Simulations of emerging magnetic flux. II. The formation of unstable coronal flux ropes and the initiation of coronal mass ejections. *Astrophys. J.* **787**, 46. ADS. DOI.
Leake, J.E., Linton, M.G., Török, T.: 2013, Simulations of emerging magnetic flux. I. The formation of stable coronal flux ropes. *Astrophys. J.* **778**, 99. ADS. DOI.
Lepping, R.P., Burlaga, L.F., Jones, J.A.: 1990, Magnetic field structure of interplanetary magnetic clouds at 1 AU. *J. Geophys. Res.* **95**, 11957. ADS. DOI.
Li, H., Schmieder, B., Aulanier, G., Berlicki, A.: 2006, Is pre-eruptive null point reconnection required for triggering eruptions? *Solar Phys.* **237**, 85. ADS. DOI.
Li, H., Schmieder, B., Song, M.T., Bommier, V.: 2007, Interaction of magnetic field systems leading to an X1.7 flare due to large-scale flux tube emergence. *Astron. Astrophys.* **475**, 1081. ADS. DOI.
Lin, J., Forbes, T.G.: 2000, Effects of reconnection on the coronal mass ejection process. *J. Geophys. Res.* **105**, 2375. ADS. DOI.
Lites, B.W.: 2005, Magnetic flux ropes in the solar photosphere: The vector magnetic field under active region filaments. *Astrophys. J.* **622**, 1275. ADS. DOI.
Liu, C., Deng, N., Lee, J., Wiegelmann, T., Moore, R.L., Wang, H.: 2013, Evidence for solar Tether-cutting magnetic reconnection from coronal field extrapolations. *Astrophys. J. Lett.* **778**, L36. ADS. DOI.
Liu, Y., Hoeksema, J.T., Sun, X.: 2014, Test of the hemispheric rule of magnetic helicity in the Sun using the Helioseismic and Magnetic Imager (HMI) data. *Astrophys. J. Lett.* **783**, L1. ADS. DOI.
Liu, Y., Luhmann, J.G., Bale, S.D., Lin, R.P.: 2011, Solar source and heliospheric consequences of the 2010 April 3 coronal mass ejection: A comprehensive view. *Astrophys. J.* **734**, 84. ADS. DOI.
López Ariste, A., Aulanier, G., Schmieder, B., Sainz Dalda, A.: 2006, First observation of bald patches in a filament channel and at a barb endpoint. *Astron. Astrophys.* **456**, 725. ADS. DOI.
López Fuentes, M.C., Démoulin, P., Mandrini, C.H., van Driel-Gesztelyi, L.: 2000, The counterkink rotation of a non-hale active region. *Astrophys. J.* **544**, 540. ADS. DOI.
Low, B.C., Zhang, M.: 2002, The hydromagnetic origin of the two dynamical types of solar coronal mass ejections. *Astrophys. J. Lett.* **564**, L53. ADS. DOI.
Lugaz, N., Roussev, I.I.: 2011, Numerical modeling of interplanetary coronal mass ejections and comparison with heliospheric images. *J. Atmos. Solar-Terr. Phys.* **73**, 1187. ADS. DOI.
Luoni, M.L., Mandrini, C.H., Démoulin, P., van Driel-Gesztelyi, L., Kövári, Z.: 2004, Can we determine the magnetic helicity sign of the solar active regions? *Bol. Asoc. Argent. Astron.* **47**, 14. ADS.
Luoni, M.L., Démoulin, P., Mandrini, C.H., van Driel-Gesztelyi, L.: 2011, Twisted flux tube emergence evidenced in longitudinal magnetograms: Magnetic tongues. *Solar Phys.* **270**, 45. DOI.
Lynch, B.J., Antiochos, S.K., DeVore, C.R., Luhmann, J.G., Zurbuchen, T.H.: 2008, Topological evolution of a fast magnetic breakout CME in three dimensions. *Astrophys. J.* **683**, 1192. ADS. DOI.
MacQueen, R.M., Fisher, R.R.: 1983, The kinematics of solar inner coronal transients. *Solar Phys.* **89**, 89. ADS. DOI.
MacTaggart, D., Hood, A.W.: 2010, Simulating the "Sliding Doors" effect through magnetic flux emergence. *Astrophys. J. Lett.* **716**, L219. DOI.
Maia, D., Vourlidas, A., Pick, M., Howard, R., Schwenn, R., Magalhães, A.: 1999, Radio signatures of a fast coronal mass ejection development on November 6, 1997. *J. Geophys. Res.* **104**, 12507. ADS. DOI.
Manchester, W. IV, Gombosi, T., DeZeeuw, D., Fan, Y.: 2004, Eruption of a buoyantly emerging magnetic flux rope. *Astrophys. J.* **610**, 588. ADS. DOI.
Manchester, W.B. IV, Vourlidas, A., Tóth, G., Lugaz, N., Roussev, I.I., Sokolov, I.V., Gombosi, T.I., De Zeeuw, D.L., Opher, M.: 2008, Three-dimensional MHD Simulation of the 2003 October 28 coronal mass ejection: Comparison with LASCO coronagraph observations. *Astrophys. J.* **684**, 1448. ADS. DOI.
Mandrini, C.H., Démoulin, P., Schmieder, B., Deluca, E.E., Pariat, E., Uddin, W.: 2006, Companion event and precursor of the X17 flare on 28 October 2003. *Solar Phys.* **238**, 293. ADS. DOI.
Mandrini, C.H., Schmieder, B., Démoulin, P., Guo, Y., Cristiani, G.D.: 2014, Topological analysis of emerging bipole clusters producing violent solar events. *Solar Phys.* **289**, 2041. ADS. DOI.

Maričić, D., Vršnak, B., Stanger, A.L., Veronig, A.: 2004, Coronal mass ejection of 15 May 2001: I. Evolution of morphological features of the eruption. *Solar Phys.* **225**, 337. ADS. DOI.

Maričić, D., Vršnak, B., Stanger, A.L., Veronig, A.M., Temmer, M., Roša, D.: 2007, Acceleration phase of coronal mass ejections: II. Synchronization of the energy release in the associated flare. *Solar Phys.* **241**, 99. DOI.

Marqué, C., Lantos, P., Delouis, J.M., Alissandrakis, C.E.: 2001, Radio observations of filaments at metric and decimetric wavelengths. *Astrophys. Space Sci.* **277**, 329. ADS. DOI.

Martin, S.F.: 1998, Filament chirality: A link between fine-scale and global patterns (review). In: Webb, D.F., Schmieder, B., Rust, D.M. (eds.) *IAU Colloq. 167: New Perspectives on Solar Prominences, Astron. Soc. Pacific CS* **150**, 419. ADS.

Masson, S., Antiochos, S.K., DeVore, C.R.: 2013, A model for the escape of solar-flare-accelerated particles. *Astrophys. J.* **771**, 82. ADS. DOI.

McCauley, P.I., Su, Y.N., Schanche, N., Evans, K.E., Su, C., McKillop, S., Reeves, K.K.: 2015, Prominence and filament eruptions observed by the Solar Dynamics Observatory: Statistical properties, kinematics, and online catalog. *Solar Phys.* DOI.

Melrose, D.B.: 1991, Neutralized and unneutralized current patterns in the solar corona. *Astrophys. J.* **381**, 306. ADS. DOI.

Mikic, Z., Linker, J.A.: 1994, Disruption of coronal magnetic field arcades. *Astrophys. J.* **430**, 898. ADS. DOI.

Miklenic, C.H., Veronig, A.M., Vršnak, B.: 2009, Temporal comparison of nonthermal flare emission and magnetic-flux change rates. *Astron. Astrophys.* **499**, 893. ADS. DOI.

Miklenic, C.H., Veronig, A.M., Vršnak, B., Hanslmeier, A.: 2007, Reconnection and energy release rates in a two-ribbon flare. *Astron. Astrophys.* **461**, 697. ADS. DOI.

Moon, Y.-J., Choe, G.S., Wang, H., Park, Y.D., Gopalswamy, N., Yang, G., Yashiro, S.: 2002, A statistical study of two classes of coronal mass ejections. *Astrophys. J.* **581**, 694. ADS. DOI.

Moon, Y.-J., Choe, G.S., Wang, H., Park, Y.D., Cheng, C.Z.: 2003, Relationship between CME kinematics and flare strength. *J. Korean Astron. Soc.* **36**, 61. ADS.

Moore, R.L., Sterling, A.C., Hudson, H.S., Lemen, J.R.: 2001, Onset of the magnetic explosion in solar flares and coronal mass ejections. *Astrophys. J.* **552**, 833. DOI.

Mouschovias, T.C., Poland, A.I.: 1978, Expansion and broadening of coronal loop transients – A theoretical explanation. *Astrophys. J.* **220**, 675. ADS. DOI.

Musset, S., Vilmer, N., Bommier, V.: 2015, Energetic electrons and electric currrent in the flare X2.2 of February 2011. *Astron. Astrophys.*, in press.

Nakagawa, Y., Raadu, M.A., Billings, D.E., McNamara, D.: 1971, On the topology of filaments and chromospheric fibrils near sunspots. *Solar Phys.* **19**, 72. ADS. DOI.

Neupert, W.M.: 1968, Comparison of solar X-ray line emission with microwave emission during flares. *Astrophys. J. Lett.* **153**, L59. ADS. DOI.

Neupert, W.M., Thompson, B.J., Gurman, J.B., Plunkett, S.P.: 2001, Eruption and acceleration of flare-associated coronal mass ejection loops in the low corona. *J. Geophys. Res.* **106**, 25215. DOI.

Nindos, A., Patsourakos, S., Wiegelmann, T.: 2012, On the role of the background overlying magnetic field in solar eruptions. *Astrophys. J. Lett.* **748**, L6. ADS. DOI.

Okamoto, T.J., Tsuneta, S., Lites, B.W., Kubo, M., Yokoyama, T., Berger, T.E., Ichimoto, K., Katsukawa, Y., Nagata, S., Shibata, K., Shimizu, T., Shine, R.A., Suematsu, Y., Tarbell, T.D., Title, A.M.: 2008, Emergence of a helical flux rope under an active region prominence. *Astrophys. J. Lett.* **673**, L215. ADS. DOI.

Pariat, E., Démoulin, P., Nindos, A.: 2007, How to improve the maps of magnetic helicity injection in active regions? *Adv. Space Res.* **39**, 1706. ADS. DOI.

Pevtsov, A.A.: 2002, Active-region filaments and X-ray sigmoids. *Solar Phys.* **207**, 111. ADS. DOI.

Pneuman, G.W.: 1980, The physical structure of coronal holes – Influence of magnetic fields and coronal heating. *Astron. Astrophys.* **81**, 161. ADS.

Priest, E.R.: 1981, *Solar Magneto-Hydrodynamics*, Gordon and Breach Science Publishers, London. ADS

Pulkkinen, T.: 2007, Space weather: Terrestrial perspective. *Living Rev. Solar Phys.* **4**, 1. ADS. DOI.

Qiu, J., Yurchyshyn, V.B.: 2005, Magnetic reconnection flux and coronal mass ejection velocity. *Astrophys. J. Lett.* **634**, L121. ADS. DOI.

Qiu, J., Wang, H., Cheng, C.Z., Gary, D.E.: 2004, Magnetic reconnection and mass acceleration in flare-coronal mass ejection events. *Astrophys. J.* **604**, 900. ADS. DOI.

Régnier, S., Walsh, R.W., Alexander, C.E.: 2011, A new look at a polar crown cavity as observed by SDO/AIA. Structure and dynamics. *Astron. Astrophys.* **533**, L1. ADS. DOI.

Rust, D.M.: 1983, Coronal disturbances and their terrestrial effects. *Space Sci. Rev.* **34**, 21. ADS. DOI.

Rust, D.M., Kumar, A.: 1996, Evidence for helically kinked magnetic flux ropes in solar eruptions. *Astrophys. J. Lett.* **464**, L199. ADS. DOI.

Sakurai, T.: 1976, Magnetohydrodynamic interpretation of the motion of prominences. *Publ. Astron. Soc. Japan* **28**, 177. ADS.

Savcheva, A., Pariat, E., van Ballegooijen, A., Aulanier, G., DeLuca, E.: 2012, Sigmoidal active region on the Sun: Comparison of a magnetohydrodynamical simulation and a nonlinear force-free field model. *Astrophys. J.* **750**, 15. ADS. DOI.

Savcheva, A.S., McKillop, S.C., McCauley, P.I., Hanson, E.M., DeLuca, E.E.: 2014, A new sigmoid catalog from hinode and the Solar Dynamics Observatory: Statistical properties and evolutionary histories. *Solar Phys.* **289**, 3297. ADS. DOI.

Schmieder, B., Aulanier, G.: 2012, What are the physical mechanisms of eruptions and CMEs? *Adv. Space Res.* **49**, 1598. ADS. DOI.

Schmieder, B., Archontis, V., Pariat, E.: 2014, Magnetic flux emergence along the solar cycle. *Space Sci. Rev.* **186**, 227. ADS. DOI.

Schmieder, B., Aulanier, G., Démoulin, P., van Driel-Gesztelyi, L., Roudier, T., Nitta, N., Cauzzi, G.: 1997, Magnetic reconnection driven by emergence of sheared magnetic field. *Astron. Astrophys.* **325**, 1213. ADS.

Schmieder, B., Mein, N., Deng, Y., Dumitrache, C., Malherbe, J.-M., Staiger, J., Deluca, E.E.: 2004, Magnetic changes observed in the formation of two filaments in a complex active region: TRACE and MSDP observations. *Solar Phys.* **223**, 119. ADS. DOI.

Schmieder, B., Bommier, V., Kitai, R., Matsumoto, T., Ishii, T.T., Hagino, M., Li, H., Golub, L.: 2008, Magnetic causes of the eruption of a quiescent filament. *Solar Phys.* **247**, 321. ADS. DOI.

Schmieder, B., Roudier, T., Mein, N., Mein, P., Malherbe, J.M., Chandra, R.: 2014, Proper horizontal photospheric flows in a filament channel. *Astron. Astrophys.* **564**, A104. ADS. DOI.

Schrijver, C.J., Elmore, C., Kliem, B., Török, T., Title, A.M.: 2008, Observations and modeling of the early acceleration phase of erupting filaments involved in coronal mass ejections. *Astrophys. J.* **674**, 586. ADS. DOI.

Schrijver, C.J., Aulanier, G., Title, A.M., Pariat, E., Delannée, C.: 2011, The 2011 February 15 X2 flare, ribbons, coronal front, and mass ejection: Interpreting the three-dimensional views from the Solar Dynamics Observatory and STEREO guided by magnetohydrodynamic flux-rope modeling. *Astrophys. J.* **738**, 167. ADS. DOI.

Schwenn, R.: 2006, Solar wind sources and their variations over the solar cycle. *Space Sci. Rev.* **124**, 51. ADS. DOI.

Shanmugaraju, A., Moon, Y.-J., Dryer, M., Umapathy, S.: 2003, On the kinematic evolution of flare-associated CMEs. *Solar Phys.* **215**, 185. ADS.

Sheeley, N.R., Walters, J.H., Wang, Y.-M., Howard, R.A.: 1999, Continuous tracking of coronal outflows: Two kinds of coronal mass ejections. *J. Geophys. Res.* **104**, 24739. ADS. DOI.

Shibata, K.: 1996, New observational facts about solar flares from Yohkoh studies – Evidence of magnetic reconnection and a unified model of flares. *Adv. Space Res.* **17**, 9. ADS.

Shibata, K., Magara, T.: 2011, Solar flares: Magnetohydrodynamic processes. *Living Rev. Solar Phys.* **8**, 6. ADS. DOI.

Shiota, D., Kusano, K., Miyoshi, T., Shibata, K.: 2010, Magnetohydrodynamic modeling for a formation process of coronal mass ejections: Interaction between an ejecting flux rope and an ambient field. *Astrophys. J.* **718**, 1305. ADS. DOI.

Steele, C.D.C., Priest, E.R.: 1989, The eruption of a prominence and coronal mass ejection which drive reconnection. *Solar Phys.* **119**, 157. ADS. DOI.

Sturrock, P.A.: 1966, Model of the high-energy phase of solar flares. *Nature* **211**, 695. ADS. DOI.

Subramanian, P., Dere, K.P.: 2001, Source regions of coronal mass ejections. *Astrophys. J.* **561**, 372. ADS. DOI.

Sun, J.Q., Cheng, X., Guo, Y., Ding, M.D., Li, Y.: 2014, Dynamic evolution of an X-shaped structure above a trans-equatorial quadrupole solar active region group. ArXiv e-prints. arXiv.

Sun, X., Hoeksema, J.T., Liu, Y., Aulanier, G., Su, Y., Hannah, I.G., Hock, R.A.: 2013, Hot spine loops and the nature of a late-phase solar flare. *Astrophys. J.* **778**, 139. ADS. DOI.

Temmer, M., Veronig, A.M., Vršnak, B., Rybák, J., Gömöry, P., Stoiser, S., Maričić, D.: 2008, Acceleration in fast halo CMEs and synchronized flare HXR bursts. *Astrophys. J. Lett.* **673**, L95. DOI.

Temmer, M., Veronig, A.M., Kontar, E.P., Krucker, S., Vršnak, B.: 2010, Combined STEREO/RHESSI study of coronal mass ejection acceleration and particle acceleration in solar flares. *Astrophys. J.* **712**, 1410. ADS. DOI.

Titov, V.S., Démoulin, P.: 1999, Basic topology of twisted magnetic configurations in solar flares. *Astron. Astrophys.* **351**, 707. ADS.

Titov, V.S., Démoulin, P., Hornig, G.: 2003, Hyperbolic flux tubes in flaring magnetic configurations. *Astron. Nachr.* **324**, 17. ADS.

Toriumi, S., Iida, Y., Bamba, Y., Kusano, K., Imada, S., Inoue, S.: 2013, The magnetic systems triggering the M6.6 class solar flare in NOAA active region 11158. *Astrophys. J.* **773**, 128. ADS. DOI.
Török, T., Kliem, B.: 2003, The evolution of twisting coronal magnetic flux tubes. *Astron. Astrophys.* **406**, 1043. DOI.
Török, T., Kliem, B.: 2005, Confined and ejective eruptions of kink-unstable flux ropes. *Astrophys. J. Lett.* **630**, L97. DOI.
Török, T., Kliem, B.: 2007, Numerical simulations of fast and slow coronal mass ejections. *Astron. Nachr.* **328**, 743. DOI.
Török, T., Chandra, R., Pariat, E., Démoulin, P., Schmieder, B., Aulanier, G., Linton, M.G., Mandrini, C.H.: 2011, Filament interaction modeled by flux rope reconnection. *Astrophys. J.* **728**, 65. DOI.
Ugarte-Urra, I., Warren, H.P., Winebarger, A.R.: 2007, The magnetic topology of coronal mass ejection sources. *Astrophys. J.* **662**, 1293. ADS. DOI.
van Ballegooijen, A.A., Martens, P.C.H.: 1989, Formation and eruption of solar prominences. *Astrophys. J.* **343**, 971. DOI.
van der Holst, B., Jacobs, C., Poedts, S.: 2007, Simulation of a breakout coronal mass ejection in the solar wind. *Astrophys. J. Lett.* **671**, L77. ADS. DOI.
van Driel-Gesztelyi, L., Démoulin, P., Mandrini, C.H., Harra, L., Klimchuk, J.A.: 2003, The long-term evolution of AR 7978: The scalings of the coronal plasma parameters with the mean photospheric magnetic field. *Astrophys. J.* **586**, 579. DOI.
van Tend, W.: 1979, The onset of coronal transients. *Solar Phys.* **61**, 89. DOI.
van Tend, W., Kuperus, M.: 1978, The development of coronal electric current systems in active regions and their relation to filaments and flares. *Solar Phys.* **59**, 115. ADS. DOI.
Vargas Domínguez, S., MacTaggart, D., Green, L., van Driel-Gesztelyi, L., Hood, A.W.: 2012, On signatures of twisted magnetic flux tube emergence. *Solar Phys.* **278**, 33. ADS. DOI.
Veronig, A., Vršnak, B., Dennis, B.R., Temmer, M., Hanslmeier, A., Magdalenić, J.: 2002, Investigation of the Neupert effect in solar flares. I. Statistical properties and the evaporation model. *Astron. Astrophys.* **392**, 699. ADS. DOI.
Vourlidas, A., Buzasi, D., Howard, R.A., Esfandiari, E.: 2002, Mass and energy properties of LASCO CMEs. In: Wilson, A. (ed.) *Solar Variability: From Core to Outer Frontiers, ESA SP* **506**, 91. ADS.
Vourlidas, A., Lynch, B.J., Howard, R.A., Li, Y.: 2013, How many CMEs have flux ropes? Deciphering the signatures of shocks, flux ropes, and prominences in coronagraph observations of CMEs. *Solar Phys.* **284**, 179. ADS. DOI.
Vršnak, B.: 2008, Processes and mechanisms governing the initiation and propagation of CMEs. *Ann. Geophys.* **26**, 3089.
Vršnak, B.: 1990, Eruptive instability of cylindrical prominences. *Solar Phys.* **129**, 295. ADS.
Vršnak, B.: 2001, Deceleration of coronal mass ejections. *Solar Phys.* **202**, 173. ADS.
Vršnak, B.: 2008, Processes and mechanisms governing the initiation and propagation of CMEs. *Ann. Geophys.* **26**, 3089. ADS.
Vršnak, B., Ruždjak, V., Rompolt, B.: 1991, Stability of prominences exposing helical-like patterns. *Solar Phys.* **136**, 151. ADS.
Vršnak, B., Sudar, D., Ruždjak, D.: 2005, The CME-flare relationship: Are there really two types of CMEs? *Astron. Astrophys.* **435**, 1149. ADS. DOI.
Vršnak, B., Ruždjak, V., Brajša, R., Zloch, F.: 1990, Oscillatory motions in an active prominence. *Solar Phys.* **127**, 119. ADS. DOI
Vršnak, B., Maričić, D., Stanger, A.L., Veronig, A.: 2004, Coronal mass ejection of 15 May 2001: II. Coupling of the CME acceleration and the flare energy release. *Solar Phys.* **225**, 355. ADS. DOI.
Wang, Y.-M.: 2013, *Astrophys. J. Lett.* **775**, L46. ADS. DOI.
Wang, Y.-M., Sheeley, N.R. Jr.: 2002, Observations of core fallback during coronal mass ejections. *Astrophys. J.* **567**, 1211. ADS. DOI.
Webb, D.F., Howard, T.A.: 2012, Coronal mass ejections: Observations. *Living Rev. Solar Phys.* **9**, 3. ADS. DOI.
Webb, D.F., Hundhausen, A.J.: 1987, Activity associated with the solar origin of coronal mass ejections. *Solar Phys.* **108**, 383. ADS. DOI.
Williams, D.R., Török, T., Démoulin, P., van Driel-Gesztelyi, L., Kliem, B.: 2005, Eruption of a kink-unstable filament in NOAA active region 10696. *Astrophys. J. Lett.* **628**, L163. ADS. DOI.
Wu, S.T., Zhang, T.X., Tandberg-Hanssen, E., Liu, Y., Feng, X., Tan, A.: 2004, Numerical magnetohydrodynamic experiments for testing the physical mechanisms of coronal mass ejections acceleration. *Solar Phys.* **225**, 157. ADS. DOI.
Xia, C., Keppens, R., Guo, Y.: 2014, Three-dimensional prominence-hosting magnetic configurations: Creating a helical magnetic flux rope. *Astrophys. J.* **780**, 130. ADS. DOI.

Yan, X.L., Xue, Z.K., Liu, J.H., Ma, L., Kong, D.F., Qu, Z.Q., Li, Z.: 2014, Kink instability evidenced by analyzing the leg rotation of a filament. *Astrophys. J.* **782**, 67. ADS. DOI.

Yashiro, S., Gopalswamy, N., Michalek, G., Cyr, O.C.St., Plunkett, S.P., Rich, N.B., Howard, R.A.: 2004, A catalog of white light coronal mass ejections observed by the SOHO spacecraft. *J. Geophys. Res.* **109**, 7105. ADS. DOI.

Yashiro, S., Akiyama, S., Gopalswamy, N., Howard, R.A.: 2006, Different power-law indices in the frequency distributions of flares with and without coronal mass ejections. *Astrophys. J. Lett.* **650**, L143. ADS. DOI.

Yeates, A.R.: 2014, Coronal magnetic field evolution from 1996 to 2012: Continuous non-potential simulations. *Solar Phys.* **289**, 631. ADS. DOI.

Zhang, J., Dere, K.P.: 2006, A statistical study of main and residual accelerations of coronal mass ejections. *Astrophys. J.* **649**, 1100. ADS. DOI.

Zhang, J., Dere, K.P., Howard, R.A., Kundu, M.R., White, S.M.: 2001, On the temporal relationship between coronal mass ejections and flares. *Astrophys. J.* **559**, 452. DOI.

Zhang, J., Dere, K.P., Howard, R.A., Vourlidas, A.: 2004, A study of the kinematic evolution of coronal mass ejections. *Astrophys. J.* **604**, 420. ADS. DOI.

Zhao, J., Li, H., Pariat, E., Schmieder, B., Guo, Y., Wiegelmann, T.: 2014, Temporal evolution of the magnetic topology of the NOAA active region 11158. *Astrophys. J.* **787**, 88. ADS. DOI.

Žic, T., Vršnak, B., Skender, M.: 2007, The magnetic flux and self-inductivity of a thick toroidal current. *J. Plasma Phys.* **73**, 741. DOI.

Zuccarello, F.P., Seaton, D.B., Mierla, M., Poedts, S., Rachmeler, L.A., Romano, P., Zuccarello, F.: 2014, Observational evidence of torus instability as trigger mechanism for coronal mass ejections: The 2011 August 4 filament eruption. *Astrophys. J.* **785**, 88. ADS. DOI.

Reprinted from *Solar Physics* Journal, DOI 10.1007/s11207-015-0708-x

SOLAR AND STELLAR FLARES

New Insights into White-Light Flare Emission from Radiative-Hydrodynamic Modeling of a Chromospheric Condensation

Adam F. Kowalski[1,2] · S.L. Hawley[3] · M. Carlsson[4] · J.C. Allred[2] · H. Uitenbroek[5] · R.A. Osten[6] · G. Holman[2]

Received: 30 January 2015 / Accepted: 18 May 2015 / Published online: 10 June 2015
© Springer Science+Business Media Dordrecht 2015

Abstract The heating mechanism at high densities during M-dwarf flares is poorly understood. Spectra of M-dwarf flares in the optical and near-ultraviolet wavelength regimes have revealed three continuum components during the impulsive phase: 1) an energetically dominant blackbody component with a color temperature of $T \approx 10^4$ K in the blue-optical, 2) a smaller amount of Balmer continuum emission in the near-ultraviolet at $\lambda \leq 3\,646$ Å, and 3) an apparent pseudo-continuum of blended high-order Balmer lines between $\lambda = 3\,646$ Å and $\lambda \approx 3\,900$ Å. These properties are not reproduced by models that employ a typical "solar-type" flare heating level of $\leq 10^{11}\ \mathrm{erg\,cm^{-2}\,s^{-1}}$ in nonthermal electrons, and therefore our understanding of these spectra is limited to a phenomenological three-component interpretation. We present a new 1D radiative-hydrodynamic model of an M-dwarf flare from precipitating nonthermal electrons with a high energy flux of $10^{13}\ \mathrm{erg\,cm^{-2}\,s^{-1}}$. The simulation produces bright near-ultraviolet and optical continuum emission from a dense ($n > 10^{15}\ \mathrm{cm^{-3}}$), hot ($T \approx 12\,000-13\,500$ K) chromospheric condensation. For the first time, the observed color temperature and Balmer jump ratio are produced self-consistently in a radiative-hydrodynamic flare model. We find that a $T \approx 10^4$ K blackbody-like continuum component and a low Balmer jump ratio result from optically thick Balmer ($\infty \rightarrow n = 2$) and Paschen recombination ($\infty \rightarrow n = 3$) radiation, and thus the properties of the flux spectrum are caused by blue ($\lambda \approx 4\,300$ Å) light escaping over a

Solar and Stellar Flares: Observations, Simulations, and Synergies
Guest Editors: Lyndsay Fletcher and Petr Heinzel

✉ A.F. Kowalski
adam.f.kowalski@nasa.gov

[1] Department of Astronomy, University of Maryland, College Park, MD 20742, USA

[2] NASA Goddard Space Flight Center, Heliophysics Science Division, Code 671, 8800 Greenbelt Rd., Greenbelt, MD 20771, USA

[3] Department of Astronomy, University of Washington, Box 351580, Seattle, WA 98195, USA

[4] Institute of Theoretical Astrophysics, University of Oslo, P.O. Box 1029, Blindern, 0315, Oslo, Norway

[5] National Solar Observatory, Sacramento Peak, P.O. Box 62, Sunpsot, NM 88349, USA

[6] Space Telescope Science Institute, 3700 San Martin Drive, Baltimore, MD 21218, USA

larger physical depth range than by red ($\lambda \approx 6\,700$ Å) and near-ultraviolet ($\lambda \approx 3\,500$ Å) light. To model the near-ultraviolet pseudo-continuum previously attributed to overlapping Balmer lines, we include the extra Balmer continuum opacity from Landau–Zener transitions that result from merged, high-order energy levels of hydrogen in a dense, partially ionized atmosphere. This reveals a new diagnostic of ambient charge density in the densest regions of the atmosphere that are heated during dMe and solar flares.

Keywords Flares, dynamics · Flares, energetic particles · Flares, impulsive phase · Flares, models · Flares, spectrum · Flares, white-light

1. Introduction

It is notoriously difficult to explain the origin of heating at high atmospheric densities during solar and stellar flares. The broadband color (white-light) distribution of the ultraviolet and optical emission during large and small flares in chromospherically active M-dwarf (dMe) stars exhibits the general shape of a $T = 8\,500-9\,500$ K blackbody (Hawley and Fisher, 1992; Hawley *et al.*, 2003) without an indication of a significant Balmer jump at $\lambda = 3\,646$ Å. A recent homogeneous analysis of flare spectra around the Balmer jump has confirmed the presence of a hot blackbody (or blackbody-like) component with a color temperature of $T \gtrsim 10\,000$ K that contributes most of the radiated near-ultraviolet (NUV) and optical energy in the impulsive phase (Hawley and Pettersen, 1991; Kowalski *et al.*, 2013, hereafter, K13). Intriguingly, the presence of hot blackbody emission with $T \approx 9\,000$ K has recently also been inferred from Sun-as-a-star observations during large (X-class) and small (C-class) solar flares (Kretzschmar, 2011). The hot (hereafter, "hot" refers to temperatures between 8 500 and 15 000 K) blackbody continuum component is often interpreted as evidence of impulsive heating in a region with a high density of $n_e \approx 10^{15}$ cm^{-3} or more, which has been inferred from phenomenological modeling of dMe flares (Cram and Woods, 1982; Houdebine, 1992; Christian *et al.*, 2003). A recent observation that directly supports this idea is the Vega-like flare spectrum detected during the megaflare event on 2009 January 16, in the dM4.5e star YZ CMi (Kowalski *et al.*, 2010, 2013), which indicates that photospheric-level densities can be produced and maintained at sufficiently high temperatures during dMe flares. However, heating at high densities requires a large amount of volumetric energy deposition such that the heating per unit mass is non-negligible. Direct heating of the photosphere by nonthermal (NT) deka-keV electrons is problematic because the electrons experience rapid energy loss as soon as they impact the top of the thick-target chromosphere.

A possible mechanism for increasing the density in the upper chromosphere was proposed in the early gas dynamic simulations of Livshits *et al.* (1981). These authors showed that a shock from precipitating nonthermal electrons (with an energy flux of 10^{12} erg cm^{-2} s^{-1}) produces a rapid downflow and compression of the material to a very high density ($> 10^{15}$ cm^{-3}). The development of downward-moving "chromospheric condensations" (hereafter, CC) is accompanied by an upward-moving "evaporation" of chromospheric material as it is heated beyond a million degrees (Fisher, Canfield, and McClymont, 1985). Chromospheric evaporation and CCs are thought to be the fundamental processes in solar and stellar flares driving the frequently observed Neupert effect (Dennis and Zarro, 1993; Hawley *et al.*, 1995; Guedel *et al.*, 1996). Several studies have explored the role of explosive CCs as the source of white-light emission in solar and dMe flares (Livshits *et al.*, 1981; Gan *et al.*, 1992; Katsova, Boiko, and Livshits, 1997), but the formation threshold of CCs is strongly dependent on the initial atmospheric structure, the energy distribution (power-law index, low-energy cutoff, and flux) of the precipitating NT electrons (Fisher, 1989), and the time-dependent ionization of helium (Abbett and Hawley, 1999; Allred *et al.*, 2005, 2006), thus necessitating an accurate and realistic description of the NT electron distribution, the pre-flare atmosphere structure, and the level populations of important cooling agents.

Recent radiative-hydrodynamic (RHD) simulations of dMe (Allred *et al.*, 2006) flares with the RADYN code (Carlsson and Stein, 1995, 1997) explored the atmospheric response to an actual NT electron spectrum that was obtained from the peak of an X-class solar flare inferred from *Reuven Ramaty High Energy Solar Spectroscopic Imager* (RHESSI) data (Holman *et al.*, 2003). A constant, solar-type flare heating level of the NT electrons was employed with injected energy fluxes of 10^{10} erg cm^{-2} s^{-1} (F10) and 10^{11} erg cm^{-2} s^{-1} (F11). These simulations produced explosive mass motions indicative of chromospheric evaporation and a CC, but with densities far too low to produce strong, blackbody-like white-light emission. Moreover, the prominent spectral component in these simulations was a hydrogen recombination spectrum with strong Balmer recombination continuum emission and a smaller amount of Paschen recombination continuum emission. In these models, the heating at high density was achieved through Balmer and Paschen continuum backwarming of the photosphere, which increased in temperature by 1 200 K in the F11 run and produced additional amounts of optical/red (blackbody-like) emission. For the range of nonthermal electron energy fluxes thus far explored in RHD models (F9 – F11; see also Hawley and Fisher, 1994; Abbett and Hawley, 1999), a flare spectrum with hot blackbody emission has not been produced through the development of a CC, through photospheric heating from Balmer and Paschen continuum backwarming (Allred *et al.*, 2006), or through photospheric heating from X-ray and EUV backwarming (Hawley and Fisher, 1992; Allred *et al.*, 2006, see also the extensive comparison of RHD model predictions and recent observations in Section 9 of Kowalski *et al.*, 2013).

In addition to hot blackbody emission, the recent spectral observations of dMe flares also show evidence for a secondary continuum component in the NUV at $\lambda \lesssim 3700$ Å. A jump in flux at these wavelengths is not accounted for by an extrapolation of a single hot blackbody component that can be fit to the redder wavelengths at $\lambda > 4000$ Å. The jump in flux relative to the blackbody flux is smaller in the peak phase than in the gradual decay phase, when the color temperature of the hot blackbody component also decreases. A comparison to the F11 RHD model spectra from Allred *et al.* (2006) led to the conclusion that Balmer recombination continuum emission was needed to account for the additional flux in the NUV (Kowalski *et al.*, 2010). However, high-resolution ($R \approx 40000$) Echelle flare spectra show that the NUV continuum from $\lambda \leq 3700$ Å can exhibit the color temperature of a hot blackbody (≈ 11000 K), while not exhibiting any signature of a Balmer discontinuity or edge feature (Fuhrmeister *et al.*, 2008) that is indicative of Balmer recombination continuum emission. Instead, the highest order Balmer lines are broadened and appear to diminish as they converge into a pseudo-continuum at wavelengths between $\lambda = 3646$ Å and the bluest identifiable Balmer line, which is typically H15 λ3 712 or H16 λ3 704 for both large solar and dMe flares (Donati-Falchi, Smaldone, and Falciani, 1984; Hawley and Pettersen, 1991; Kowalski *et al.*, 2010).

The broadening and merging of the high-order Balmer lines in flares is thought to be due to the well-known Stark effect from ambient charges (Švestka, 1963; Donati-Falchi, Falciani, and Smaldone, 1985). The Balmer edge itself is also thought to be broadened, but thermodynamically self-consistent modeling (where the internal partition function converges) of this effect was not available until the occupational probability theory of pressure ionization was developed by Hummer and Mihalas (1988). This improved description of pressure ionization from Stark broadening at the Balmer edge is currently employed in white dwarf atmospheric modeling codes (Tremblay and Bergeron, 2009), but has not yet been implemented in models of solar and stellar flares to determine if the theory can explain the NUV/blue continuum, and in particular, the lack of a Balmer edge. The Stark broadening of the higher order Balmer lines is potentially a useful tool for constraining the environment (*i.e.*, charge density) where the (blackbody) continuum emission is produced in stellar flares (Cram and Woods, 1982).

What atmospheric conditions and atomic processes at high density produce a flare radiation field that self-consistently exhibits all the important continuum components and line properties in the blue and NUV? First, we must relate the observables in blue/NUV spectra to physical parameters of the atmosphere. Then, the spectra can be used as powerful tools to constrain the heating mechanisms that produce these atmospheric conditions. In this article, we explore the possible role of a very dense CC in producing white-light emission during dMe flares, which reveals a meaningful interpretation of the spectra that will be used to constrain targeted modeling of individual solar and stellar flares. These simulations are timely because of new spectral constraints in the blue/NUV and optical wavelength regimes (Kowalski *et al.*, 2013) and modern computational facilities enabling calculation of detailed spectra with sophisticated radiative-transfer codes. Section 2 introduces our method of solution and the starting atmospheric conditions, Section 3 describes the atmospheric dynamics and a new interpretation of the 10^4 K blackbody color temperature, Section 4 reports on the improved modeling of the Balmer edge, Section 5 explores the viability of the model, Section 6 summarizes the study and presents several conclusions, and Section 7 describes future modeling work.

2. Method of Solution and Initial Atmosphere

2.1. NLTE Radiative-Hydrodynamic Modeling with RADYN

We performed 1D radiative-hydrodynamic modeling using the RADYN code (Carlsson and Stein, 1992, 1994, 1995, 1997, 2002). RADYN solves the equations of mass, momentum, internal energy, non-equilibrium level populations, charge conservation, and radiative transfer (see also Abbett and Hawley, 1999). The electron and ion temperatures are equal, and an adaptive grid (Dorfi and Drury, 1987) with 250 grid points resolves shocks in the atmosphere. The NLTE (non-LTE) problem is solved in RADYN using the technique of Scharmer (1981) and Scharmer and Carlsson (1985). The NLTE level populations for three atoms are calculated: a hydrogen atom (six levels including H II), a helium atom (five levels for He I, three levels for He II, and He III), and a singly ionized calcium ion (six levels including the Ca III ground state), giving a total of 22 bound–bound (b–b) and 19 bound–free (b–f) transitions calculated in detail (see Allred *et al.*, 2005). Continua from elements other than H, He, and Ca II are treated as background continua in LTE using the Uppsala atmospheres program (Gustafsson, 1973). A radiative cooling function from CHIANTI (Dere *et al.*, 1997; Young *et al.*, 2003) at 42 temperature points between 10^4 K and 10^8 K is included to account for optically thin cooling. Radiation from several ionization states (*e.g.*, Mg II, Fe II, and Si II) that are most likely optically thick at 10^4 K (Hansteen, 2012) may be important in flares (Avrett, Machado, and Kurucz, 1986), but is not yet included in the models (see Section 7).

Each line profile is calculated with up to 100 frequency points across the transition using a Voigt profile and complete redistribution. For Lyman transitions, the effects of partial redistribution are approximated by truncating the line profiles at ten Doppler widths. A change to the line broadening that is due to the Stark effect is implemented in RADYN compared to the models of Allred *et al.* (2006). Here, we used the analytic formulae (Method 1) from Sutton (1978) to add the Stark damping parameter to the total damping parameter in the Voigt profiles for hydrogen. The Stark broadening of calcium and helium transitions remains the same as in previous versions of the code.

Backwarming from coronal X-ray and extreme-ultraviolet (XEUV) radiation is treated according to the method of Hawley and Fisher (1994), Gan and Fang (1990), and Abbett and Hawley (1999), which approximates the radiation source using a plane-parallel geometry within 1 Mm and as a point source beyond this distance. The emissivity at 41 temperature points from $10^4 - 10^8$ K in 14 wavelength bins from $\lambda = 1 - 2500$ Å is generated with the Astrophysical Plasma Emission Code (APEC; Smith *et al.*, 2001). The hydrogen, helium, and metal cross sections are obtained from CHIANTI.

We ran flare simulations with the starting atmospheric parameters and boundary conditions similar to those used in Allred *et al.* (2006) to facilitate the comparison. A preflare, plane-parallel 1D atmosphere with $\log g = 4.75$ was relaxed to have velocities $\lesssim 10$ m s^{-1} and converged NLTE populations. The upper temperature boundary was initially fixed at 6×10^6 K ($n_e \approx 2.6 \times 10^{10}$ cm^{-3}), and constant non-radiative heating was applied at a column mass m greater than 3.16 g cm^{-2} ($\tau_{5\,000} = 0.06$) to emulate a convective energy flux. The atmosphere was relaxed using a plane-parallel approximation of the X-ray backwarming, as would be appropriate for extended overlying active region loops. The effective temperature of the relaxed atmosphere is 3 630 K, corresponding to that of an early- to mid-M dwarf. The height coordinate (z) of the atmosphere lies along the length of a semicircular magnetic loop of half-length $L_{1/2} = 10^9$ cm (10 Mm), with $z = 0$ at the photosphere ($\tau_{5\,000} = 1.0$, $\log m = 1.14$ g cm^{-2}). A modification to the gravitational acceleration is incorporated near the top of the loop (Abbett and Hawley, 1999); otherwise, the geometry is assumed to be a vertical tube of constant cross-sectional area. The imposed 1D geometry assumes that the plasma motion is confined to the direction of the magnetic field, as would be appropriate for a low plasma β (plasma $\beta = \frac{P_{\mathrm{gas}}}{P_{\mathrm{mag}}}$), but is not necessarily satisfied in the photosphere of the pre-flare atmosphere (the equipartition magnetic field strength is 4.2 kG at the photosphere, consistent with observations Johns-Krull and Valenti, 1996) or in the chromosphere during the flare. Detailed molecular transitions that are important for flux redistribution in an M-dwarf photosphere have not yet been included in the opacity.

2.2. Flare Heating Prescription

We modeled the atmospheric heating from a double power-law, NT electron beam injected at the top of the atmosphere. In this study, we seek to directly connect to the work of Allred *et al.* (2006), who used the thick-target parameters inferred from RHESSI (Lin *et al.*, 2002) hard X-ray spectra during the peak of the 2002 July 23 X4.8 GOES class solar flare (Holman *et al.*, 2003, see also Ireland *et al.*, 2013). A double power-law of electron energies was parameterized by $\delta_l = 3.0$ (the power-law index from the cutoff energy to the break energy), $\delta_u = 4.0$ (the power-law index at energies greater than the break energy), $E_c = 37$ keV (the cutoff energy or lowest energy of the NT electron spectrum), and $E_B = 105$ keV (the break energy between δ_l and δ_u). The value of E_c derived from observations is an upper limit, and $E_c = 37$ keV is moderately high compared to typically assumed or inferred values on the order of 20 keV. We also note that the break in hard X-ray photon spectra (and thus also the inferred double power-law shape of the NT electron spectrum) can be the result of the nonuniform ionization of the ambient atmosphere (Su, Holman, and Dennis, 2011) or from energy losses to the return-current electric field (Holman, 2012). The parameters of the NT electron spectrum are poorly constrained during dMe flares because the hard X-ray emission is too faint to be observed except during the largest flares (Osten *et al.*, 2010). A powerful superflare on the RS CVn HR 1099 is consistent with a hard NT electron spectrum with a single power-law index $\delta \approx 3$ (Osten *et al.*, 2007), similar to the relatively hard double power-law spectrum employed in our simulations.

Free parameters that may not be well-constrained by RHESSI observations alone are the energy flux (Krucker *et al.*, 2011) and the precise time profile (duration) of the flare heating in any given magnetic loop. We extended the range of heating fluxes from Allred *et al.* (2006) to also consider those with 10^{12} erg cm^{-2} s^{-1} (F12) and 10^{13} erg cm^{-2} s^{-1} (F13), in addition to 10^{11} erg cm^{-2} s^{-1} (F11). The F13 flux is not much higher than the highest values that have been derived during solar flares (Neidig, Grosser, and Hrovat, 1994), and such a level has recently been suggested to be relevant to the brightest solar flare kernels (Krucker *et al.*, 2011; Milligan *et al.*, 2014; Gritsyk and Somov, 2014). The duration of the heating in a single magnetic loop in solar flares is constrained in some cases to $\leq 25-60$ s (Wang, Fang, and Ding, 2007; Schrijver *et al.*, 2006) and in other cases may be several minutes (Qiu *et al.*, 2010; Warren, 2006). A wavelet analysis of high-time resolution hard X-ray light curves for 647 solar flares observed with the *Compton Gamma Ray Observatory* (CGRO) indicates a range in the duration of individual bursts from 0.1 s to 50 s (Aschwanden *et al.*, 1998). The shortest burst durations from high-time resolution constraints observed in the NUV in dMe flares range from ≈ 2 s (Robinson *et al.*, 1995) to 10 s (Mathioudakis *et al.*, 2006). For our heating profile, we used a constant energy flux for a duration of 2.3 s, after which the NT electron energy deposition decreases to 0, and the cooling phase is modeled up to 5.1 s for the F13 model (and 30 s for the models with lower beam fluxes). Short heating bursts represent the heating of a single loop, which can be superposed in sequence to emulate an arcade development, as has been done for soft X-ray data of solar flares (Warren, 2006).

The elastic collisions between NT electrons and the ambient charged and neutral components of the atmosphere heat the gas through Coulomb interactions. Following Hawley and Fisher (1994), Abbett and Hawley (1999), and Allred *et al.* (2005, 2006), the NT electron energy deposition rate was included as an additional term in the conservation of energy equation. The Coulomb heating rate as a function of height was modeled here using the analytic, nonrelativistic test-particle approach of Emslie (1978), which was modified to include the height-dependent ionization fraction of the atmosphere in Hawley and Fisher (1994). This prescription is based on the collisional thick-target model (Brown, 1971), but it also includes an approximation for pitch angle scattering of the NT electrons (Emslie, 1978). The inelastic collisions between NT electrons and the ambient neutral hydrogen component of the atmosphere cause direct excitation and ionization (Mott and Massey, 1965); we followed Kašparová *et al.* (2009) and added these NT collisional rates to the thermal rates for hydrogen according to the prescription in Fang, Hénoux, and Gan (1993).[1] The results from Dalgarno and Griffing (1958) were included in these rates to account for secondary ionizations of hydrogen by electrons that are ionized by the beam from the ground state of hydrogen.

3. Properties of the NUV and Optical Continuum Predictions

3.1. Comparison to Observables

Continuum predictions from radiative-transfer modeling are constrained with two simple observational parameters: 1) the Balmer jump ratio and 2) the color temperature of the con-

[1] A comprehensive study of NT rates was presented in the calculations of Ricchiazzi (1982) and Ricchiazzi and Canfield (1983), who found that the H I $n = 1 \rightarrow \infty$ and $n = 1 \rightarrow n = 2$ rates are important, but that the $n = 2 \rightarrow \infty$ rate was negligible compared to the $n = 2 \rightarrow \infty$ thermal rates for their range of beam fluxes.

Table 1 Observables calculated from the model flux spectra. These values have been obtained without subtracting the preflare spectrum.

Model (time step; phase)	χ_{flare}	T_{BB} (K)	Hγ_{EW4170}
F11 ($t = 2.2$ s; impulsive peak)	9.2	5 300	150
F12 ($t = 2.2$ s; impulsive peak)	7.4	5 700	85
F13 ($t = 0.2$ s; impulsive early, rise 1)	6.0	6 300	70
F13 ($t = 1.2$ s; impulsive late, rise 2)	3.3	7 400	40
F13 ($t = 2.2$ s; impulsive peak)	2.0	9 300	20
F13 ($t = 4$ s; gradual decay)	3.3	5 400	60
F13 (average burst)	2.7	7 500	30

tinuum at wavelengths longer than the Balmer edge. These two parameters were referred to as χ_{flare} and T_{BB}, respectively, in a recent homogeneous analysis of 20 flares on dMe stars from K13. A blackbody function was fit to the blue window (BW) wavelength ranges (see K13) of the flux spectrum from $\lambda = 4\,000-4\,800$ Å (the blue-optical wavelength regime) to obtain T_{BB}, while χ_{flare} is the ratio of the flux averaged over 30 Å centered on $\lambda = 3\,615$ Å (blueward of the Balmer jump) to the flux averaged over 30 Å centered on $\lambda = 4\,170$ Å (redward of the Balmer jump). These wavelength ranges are least affected by emission lines while also being near the Balmer jump (see K13). We note that the detected continuum emission shortward of the Balmer jump in K13 also included wavelengths between $\lambda = 3\,420$ Å and $\lambda = 3\,600$ Å. A more secure detection and characterization of this continuum emission would include shorter wavelengths (such as in Heinzel and Kleint, 2014, for a solar flare observed from space), but the signal-to-noise ratio decreases at $\lambda < 3\,600$ Å as a result of the atmospheric cutoff. The distribution of these parameters was found to fall between $\chi_{\text{flare}} = 1.5-4$ and $T_{\text{BB}} = 9\,000-14\,000$ K during the peak phases of nearly all of the flares in the sample from K13. The value of $\chi_{\text{flare,peak}}$ varied with flare type, with the most impulsive flares exhibiting a narrow range of 1.5–2.2 and the gradual flares exhibiting higher values. We measured these same parameters from the flux spectra for the peak phases of the model continua for the range of beam fluxes studied here. The model continua were calculated at a fixed number of points across the NUV and optical, and so we interpolated the continua to the wavelengths used in the observations. We also computed the ratio of the line-integrated, continuum-subtracted Hγ emission line flux to the value of the continuum flux at $\lambda = 4\,170$ Å (hereafter, Hγ_{EW4170}; note that this was referred to as Hγ/C4170 in K13).

The model predictions for χ_{flare}, T_{BB}, Hγ_{EW4170} are given for each NT electron flux in Table 1. For the F13 simulation, the table gives the coarse time evolution of these parameters for the impulsive and gradual phases. The evolution of the flux spectrum for F13 is shown in Figure 1.[2] At early times, the F13 flux spectrum exhibits a high Balmer jump ratio and a cool color temperature in the blue-optical. However, from $t = 1.2$ s to 2.2 s, the continuum at $\lambda > 3\,646$ Å becomes bluer as the flux at $\lambda = 4\,300$ Å increases significantly, while the flux at $\lambda \approx 6\,690$ Å increases only marginally. This results in an increase in the blue-optical color temperature, T_{BB}, of 2 000 K. The Balmer jump ratio also decreases from 3.3 to 2 as the flux at $\lambda \approx 3\,550$ Å decreases, while the flux at $\lambda = 4\,300$ increases. The continuum at wavelengths shorter than the Balmer edge also becomes bluer over this time interval as the flux at $\lambda < 3\,000$ Å increases, while the flux at $\lambda \approx 3\,550$ Å decreases. In Section 3.3, we explain how these changes in the flux spectrum are related to the increasing hydrogen b–f opacities in the atmosphere.

[2] The flux spectra for the F13 simulation are available upon request.

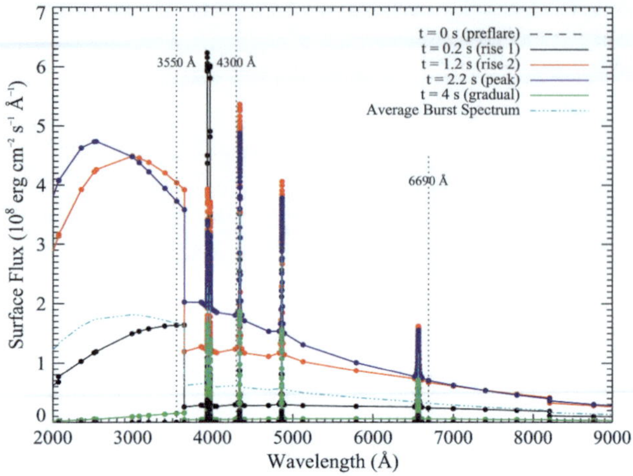

Figure 1 The evolution of the NUV and optical flux spectra during the F13 simulation. The observable parameters extracted from these spectra are given in Table 1. The coadded (average) burst spectrum is discussed in Section 5. The preflare spectrum is not visible on this scale. The vertical dotted lines indicate the NUV ($\lambda = 3\,550$ Å), blue-optical ($\lambda = 4\,300$ Å), and red-optical ($\lambda = 6\,690$ Å) wavelengths used in the continuum analysis (Section 3.3).

At $t = 2.2$ s, the F13 simulation produces a flux spectrum that falls into the regime of the observations for the peak phase of dMe flares with a relatively low Balmer jump ratio, $\chi_{\text{flare}} = 2.0$, and a blue-optical color temperature of $T_{\text{BB}} = 9\,300$ K. The F13 model NUV flux between $\lambda = 3\,000 - 3\,646$ Å also exhibits the color temperature of a hot blackbody with $T = 11\,500$ K. Although the signal-to-noise ratio of spectra at NUV wavelengths is lower than that of the data at blue-optical wavelengths (due to the atmospheric cutoff), the color temperature in the NUV is generally consistent with the value derived at blue-optical wavelengths (*cf.* Figures 8 – 9 of K13). The $\text{H}\gamma_{\text{EW4170}}$ also falls within the range of the observed values at flare peak (*cf.* Figures 11 – 12 of K13). Blue-optical emission with a color temperature higher than $\approx 6\,000$ K has not previously been self-consistently reproduced in RHD models of dMe flares (K13). The F11 and F12 simulations do not produce properties that are consistent with observations of the impulsive phase: the Balmer jump ratios are too high (> 7) and the color temperatures are too low ($< 6\,000$ K). From Table 1, it can be seen that the time-evolution of the characteristics of the F13 simulation is consistent with the general time-evolution observed in dMe flares: the Balmer jump and $\text{H}\gamma_{\text{EW4170}}$ decrease in the rise phase and increase in the decay phase. The timescale of the burst (5 s), however, is much shorter than the timescales of most dMe flares (minutes to hours). Adding up sequentially heated magnetic strands as in the spatial development of a flare arcade (*e.g.*, Warren, 2006; Qiu *et al.*, 2010) is beyond the scope of this work, but we intend to pursue this in a future paper.

3.2. Dynamics and Energy Balance in the F13 Atmosphere

The evolution of the F13 flux spectrum in Figure 1 is tied to the dynamical evolution of the atmosphere. We discuss here general properties of the hydrodynamics.

The NT electrons initially deposit their energy over a broad extent, with significant volumetric heating from 275 km ($\log m = -2.1$ g cm^{-2}) to the top of the chromosphere at 450 km ($\log m = -4.05$ g cm^{-2}), and there is non-negligible NT electron energy deposition (300 erg s^{-1} cm^{-3}) directly in the corona. The maximum energy deposition from the NT electrons (8×10^5 erg s^{-1} cm^{-3}) initially occurs at $z_{\text{o}} = 390$ km ($\log m = -3.05$ g cm^{-2}), which is 99 % neutral, but has a significant electron density $n_e = 10^{12}$ cm^{-3}. Most of the NT energy is initially deposited into NT hydrogen ionization at z_{o}. As the electron density increases at z_{o}, an increasing fraction of the NT energy goes into raising the temperature since heating of electrons occurs more efficiently than heating of neutrals (note that the thermalization of electrons, protons, and neutrals is assumed to occur instantaneously in these models).

Hydrogen becomes completely ionized by $t = 0.002$ s and thermal ionization increases over NT ionization as the temperature exceeds 20 000 K. He I then becomes the primary cooling agent as the temperature reaches 40 000 K. At $t = 0.003$ s, the temperature approaches 65 000 K and most helium at this height is now singly ionized. From $t = 0.003$ s to $t = 0.01$ s the He III fraction increases from 2 % to 99.6 % at $z_X = 425$ km as the temperature rapidly increases to 140 000 K. At $t = 0.016$ sec, a localized temperature maximum of 180 000 K forms at z_X, and the thermal pressure generates upward velocities of 4 $\mathrm{km\,s^{-1}}$ and downward velocities of 2 $\mathrm{km\,s^{-1}}$. These mass motions are the seeds to the explosive condensation and evaporation that follow from runaway heating as material is evacuated from this region of the atmosphere.

Meanwhile at $t = 0.016$ s, between $z = 450$ and $z = 460$ km (at the preflare transition region heights), there is a strong upward flow of 30 $\mathrm{km\,s^{-1}}$ resulting from the increase in temperature to 90 000 K of lower density material (previously the upper chromospheric material) between the site of maximum beam heating (425 km) and the transition region (450 km). At $t = 0.2$ s, the evaporation consists of two fronts: a high-density chromospheric front with a maximum velocity of 125 $\mathrm{km\,s^{-1}}$ and a low-density transition region front with a maximum velocity of 250 $\mathrm{km\,s^{-1}}$. The downward front is a high-density chromospheric condensation (CC), with a maximum velocity of 95 $\mathrm{km\,s^{-1}}$. The maximum densities of the chromospheric fronts have increased from $n_H \approx 2 \times 10^{14}$ $\mathrm{cm^{-3}}$ ($t = 0.2$ s) to $n_H \approx 5 \times 10^{14}$ $\mathrm{cm^{-3}}$ ($t = 0.4$ s) and the temperatures have decreased from 50 000 K to 37 000 K. At the location of z_X, the temperature has reached nearly 10 MK by $t = 0.4$ s from the runaway NT electron heating as the density has decreased (material is evacuated) from $n_H = 5 \times 10^{13}$ $\mathrm{cm^{-3}}$ ($t = 0$ s) to $n_H = 6 \times 10^{12}$ $\mathrm{cm^{-3}}$ ($t = 0.4$ s). At the upper and lower ends of the transition from very hot ($T \approx 400\,000$ K – 10 MK) plasma to $T \approx 35\,000$ K plasma, upward and downward propagating *shock fronts* are located at the very steep (nearly discontinuous) jumps in gas pressure, where the He III fraction changes from 100 % to < 30 %.[3]

At $t = 0.33$ s, the time steps of the simulation become unmanageably small ($\Delta t \approx$ 10 – 100 ns) due to the precision requirements of RADYN in calculating the population levels of He I and He II in regions of strong temperature gradient, especially at the upper evaporation fronts. At $t = 0.478$ s, we decrease the accuracy of the minor level populations, with accurate solutions provided only for level population densities that are $> 10^{-9}$ of the total elemental population density (initially, the parameter is 10^{-12}), while the second derivative of the grid weights is increased to weight more toward the higher density (CC) front. The result is a dramatic increase in the time-step size until $t = 5.1$ s as the low-density (upper) chromospheric evaporation front is smoothed over (the number of grid points here drastically decreases and the temperature is no longer maintained at 35 000 – 50 000 K). The results for the simulation at $z > 450$ km are thus less accurate, but this approximation is required to compute 1D simulations with very steep and dynamic temperature and density gradients. We focus only on the radiating material in the CC and below, since this higher density material produces most of the optical emission; the model above 450 km should not affect the analysis at the lower heights.

Figure 2 summarizes the dynamical evolution of the lower atmosphere at $t = 2.2$ s, at the maximum white-light emission. The top panel shows the velocity of the downward-moving CC. At this time in the CC, the maximum hydrogen density is 7.6×10^{15} $\mathrm{cm^{-3}}$,

[3]The speed of sound in the CC at $t = 0.4$ s is 30 $\mathrm{km\,s^{-1}}$, whereas material is traveling at speeds of nearly 100 $\mathrm{km\,s^{-1}}$.

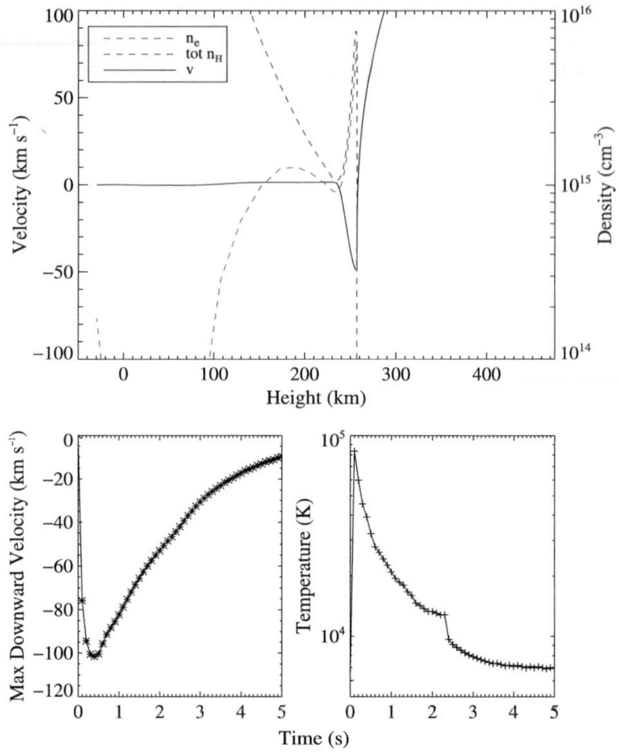

Figure 2 Top: The velocity field in the lower atmosphere at $t = 2.2$ s shows a chromospheric condensation (CC) with a high hydrogen and electron density exceeding 5×10^{15} cm^{-3}. Bottom left: The time-dependence of the velocity in the highest density zone in the CC. Bottom right: The time-dependence of the temperature in the highest density zone in the CC. The velocity of the CC changes smoothly through the burst, while the temperature reflects the cutoff of the NT electron energy deposition at $t = 2.3$ s.

or 17.4×10^{-9} g cm^{-3}, which is over a factor of 12 greater than the hydrogen density at this height ($z = 255.4$ km) before the flare. This mass density is comparable to the preflare density at $z \approx 145$ km, which lies between the temperature minimum and the lower chromosphere. The evolution of the speed and temperature of the maximum density zone in the CC is shown in the bottom panels. By $t = 2.2$ s, the speed of the downflowing material has decreased to ≈ 50 km s^{-1} as it encounters denser material and higher gas pressure lower in the atmosphere, the equation of motion for which was derived analytically by Fisher (1989). During the simulation, the temperature of the region of maximum density in the CC decreases steadily; this is a result of the higher hydrogen emissivity for lower temperatures, while increasing the density in the CC gives a higher volumetric cooling rate. Eventually, the temperature nearly settles between 12 500 and 13 000 K at $t = 2.2$ s as continuous injection of NT electrons deposit energy in the CC. After the NT electron energy deposition decreases to zero at $t = 2.3$ s, the CC continues to propagate downward, increasing its density to 1.3×10^{16} cm^{-3} at $t = 4.0$ s. The temperature drops to 7 000 K at this time (when the NT energy input is cut off), which is much higher than the preflare temperature of $\approx 5\,000$ K at this height (196.5 km).

At $t = 2.2$ s, the $E \approx 65-200$ keV NT electrons encounter a thick-target wall at the CC and produce a maximum energy deposition rate of 5×10^7 erg cm^{-3} s^{-1} (compared to a maximum of $< 10^6$ erg cm^{-3} s^{-1} initially) at $z = 256.9$ km; the full width at half maximum (FWHM) of the energy deposition rate is only 0.46 km (compared to 75 km initially). The simulation becomes computationally problematic shortly after this time, and therefore the NT heating rate is cut off at 2.3 s. The energy budget around the CC is shown in Figure 3, which gives the individual contributions to the change in total internal energy *per* unit mass.

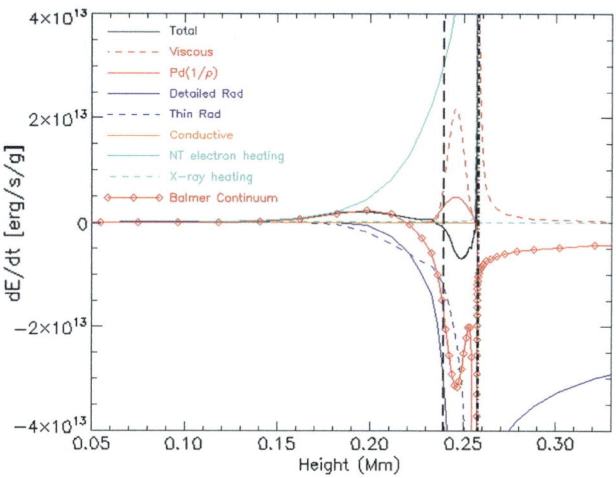

Figure 3 Contributions to the energy balance for $t = 2.2$ s in the F13 simulation. These curves show the *net* amount for each component. For the detailed radiation, the heating and cooling are integrated over the wavelengths of the transitions. The Balmer continuum (red diamonds) heating/cooling is shown as a separate curve from the total detailed radiation curve (which includes the Balmer continuum). The CC, with velocities lower than $5\ \mathrm{km\,s^{-1}}$, is material between $z = 239.4$ km (long dashed vertical black line) and 257.4 km (thick vertical dash-dotted line).

Positive quantities contribute toward heating, excitation, and ionization; negative quantities contribute toward cooling, recombination, and de-excitation. The lower end of the CC from Figure 2 at 239.4 km is indicated by a vertical long-dashed line in Figure 3. In the CC, the NT electron energy deposition is balanced by the thin and thick (detailed) radiative losses, resulting in a net decrease in the internal energy (recombination and cooling). At deeper layers than the CC, $E \gtrsim 200$ keV NT electrons can still penetrate and heat the atmosphere. At $z \approx 200$ km, there is net ionization and heating from both Balmer continuum backwarming (red diamonds) and NT electron energy deposition. At these depths, radiation from the optically thin loss function and from the detailed transitions does not counterbalance these sources of energy. There are also some contributions from compressional and viscous work where the CC impacts and accretes the stationary material below it.

3.3. Properties of the Contribution Function for Continuum Emission in the Flare Atmosphere

The CC decreases in temperature and increases in density over time (Figure 2). By $t = 2.2$ s the CC has attained temperatures $> 12\,000$ K and increased in density over 12 times the preflare density, thus producing a large amount of radiative cooling from hydrogen and other elements (Figure 3). In this section, we explore the height-dependent properties of the emergent intensity from the CC at three continuum wavelengths calculated in detail: $\lambda = 3\,550$ Å in the NUV, $\lambda = 4\,300$ Å in the blue-optical, and $\lambda = 6\,690$ Å in the red-optical, which are indicated in Figure 1. In Section 3.1, we found that the NUV and blue-optical F13 model peak spectrum (blue spectrum in Figure 1) is described by two blackbody spectra with color temperatures of $T = 11\,500$ K between $\lambda = 3\,000 - 3\,646$ Å and $T = 9\,300$ K at $\lambda > 4\,000$ Å in relative amounts given by the value of the Balmer jump ratio (≈ 2). In this section, we use the contribution function to the emergent intensity (Magain, 1986; Carlsson, 1998) to explain the origin and characteristics of the white-light continuum emission in

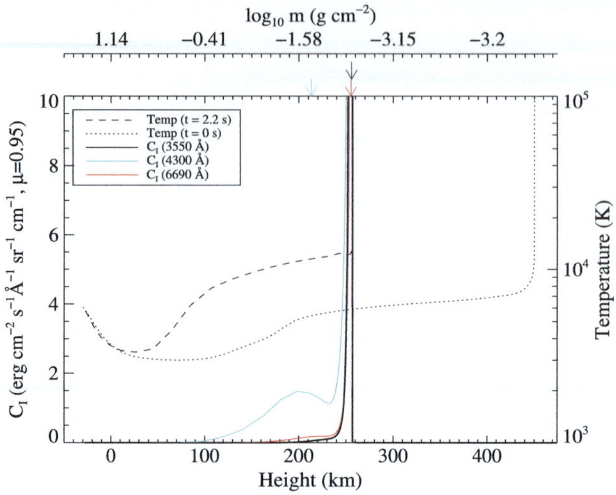

Figure 4 The contribution to the emergent intensity for three continuum wavelengths in the NUV, blue, and red at $t = 2.2$ s compared to the temperature structure (for the same height range as in Figure 2). The column mass scale for $t = 2.2$ s is shown at the top axis. The arrows indicate where $\tau = 1$ occurs for each wavelength. The contribution functions in the CC ($z \approx 239-257$ km) are extremely thin and extend off the top of the plot (see text).

this model spectrum. We use the following form of the contribution function, $C_I = \mathrm{d}I_\lambda/\mathrm{d}z$ (Leenaarts *et al.*, 2013), for a plane-parallel atmosphere:[4]

$$C_I = \frac{\mathrm{d}I_\lambda}{\mathrm{d}z} = \frac{\chi_\nu}{\mu} S_\nu \mathrm{e}^{-\tau_\nu/\mu} \frac{c}{\lambda^2}, \tag{1}$$

where χ_ν is the height-dependent extinction coefficient (in units of cm^{-1}; not to be confused with χ_{flare} the Balmer jump ratio), τ_ν is the optical depth ($\int_z^{10^9\ \mathrm{cm}} \chi_\nu\, \mathrm{d}z$), and S_ν is the height-dependent source function. More simply, this equation is $\frac{j_\nu}{\mu}\mathrm{e}^{-\tau_\nu/\mu}\frac{c}{\lambda^2}$, or the plasma emissivity (j_ν) at a given height and frequency multiplied by the attenuation of the radiation (determined by $\mathrm{e}^{-\tau_\nu}$) as it propagates outward in the direction of μ from that height.[5] We discuss the atomic processes that contribute to the continuum emissivity in Section 3.4.

In Figure 4, we show the contribution function to the emergent vertical intensity ($\mu \approx 1$) for $\lambda = 3\,550$ Å, $4\,300$ Å, and $6\,690$ Å at $t = 2.2$ s. The temperature profile (dashed line) of the flare atmosphere is qualitatively similar to that of the quiescent atmosphere, except that the flare chromosphere is hotter and has moved to higher column mass. Radiation escapes from two principal regions of the flare atmosphere: a stationary component (see Figure 2) produces a wide maximum in C_I at $z \approx 200$ km in the mid-/upper- flare chromosphere below the CC, while the CC produces a very thin component peaking at $z \approx 256.3$ km with maxima in C_I of 950, 140, and 120 erg cm^{-2} s^{-1} Å^{-1} sr^{-1} cm^{-1} (extending above the top of the figure) for $\lambda = 3\,550$, $4\,300$, and $6\,690$ Å, respectively. Lower values of z (defined in Section 2) refer to larger physical depths, whereas larger values of Δz refer to larger physical depth *ranges* in the atmosphere. We measured the FWHM of the narrow component of C_I to compare the physical depth ranges (Δz_{FWHM}) over which radiation escapes from the CC. The Δz_{FWHM} is the largest for $\lambda = 4\,300$ Å: the Δz_{FWHM} values are 1 km, 1.75 km,

[4]The differential $\mathrm{d}I_\lambda/\mathrm{d}z$ in the definition of the contribution function is the same as the differential describing the variation of the specific intensity with height in the radiative-transfer equation; however, these differentials are not equivalent.

[5] j_ν is often used to describe the emissivity in units of photons cm^{-3} s^{-1} sr^{-1} Hz^{-1}, whereas η_ν is used for the emissivity in units of ergs cm^{-3} s^{-1} sr^{-1} Hz^{-1}; here, we use j_ν to describe the emissivity in units of ergs cm^{-3} s^{-1} sr^{-1} Hz^{-1}.

Figure 5 The evolution of column density for $n = 2$ and $n = 3$ of hydrogen (left axis, dashed lines) and optical depth (right axis, solid lines) for atmospheric heights in the CC with velocities $< -5\ \mathrm{km\,s^{-1}}$. At $t = 2.2$ s, the CC has an extent of 18 km from 239 km to 257 km and has temperatures > 12 000 K. Approximately $3.5 \times 10^{12}\ \mathrm{erg\,cm^{-2}\,s^{-1}}$ of NT electron energy flux is being deposited within the CC at this time.

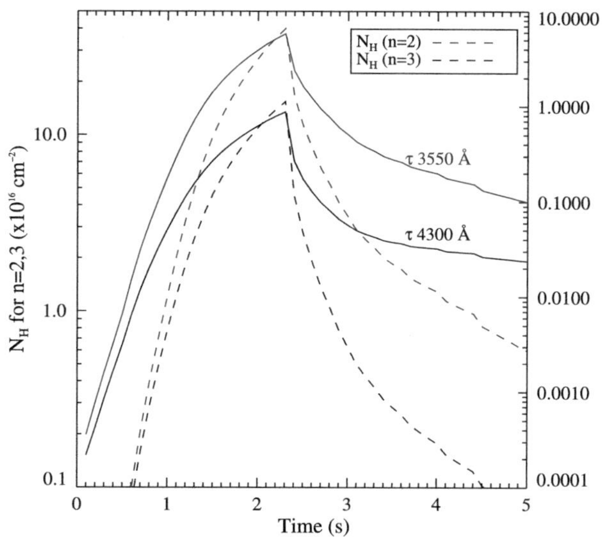

and 1.25 km for $\lambda = 3\,550$, $4\,300$, and $6\,690$ Å, respectively. The contribution function in the stationary component of the flare atmosphere is also markedly different among these three wavelengths. Whereas $\lambda = 4\,300$ Å has a local maximum at 200 km and 20–25 % of the emergent intensity originates from layers below the CC at $z < 230$ km, the other two wavelengths have very little emergent intensity originating from these depths, with $\lambda = 3\,550$ Å showing the least. The arrows on the top of the figure indicate the layers at which $\tau_\lambda = 1$ for each wavelength. The $\tau_\lambda = 1$ layer occurs in the CC for $\lambda = 3\,550$ and $6\,690$ Å, and in the stationary flare layers for $\lambda = 4\,300$ Å. Therefore, the intensity at $\lambda = 4\,300$ Å originates over the largest physical depth range (largest Δz) and from the lowest atmospheric heights due to the smallest optical depth in the CC.

At times earlier than 2.2 s, a larger fraction of the emergent intensity at $\lambda = 3\,550$ Å and $6\,690$ Å originates from the stationary component of the flare atmosphere because the CC has a lower density and optical depth. The evolution of the optical depth for $\lambda = 4\,300$ Å and $3\,550$ Å and the evolution of the column densities (N) for hydrogen in $n = 2$ and $n = 3$ in the CC (where $v < -5\ \mathrm{km\,s^{-1}}$) are shown in Figure 5. As the CC becomes denser and attains temperatures that excite hydrogen to $n = 2$ and $n = 3$, the probability of photoionization by light with these wavelengths increases. The values of the column densities at $t = 2.2$ s of $N_\mathrm{H}(n = 2) > 3 \times 10^{17}\ \mathrm{cm^{-2}}$ and $N_\mathrm{H}(n = 3) > 10^{17}\ \mathrm{cm^{-2}}$ define useful requirements[6] for large optical depths ($\tau_{4\,300} = 0.8$, $\tau_{3\,550} = 5$) in any future flare model of CCs with material near temperatures of 12 000 K.

The evolution of the contribution function results from increasing, wavelength-dependent optical depths at optical/NUV wavelengths and can explain the evolution of the flux spectrum in Figure 1 (Section 3.1). The very large and continuously increasing optical depth at $\lambda = 3\,550$ Å causes a *decrease* in the flux at this wavelength between $t = 1.2$ s and $t = 2.2$ s because the radiation can only escape from a very thin layer. At $t = 2.2$ s, the radiation at $\lambda = 4\,300$ Å is still relatively optically thin in the CC ($\tau < 0.8$) and therefore has a higher

[6] See Section 3.4, there are additional opacities from photoionization of the higher levels of hydrogen, hydrogen free–free absorption, and opacity from the $\mathrm{H^-}$ ion that can also contribute towards the optical depth at these wavelengths.

probability of escape, giving rise to a larger FWHM of the contribution function in the CC (Δz_{FWHM}) and also a larger fraction that can escape from the stationary component of the flare atmosphere. The flux at 4 300 Å increases from $t = 1.2$ s to 2.2 s in Figure 1 and results from the increasing density but a lower optical depth due to the lower photoionization cross-section and lower population density for $n = 3$ than for $n = 2$. The greater optical depth at $\lambda = 6\,690$ Å relative to $\lambda = 4\,300$ Å is caused by a larger photoionization cross-section (but same population density), which leads to a much smaller increase in flux at this wavelength in Figure 1 from $t = 1.2-2.2$ s. The NUV flux significantly increases at $\lambda < 3\,000$ Å, and the color temperature at $\lambda = 3\,000-3\,646$ Å becomes higher over this time interval due to similar reasoning for the increase of the flux at $\lambda = 4\,300$ Å and the increase of the color temperature at $\lambda > 4\,000$ Å. Flare spectra with higher signal-to-noise ratios at $\lambda < 3\,600$ Å would help to rigorously test the slope changes predicted by these models at NUV wavelengths.

3.4. Opacity Components in the Optical Depth

3.4.1. *The Optically Thick Hydrogen-Recombination Flare Spectrum for the F13 Model*

In this section, we quantitatively demonstrate that the properties of the flux spectrum at $t = 2.2$ s in Figure 1 primarily arise from spontaneous hydrogen-recombination continuum emission by reconstructing the contribution function to the emergent intensity from atomic emissivity processes. The b–f opacity (χ_ν) corrected for stimulated emission is given by Mihalas (1978, Equation 7-1)

$$\chi_\nu = \Sigma_i \bigl\{ \bigl[n_i - n_i^* \mathrm{e}^{-h\nu/kT} \bigr] \sigma_i(\nu) \bigr\}, \tag{2}$$

where σ_i is the photoionization cross-section of an atom with an electron in level i for light of frequency ν; $n_i^* = n_c \times (n_i/n_c)^*$ where n_c is the number density of protons and $(n_i/n_c)^*$ is the density ratio in LTE, which can be calculated using the Saha equation. We calculated the hydrogen b–f and hydrogen free–free (f–f, bremsstrahlung) opacity at $\lambda = 3\,550$ Å, 4 300 Å, and 6 690 Å for the flare atmosphere at $t = 2.2$ s. The total hydrogen continuum (b–f + f–f) opacity values for $\lambda = 4\,300$ and 3 550 Å are shown in the top left panel of Figure 6. The hydrogen b–f opacity dominates for both wavelengths, but the hydrogen f–f is non-negligible in the CC at optical wavelengths. At $\lambda = 4\,300$ Å, the Paschen b–f ($n = 3 \rightarrow \infty$) continuum opacity is highest, whereas at $\lambda = 3\,550$ Å the Balmer b–f ($n = 2 \rightarrow \infty$) continuum opacity dominates. The b–f and f–f opacities from the H^- ion are not shown explicitly in Figure 6, but are included in RADYN and in the contribution function (Figure 4). The H^- opacity is low (≈ 3.5 %) compared to the hydrogen b–f and f–f opacities in the CC, but dominates at the lower temperatures ($T < 9\,000$ K) in the flare atmosphere at $z \lesssim 130$ km. Although the electron density is high in the CC, Thomson scattering is relatively very small.

Most of the CC is so dense that the LTE condition $S = B$ (where B is the Planck function) applies and the source function is only dependent on local quantities. The level populations at $t = 2.2$ s for $n = 2$ and $n = 3$ are equal to the LTE populations, and collisional rates are generally higher than radiative rates for most of the flare atmosphere ($z = 100-256.6$ km) except at the surface of the CC. The level population n_p is the LTE value through all of the CC. The hydrogen b–f and f–f continuum emissivity, j_λ, is shown in the top right panel of Figure 6. For simplicity, only the spontaneous recombination emission is shown. The relative rates of induced to spontaneous recombination are very low at these wavelengths, but can become higher at redder wavelengths. The hydrogen emissivity can be approximated as

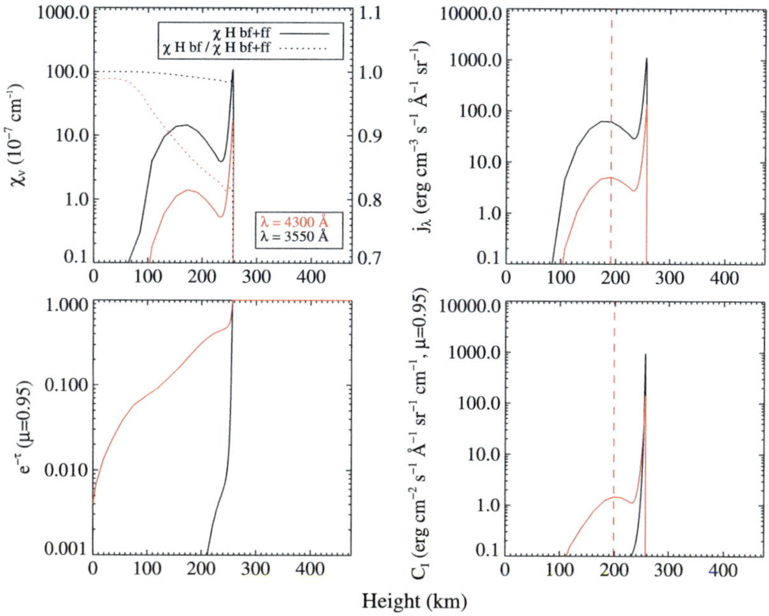

Figure 6 The contribution function (bottom right) for 4 300 Å and 3 550 Å is broken up into three parts: opacity (top left), emissivity (top right), and the attenuation (bottom left). The ratios of hydrogen b–f opacity to total hydrogen (b–f and f–f) opacity are shown in the top left panel (right axis). In the stationary flare layers ($z \approx 190$ km), the emissivity at $\lambda = 3\,550$ Å is high, but the attenuation is very high as well due to photoionizations of hydrogen from $n = 2$ in the higher layers of the CC. The peaks of the emissivity and contribution function in the stationary layers for $\lambda = 4\,300$ Å are indicated by vertical dashed lines (see text).

$\{\chi_{\nu,\text{b–f}} + \chi_{\nu,\text{f–f}}\} B_\nu \frac{c}{\lambda^2}$ through most of the CC; but to account for departures from LTE at the surface of the CC, we used the NLTE hydrogen b–f emissivity from Equation 7-2 of Mihalas (1978). The ratio of emissivities ($j_{\lambda=3\,550}/j_{\lambda=4\,300}$) in the CC is about 9.5, as we expect for hydrogen-recombination (Balmer and Paschen) continuum emission at these temperatures. When τ is non-negligible, the wavelength-dependent attenuation becomes important in the contribution function (Equation (1)). The attenuation factor, $e^{-\tau_\lambda}$, is shown in the bottom left panel of Figure 6 and illustrates that the $\lambda = 3\,550$ Å emissivity is preferentially attenuated at $z = 180-200$ km, where $\tau_{3\,550} = 20$. Here, $\tau_{4\,300} \approx 1$ which is large enough to attenuate some of the radiation from these heights (compare the shift in peaks at this height indicated by the vertical dashed lines in the top right and bottom right panels). The $e^{-\tau_\lambda}$ factor gives rise to the wavelength-dependent changes in the FWHM of C_I (Δz_{FWHM}) and in the fraction of the emergent intensity that originates from the stationary component (Section 3.3 and Figure 4).

The contribution function, C_I, is shown again in the bottom right panel of Figure 6 (note the logarithmic y-scale as compared to Figure 4, which is on a linear scale). Integrating C_I over height gives the emergent intensity at $\mu = 0.95$. The integration over height reveals that the Balmer jump intensity ratio ($I_{\lambda=3\,550}/I_{\lambda=4\,300}$, similar to χ_{flare}) is $\approx$2.2, a value indicative of the small Balmer jump in emission in the flux spectrum in Figure 1. The emergent optical intensity ratio $I_{\lambda=4\,300}/I_{\lambda=6\,690}$ of 2.3 is a value consistent with the color temperature of a hot ($\approx$9 000 K) blackbody. We have thus shown that hydrogen-recombination (and bremsstrahlung) continuum emissivity between 11 000 – 14 000 K integrated over heights where optical depths become large ($\tau_{4\,300} \gtrsim 0.5$) and are wavelength de-

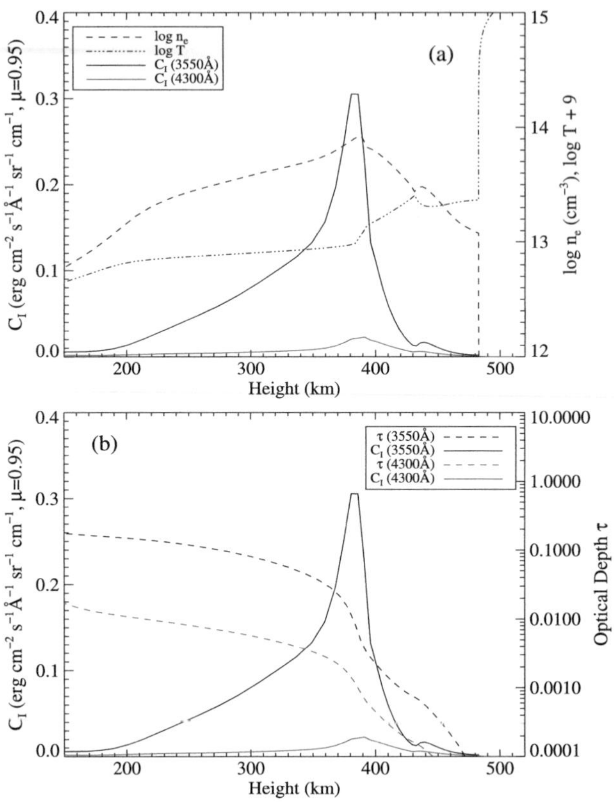

Figure 7 The contribution function for the F11 model at $\mu = 0.95$ and $t = 2.2$ s compared to the (a) electron density and temperature and (b) optical depth. We only show heights corresponding to the flare chromosphere, transition region, and lower corona.

pendent (from the hydrogen cross-section dependence and thermal population density variation) can self-consistently generate the color temperature (T_{BB}) of the blue-optical emission and the Balmer jump ratio (χ_{flare}), in agreement with observations of the peak phases of dMe flares.

3.4.2. The Optically Thin Hydrogen-Recombination Flare Spectrum for the F11 Model

The contribution functions for the F11 model at $t = 2.2$ s are shown in Figure 7(a–b). At this time, the atmosphere has not yet experienced explosive evaporation, which is consistent with the F11 simulation of Allred *et al.* (2006). In contrast to the F13 simulation, the C_I profiles for $\lambda = 3\,550$ Å and 4 300 Å during the F11 simulation extend over a large portion of the flare chromosphere. The C_I profiles also have second peaks near the preflare photosphere, which is not shown here because the important molecular species and opacities in M-dwarf photospheres have not yet been included in RADYN (Section 2). The optical depths for 4 300 Å and 3 550 Å do not reach values of 1 until heights near the pre-flare photosphere; the values of $\tau_{3\,550}$ in the flare chromosphere range between 0.001 and 0.1, with $\tau_{4\,300}$ about an order of magnitude lower. Therefore, the NUV and optical radiation from the flare chromosphere is dominated by optically thin hydrogen Balmer and Paschen continua emission from recombination due to the increased electron density (and temperature). The emergent F11 NUV and optical spectrum is a combination of mostly unattenuated hydrogen-recombination emission from $z = 250-400$ km, combined with any increased photospheric radiation from backwarming (Allred *et al.*, 2006). The value of χ_{flare} is high (≈ 9; Table 1) and T_{BB} is low ($\approx 5\,300$ K; Table 1), as expected for hydrogen recombination at $T \approx 10\,000$ K with negligible wavelength-dependent attenuation.

If we use the formula for spontaneous thermal hydrogen recombination emissivity (Equation 7-2 of Mihalas, 1978, see also Aller, 1963 and Wiese, Kelleher, and Paquette, 1972) in Equation (1), we obtain an approximate formula for the ratio of the contribution function at two wavelengths, λ_1 and λ_2:

$$\frac{C_I(\lambda_1)}{C_I(\lambda_2)} \approx \frac{\lambda_2^2 n_2^3 g_{\text{bf},1}}{\lambda_1^2 n_1^3 g_{\text{bf},2}} \mathrm{e}^{\frac{hc}{k_{\text{B}} T_e}(\frac{1}{\lambda_2} - \frac{1}{\lambda_1})} \mathrm{e}^{1.578\times 10^5 / T_e (\frac{1}{n_1^2} - \frac{1}{n_2^2})} \mathrm{e}^{(\tau_2 - \tau_1)/\mu} \tag{3}$$

at height z with (T_e, n_e), where g_{bf} are the Gaunt factors and n_1 and n_2 are the principal quantum numbers that dominate the opacity in Equation (2) for λ_1 and λ_2, respectively.[7] When $\tau_{1,2}$ are negligible (as for the F11 in the flare chromosphere), $C_I(\lambda_1 = 4\,300)/C_I(\lambda_2 = 6\,690)$ for $\mu = 1$ is ≈ 0.8 at $T = 10\,000$ K and $\mathrm{e}^{\tau_2 - \tau_1}$ is not much larger than 1, which causes an optically thin Paschen recombination continuum to exhibit a color temperature of $\approx 5\,000$ K (Kunkel, 1970; Kerr and Fletcher, 2014, ; Section 9 of K13). When $\tau_{1,2}$ are non-negligible (as for the F13 at $t = 2.2$ s, Section 3.4.1), the ratio of the attenuation factors is $\mathrm{e}^{\tau_2 - \tau_1} > 1$, which causes the optical continuum to appear hotter. Therefore, increasing the optical depth makes the Paschen continuum bluer with a higher value of T_{BB} (apparently hotter), as can be seen by approximating the atmosphere as a slab with constant (T_e, n_e) and integrating over height. The correction of the b–f opacity for stimulated emission (second term in Equation (2)) also affects the color temperature because it decreases the opacity at $\lambda = 4\,300$ Å by a smaller factor than at $\lambda = 6\,690$ Å. However, the actual emergent intensity is an integration over height with varying (T_e, n_e), which is why we modeled the atmosphere gradients, which are shown in Figure 2. The emergent radiative flux is a Gaussian integration of intensity over five μ values: the intensity at a lower value of μ gives an even bluer optical continuum and lower Balmer jump ratio because of the $\mathrm{e}^{(\tau_2 - \tau_1)/\mu}$ factor.

4. Stark Broadening at the Balmer Edge

Stark broadening is thought to be an important symmetric broadening mechanism of the hydrogen Balmer lines during solar and stellar flares (Švestka, 1963; Worden *et al.*, 1984; Hawley and Pettersen, 1991; Johns-Krull *et al.*, 1997; Paulson *et al.*, 2006; Allred *et al.*, 2006), in addition to possible turbulent (Eason *et al.*, 1992; Doyle *et al.*, 1988) and thermal contributions. The linear (first order) Stark effect is the splitting of the degenerate orbital angular momentum (l) states of each principal quantum state (n) of hydrogen, due to a net electric microfield from the surrounding distribution of charges. To date, the most complete theories of the linear Stark effect are those of Vidal, Cooper, and Smith (1970, 1971, 1973), Seaton (1990), Stehle and Jacquemot (1993), and Kepple and Griem (1968), which take into account the perturbations from the slowly moving (quasi-static) protons and from the rapidly moving (dynamic) electrons. The Vidal *et al.* treatment is known as the unified theory because it provides an accurate treatment of the electron perturbations from the line core through the far wings. Theories of Stark broadening typically only present results for the lowest order Balmer transitions, but Lemke (1997) extended the results from the Vidal *et al.* unified theory to $n = 22$. Therefore, this Vidal *et al.* unified theory (which is considered the

[7] The value 1.578×10^5 in the second exponential is 13.598 eV/k_{B}.

best treatment of the non-overlapping far line wings and of the higher order Balmer lines; see the discussion in Johns-Krull *et al.*, 1997, and in Vidal, Cooper, and Smith, 1970) is most useful to the interpretation of flare spectra in the Balmer jump region.[8]

In future work (Section 7), we will implement the Lemke (1997) extension of Vidal *et al.* results (see Section 7). In RADYN, the analytic approximations of the Stark damping from Sutton (1978) are added to the total damping in the Voigt profile (Section 2). Solar modeling of infrared hydrogen lines has shown that this is adequate (Carlsson and Rutten, 1992).

4.1. Occupational Probability of Level *n*

Here, we address the atomic physics necessary to interpret the merging of the rapidly converging higher order Balmer lines and the apparent lack of a Balmer discontinuity (or edge) at $\lambda = 3\,646$ Å in flare spectra.

To model the Stark broadening of the Balmer edge, we applied the occupational probability theory developed in Hummer and Mihalas (1988) and Dappen, Anderson, and Mihalas (1987), which is summarized thoroughly in the recent modeling study of cool DA white-dwarf atmospheres of Tremblay and Bergeron (2009). This theory provides a thermodynamically self-consistent approach to pressure ionization, which allows the internal partition function to converge. Unlike procedures that truncate the internal partition function, occupational probability theory includes a continuous transition from bound to free states as upper levels of hydrogen converge and Stark energy shifts cause them to overlap. Electric field ionization experiments with sodium have shown that enhanced ionization rates of electrons in level n occur when there is a *critical* electric microfield strength causing the highest Stark state of n to cross with the lowest Stark state of $n+1$ (Pillet *et al.*, 1984). At a level crossing, electrons can undergo a Landau–Zener transition from n to $n+1$ as the electric field fluctuates with time above and then below the critical value (Zener, 1932). For electric field perturbations that cause level crossing of n and $n+1$, all states $\geq n$ become *dissolved* or *destroyed*, and an electron excited to n can undergo a reverse-cascade to the continuum through a series of Landau–Zener transitions.

The occupational probability, w_n, is the probability of an atom being perturbed by a net electric microfield that is below the critical field strength that would dissolve/destroy level n:

$$w_n(\text{charged}) = \int_0^{\beta_{crit}} P(\beta)\,\mathrm{d}\beta, \tag{4}$$

where $P(\beta)$ is the probability distribution of net electric microfield strength (β) and β_{crit} is the critical net electric microfield that dissolves/destroys level n (Seaton, 1990). Both the high-frequency electron collisions and quasi-static proton collisions contribute to level dissolution, but w_n is primarily determined from the ambient protons (Hummer and Mihalas, 1988; Tremblay and Bergeron, 2009). The ambient electrons contribute to the cascade plus ionization process through their high-frequency collisions with the electrons that are excited to the dissolved states and through the mixing of Stark states allowing the reverse-cascade of Landau–Zener transitions to proceed (see Tremblay and Bergeron, 2009). Following Tremblay and Bergeron (2009), we adopted the *Q-fit* electric microfield model (Nayfonov *et al.*,

[8]Bengtson, Tannich, and Kepple (1970) provided results up to H12 using the theory of Kepple and Griem (1968) for one temperature and electron density. Bengtson (1996) provided higher order line profiles for the study of Johns-Krull *et al.* (1997), but these results have not been made publicly available to our knowledge.

1999) of the Hooper distribution to represent the probability distribution $P(\beta)$. In our calculations, we used the *total* occupational probability obtained by multiplying Equation (4) by the probability of level destruction from neutral collisions, which are modeled using the same method described in Tremblay and Bergeron (2009) and Hummer and Mihalas (1988). For the conditions in flare atmospheres, the total occupational probability is approximately equal to w_n(charged), given by Equation (4).

4.2. Continuous Opacities at the b–f Edge

Dappen, Anderson, and Mihalas (1987) developed the method for incorporating the occupational probability theory into b–f and b–b opacities for hydrogen to account for the effects of Landau–Zener transitions and the subsequent cascade to the continuum.

The Landau–Zener b–b opacity for a Balmer line, Hn, is the nominal b–b opacity multiplied by the occupational probability $w_n/w_{n=2}$, where $w_2 = 1$ for the conditions in flare atmospheres. The Landau–Zener b–f opacity is the nominal b–f opacity (Equation (2)) extrapolated to wavelengths longer than the standard b–f edge (λ_o), multiplied by the probability that an atom experiences a critical microfield strength or greater, causing the upper level of the transition to be dissolved. For any wavelength $\lambda > \lambda_o$, this probability is obtained by using a pseudo-quantum number n^*:

$$n^* = \left(\frac{1}{n_i^2} - \frac{hc}{\lambda Z^2 E}\right)^{-1/2}, \tag{5}$$

where $E = 13.598$ eV $= 2.1787102 \times 10^{-11}$ erg and Z is the atomic number ($Z = 1$ for hydrogen). The *dissolved probability* (or *dissolved fraction*) is

$$D(\lambda) = 1 - \frac{w_{n^*}}{w_i}, \tag{6}$$

where w_{n^*} is the occupational probability for level n^* and $i = 2$ for the Balmer lines. As discussed by Dappen, Anderson, and Mihalas (1987), the pseudo-levels (n^*) are simple extrapolation tools between the limits of $w_n \approx 1$ and $w_n \ll 1$ that are useful for estimating the critical microfield strength (where the highest pseudo-stark state of n^* and lowest of $n^* + 1$ overlap) that can cause level dissolution through a cascade of Landau–Zener transitions. We also note that there are several different ways to estimate the critical field strength, and we used the popular method from Seaton (1990).

Now accounting for Landau–Zener transitions, wavelengths longer than the Balmer edge ($\lambda_o = 3\,646$ Å) allow there to be a fraction of atoms in $n = 2$ that can be photoionized by $\lambda > 3\,646$ Å, whereas without Landau–Zener transitions, photons with $\lambda > 3\,646$ Å could photoionize atoms only with electrons in $n > 2$. The total b–f opacity, with and without Landau–Zener transitions, in a representative layer of the CC in the F13 model is shown in Figure 8. As a result of the high ambient proton density (5.6×10^{15} cm^{-3}), the dissolved fraction is > 99 % (the occupational probability, w_n, is < 1 %) for the upper levels that contribute opacity at $\lambda \lesssim 3\,730$ Å. The nearly complete level dissolution results in a direct extrapolation of the Balmer b–f opacity to these wavelengths. As λ increases, the fraction of $n = 2$ hydrogen atoms that can be photoionized decreases because larger critical microfield is required to cause destruction of the upper level of an atomic transition. This leads to a continuous transition from the Balmer continuum opacity into the Paschen continuum opacity, which dominates at $\lambda > 4\,200$ Å. The hydrogen f–f opacity (without Landau–Zener transitions) is also shown in this figure, illustrating that it is small but non-negligible ($\chi_{\text{f–f}}/\chi_{\text{tot}} \approx 0.2$) at redder wavelengths.

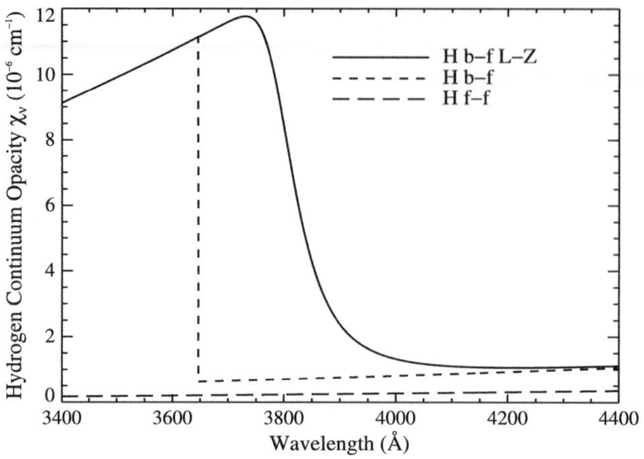

Figure 8 The hydrogen b–f continuum opacity accounting for Landau–Zener (L–Z) transitions, at $t = 2.2$ s in the F13 simulation, at a height of 255.4 km ($\log_{10} m = -2.57$ g cm^{-2}), which is the maximum of $C_I(\lambda = 3\,550)\Delta z \times C_I(\lambda = 4\,300)\Delta z$, where Δz is the vertical extent of a grid cell. The values of n_p and T_e are 5.6×10^{15} cm^{-3} and 12 770 K. At this height, hydrogen is nearly 75 % ionized. Also shown are the hydrogen f–f opacity and the hydrogen b–f opacity without accounting for opacity effects from Landau–Zener transitions at this height. At this layer, $\tau_{3\,550} = 1.1$ and $\tau_{4\,300} = 0.2$.

Hereafter, we refer to these new opacities as Landau–Zener opacities, and the pseudo-continuum resulting from this opacity as the Landau Zener continuum. We note that Dappen, Anderson, and Mihalas (1987) referred to this as the "dissolved level pseudo-continuum." In Figure 9, we show how the Landau–Zener b–f opacity from $\lambda = 3\,646$ Å to $\lambda = 4\,200$ Å changes as a function of n_p ($\approx n_e$). Self-consistent atmospheric parameters were obtained from representative heights in the F11 and F13 models, at various times, and covering a range of proton densities, $n_p = 10^{13}$–5×10^{15} cm^{-3}. As the ambient charge density increases, the Landau–Zener b–f opacity maximum shifts from $\lambda = 3\,660$ Å to $\lambda = 3\,730$ Å. We also show the occupational probabilities, w_n/w_2 (open circles, right axis, $w_2 = 1$), that are multiplied by the nominal b–b opacity for each of the Balmer b–b transitions. Thus the opacity from the lines is transferred to the Landau–Zener continuum.

4.3. Landau–Zener Opacity Modeling Using the RH Code

We used the NLTE RH code (Uitenbroek, 2001) to model the Landau–Zener b–f and b–b opacities and emissivities between $\lambda = 3\,646$ Å and $\lambda = 5\,000$ Å for the F13 model at $t = 2.2$ s. The RH code is well-suited for modeling the Balmer jump region because it can calculate overlapping transitions (Rybicki and Hummer, 1992) with more wavelength points than in the original dynamic (RADYN) simulation. By assuming statistical equilibrium and hydrogen number conservation, we can use an arbitrarily large hydrogen atom to model the merging of higher order lines near the Balmer b–f edge. Statistical equilibrium is a poor assumption during the early phases of the flare simulation, but at $t = 2.2$ s, the resulting error is small.[9] As in RADYN, the linear Stark broadening in RH is treated with a Voigt profile and the approximate formulae from Sutton (1978) are used; a prescription for quadratic Stark broadening is also included.

[9] We compared the RH and RADYN predictions at $t = 2.2$ s for the Balmer lines and continua that overlap, and we found satisfactory agreement.

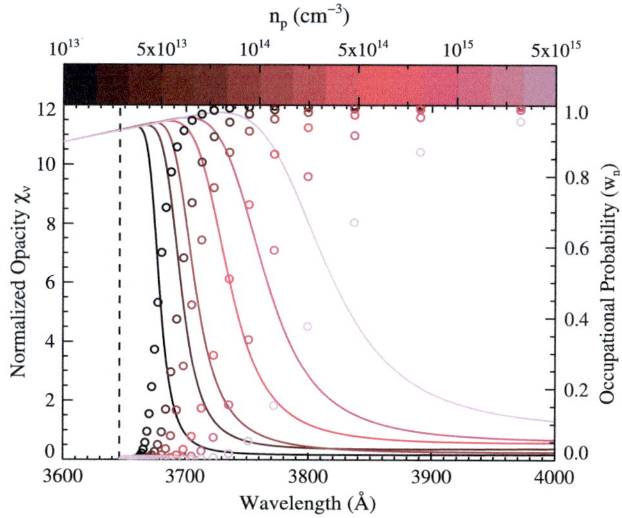

Figure 9 The hydrogen b–f opacity accounting for Landau–Zener transitions at representative times and heights in the F11 and F13 simulations. Each curve corresponds to approximately the ambient charge density at each of the tick marks in the color bar. The opacity curves have been scaled to a common value at $\lambda = 3\,600$ Å to illustrate the wavelength shift of the opacity maximum compared to the Balmer b–f edge (vertical dashed line) without accounting for Landau–Zener transitions. The occupational probabilities (right axis) of the upper levels of the Balmer transitions are shown as circles (for reference H10 is at $\lambda \approx 3\,800$ Å).

We obtained the atmospheric parameters (column mass, temperature, electron density, and velocity) at $t = 2.2$ s from the RADYN simulation with a six-level hydrogen atom, and we calculated the spectrum with the RH code using a 20-level hydrogen atom. In the RH calculation, Ca II and He are treated as background (LTE) elements. At each depth in the atmosphere, the effect of Landau–Zener transitions is modeled by multiplying the Balmer b–f cross section by the dissolved fraction (Equation (6)) and the Balmer b–b line profile by the occupational probability (w_n/w_2), as in Tremblay and Bergeron (2009). The NLTE opacity and emissivity is then calculated in RH with these modifications. We show the result of the RH flux calculation in Figure 10 compared to several observed flare spectra during the mid rise (S#27),[10] late rise (S#28), and peak phases (S#31) of a large flare on the dM4.5e star YZ CMi from K13. The observed spectra have been normalized to the flux at $\lambda = 4\,700$ Å to show the range of properties at $\lambda < 3\,900$ Å and the lack of color temperature evolution at $\lambda > 4\,000$ Å (see K13 for the evolution of the flare flux without scaling). Because the flare area is unresolved, the continuum brightness is not constrained by observations. Thus, the RH model surface-flux spectrum has been scaled by the factor $R_{\rm flare}^2/d^2$ (where $R_{\rm flare}$ is the projected radius of a circular flare area, and d is the distance to the star) to match the $\lambda = 4\,700$ Å flux of each observed flare spectrum. An instrumental convolution with a Gaussian of FWHM $= 13.5$ Å has also been applied to the model. We are not attempting to model this flare here, but we use these spectra to illustrate that the model spectrum characteristics are similar to the data at least at some phases of the flare evolution.

First, the blue-optical color temperature between the Balmer lines at $\lambda > 4\,000$ Å is well fit to a blackbody with a temperature of $T_{\rm BB} = 10\,400$ K, which is shown in Figure 10 as the dashed line. Note that this color temperature is higher by 1 000 K than that in Table 1 because

[10]The S# refer to the time-sequential spectrum numbers over each flare analyzed in K13.

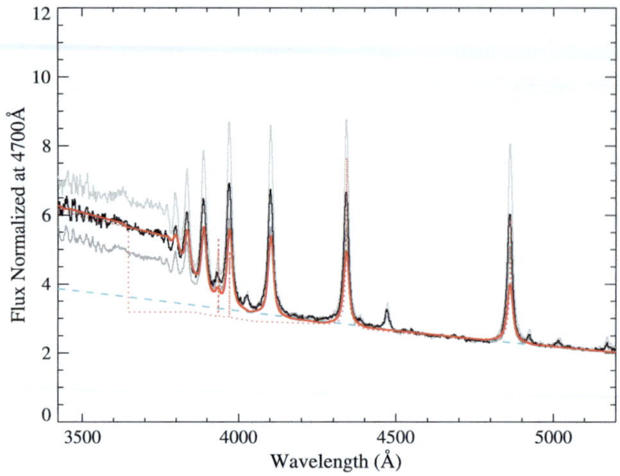

Figure 10 RH calculation with a 20-level hydrogen atom and the b–f and b–b opacities including Landau–Zener transitions, during a snapshot of the F13 simulation at $t = 2.2$ s (solid red). A $T = 10400$ K blackbody fit to continuum windows at $\lambda = 4000-4800$ Å is also shown (dashed light blue), along with the RADYN prediction with a 6-level hydrogen atom (red dotted line). Instrumental broadening has been applied to the red spectrum. Several observations during the rise and peak phases of the IF3 flare of K13 are shown: peak spectrum is S#31 from K13 (dark gray), middle rise phase is S#27 from K13 (light gray), and late rise phase is S#28 from K13 (black). All spectra are scaled to the flux at $\lambda = 4690-4710$ Å.

of the extension of the Landau–Zener continuum and Balmer wings between $\lambda = 4000$ Å and $\lambda = 4060$ Å.[11] Although the color temperature is represented by a blackbody with $T = 10400$ K, the F13 model radiative flux is higher by a factor of 1.7 than the surface flux of a blackbody with this temperature. The radiation temperature ($T_{\mathrm{rad},\mu=1,\lambda=4300} \approx 11600$ K) is also significantly higher than the color temperature ($T_{\mathrm{BB}} = 9300-10400$ K). Hereafter, we refer to the hot blackbody component as the hot blackbody-like continuum component, since we have identified the emission as predominantly due to optically thick hydrogen (Paschen) recombination emission in Section 3.4.1. Second, a jump in flux occurs above an extrapolation of the blackbody to the bluest wavelengths, and this jump in flux is the same in both the F13 model and the observations of the late rise-phase spectrum. Third, there is no Balmer discontinuity feature present in the model spectrum at $\lambda = 3646$ Å: the model calculation produces a featureless Landau–Zener continuum connecting the Balmer continuum emission at $\lambda < 3646$ Å to the bluest significant Balmer line in emission, H10 λ3798. This effect originates from the direct extrapolation of Balmer continuum opacity in Figure 8 to these wavelengths where the opacity is due to the upper levels of hydrogen that have nearly 100 % probability of being dissolved.

The RH flux spectrum with the opacity effects from Landau–Zener transitions (and without instrumental convolution) is shown in Figure 11 around the Balmer edge region; no Balmer lines at all are apparent between $\lambda = 3646-3740$ Å, and H11 is barely identifiable. The RH prediction using a 20-level hydrogen atom *without* the effects from Landau–Zener transitions (gray line, "no L–Z") is shown and definitively illustrates that the overlap or merging of the Balmer wings indeed form a pseudo-continuum between the Balmer lines at wavelengths shorter than 4 100 Å, but less so than derived from the calculation with the opacity effects from Landau–Zener transitions (black line, "L–Z"). The two calculations

[11] We used the same continuum blue window (BW) wavelength regions and weighting as in K13.

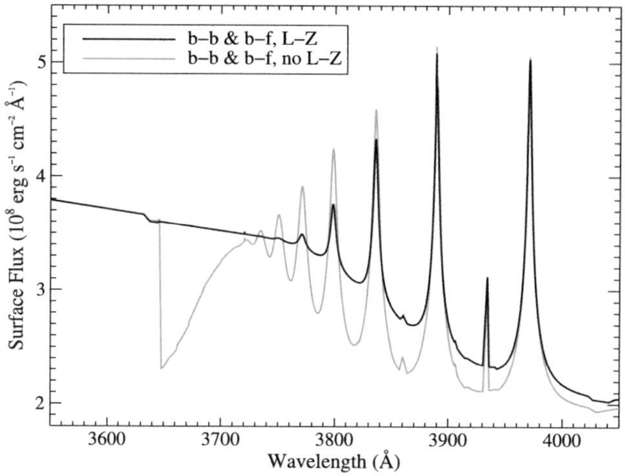

Figure 11 RH spectra of the Balmer jump region without instrumental convolution applied and with a magnified flux scale. *Black line:* Electric field effects on b–b opacities and Landau–Zener transition effects on b–b and b–f opacities from critical electric fields. H11 is the bluest identifiable Balmer line. *Gray line:* Electric field effects on b–b opacities and Landau–Zener transitions not included. H13 is the bluest identifiable Balmer line. Including Landau–Zener transitions in the b–b and b–f opacities predicts more continuum emission between 3 646 Å and 3 720 Å, whereas not accounting for Landau–Zener transitions produces a "blue continuum bump" at $\lambda \approx 3\,720$ Å from the blending of Stark-broadened Balmer lines. Instrumental broadening is not shown; it broadens the Balmer lines, but does not have a strong effect on the amount of continuum between the lines.

disagree most for the appearance and nature of the pseudo-continuum between 3 646 Å and 3 720 Å. In particular, the "no L–Z" calculation exhibits a "blue continuum bump" at $\lambda \approx 3\,720$ Å and a large positive slope at wavelengths between 3 646 Å and 3 720 Å, whereas the "L–Z" calculation predicts a continuum with the same slope as the observed Balmer continuum at all $\lambda < 3\,720$ Å. The "no L–Z" calculation also shows that the H10 – H19 lines do not merge enough to conceal the Balmer discontinuity, which is just as apparent if we apply a large instrumental broadening. Using a hydrogen atom with more levels (we also tried a 35-level hydrogen atom) does not improve the modeling result: the opacity effects from Landau–Zener transitions are necessary to completely conceal the Balmer edge at these densities. As discussed by Dappen, Anderson, and Mihalas (1987), the slope of the pseudo-continuum in the "no L–Z" calculation between 3 646 Å and the blue continuum bump is the result of the rapidly decreasing oscillator strength *density* of the higher order transitions. In the "L–Z" calculation, Stark broadening decreases the oscillator strength density of the b–b transitions while replacing the oscillator strength density with the Landau–Zener continuum.

In Figure 12, we show how the contribution function at $\lambda = 3\,870$ Å (between H8 λ3889 and H9 λ3835) and $\lambda = 4\,300$ Å compare with and without opacity effects from Landau–Zener transitions. As expected, the contribution at $\lambda = 4\,300$ Å does not change. For $\lambda = 3\,870$ Å without Landau–Zener transitions, the opacity is dominated by Paschen b–f opacity. This wavelength is more optically thin than $\lambda = 4\,300$ Å. Hence there is more contribution from the stationary component of the flare atmosphere at $z = 150 - 230$ km. In the case with Landau–Zener transitions, the Landau–Zener opacity increases the optical depth beyond that of the optical depth at 4 300 Å (the contribution at $z = 150 - 230$ km is less than in the calculation without Landau–Zener transitions) and more $\lambda = 3\,870$ Å photons escape from higher in the atmosphere in the CC (at $z = 251$ km) as a result of the recombination of electrons from the $n^* = 8.3$ pseudo-level (Equation (5)) to the $n = 2$ level of hydrogen.

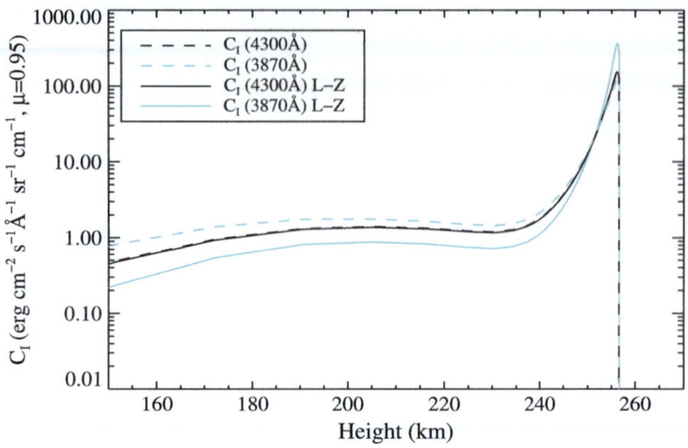

Figure 12 The contribution function at $\lambda = 3\,870$ Å and $4\,300$ Å accounting for opacity effects from Landau–Zener transitions (solid) compared to the original calculation (dashed). With L–Z transitions, the additional Balmer opacity at $\lambda = 3\,870$ Å causes more emission to escape from the CC and less from the stationary flare layers than without the L–Z transitions. The contribution function at $\lambda = 4\,300$ Å is not affected.

We can now associate the model flux spectrum in Figure 10 with several atomic emission processes. At $\lambda < 3\,646$ Å the flux is almost entirely Balmer continuum emission from free electrons recombining to $n = 2$. This part of the spectrum also exhibits a color temperature $T_{\mathrm{BB}} > 10\,000$ K, with a peak in the NUV between $\lambda = 2\,500-2\,600$ Å. The Landau–Zener continuum flux between $\lambda = 3\,646-3\,760$ Å is *bona-fide* Balmer continuum recombination radiation resulting from free electrons recombining from the n^* pseudo (dissolved) upper levels of hydrogen to $n = 2$ (Dappen, Anderson, and Mihalas, 1987); *i.e.*, undergoing reverse Landau–Zener transitions from the continuum to $n = n^*$, at which point they transition to $n = 2$. The flux between the rapidly converging Balmer lines from H10 to Hε is mostly Landau–Zener continuum emission from electrons recombining from the dissolved upper levels to $n = 2$, but at these wavelengths there is a smaller fraction of atoms in the atmosphere that experience a critical field strength, and thus the Landau–Zener continuum emission is lower than at $\lambda = 3\,646-3\,760$ Å. There is also some contribution from the far wings of the Balmer b–b transitions. The far wings of Hδ and Hε and the redmost extent of the Landau–Zener continuum emission produce additional emission above the nominal Paschen recombination spectrum as red as $\lambda = 4\,200$ Å (compare solid and dotted red lines). At $\lambda > 4\,200$ Å, Paschen continuum emission from recombination of free electrons to $n = 3$ dominates, with a smaller (≈ 20 %) contribution from thermal bremsstrahlung (free–free) emission. Paschen recombination radiation also contributes toward the radiation at $\lambda < 4\,200$ Å, but to a much lower degree than the calculation without the Landau–Zener continuum, as can be seen by the attenuation of radiation at $z = 160-230$ km in Figure 12. An interesting observational effect of including the Landau–Zener continuum emission is that the measured color temperature of the flux spectrum increases from 9 300 K (6 levels, without Landau–Zener transitions, dotted red line, Table 1) to 10 400 K (20 levels, with Landau–Zener transitions, thick red line), which further improves the applicability of the F13 model to the measured T_{BB} distribution at peak for a variety of flares (K13).

We note that the lower order Balmer lines (Hα – Hδ) can be used to constrain the models. The CC in the F13 model generates broad profiles, high surface fluxes, and a reverse decrement (more Hγ flux than Hα flux) in the lower order Balmer lines. The large widths of the lower order Balmer lines in Figure 10 result from Stark broadening. However, improvements to the implementation of the Stark effect (see Section 7) are required for a detailed comparison. Although the F13 model produces approximately 18 times the surface flux in Hγ compared to the F11 model (and approximately six times that in the F12 model), the ratio of Balmer line to continuum values is notably lower than in the simulations with a lower NT electron flux (Table 1). At $t = 2.2$ s, the relative amount of Balmer line to continuum emission produced by the CC is consistent with typical observed values in the impulsive phase: $\mathrm{H}\gamma_{\mathrm{EW4170}} \approx 20$ in the F13 model compared to $\mathrm{H}\gamma_{\mathrm{EW4170}} \approx 15-30$ in the late rise and peak phase spectra in Figure 10. Additional flaring areas that are either impulsively heated with a lower NT electron flux or are gradually decaying from a high electron flux may be required to more accurately reproduce the time evolution of the emission lines in these individual spectra.

4.4. A New Diagnostic of Ambient Charge Density in Chromospheric Flares

The opacity effects from Landau–Zener transitions result in Balmer recombination radiation longward of 3 646 Å and explain the NUV/blue pseudo-continuum previously attributed to the merging of Stark-broadened Balmer lines. The occupational probability theory as employed in white-dwarf atmospheric models provides a self-consistent quantum mechanical method of modeling the opacity effects from cascading Landau–Zener transitions. This new modeling explains the lack of an observed Balmer discontinuity (or edge) in flare spectra while still allowing a *Balmer jump* in flux above an extrapolation of a blackbody that is fit to the blue-optical wavelengths in the Paschen continuum ($\lambda > 4\,000$ Å). Accounting for Landau–Zener transitions gives an emergent flux spectrum from the F13 model at $t = 2.2$ s with continuum emission bridging Balmer continuum emission at $\lambda \leq 3\,646$ Å to the highest order Balmer line that is detectable, which is H10 or H11 in the case of the F13 model. The microscopic picture that is reinforced here is that slowly moving thermal electrons are able to recombine to $n = 2$ and produce a photon with wavelength $> 3\,646$ Å (a Balmer continuum photon) as a result of the nearly complete dissolution of the upper levels: $1 - w_n/w_2 \geq 0.85$ for levels $n \geq 11$, in a dense plasma with $n_p = 5 \times 10^{15}$ cm^{-3}. The success of modeling the pseudo-continuum emission from $\lambda = 3\,646-3\,760$ Å in Figure 10 as a red-wavelength extension of Balmer continuum emission is consistent with our conclusion that the white-light emission, in this RHD model, is dominated by a combination of hydrogen Balmer b–f, hydrogen Paschen b–f, and hydrogen f–f emission.

In solar flares, a pseudo-continuum component between the Balmer lines from $\approx 3\,700-3\,870$ Å was first noted by Zirin (1980) and Zirin and Neidig (1981). The current, best treatment of the modeling of the Balmer jump region and pseudo-continuum was applied to several solar flares by Donati-Falchi, Falciani, and Smaldone (1985), who used static, isothermal models with $n_e = 10^{13.2} - 10^{13.6}$ cm^{-3} and $T_e = 7000 - 10\,000$ K. These models produced a "blue continuum bump" at $\lambda = 3\,675$ Å that shifts to redder wavelengths and increases in brightness for higher electron density and has been discussed in light of a new solar flare observation with broad wavelength coverage spectra in the NUV (Kowalski, Cauzzi, and Fletcher, 2015). The pseudo-continuum bump at $\lambda \approx 3\,720$ Å in Figure 11 produced in our calculation without Landau–Zener opacity effects (Section 4.3) is very likely analogous to the NUV/blue bump feature at $\lambda \approx 3\,675$ Å produced in the lower density slab models of Donati-Falchi, Falciani, and Smaldone (1985). As noted by these authors, this feature is due to the merging of Stark-broadened high order Balmer lines. Another feature in the models of Donati-Falchi, Falciani, and Smaldone (1985) is a continuum dip at 3 646 Å that is due to the smearing of the Balmer edge. Donati-Falchi, Falciani, and Smaldone (1985) truncated the model hydrogen atom at $n = 84$ and broadened the Balmer edge using the Stark width of the highest Balmer line. For $n = 84$, the Stark damping parameter is 30 Å (Švestka, 1963; Johns-Krull *et al.*, 1997), which results in a flattening and smearing of the edge in their models. Since the Landau–Zener continuum modeling from occupational probability theory does not predict a continuum bump at $\lambda \approx 3\,720$ Å or a dip near the Balmer edge (for very high proton densities), this modeling may also provide a more self-consistent explanation for the range in Balmer jump characteristics observed in solar flare spectra (as in the compilation by Neidig, 1983), while also improving the diagnostic capabilities of these spectra.

Intuitively, occupational probability theory provides insight beyond a model that considers pseudo-continuum from merged lines as just a superposition of b–b opacities. In the classical framework (Donati-Falchi, Falciani, and Smaldone, 1985), the opacity of a b–b transition is spread over a wider spectral range by the Stark broadening and is then added to the opacity of other Stark-broadened b–b opacities at a given wavelength. In the occupational probability framework, the Stark-broadened b–b opacity is replaced by a higher b–f opacity at a given wavelength when hydrogen atoms experience a critical microfield perturbation. The higher b–f opacity results from the fact that free states are not degenerate (in orbital angular momentum), and therefore the oscillator strength density of the continuum is not re-distributed over a superposition of Stark states as for b–b transitions. This physical difference explains the large discrepancy between the two model spectra in Figure 11 from $\lambda = 3\,646-3\,730$ Å (see also the discussion in Dappen, Anderson, and Mihalas, 1987). In summary, the opacity sources from Donati-Falchi, Falciani, and Smaldone (1985) are *continuum opacity from a Stark-broadened b–f edge and a superposition of Stark-broadened b–b opacities,* whereas in occupational probability theory, the opacity sources are *continuum opacity from Stark-dissolved levels and a superposition of the remaining undissolved Stark-broadened b–b opacities.*

Modeling the opacity effects from Landau–Zener transitions provides a new physical interpretation of several dMe flare spectra in the literature. We have already shown how this modeling with a very high density ($\approx 5 \times 10^{15}$ cm^{-3}) can account for the continuum and line properties in a flare with H10 or H11 as the bluest identifiable Balmer line (Figure 10). In the continuum peak spectrum (1:15 UT) from the dMe flare studied in García-Alvarez *et al.* (2002), H9 or H10 also appear as the bluest Balmer line; however, their gasdynamic modeling using the Jevremovic *et al.* (1998) fit to the Balmer decrement indicates densities nearly two orders of magnitude lower than the occupational probability modeling indicates based on the bluest identifiable Balmer line. The bluest identifiable Balmer lines in most dMe flares with moderate spectral resolution are typically H15 and H16 both in the impulsive phase (Hawley and Pettersen, 1991) and in the gradual decay phase (Kowalski *et al.*, 2010, 2013). Even $R = 40\,000$ Echelle spectra of dMe flares sometimes do not show emission in Balmer lines bluer than about H15 and instead exhibit a nearly featureless[12] continuum with a color temperature of $9\,000-11\,000$ K at $\lambda \lesssim 3\,700$ Å (Figures 2 and 5 of Fuhrmeister *et al.*, 2008). It is also notable that one peak flare spectrum of a dMe flare showed Balmer lines as blue as H19 (Fuhrmeister *et al.*, 2011), which the authors concluded was a result of only a small enhancement of the chromospheric electron density in the flare. Since the Landau–Zener continuum resulting from occupational probability theory can fully account for similar characteristics of this spectral region in the flare spectra in Figure 10, we suggest the *range* of characteristics observed around the Balmer jump in stellar flares may be explained with the range of Landau–Zener opacities in Figure 9.

Future modeling of particular flares using occupational probability theory can be used to infer the magnitude and gradient of the ambient charge (proton) density in the densest regions of the flare atmosphere that are significantly heated and produce white-light emission. In Section 4.3 and Figure 9, we showed how the Landau–Zener continuum opacity maximum changes and occupational probabilities dissolve the hydrogen Balmer transitions over 2.5 orders of magnitude in charge density. We presented a simple method using the Balmer

[12] In addition to many faint metallic and helium emission lines.

Table 2 The Landau–Zener Balmer decrement (H11/Hγ) in the optically thick, isothermal LTE limit. Here, we give only the charged component of w_n (Equation (4), averaged for the range of temperatures), ignoring the minor dependence on the destruction of levels from collisions with other neutrals.

n_p (cm^{-3})	$w_{n=11}$	$T = 10\,000$ K	$T = 12\,500$ K	$T = 15\,000$ K	$T = 20\,000$ K
10^{13}	1.00	0.30	0.33	0.35	0.37
10^{14}	0.98	0.30	0.32	0.34	0.37
5×10^{14}	0.85	0.26	0.28	0.30	0.32
10^{15}	0.69	0.21	0.23	0.24	0.25
2.5×10^{15}	0.35	0.11	0.12	0.12	0.12
5×10^{15}	0.15	0.05	0.05	0.05	0.05

decrement to estimate the charge density for an optically thick flare atmosphere in LTE (as for the F13 atmosphere studied here). We introduced the Landau–Zener Balmer decrement, which is the flux in a Balmer line transition (*e.g.*, H10 or H11, which are relatively unaffected by strong metallic or helium lines) that experiences a large change in the dissolved fraction of the upper level of this transition, divided by the flux of a lower order line that is not affected by level dissolution (*e.g.*, Hγ).[13] In Table 2, we showed how the Landau–Zener Balmer decrement H11/Hγ changes for different temperatures and over the density range of $n_p = 10^{13}$ to $n_p = 5 \times 10^{15}$ cm^{-3}. This Balmer decrement was obtained by multiplying $w_{n=11}/w_{n=5}$ (where $w_5 \approx 1$) by the Balmer decrement formula for the optically thick, isothermal, LTE limit in Appendix B of Drake and Ulrich (1980). The decrement sharply drops for $n_p > 10^{15}$ cm^{-3} as $n = 11$ becomes dissolved (see also Figure 9). The observed decrements for H11/Hγ in the two peak spectra from Hawley and Pettersen (1991) are 0.26 and 0.22, which suggests a charge density of not much higher than 10^{15} cm^{-3}, but also not lower than 10^{14} cm^{-3}. This method might be used as an alternative to the Inglis–Teller relation,[14] which requires data with high spectral resolution. However, since lines and continua are formed at different physical depths and depth ranges (as for the F13 atmosphere studied here), we plan to improve the accuracy of this spectral diagnostic using a grid of NLTE flare atmospheres in future works.

5. The Viability of the F13 Model

M-dwarf flares are more energetic than solar flares (in total). Typical white-light energies of dMe flares are $10^{30} - 10^{32}$ erg (Hawley *et al.*, 2014), with occasional events exceeding 10^{34} erg in the U band alone (Kowalski *et al.*, 2010). The areal extent of flare footpoints cannot be measured directly,[15] and hard X-ray emission is far too faint except during the largest dMe flares (Osten *et al.*, 2010). Although we cannot determine if NT electron fluxes on M-dwarfs are indeed as high as 10^{13} erg cm^{-2} s^{-1}, typical values of $\approx 10^{11}$ erg cm^{-2} s^{-1} in strong solar flares cannot explain the properties of white-light emission during dMe flares,

[13] Assuming that the Balmer line emission does not originate at a different spatial location, for example from a larger area with a lower density, than the continuum emission.

[14] See the discussion in Hummer and Mihalas (1988) about how the theory of occupational probability differs from the Inglis–Teller relationship: the Landau–Zener transitions require *fluctuations* of the microfield around the critical field value.

[15] Emission model fits, *e.g.*, with a blackbody function or RHD model spectra, give an indirect estimate of the flare area and energy flux requirements.

and thus we infer that the energy flux is higher. The atmospheric response to an F13 NT electron burst provides insight into the radiative transfer processes that could explain white-light emission, but is this very high flux a realistic model for dMe flares? In this section, we evaluate its viability and conclude that some issues still remain.

First, is an F13 beam even possible during an M-dwarf flare? The energy flux available to accelerate particles at the reconnection site is limited by the free magnetic energy in the corona, which can be estimated by $\frac{B_{\mathrm{cor}}^2}{8\pi} v_{\mathrm{inflow}} = 2 \times 10^{13}$ erg cm^{-2} s^{-1}, where v_{inflow} is the rate at which a magnetic field flows into the reconnection region and may be between $0.01-0.1$ times the Alfvén velocity. The energy flux requirement of 2×10^{13} erg cm^{-2} s^{-1} into each magnetic footpoint assumes equipartition of energy among protons and electrons and that all free magnetic energy goes into accelerating particles. Using the higher estimate for inflow velocity, $0.1 \times v_{\mathrm{Alfven}}$ gives $B_{\mathrm{cor}} \approx 1.5$ kG as the coronal magnetic field (assuming $v_{\mathrm{inflow}} = 0.01 v_{\mathrm{Alfven}}$ gives 3.3 kG as the coronal magnetic field) necessary to supply this energy flux. Direct (Zeeman broadening) measurements of small-scale magnetic fields of M-dwarfs reveal average photospheric magnetic flux density of $3-4.5$ kG (Johns-Krull and Valenti, 1996; Phan-Bao *et al.*, 2009), but the coronal magnetic field above these starspots has not been directly measured. The most well-studied active corona is that of the dM3.5e star EV Lac (Osten *et al.*, 2006), which exhibits a very hot (10 MK), confined ($L \approx 10^8$ cm), and dense ($n_e \approx 10^{13}$ cm^{-3}) component in quiescence. For these structures, the minimum magnetic field required for confinement is estimated to be ≈ 1 kG. Other values from microwave data of flares and quiescence have suggested coronal magnetic fields on M-dwarfs exceeding 1 kG (Osten *et al.*, 2006; Smith, Güdel, and Audard, 2005), which have also been inferred from magnetic confinement estimates using the Haisch Simplified Analysis of data from the *Extreme Ultraviolet Explorer* (EUVE) of many M-dwarf flares (Mullan *et al.*, 2006). Thus, the energy requirement for accelerating an F13 NT electron flux requires relatively high (kilo-Gauss) coronal magnetic fields, but they are not outside the realm of current measurements.

Although the energy flux requirements may be plausible (given strong M-dwarf coronal magnetic fields), the electron number flux for the F13 model may be problematic due to return-current effects. The electron number flux corresponding to our double power-law F13 spectrum is 10^{20} electrons cm^{-2} s^{-1}. A return-current electric field is established from the charge displacement of electrons and protons, given that both particles are energized equally in the process of magnetic reconnection and retraction. The initial energy lost *per* NT electron from the return-current electric field is 13 keV Mm^{-1} at the apex of our model loop (Equation 29 of Holman, 2012). The NT electrons with the cutoff energy (37 keV) lose all of their energy and are thermalized after penetrating only 3 Mm into the atmosphere, after which the flux of ongoing NT electrons decreases with distance as more NT electrons are thermalized. Holman (2012) derived the evolution of the NT electron flux distribution accounting for losses by the return-current electric field. We used these results and found that only NT electrons with $E_{\mathrm{initial}} > 85$ keV penetrate the top of the chromosphere without thermalizing. Furthermore, the F13 NT electron density is comparable to the ambient density ($n_e = 2.6 \times 10^{10}$ cm^{-3}) at the apex of our model coronal loop; this condition would cause the ambient electrons to be accelerated and would result in instabilities and other effects that are not modeled here.

A more basic issue is that the number flux of electrons available in the corona flowing into the reconnection site is only $\approx 5 \times 10^{18}$ electrons cm^{-2} s^{-1} (using $n_e \times 0.1 v_{\mathrm{Alfven}}$, where $B = 1.5$ kG is obtained from the energy flux), whereas the number flux in the F13 beam is 10^{20} electrons cm^{-2} s^{-1}. Even if electrons in the coronal loop were globally accelerated, there are only an additional 4×10^{19} electrons cm^{-2} s^{-1} available. However, this

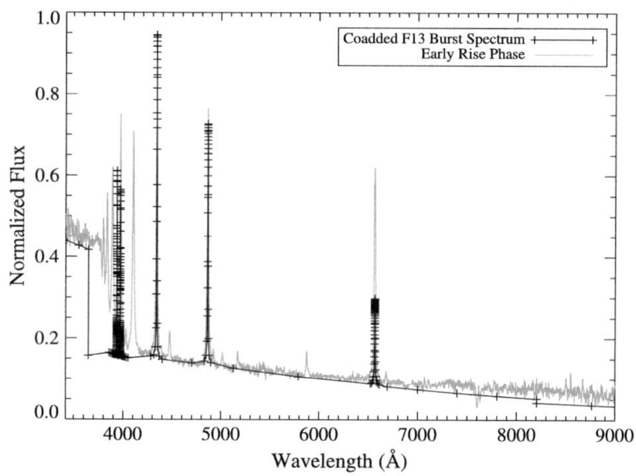

Figure 13 The coadded (average) model F13 spectra (from RADYN; instrumental broadening is not applied) over 5 s compared to the early rise spectrum (S#26) during the IF3 flare from K13. The coadded model spectrum has a larger Balmer jump and lower optical color temperature than the instantaneous model spectrum at $t = 2.2$ s in Figure 10. The flux at $\lambda < 3\,646$ Å increases towards bluer wavelengths in the model and observation. Note that the opacity effects from Landau–Zener transitions are not included here because statistical equilibrium is a poor approximation during the early time steps in the model (Section 4.3).

simple assessment does not account for processes that may increase electron density during reconnection and retraction (*e.g.*, Longcope and Guidoni, 2011). Furthermore, high densities between $10^{12}-10^{13}$ cm^{-3} have been inferred for some (non-flaring) coronal structures on the Sun (Feldman *et al.*, 1994) and on the flare star EV Lac (Osten *et al.*, 2006). In summary, the energy and number flux requirements for an F13 beam could be met under certain conditions (a dense corona with a high magnetic field), but the return-current losses would be significant.

A final issue with the F13 model is the short presence of the hot, blackbody-like emission, which only appears at the end of the impulsive phase between $t \approx 1.8$ s and $t = 2.3$ s. We expect a coadded burst spectrum to be more directly comparable to spectral observations, which have integration times of 10 s or more. The coadded spectrum over the 5.1 s duration of the impulsive heating and the decay phases (dot-dashed light blue line in Figure 1) exhibits a lower color temperature (7 500 K) and higher Balmer jump ratio (2.7) than the instantaneous model spectrum at $t = 2.2$ s (Table 1). The coadded F13 spectrum is not as representative of the peak phases of the M-dwarf flares as the instantaneous $t = 2.2$ s spectrum, which exhibits prominent $T_{\mathrm{BB}} \approx 10^4$ K blackbody-like emission, as observed. A longer duration of NT electron energy deposition for $5-10$ s may produce similar coadded and instantaneous spectra, but the simulation becomes very computationally intensive past 2.3 s with sustained heating (Section 3.2).

The coadded 5 s F13 spectrum is more representative of the atmospheric conditions in the gradual decay phase and the very early rise phase. The gradual decay phase of flares exhibits a Balmer jump ratio of $2.6-3.3$ and a blue-optical color temperature of about 8 000 K (K13), which are similar to the observables obtained from the coadded F13 spectrum (Table 1). The coadded F13 spectrum is also consistent with a spectral observation from the very early rise phase of a flare from K13, which we show in Figure 13. K13 characterized two additional properties of flare spectra that can constrain models of the time-evolution: the slope in the NUV at $\lambda < 3\,646$ Å and the amount of deviation at red wavelengths (at $\lambda > 5\,000$ Å) from a blackbody fit to the blue-optical range at $\lambda = 4\,000-4\,800$ Å. Although observations of both the early rise phase and gradual decay phase exhibit a Balmer jump ratio that are higher than during the peak phase, only the early rise exhibits a $T \approx 10\,000$ K color temperature at $\lambda < 3\,646$ Å and a small deviation from the blackbody at redder wavelengths. These two common properties of the average burst model and the early rise phase observation can be seen in Figure 13. However, we expect that flare emission at any particular time (even in the very early rise phase) may consist of a superposition of many individual bursts. In order to conclude that any individual model can explain the time-evolution at any particular time, future modeling work should incorporate a treatment of the spatial evolution of flares, such as the development of two ribbon arcades and sequentially heated white-light kernels.

6. Summary and Conclusions

We have simulated the dynamic and radiative response of an M-dwarf atmosphere subject to a short burst of a constant, high flux (10^{13} erg cm^{-2} s^{-1}) of NT electrons with a moderately high low-energy cutoff (37 keV). This simulation produced a heated ($T = 12\,000-13\,500$ K), high-density ($n_H > 10^{15}$ cm^{-3}) chromospheric condensation with $N_H(n=2) > 3 \times 10^{17}$ cm^{-2}, $N_H(n=3) > 10^{17}$ cm^{-2}, and a total physical depth range of ≈ 20 km, resulting in high hydrogen b–f and f–f opacities within a short amount of time (2.2 s) after the start of non-thermal electron heating. At this time, the flare photosphere ($\tau_{5\,000} = 1$) is located in the CC at a height that is nearly 250 km higher than the pre-flare photosphere, and the density of the CC is over an order of magnitude higher than the material at this height in the pre-flare. Significant heating to $T > 10\,000$ K at high densities (comparable to the density in the very low chromosphere of the pre-flare) was achieved and maintained in the CC and also in the stationary, slightly less dense regions of the atmosphere ≈ 50 km below the CC (Figures 2, 4).

The atmospheric response in the F13 model provides a new, accurate description of the radiation field in a flare atmosphere with a dense CC and moreover shows that such an atmosphere can produce bright white-light emission with properties that agree well with the observations. The flux spectrum from the F13 model at $t = 2.2$ s exhibits a blue-optical color temperature of $T_{BB} \approx 10^4$ K between $\lambda = 4\,000-4\,800$ Å and a relatively low Balmer jump ratio ($\chi_{flare} \approx 2.0$; Table 1), both of which are large improvements compared to the range of these parameters in lower beam flux models (F10, F11, and F12 keeping all else the same with the NT electron spectrum). These properties describe typical peak spectra of dMe flares, and we showed the striking match of the instantaneous F13 model at $t = 2.2$ s to the observations of one large dMe flare with time-resolved spectra in the impulsive phase (Figure 10). The F13 model is the first RHD model to self-consistenly reproduce three generally observed characteristics (T_{BB}, χ_{flare}, and $H\gamma_{EW4170}$; Table 1).

The pioneering gasdynamic simulations of Livshits *et al.* (1981) and Katsova, Boiko, and Livshits (1997) produced a CC with similar characteristics to those in the F13 CC, and also $\Delta\tau \approx 1$ at $\lambda = 4\,500$ Å and $T \approx 9\,000$ K, albeit with 10–30 times lower NT electron energy fluxes. It is interesting that at least an order of magnitude difference in heating rates gives similar dynamic responses in gasdynamic modeling and radiative-hydrodynamic modeling. Our new radiative-hydrodynamic models use a more realistic starting dMe atmosphere with an NLTE treatment of the entire atmosphere (photosphere, chromosphere, transition region, and corona), a detailed treatment of the helium-level populations (which are important for regulating the onset of explosive evaporation; see Section 3.2), and a modern parameterization of the NT electron spectrum from RHESSI data. However, a higher cutoff of the NT electron spectrum (as employed in our model) is expected to require a higher flux in order to produce similar dynamics (Fisher, 1989). Finally, the RADYN and RH codes provide a detailed spectrum, which we have shown is important for critically testing the models against the observations, for revealing new atomic physics in flares around the Balmer jump, for constraining the charge density in the densest atmospheric regions that are heated during flares (Sections 4.3, 4.4), and for understanding how hydrogen recombination appears as a blackbody-like spectrum with $T \approx 10^4$ K (Section 3.4).

By analyzing the contribution function for the continuum intensity, we conclude that the values of the slope (T_{BB}) and Balmer jump ratio (χ_{flare}) of the emergent optical and NUV continuum intensity in the F13 flare atmosphere are due primarily to hydrogen recombination radiation subject to wavelength-dependent attenuation from Balmer and Paschen b–f opacities. Laboratory spectra of hydrogen at high density ($n \approx 10^{16}$ cm^{-3}) and temperatures of $T \approx 10^4$ K (Wiese, Kelleher, and Paquette, 1972) do not exhibit the blue-optical color temperature of a hot blackbody with $T \approx 10^4$ K. The necessary ingredient to build up optical depth at this density and temperature, and hence produce a $T \approx 10^4$ K *blackbody-like* spectrum with a small Balmer jump, is a large physical depth range (Δz) of emitting material along the line of sight. Thus, the color temperature in flare spectra does not directly relate to the atmospheric temperature so much as it is an indication of how the amount of emitting material along the line of sight varies as a function of wavelength (indeed, material must be heated to $T \approx 10^4$ K for hydrogen to be significantly populated out of the ground state into $n = 2$ and $n = 3$ and generate optical depth). In the CC of the F13 atmosphere, the physical depth range over which continuum emission escapes is $\Delta z_{FWHM} \approx 1-2$ km, which is not present in laboratory experiments. At the blue continuum wavelengths with the lowest opacity (*e.g.*, at $\lambda = 4300$ Å), the physical depth range over which continuum emission escapes is relatively large in the CC and extends to lower heights at $z \approx 200$ km (50 km below the CC). The amount of emitting material along a line of sight can also change due to variations in the density at the heights where continuum emission escapes; in the F13 flare atmosphere, the density is lower at $z \approx 200$ km than the maximum density in the CC (Figure 2). Thus, the integration of the *height-dependent* properties (as for the contribution function in Figure 4) is important for determining the emergent intensity and flux spectra.

Another important result is that by applying the occupational probability formulae (Hummer and Mihalas, 1988; Dappen, Anderson, and Mihalas, 1987; Tremblay and Bergeron, 2009) to NLTE b–f and b–b Balmer opacities, we can explain the pseudo-continuum properties in the spectral region redward of the Balmer edge ($\lambda > 3646$ Å) and between the Balmer lines from $\lambda = 3750-3900$ Å. These formulae approximate how wavelengths redward of the Balmer jump experience Balmer b–f opacity in addition to Paschen b–f opacity (and f–f opacity); the atomic processes that allow for this additional Balmer continuum opacity are dissolved upper levels of hydrogen from Stark perturbations by ambient protons and Landau–Zener transitions of electrons between dissolved levels of hydrogen. The amount of Balmer continuum emission redward of the Balmer jump and the amount of high order b–b Balmer emission vary inversely, and therefore the relative amounts of dissolved and undissolved components in spectra provide a new diagnostic of ambient charge density in the white-light emitting, partially ionized regions in the lower flare atmosphere (Section 4.4, Table 2).

The heating mechanism at high density in dMe flares is still unresolved since the CC required to produce the observed properties of the white-light emission results from a NT electron beam that would experience a considerable loss of energy from the return-current electric field. The main results of this work are therefore the new interpretation of the blue-optical color temperature and Balmer jump ratio (as the variations in the physical depth range and optical depth as a function of wavelength) and of the apparent pseudo-continuum between the higher order Balmer lines (as Balmer continuum emission due to Landau–Zener transitions of electrons between dissolved levels). Combined with future modeling efforts using a more plausible heating mechanism (Section 7), these insights will allow the spectral observables from individual flares to directly constrain the atmospheric properties in the dense, lower atmosphere.

7. Future Modeling Improvements

We plan to develop a grid of short flare bursts using a higher low-energy cutoff of the NT electron spectrum, thereby reducing the number flux of NT electrons (and thus return-current effects) while allowing a larger fraction of energy to penetrate deeper in the atmosphere. A possible alternative is to employ lower NT electron fluxes and explore possible re-acceleration mechanisms in the chromosphere in a similar way as done recently by (Varady *et al.*, 2014). Including additional energy deposition mechanisms, such as NT proton energy deposition (*e.g.*, a neutral beam simulation such as in Karlický *et al.*, 2000) or a combination of energy deposition by other means (*e.g.*, Fletcher and Hudson, 2008) may also be important for producing a similar atmospheric response with a lower flux of NT electrons. In addition, we plan to incorporate the spatial development of a flare arcade of sequentially heated footpoints (*e.g.*, as in the observations of the Bastille Day solar flare from Qiu *et al.*, 2010 or the cartoon two-ribbon model of the megaflare event in Kowalski *et al.*, 2012) to reproduce the observed timescales of dMe flares, which are longer than the 5 s burst modeled here.

In future work with the RH code, we intend to improve the implementation of occupational probability theory and Stark broadening. Hubeny, Hummer, and Lanz (1994) fully extended the occupational probability formalism to NLTE, and we suggest that flare modeling codes developed in the future should similarly include occupational probability formalism in the non-equilibrium rate equation (following Equation 2.23 in Hubeny, Hummer, and Lanz, 1994) in addition to the NLTE b–b and b–f opacity and emissivity. Here, our goal was to demonstrate that large improvements in the NUV model spectrum result from implementing the simple modifications to the opacity and emissivity given in Dappen, Anderson, and Mihalas (1987) and Tremblay and Bergeron (2009). We will also follow the work of Tremblay and Bergeron (2009) and Seaton (1990) and *renormalize* the Stark profiles of Vidal, Cooper, and Smith (1973) using β_{crit} as an integration limit, which leads to significantly narrower higher order Balmer lines (*cf.* Figure 5 of Tremblay and Bergeron, 2009) The theory of Vidal, Cooper, and Smith (1973) is only valid to first order when the line wings do not overlap, since the Landau–Zener continuum accounts for larger perturbations. Therefore, too much emission is probably produced between the Balmer lines between $\lambda = 3\,800-4\,000$ Å in the model spectrum in Figure 11. Between $\lambda = 3\,646-3\,800$ Å, the low values of w_n (that multiply into the b–b opacity) lead to only a small amount of blended Balmer line wings compared to the Landau–Zener continuum, so the error is not important. Finally, the linear Stark broadening theory of the Balmer lines using the analytic approximations of Sutton (1978) as Voigt damping parameters should be replaced by the full Stark profiles from Vidal, Cooper, and Smith (1973) and Lemke (1997) for the conditions present in flare atmospheres, in order to give accurate diagnostics on the electron density for unblended Balmer lines.

In future work with the RADYN code Allred, Kowalski, and Carlsson (2015), we intend to include the following improvements to the modeled physics: the cooling from Mg II and Fe II ions, an improved optically thin radiative loss function, non-thermal ionization of He

I and He II, return-current heating, magnetic mirroring of the NT electrons, and the fully relativistic Fokker–Planck solution to energy deposition and pitch angle scattering of NT electrons.

Acknowledgements Adam F. Kowalski thanks the Science Organizing Committee of the *Solar and Stellar Flares* meeting in Prague, Czech Republic for the opportunity to present this work. We thank an anonymous referee for clarifications and comments that helped improve this work. Adam F. Kowalski thanks Petr Heinzel and Hans Ludwig for bringing hot star modeling articles of Stark broadening to his attention. We thank Adrian Daw, Eric Agol, Ellen Zweibel, and Jeremiah Murphy for helpful discussions and William Abbett for several IDL routines used in the analysis. AFK also acknowledges helpful discussions at the International Space Science Institute with Lyndsay Fletcher's Solar and Stellar Flares team and with Sven Wedemeyer-Bohm's Magnetic Activity of M-type Dwarf Stars and the Influence on Habitable Extra-solar Planets team. This research was supported by an appointment to the NASA Postdoctoral Program at the Goddard Space Flight Center, administered by Oak Ridge Associated Universities through a contract with NASA, and by the University of Maryland Goddard Planetary Heliophysics Institute (GPHI) Task 132.

References

Abbett, W.P., Hawley, S.L.: 1999, Dynamic models of optical emission in impulsive solar flares. *Astrophys. J.* **521**, 906. DOI.

Aller, L.H.: 1963, *Astrophysics. The Atmospheres of the Sun and Stars*, 2nd edn. Ronald Press, New York.

Allred, J.C., Hawley, S.L., Abbett, W.P., Carlsson, M.: 2005, Radiative hydrodynamic models of the optical and ultraviolet emission from solar flares. *Astrophys. J.* **630**, 573. DOI.

Allred, J.C., Hawley, S.L., Abbett, W.P., Carlsson, M.: 2006, Radiative hydrodynamic models of optical and ultraviolet emission from M dwarf flares. *Astrophys. J.* **644**, 484. DOI.

Allred, J.C., Kowalski, A.F., Carlsson, M.: 2015, submitted, A unified computational model for solar and stellar flares. *Astrophys. J.*

Aschwanden, M.J., Kliem, B., Schwarz, U., Kurths, J., Dennis, B.R., Schwartz, R.A.: 1998, Wavelet analysis of solar flare hard X-rays. *Astrophys. J.* **505**, 941. DOI.

Avrett, E.H., Machado, M.E., Kurucz, R.L.: 1986, Chromospheric flare models. In: Neidig, D.F. (ed.) *The Lower Atmosphere of Solar Flares*, National Solar Observatory, 216.

Bengtson, R.D.: 1996, private communication.

Bengtson, R.D., Tannich, J.D., Kepple, P.: 1970, Comparison between measured and theoretical Stark-broadened profiles of H_6-H_{12} emitted from a low-density plasma. *Phys. Rev. A* **1**, 532. DOI.

Brown, J.C.: 1971, The deduction of energy spectra of non-thermal electrons in flares from the observed dynamic spectra of hard X-ray bursts. *Solar Phys.* **18**, 489. DOI.

Carlsson, M.: 1998, Radiative transfer and radiation hydrodynamics. In: Vial, J.C., Bocchialini, K., Boumier, P. (eds.) *Space Solar Physics: Theoretical and Observational Issues in the Context of the SOHO Mission*, *Lecture Notes in Physics* **507**, Springer, Berlin, 163. DOI.

Carlsson, M., Rutten, R.J.: 1992, Solar hydrogen lines in the infrared. *Astron. Astrophys.* **259**, L53.

Carlsson, M., Stein, R.F.: 1992, Non-LTE radiating acoustic shocks and CA II K2V bright points. *Astrophys. J. Lett.* **397**, L59. DOI.

Carlsson, M., Stein, R.F.: 1994, Radiation shock dynamics in the solar chromosphere – Results of numerical simulations. In: Carlsson, M. (ed.) *Chromospheric Dynamics*, University of Oslo, Oslo, 47.

Carlsson, M., Stein, R.F.: 1995, Does a nonmagnetic solar chromosphere exist? *Astrophys. J. Lett.* **440**, L29. DOI.

Carlsson, M., Stein, R.F.: 1997, Formation of solar calcium H and K bright grains. *Astrophys. J.* **481**, 500. DOI.

Carlsson, M., Stein, R.F.: 2002, Dynamic hydrogen ionization. *Astrophys. J.* **572**, 626. DOI.

Christian, D.J., Mathioudakis, M., Jevremović, D., Dupuis, J., Vennes, S., Kawka, A.: 2003, The extreme-ultraviolet continuum of a strong stellar flare. *Astrophys. J. Lett.* **593**, L105. DOI.

Cram, L.E., Woods, D.T.: 1982, Models for stellar flares. *Astrophys. J.* **257**, 269. DOI.

Dalgarno, A., Griffing, G.W.: 1958, Energy per ion pair for electron and proton beams in atomic hydrogen. *Proc. Roy. Soc. London Ser. A* **248**, 415. DOI.

Dappen, W., Anderson, L., Mihalas, D.: 1987, Statistical mechanics of partially ionized stellar plasma – The Planck–Larkin partition function, polarization shifts, and simulations of optical spectra. *Astrophys. J.* **319**, 195. DOI.

Dennis, B.R., Zarro, D.M.: 1993, The Neupert effect – What can it tell us about the impulsive and gradual phases of solar flares? *Solar Phys.* **146**, 177. DOI.

Dere, K.P., Landi, E., Mason, H.E., Monsignori Fossi, B.C., Young, P.R.: 1997, CHIANTI – An atomic database for emission lines. *Astron. Astrophys. Suppl.* **125**, 149. DOI.

Donati-Falchi, A., Falciani, R., Smaldone, L.A.: 1985, Analysis of the optical spectra of solar flares. IV. The 'blue' continuum of white light flares. *Astron. Astrophys.* **152**, 165.

Donati-Falchi, A., Smaldone, L.A., Falciani, R.: 1984, Analysis of the optical spectra of solar flares. II. The energetics of the June 4, 1980 white light flare. *Astron. Astrophys.* **131**, 256.

Dorfi, E.A., Drury, L.O.: 1987, Simple adaptive grids for 1-D initial value problems. *J. Comput. Phys.* **69**, 175. DOI.

Doyle, J.G., Butler, C.J., Bryne, P.B., van den Oord, G.H.J.: 1988, Rotational modulation and flares on RS CVn and BY DRA systems. *Astron. Astrophys.* **193**, 229.

Drake, S.A., Ulrich, R.K.: 1980, The emission-line spectrum from a slab of hydrogen at moderate to high densities. *Astrophys. J. Suppl. Ser.* **42**, 351. DOI.

Eason, E.L.E., Giampapa, M.S., Radick, R.R., Worden, S.P., Hege, E.K.: 1992, Spectroscopic and photometric observations of a five-magnitude flare event on UV Ceti. *Astron. J.* **104**, 1161. DOI.

Emslie, A.G.: 1978, The collisional interaction of a beam of charged particles with a hydrogen target of arbitrary ionization level. *Astrophys. J.* **224**, 241. DOI.

Fang, C., Hénoux, J.C., Gan, W.Q.: 1993, Diagnostics of non-thermal processes in chromospheric flares. 1. Hoe and CaII K line profiles of an atmosphere bombarded by 10– 500 keV electrons. *Astron. Astrophys.* **274**, 917.

Feldman, U., Hiei, E., Phillips, K.J.H., Brown, C.M., Lang, J.: 1994, Very impulsive solar flares observed with the YOHKOH spacecraft. *Astrophys. J.* **421**, 843. DOI.

Fisher, G.H.: 1989, Dynamics of flare-driven chromospheric condensations. *Astrophys. J.* **346**, 1019. DOI.

Fisher, G.H., Canfield, R.C., McClymont, A.N.: 1985, Flare loop radiative hydrodynamics. Part Six. Chromospheric evaporation due to heating by nonthermal electrons. *Astrophys. J.* **289**, 425. DOI.

Fletcher, L., Hudson, H.S.: 2008, Impulsive phase flare energy transport by large-scale Alfvén waves and the electron acceleration problem. *Astrophys. J.* **675**, 1645. DOI.

Fuhrmeister, B., Lalitha, S., Poppenhaeger, K., Rudolf, N., Liefke, C., Reiners, A., Schmitt, J.H.M.M., Ness, J.U.: 2011, Multi-wavelength observations of Proxima Centauri. *Astron. Astrophys.* **534**, A133. DOI.

Fuhrmeister, B., Liefke, C., Schmitt, J.H.M.M., Reiners, A.: 2008, Multiwavelength observations of a giant flare on CN Leonis. I. The chromosphere as seen in the optical spectra. *Astron. Astrophys.* **487**, 293. DOI.

Gan, W.Q., Fang, C.: 1990, A hydrodynamic model of the gradual phase of the solar flare loop. *Astrophys. J.* **358**, 328. DOI.

Gan, W.Q., Rieger, E., Zhang, H.Q., Fang, C.: 1992, The role of chromospheric condensations in the continuum emission of white-light flares. *Astrophys. J.* **397**, 694. DOI.

García-Alvarez, D., Jevremović, D., Doyle, J.G., Butler, C.J.: 2002, Observations and modelling of a large optical flare on AT Microscopii. *Astron. Astrophys.* **383**, 548. DOI.

Gritsyk, P.A., Somov, B.V.: 2014, Reverse-current effect in present-day models of solar flares: Theory and high-accuracy observations. *Astron. Lett.* **40**, 499. DOI.

Guedel, M., Benz, A.O., Schmitt, J.H.M.M., Skinner, S.L.: 1996, The Neupert effect in active stellar coronae: Chromospheric evaporation and coronal heating in the dMe flare star binary UV ceti. *Astrophys. J.* **471**, 1002. DOI.

Gustafsson, B.: 1973, *A FORTRAN Program for Calculating "Continuous" Absorption Coefficients of Stellar Atmospheres, Uppsala Astr. Obs. Ann.* **5**.

Hansteen, V.: 2012, private communication.

Hawley, S.L., Allred, J.C., Johns-Krull, C.M., Fisher, G.H., Abbett, W.P., Alekseev, I., Avgoloupis, S.I., Deustua, S.E., Gunn, A., Seiradakis, J.H., Sirk, M.M., Valenti, J.A.: 2003, Multiwavelength observations of flares on AD Leonis. *Astrophys. J.* **597**, 535. DOI.

Hawley, S.L., Davenport, J.R.A., Kowalski, A.F., Wisniewski, J.P., Hebb, L., Deitrick, R., Hilton, E.J.: 2014, Kepler flares. I. Active and inactive M dwarfs. *Astrophys. J.* **797**, 121. DOI.

Hawley, S.L., Fisher, G.H.: 1992, X-ray-heated models of stellar flare atmospheres – Theory and comparison with observations. *Astrophys. J. Suppl. Ser.* **78**, 565. DOI.

Hawley, S.L., Fisher, G.H.: 1994, Solar flare model atmospheres. *Astrophys. J.* **426**, 387. DOI.

Hawley, S.L., Fisher, G.H., Simon, T., Cully, S.L., Deustua, S.E., Jablonski, M., Johns-Krull, C.M., Pettersen, B.R., Smith, V., Spiesman, W.J., Valenti, J.: 1995, Simultaneous extreme-ultraviolet explorer and optical observations of AD Leonis: Evidence for large coronal loops and the Neupert effect in stellar flares. *Astrophys. J.* **453**, 464. DOI.

Hawley, S.L., Pettersen, B.R.: 1991, The great flare of 1985 April 12 on AD Leonis. *Astrophys. J.* **378**, 725. DOI.

Heinzel, P., Kleint, L.: 2014, Hydrogen Balmer continuum in solar flares detected by the Interface Region Imaging Spectrograph (IRIS). *Astrophys. J. Lett.* **794**, L23. DOI.

Holman, G.D.: 2012, Understanding the impact of return-current losses on the X-ray emission from solar flares. *Astrophys. J.* **745**, 52. DOI.

Holman, G.D., Sui, L., Schwartz, R.A., Emslie, A.G.: 2003, Electron Bremsstrahlung hard X-ray spectra, electron distributions, and energetics in the 2002 July 23 solar flare. *Astrophys. J. Lett.* **595**, L97. DOI.

Houdebine, E.R.: 1992, Investigating the spectroscopic signatures of stellar flares. *Ir. Astron. J.* **20**, 213.

Hubeny, I., Hummer, D.G., Lanz, T.: 1994, NLTE model stellar atmospheres with line blanketing near the series limits. *Astron. Astrophys.* **282**, 151.

Hummer, D.G., Mihalas, D.: 1988, The equation of state for stellar envelopes. I – An occupation probability formalism for the truncation of internal partition functions. *Astrophys. J.* **331**, 794. DOI.

Ireland, J., Tolbert, A.K., Schwartz, R.A., Holman, G.D., Dennis, B.R.: 2013, Estimating the properties of hard X-ray solar flares by constraining model parameters. *Astrophys. J.* **769**, 89. DOI.

Jevremovic, D., Butler, C.J., Drake, S.A., O'Donoghue, D., van Wyk, F.: 1998, Ultraviolet and optical flares on GL 866. *Astron. Astrophys.* **338**, 1057.

Johns-Krull, C.M., Hawley, S.L., Basri, G., Valenti, J.A.: 1997, Hamilton Echelle spectroscopy of the 1993 March 6 solar flare. *Astrophys. J. Suppl. Ser.* **112**, 221. DOI.

Johns-Krull, C.M., Valenti, J.A.: 1996, Detection of strong magnetic fields on M dwarfs. *Astrophys. J. Lett.* **459**, L95. DOI.

Karlický, M., Brown, J.C., Conway, A.J., Penny, G.: 2000, Flare hard X-rays from neutral beams. *Astron. Astrophys.* **353**, 729.

Katsova, M.M., Boiko, A.Y., Livshits, M.A.: 1997, The gas-dynamic model of impulsive stellar flares. *Astron. Astrophys.* **321**, 549.

Kašparová, J., Varady, M., Heinzel, P., Karlický, M., Moravec, Z.: 2009, Response of optical hydrogen lines to beam heating. I. Electron beams. *Astron. Astrophys.* **499**, 923. DOI.

Kepple, P., Griem, H.R.: 1968, Improved Stark profile calculations for the hydrogen lines Hα, Hβ, Hγ, and Hδ. *Phys. Rev.* **173**, 317. DOI.

Kerr, G.S., Fletcher, L.: 2014, Physical properties of white-light sources in the 2011 February 15 solar flare. *Astrophys. J.* **783**, 98. DOI.

Kowalski, A.F., Cauzzi, G., Fletcher, L.: 2015, Optical spectral observations of a flickering white-light kernel in a C1 solar flare. *Astrophys. J.* **798**, 107. DOI.

Kowalski, A.F., Hawley, S.L., Holtzman, J.A., Wisniewski, J.P., Hilton, E.J.: 2010, A white light megaflare on the dM4.5e Star YZ CMi. *Astrophys. J. Lett.* **714**, L98. DOI.

Kowalski, A.F., Hawley, S.L., Holtzman, J.A., Wisniewski, J.P., Hilton, E.J.: 2012, The multiple continuum components in the white-light flare of 16 January 2009 on the dM4.5e star YZ CMi. *Solar Phys.* **277**, 21. DOI.

Kowalski, A.F., Hawley, S.L., Wisniewski, J.P., Osten, R.A., Hilton, E.J., Holtzman, J.A., Schmidt, S.J., Davenport, J.R.A.: 2013, Time-resolved properties and global trends in dMe flares from simultaneous photometry and spectra. *Astrophys. J. Suppl. Ser.* **207**, 15. DOI.

Kretzschmar, M.: 2011, The Sun as a star: Observations of white-light flares. *Astron. Astrophys.* **530**, A84. DOI.

Krucker, S., Hudson, H.S., Jeffrey, N.L.S., Battaglia, M., Kontar, E.P., Benz, A.O., Csillaghy, A., Lin, R.P.: 2011, High-resolution imaging of solar flare ribbons and its implication on the thick-target beam model. *Astrophys. J.* **739**, 96. DOI.

Kunkel, W.E.: 1970, On the spectra of stellar flares. *Astrophys. J.* **161**, 503. DOI.

Leenaarts, J., Pereira, T.M.D., Carlsson, M., Uitenbroek, H., De Pontieu, B.: 2013, The formation of IRIS diagnostics. II. The formation of the Mg II h & k lines in the solar atmosphere. *Astrophys. J.* **772**, 90. DOI.

Lemke, M.: 1997, Extended VCS Stark broadening tables for hydrogen – Lyman to Brackett series. *Astron. Astrophys. Suppl.* **122**, 285. DOI.

Lin, R.P., Dennis, B.R., Hurford, G.J., Smith, D.M., Zehnder, A., Harvey, P.R., Curtis, D.W., Pankow, D., Turin, P., Bester, M., Csillaghy, A., Lewis, M., Madden, N., van Beek, H.F., Appleby, M., Raudorf, T., McTiernan, J., Ramaty, R., Schmahl, E., Schwartz, R., Krucker, S., Abiad, R., Quinn, T., Berg, P., Hashii, M., Sterling, R., Jackson, R., Pratt, R., Campbell, R.D., Malone, D., Landis, D., Barrington-Leigh, C.P., Slassi-Sennou, S., Cork, C., Clark, D., Amato, D., Orwig, L., Boyle, R., Banks, I.S., Shirey, K., Tolbert, A.K., Zarro, D., Snow, F., Thomsen, K., Henneck, R., McHedlishvili, A., Ming, P., Fivian, M., Jordan, J., Wanner, R., Crubb, J., Preble, J., Matranga, M., Benz, A., Hudson, H., Canfield, R.C., Holman, G.D., Crannell, C., Kosugi, T., Emslie, A.G., Vilmer, N., Brown, J.C., Johns-Krull, C., Aschwanden, M., Metcalf, T., Conway, A.: 2002, The Reuven Ramaty High-Energy Solar Spectroscopic Imager (RHESSI). *Solar Phys.* **210**, 3. DOI.

Livshits, M.A., Badalian, O.G., Kosovichev, A.G., Katsova, M.M.: 1981, The optical continuum of solar and stellar flares. *Solar Phys.* **73**, 269. DOI.
Longcope, D.W., Guidoni, S.E.: 2011, A model for the origin of high density in looptop X-ray sources. *Astrophys. J.* **740**, 73. DOI.
Magain, P.: 1986, Contribution functions and the depths of formation of spectral lines. *Astron. Astrophys.* **163**, 135.
Mathioudakis, M., Bloomfield, D.S., Jess, D.B., Dhillon, V.S., Marsh, T.R.: 2006, The periodic variations of a white-light flare observed with ULTRACAM. *Astron. Astrophys.* **456**, 323. DOI.
Mihalas, D.: 1978, *Stellar Atmospheres*, 2nd edn. Freeman, New York.
Milligan, R.O., Kerr, G.S., Dennis, B.R., Hudson, H.S., Fletcher, L., Allred, J.C., Chamberlin, P.C., Ireland, J., Mathioudakis, M., Keenan, F.P.: 2014, The radiated energy budget of chromospheric plasma in a major solar flare deduced from multi-wavelength observations. *Astrophys. J.* **793**, 70. DOI.
Mott, N.F., Massey, H.S.W.: 1965, *Theory of Atomic Collisions*, 3rd edn. Oxford University Press, Oxford.
Mullan, D.J., Mathioudakis, M., Bloomfield, D.S., Christian, D.J.: 2006, A comparative study of flaring loops in active stars. *Astrophys. J. Suppl. Ser.* **164**, 173. DOI.
Nayfonov, A., Däppen, W., Hummer, D.G., Mihalas, D.: 1999, The MHD equation of state with post-Holtsmark microfield distributions. *Astrophys. J.* **526**, 451. DOI.
Neidig, D.F.: 1983, Spectral analysis of the optical continuum in the 24 April 1981 flare. *Solar Phys.* **85**, 285. DOI.
Neidig, D.F., Grosser, H., Hrovat, M.: 1994, Optical output of the 24 April 1984 white-light flare. *Solar Phys.* **155**, 199. DOI.
Osten, R.A., Drake, S., Tueller, J., Cummings, J., Perri, M., Moretti, A., Covino, S.: 2007, Nonthermal hard X-ray emission and iron Kα emission from a superflare on II Pegasi. *Astrophys. J.* **654**, 1052. DOI.
Osten, R.A., Godet, O., Drake, S., Tueller, J., Cummings, J., Krimm, H., Pye, J., Pal'shin, V., Golenetskii, S., Reale, F., Oates, S.R., Page, M.J., Melandri, A.: 2010, The mouse that roared: A superflare from the dMe flare star EV Lac detected by Swift and Konus-Wind. *Astrophys. J.* **721**, 785. DOI.
Osten, R.A., Hawley, S.L., Allred, J., Johns-Krull, C.M., Brown, A., Harper, G.M.: 2006, From radio to X-ray: The quiescent atmosphere of the dMe flare star EV Lacertae. *Astrophys. J.* **647**, 1349. DOI.
Paulson, D.B., Allred, J.C., Anderson, R.B., Hawley, S.L., Cochran, W.D., Yelda, S.: 2006, Optical spectroscopy of a flare on Barnard's star. *Publ. Astron. Soc. Pac.* **118**, 227. DOI.
Phan-Bao, N., Lim, J., Donati, J.F., Johns-Krull, C.M., Martín, E.L.: 2009, Magnetic field topology in low-mass stars: Spectropolarimetric observations of M dwarfs. *Astrophys. J.* **704**, 1721. DOI.
Pillet, P., van Linden van den Heuvell, H.B., Smith, W.W., Kachru, R., Tran, N.H., Gallagher, T.F.: 1984, Microwave ionization of Na Rydberg atoms. *Phys. Rev. A* **30**, 280. DOI.
Qiu, J., Liu, W., Hill, N., Kazachenko, M.: 2010, Reconnection and energetics in two-ribbon flares: A revisit of the Bastille-day flare. *Astrophys. J.* **725**, 319. DOI.
Ricchiazzi, P.J.: 1982, A static model of chromospheric heating in solar flares. Ph.D. Thesis, California University, San Diego.
Ricchiazzi, P.J., Canfield, R.C.: 1983, A static model of chromospheric heating in solar flares. *Astrophys. J.* **272**, 739. DOI.
Robinson, R.D., Carpenter, K.G., Percival, J.W., Bookbinder, J.A.: 1995, A search for microflaring activity on dMe flare stars. I. Observations of the dM8e star CN Leonis. *Astrophys. J.* **451**, 795. DOI.
Rybicki, G.B., Hummer, D.G.: 1992, An accelerated lambda iteration method for multilevel radiative transfer. II. Overlapping transitions with full continuum. *Astron. Astrophys.* **262**, 209.
Scharmer, G.B.: 1981, Solutions to radiative transfer problems using approximate lambda operators. *Astrophys. J.* **249**, 720. DOI.
Scharmer, G.B., Carlsson, M.: 1985, A new approach to multi-level non-LTE radiative transfer problems. *J. Comput. Phys.* **59**, 56. DOI.
Schrijver, C.J., Hudson, H.S., Murphy, R.J., Share, G.H., Tarbell, T.D.: 2006, Gamma rays and the evolving, compact structures of the 2003 October 28 X17 flare. *Astrophys. J.* **650**, 1184. DOI.
Seaton, M.J.: 1990, Atomic data for opacity calculations. XIII. Line profiles for transitions in hydrogenic ions. *J. Phys. B, At. Mol. Opt. Phys.* **23**, 3255. DOI.
Smith, K., Güdel, M., Audard, M.: 2005, Flares observed with XMM-Newton and the VLA. *Astron. Astrophys.* **436**, 241. DOI.
Smith, R.K., Brickhouse, N.S., Liedahl, D.A., Raymond, J.C.: 2001, Collisional plasma models with APEC/APED: Emission-line diagnostics of hydrogen-like and helium-like ions. *Astrophys. J. Lett.* **556**, L91. DOI.
Stehle, C., Jacquemot, S.: 1993, Line shapes in hydrogen opacities. *Astron. Astrophys.* **271**, 348.
Su, Y., Holman, G.D., Dennis, B.R.: 2011, Evidence for the full hard X-ray spectral signature of nonuniform ionization in a solar flare. *Astrophys. J.* **731**, 106. DOI.
Sutton, K.: 1978, Approximate line shapes for hydrogen. *J. Quant. Spectrosc. Radiat. Transf.* **20**, 333. DOI.

Švestka, Z.: 1963, Spectral analysis of the moustache-like flare of August 7, 1960. *Bull. Astron. Inst. Czechoslov.* **14**, 234.

Tremblay, P.E., Bergeron, P.: 2009, Spectroscopic analysis of DA white dwarfs: Stark broadening of hydrogen lines including nonideal effects. *Astrophys. J.* **696**, 1755. DOI.

Uitenbroek, H.: 2001, Multilevel radiative transfer with partial frequency redistribution. *Astrophys. J.* **557**, 389. DOI.

Varady, M., Karlický, M., Moravec, Z., Kašparová, J.: 2014, Modifications of thick-target model: Re-acceleration of electron beams by static and stochastic electric fields. *Astron. Astrophys.* **563**, A51. DOI.

Vidal, C.R., Cooper, J., Smith, E.W.: 1970, Hydrogen Stark broadening calculations with the unified classical path theory. *J. Quant. Spectrosc. Radiat. Transf.* **10**, 1011. DOI.

Vidal, C.R., Cooper, J., Smith, E.W.: 1971, Unified theory calculations of Stark broadened hydrogen lines including lower state interactions. *J. Quant. Spectrosc. Radiat. Transf.* **11**, 263. DOI.

Vidal, C.R., Cooper, J., Smith, E.W.: 1973, Hydrogen Stark-broadening tables. *Astrophys. J. Suppl. Ser.* **25**, 37. DOI.

Wang, L., Fang, C., Ding, M.: 2007, Small-scale brightenings in the UV continuum of an M9.1 solar flare. *Chin. J. Astron. Astrophys.* **7**, 721. DOI.

Warren, H.P.: 2006, Multithread hydrodynamic modeling of a solar flare. *Astrophys. J.* **637**, 522. DOI.

Wiese, W.L., Kelleher, D.E., Paquette, D.R.: 1972, Detailed study of the Stark broadening of Balmer lines in a high-density plasma. *Phys. Rev. A* **6**, 1132. DOI.

Worden, S.P., Schneeberger, T.J., Giampapa, M.S., Deluca, E.E., Cram, L.E.: 1984, The response of chromospheric emission lines to flares on YZ Canis Minoris. *Astrophys. J.* **276**, 270. DOI.

Young, P.R., Del Zanna, G., Landi, E., Dere, K.P., Mason, H.E., Landini, M.: 2003, CHIANTI – An atomic database for emission lines. VI. Proton rates and other improvements. *Astrophys. J. Suppl. Ser.* **144**, 135. DOI.

Zener, C.: 1932, Non-adiabatic crossing of energy levels. *Proc. Roy. Soc. London Ser. A* **137**, 696. DOI.

Zirin, H.: 1980, Three flares with blue continuum, one with a D3 preflare shell. *Astrophys. J.* **235**, 618. DOI.

Zirin, H., Neidig, D.F.: 1981, Continuum emission in the 1980 July 1 solar flare. *Astrophys. J. Lett.* **248**, L45. DOI.

Reprinted from *Solar Physics* Journal, DOI 10.1007/s11207-015-0814-9

Mg II Lines Observed During the X-class Flare on 29 March 2014 by the *Interface Region Imaging Spectrograph*

W. Liu[1] · P. Heinzel[1] · L. Kleint[2] · J. Kašparová[1]

Received: 13 March 2015 / Accepted: 3 November 2015 / Published online: 16 November 2015
© Springer Science+Business Media Dordrecht 2015

Abstract Mg II lines represent one of the strongest emissions from the chromospheric plasma during solar flares. In this article, we studied the Mg II lines observed during the X1 flare on 29 March 2014 (SOL2014-03-29T17:48) by the *Interface Region Imaging Spectrograph* (IRIS). IRIS detected large intensity enhancements of the Mg II *h* and *k* lines, subordinate triplet lines, and several other metallic lines at the flare footpoints during this flare. We have used the advantage of the slit-scanning mode (rastering) of IRIS and performed, for the first time, a detailed analysis of spatial and temporal variations of the spectra. Moreover, we were also able to identify positions of strongest hard X-ray (HXR) emissions using the *Reuven Ramaty High Energy Solar Spectroscopic Imager* (RHESSI) observations and to correlate them with the spatial and temporal evolution of IRIS Mg II spectra. The light curves of the Mg II lines increase and peak contemporarily with the HXR emissions but decay more gradually. There are large red asymmetries in the Mg II *h* and *k* lines after the flare peak. We see two spatially well-separated groups of Mg II line profiles, non-reversed and reversed. In some cases, the Mg II footpoints with reversed profiles are correlated with HXR sources. We show the spatial and temporal behavior of several other line parameters (line metrics) and briefly discuss them. Finally, we have synthesized the Mg II *k* line using our non-LTE code with the Multilevel Accelerated Lambda Iteration (MALI) technique. Two kinds of models are considered, the flare model F2 of Machado *et al.* (*Astrophys. J.*

Solar and Stellar Flares: Observations, Simulations, and Synergies
Guest Editors: Lyndsay Fletcher and Petr Heinzel

✉ W. Liu
wenjuan.liu@asu.cas.cz

P. Heinzel
petr.heinzel@asu.cas.cz

L. Kleint
lucia.kleint@fhnw.ch

J. Kašparová
jana.kasparova@asu.cas.cz

1 Astronomical Institute, The Czech Academy of Sciences, 25165 Ondřejov, Czech Republic

2 University of Applied Sciences and Arts Northwestern Switzerland, 5210 Windisch, Switzerland

242, 336, 1980) and the models of Ricchiazzi and Canfield (*Astrophys. J.* **272**, 739, 1983, RC models). Model F2 reproduces the peak intensity of the non-reversed Mg II *k* profile at flare maximum, but does not account for high wing intensities. On the other hand, the RC models show the sensitivity of Mg II line intensities to various electron-beam parameters. Our simulations also show that the microturbulence produces a broader line core, while the intense line wings are caused by an enhanced line source function.

Keywords Flares, spectrum · Flares, dynamics · Radiative transfer

1. Introduction

The *Interface Region Imaging Spectrograph* (IRIS: De Pontieu *et al.*, 2014) was launched in the middle of 2013, and since then, many excellent images and spectra were acquired and analyzed, see *e.g.* a special issue of *Science* in 2014 (Vol. **346**). Among the targets of IRIS, solar flares represent a new challenge because of the unprecedented spatial resolution achieved in the UV/EUV and potentially high cadence, which allows continuous rastering of the flaring region. Many flares have already been captured by IRIS, and some of them were X-class flares. IRIS records spectra in two different wavelength channels: in the far-UV (FUV) from 1332 Å to 1358 Å and 1389 Å to 1407 Å, and in the near-UV (NUV) from 2783 Å to 2835 Å. In this article, we concentrate on NUV observations with the objective of studying the behavior of Mg II lines, which appear very bright during flares. Before the launch of IRIS, the only comprehensive study of Mg II lines in a flare, to our knowledge, was based on the *8th Orbiting Solar Observatory* (OSO-8) spectral observations obtained in hydrogen, calcium and magnesium lines by the *Laboratoire de Physique Stellaire et Planetaire* (LPSP) UV spectrometer (Lemaire, Choucq-Bruston, and Vial, 1984). Six resonance lines were detected and their temporal evolution was described: hydrogen Ly α and Ly β, Ca II H (together with the blending Hε Balmer line) and K lines, and Mg II *h* and *k* resonance lines. During the observed flare, the LPSP spectrometer was also able to detect an enhanced emission in three subordinate (intersystem) Mg II lines, which have wavelengths around the Mg II *k* line and appear as absorption lines in the quiet-Sun (QS) spectrum. With the observations from IRIS, the Mg II lines during an M-class flare were studied in detail by Kerr *et al.* (2015). They investigated the spatial and temporal behavior of Mg II lines and found redshifts of line centroids, line broadening, and blue asymmetries in flare ribbons. The formation of Mg II lines in the QS has recently been thoroughly discussed by Leenaarts *et al.* (2013), in preparation for the analysis of the new IRIS data. It is interesting to note that, according to Lemaire, Choucq-Bruston, and Vial (1984), both Mg II *h* and *k* integrated line intensities are stronger than the hydrogen Ly α line during the whole evolution of the flare.

Although the hydrogen lines, and in particular the Lyman and Balmer lines, have been extensively modeled for various flare conditions, including time-dependent models, Mg II lines are largely unexplored. They have been modeled under QS conditions (Milkey and Mihalas, 1974; Bocchialini and Gouttebroze, 1996; Uitenbroek, 1997; Leenaarts *et al.*, 2013; Avrett, Landi, and McKillop, 2013), but the only study of flare atmospheres is that of Avrett, Machado, and Kurucz (1986), who used the well-known semi-empirical flare models of Machado *et al.* (1980) and synthesized the Mg II *k* line intensities that emerge from F1, F2, and F3 model atmospheres. Lemaire, Choucq-Bruston, and Vial (1984) used models F1 and F2 and computed their own synthetic intensities to compare them with Mg II OSO-8 observations.

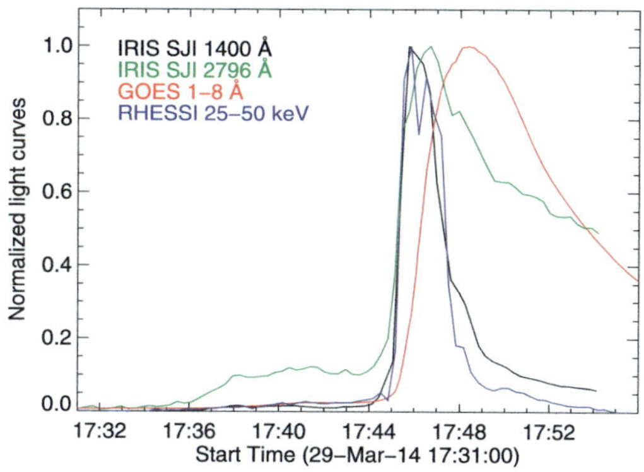

Figure 1 Normalized and pre-flare-subtracted light curves of the 29 March 2014 flare in 1400 Å and 2796 Å (the sum of counts in the images of the AR NOAA 12017 taken by the IRIS slitjaw imager 1400 Å and 2796 Å, divided by the exposure time) in comparison with GOES 1 – 8 Å and RHESSI 25 – 50 keV HXR light curves.

One of the most spectacular events detected by IRIS was the X1-class flare on 29 March 2014 (Kleint *et al.*, 2015a). Several other ground- and space-based instruments observed this flare. Strong Mg II emission was reported by Heinzel and Kleint (2014), who studied the accompanying enhancement of the nearby hydrogen Balmer continuum. In this article we perform the first detailed analysis of Mg II lines in this flare, using the temporal coverage of the flare from its pre-flare period to the decay phase shortly after the flare maximum. We define several 'metrics' quantities and show how they evolve in time. We also perform non-LTE radiative-transfer modeling using the semi-empirical model F2 (Machado *et al.*, 1980) as a representative atmospheric snapshot of the temperature and density structure and compare our synthetic Mg II line intensities with IRIS observations. We use the theoretical energy balance models of Ricchiazzi and Canfield (1983) to study how various electron-beam parameters affect the profiles of the Mg II lines.

2. Observations

The X1 flare observed on 29 March 2014 occurred in the active region (AR) NOAA 12017. It was observed by IRIS from three hours before the flare to a few minutes after the flare peak with a NUV exposure time of 8.00 s before 17:46:04.78 UT and 2.44 s after 17:46:13.98 UT, and 75 s raster cadence (see Heinzel and Kleint (2014) and Young, Tian, and Jaeggli (2015) for detailed descriptions of IRIS observations of this flare). Light curves of the flare in IRIS 1400 Å and IRIS 2796 Å channels together with X-ray light curves are plotted in Figure 1. The light curves in 1400 Å and 2796 Å are the total counts of the active region, derived from the IRIS slitjaw images, divided by the exposure time. Figure 1 shows that the GOES flux started to rise impulsively at 17:45 UT and peaked at 17:48 UT, while fluxes in 1400 Å, 2796 Å and hard X-rays (HXR) peaked 1 – 3 minutes earlier. Figure 1 also shows that during the decay phase of the flare, the light curves in the 1400 Å slitjaw image followed the 25 – 50 keV HXR fluxes observed by the *Reuven Ramaty High Energy Solar Spectroscopic Imager* (RHESSI: Lin *et al.*, 2002), while the 2796 Å SJI light curve seems to be slightly delayed and more gradual.

Images of the flare at 17:46 UT in 1400 Å and 2796 Å are presented in Figure 2. The flare ribbons are clearly seen as black patches in the intensity-reversed images. Although the contrast of the image in 2796 Å is lower, the flare ribbons look nearly identical in the two images. These images are representative of the plasma in the lower transition region (1400 Å) and upper chromosphere (2796 Å) (De Pontieu *et al.*, 2014). The IRIS slit positions

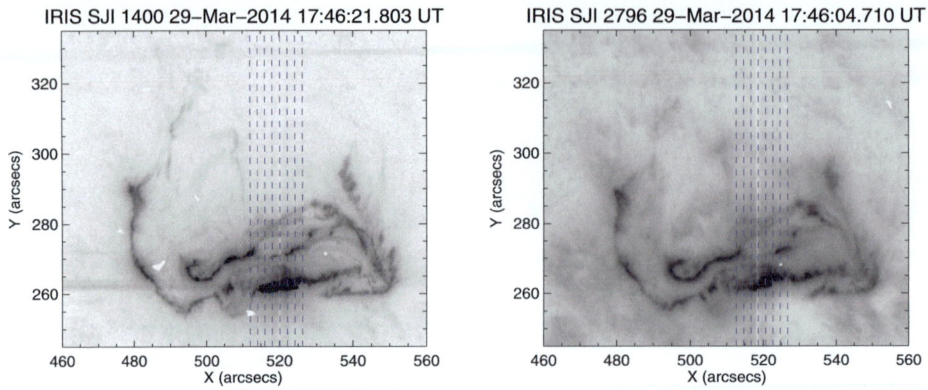

Figure 2 An IRIS 1400 Å slitjaw image taken at 17:46:22 UT (left) and a 2796 Å slitjaw image at 17:46:04 UT (right), with intensity reversed and in log scale. The IRIS slit positions are indicated by vertical dashed lines.

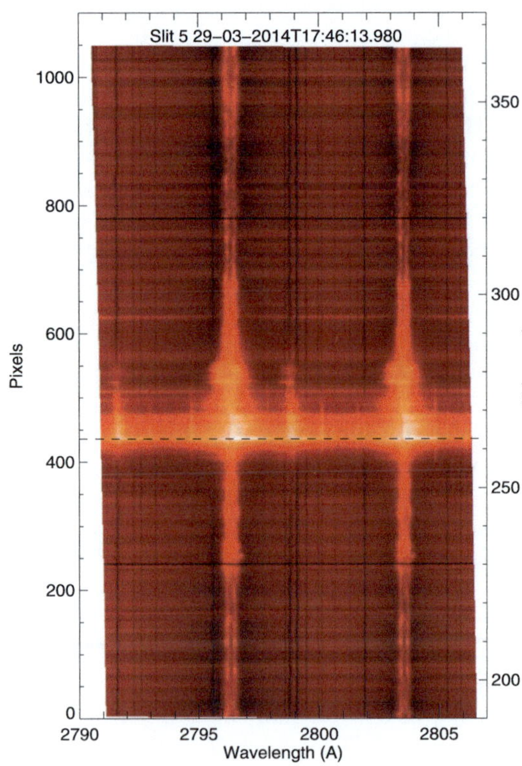

Figure 3 IRIS spectra of the Mg II lines at 17:46:13.98 UT along slit 5. The dashed horizontal line indicates the position of pixel 436, which is used to plot the temporal evolution of the line profiles in Figure 6.

are also plotted in Figure 2 as vertical dashed lines; they covered the brightest part of the southern ribbons and part of the northern ribbon during the flare.

Figure 3 shows the IRIS spectra of the Mg II lines taken at 17:46:13.98 UT along slit 5. The strong emission in Mg II *h*, *k*, and in the subordinate lines as well as in other metallic lines at 17:46 UT is clearly visible. These emitted Mg II lines are asymmetric in some of the

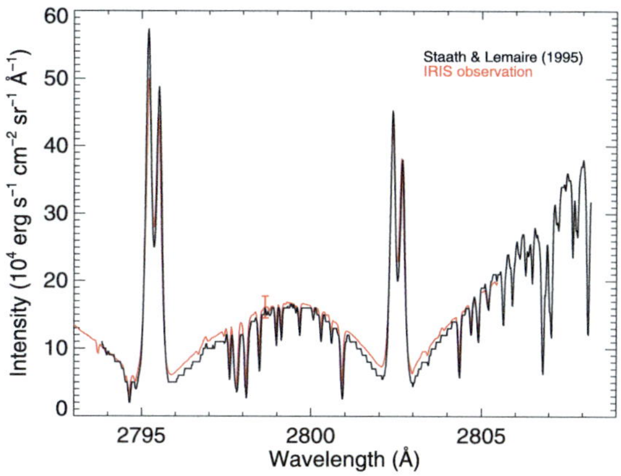

Figure 4 The Mg II spectrum averaged over 100 pixels in the QS region at $\mu = \cos\theta = 0.83$ in comparison with the QS spectrum observed by RASOLBA at disk center (Staath and Lemaire, 1995). Note that IRIS records spectra in vacuum while the RASOLBA observation was made in the air. We shifted the IRIS spectrum by 0.84 Å to the blue to align the wavelengths of these two observations. The error bar indicates the 10 % error for the IRIS calibration.

brightest pixels, with the red wings much broader and brighter than the blue wings. The red asymmetries are usually observed in chromospheric lines (*e.g.*, Hα, Ca II K) during flares (*e.g.*, Švestka, Kopecký, and Blaha, 1962; Tang, 1983; Ichimoto and Kurokawa, 1984; Asai *et al.*, 2012; Deng *et al.*, 2013) and are caused by downward motions of the chromospheric material, which is known as chromospheric condensation (Fisher, Canfield, and McClymont, 1985; Fisher, 1989; Longcope, 2014). The dashed horizontal line in Figure 3 shows the position of one footpoint pixel, which is the brightest pixel along the slit 5 at 17:46:13.98 UT and is used to plot the temporal evolution of Mg II line profiles in Figure 6.

We used calibrated IRIS level-2 data. The intensity observed by IRIS in units of data number (DN) is converted into intensities in physical units by the relation

$$I(\lambda, t) = \frac{n(\lambda, t)\ Q(\lambda)hc/\lambda}{\delta t \delta\lambda A_{\mathrm{eff}}\Omega}, \tag{1}$$

where $I(\lambda, t)$ is the specific intensity in physical units of erg s^{-1} cm^{-2} sr^{-1} Å^{-1}, $n(\lambda, t)$ is the observed intensity in units of DN, read from IRIS level-2 data, $Q(\lambda)$ is the product of the CCD gain and the number of photons needed to create one electron-hole pair on the detector, which is 18 photons DN^{-1} for the NUV band (De Pontieu *et al.*, 2014), h is the Planck constant in units of erg s, c is the speed of light in units of cm s^{-1}, λ is the wavelength in cm, δt is the exposure time, which is 8 s for this flare in the NUV channel before 17:46 UT, $\delta\lambda$ is the spectral resolution, which is 0.02546 Å for the NUV spectra, A_{eff} is the effective area for the NUV band obtained from the SolarSoftware (SSW: Freeland and Handy, 1998) IDL procedure *iris_get_response*, and Ω is the solid angle subtended by one pixel in the y-axis of the slit. For the Mg II spectrum between 2790 Å and 2806 Å, the factor to convert DN s^{-1} into physical units is in the range of 1.82×10^4 to 1.9×10^4. The new IRIS calibration by J.-P. Wülser shows that there is a difference of 10 % for the calibration.

To test the absolute radiometric calibration of the IRIS NUV spectrum, we plot the QS spectrum in the NUV band observed by IRIS in comparison with that observed by the balloon-borne telescope and spectrograph RASOLBA on 19 September 1986 (Staath and Lemaire, 1995) in Figure 4. The IRIS QS spectrum shown in Figure 4 is a spectrum averaged over 100 consecutive pixels (pixels 100 – 199, corresponding to 16.7″) and averaged over all eight slit positions in the QS region located at $\mu = \cos\theta = 0.83$ (Heinzel and Kleint, 2014) taken around 17:30 UT, while the RASOLBA spectrum was an average over 30″ slit length and over ten sets of data with alternative 30 s and 90 s exposure times taken at the

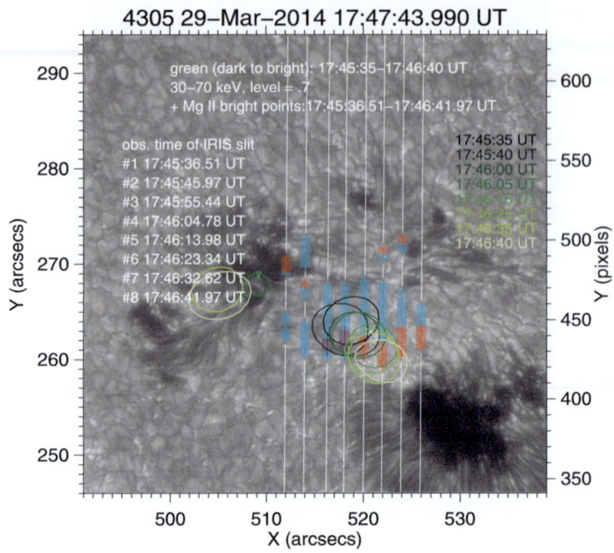

Figure 5 The Mg II bright footpoints (with the integrated intensity for the Mg II *h* line $\geqslant 10^7\ \mathrm{erg\,s^{-1}\,cm^{-2}\,sr^{-1}}$) along all eight slits and RHESSI 30 – 70 keV HXR contours (level = 0.7) overplotted on a G-band image from the *Richard B. Dunn Solar Telescope* at the National Solar Observatory (NSO/DST). The positions of slits are indicated by vertical white lines. The Mg II bright points are plotted with a blue "+" sign for a non-reversed line profile, red for a reversed line profile, and violet for saturated pixels. The RHESSI contours were created with an integration of 12 s.

center of the solar disk. Figure 4 shows that the IRIS QS spectrum in the wings of the Mg II *h* and *k* lines is very close to the one from RASOLBA. For example, the IRIS QS spectrum at 2800 Å is $1.62 \times 10^5\ \mathrm{erg\,s^{-1}\,cm^{-2}\,sr^{-1}\,Å^{-1}}$, while the RASOLBA spectrum at the same wavelength is $1.6 \times 10^5\ \mathrm{erg\,s^{-1}\,cm^{-2}\,sr^{-1}\,Å^{-1}}$. The RASOLBA spectrum at 2800 Å is higher than that from the HRTS-9 rocket (Morrill and Korendyke, 2008; Pereira *et al.*, 2013) and other previously calibrated rocket spectra at QS disk center. For example, Kohl and Parkinson (1976) obtained 14.7 ± 1.8, and Bonnet (1968) derived 15.0 ± 3.8 (see Table 2a in Lemaire *et al.*, 1981, for a list of 2800 Å absolute intensity at Sun center). In general, the difference between IRIS and RASOLBA QS spectra in the wings is less than 10 % of the RASOLBA data. The differences in the line cores are more pronounced, and they reflect the spatial and temporal variations of the quiet chromosphere.

To study the relation between the location of footpoints in Mg II and HXR sources, we plot the Mg II bright footpoints along all eight slits for one raster with RHESSI 30 – 70 keV HXR contours around the peak of HXR fluxes in Figure 5. We can see that there are two HXR sources around the peak. The IRIS slit did not cross the northern HXR source. The southern HXR source moved southwest and crossed slits 3 – 7 from 17:45:35 UT to 17:46:40 UT. Mg II bright footpoints along slits 5 and 6 with reversed profiles coincide well with HXR sources. The Mg II line profiles at these footpoints become non-reversed after the HXR peak, which is clearly seen in the plot of the center-to-peak ratio for slit 5 (Figure 13). Figure 5 also shows that only a few pixels along slits 3 and 4 (at 17:45:55 UT and 17:46:04 UT, respectively) are saturated in the Mg II lines. This shows that the majority of IRIS spectra for this flare are of excellent quality.

In Section 3 we use several parameters (metrics) to characterize the profiles and intensities of the Mg II lines and to study their variations during the flare. Since the Mg II bright footpoints along slits 5 and 6 with reversed profiles are correlated with HXR sources at the HXR peak and the temporal evolution of these metrics along slit 5 and 6 are very similar, we show the evolution of these parameters along slit 5 as an example for most cases.

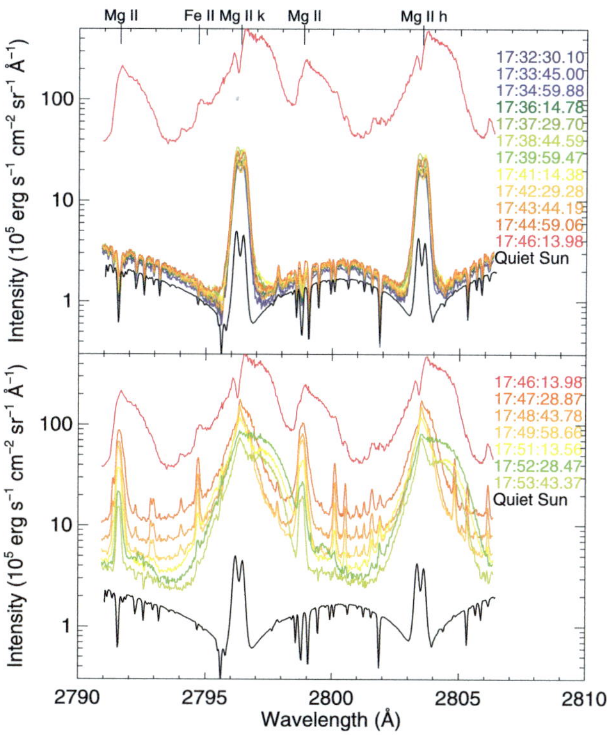

Figure 6 Temporal evolution of the Mg II spectra in one pixel (pixel 436, slit 5) at the footpoint of the flare. The upper panel shows spectra during the rising phase up to the peak time of Mg II, while the lower panel shows spectra during the decay phase (including the peak time of Mg II). The QS spectrum is plotted in both panels in black as a reference. The position of the pixel is marked by a dashed horizontal line in Figure 3 and in Figures 8 to 14.

3. Evolution of Mg II Spectra During the X1 Flare

Figure 6 shows the temporal evolution of Mg II spectra in one pixel (pixel 436, slit 5) at the footpoint of the flare. The averaged QS spectrum is also plotted in Figure 6 as a reference. Increases of line intensities due to flare energy deposition in the chromosphere are clearly seen in Figure 6. At the center of the Mg II *h* and *k* lines and the subordinate lines (the three 3p – 3d triplet lines at 2791.60 Å, 2798.75 Å, and 2798.82), intensities increase from 17:36 UT, peak at 17:46 UT, and decrease thereafter. At the peak time (17:46 UT), the intensities in the whole spectral range from 2790 Å to 2806 Å increase dramatically, and the widths of the Mg II lines and metallic lines increase as well. We note that for pixel 436 the line profiles of the Mg II *h* and *k* lines are reversed at the peak time and become non-reversed after the peak. For pixels close to pixel 436 the Mg II lines are reversed at the peak time, while for positions a few pixels north of pixel 436, the Mg II lines are in emission and more symmetric at the peak time. After the peak, the temporal evolution of the line profiles in Mg II bright pixels along slit 5 is similar to that of pixel 436 (line profiles are not shown). At 17:54 UT, the intensities in the wings of the Mg II *h* and *k* lines return to their pre-flare level, which is about 1.5 times the QS intensities. The subordinate lines change from absorption to emission at 17:44 UT and remain in emission after the peak time. All other metallic lines also go into emission at the peak time (17:46 UT) and return to absorption at 17:54 UT. Figure 6 shows that the line profiles of the *h* and *k* lines are very similar and the peak intensities in particular are almost the same. Significant red asymmetries of the Mg II *h* and *k* lines are observed at the peak time and after 17:51 UT. Subordinate lines are also significantly asymmetric at the peak time for pixel 436.

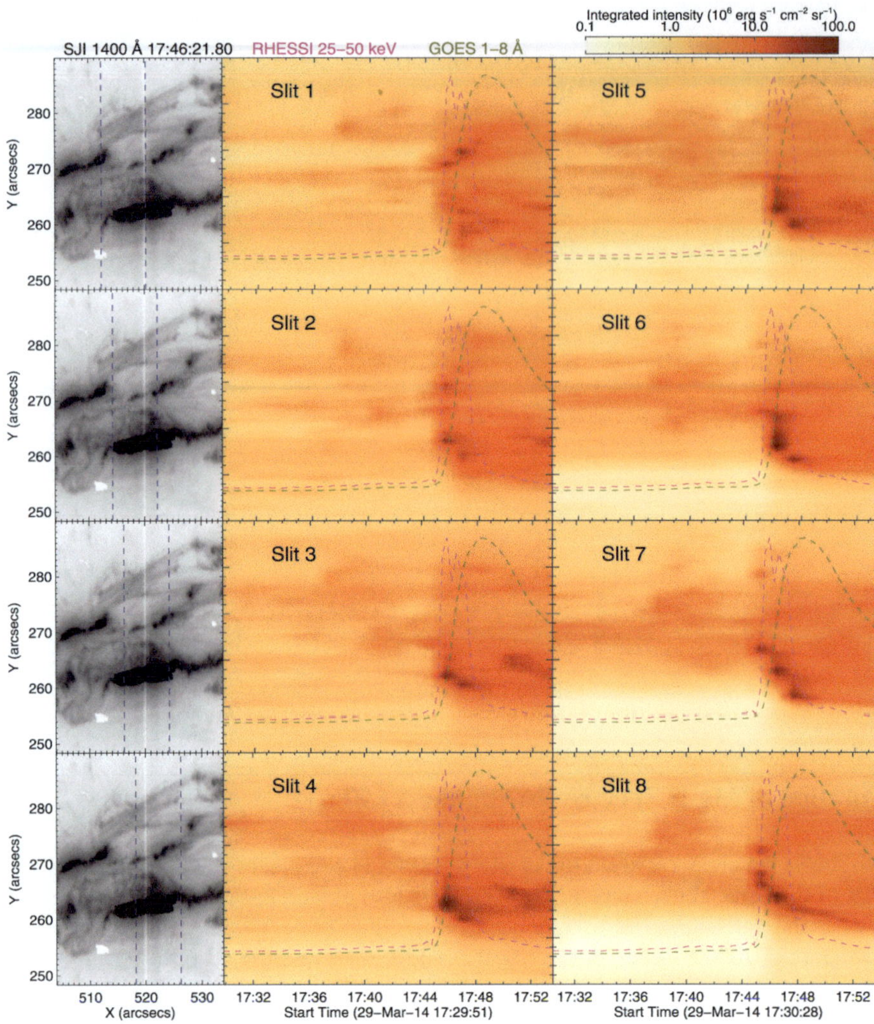

Figure 7 Left: an IRIS 1400 Å slitjaw image taken at 17:46:21 UT. The vertical dashed lines indicate the positions of the slits that were used to plot the integrated intensity in the center and right columns. Center and right: temporal evolution of the integrated intensities of the Mg II *h* line over 4 Å along all eight slits. Arbitrarily scaled RHESSI and GOES light curves are also plotted in the center and right panels. The tiny shift in the *x*-axes of the integrated intensity is due to a 9 s time step between two consecutive IRIS slit positions.

The temporal evolution of integrated line intensities for the Mg II *h* line over 4 Å along all eight slits from pixel 350 to pixel 600 is plotted in Figure 7. We also plot the arbitrarily scaled RHESSI 25 – 50 keV HXR flux with magenta dashed lines and the GOES 1 – 8 Å flux with olive dashed lines. The integrated intensities for the Mg II *h* line in Figure 7 are integrated over a 4 Å band because the *h* line from the footpoint pixels is very wide, with a full width at half maximum (FWHM) around 2 Å after 17:50 UT. Figure 7 shows that the temporal evolution of integrated Mg II *h* line intensities at footpoint pixels along all eight slits is very similar. Bright pixels with an intense Mg II *h* line appear when the HXR flux

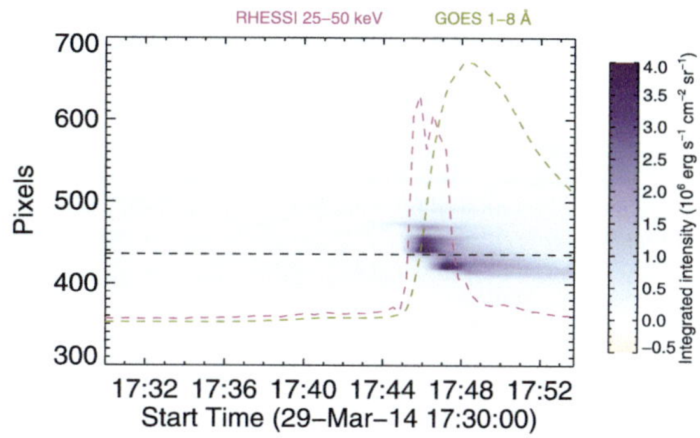

Figure 8 Temporal evolution of the integrated intensities of the Mg II subordinate line at 2791.60 Å over 0.53 Å along slit 5. The dashed horizontal line indicates the position of pixel 436, which is used to plot the temporal evolution of line profiles in Figure 6. Arbitrarily scaled RHESSI and GOES light curves are also plotted in the figure.

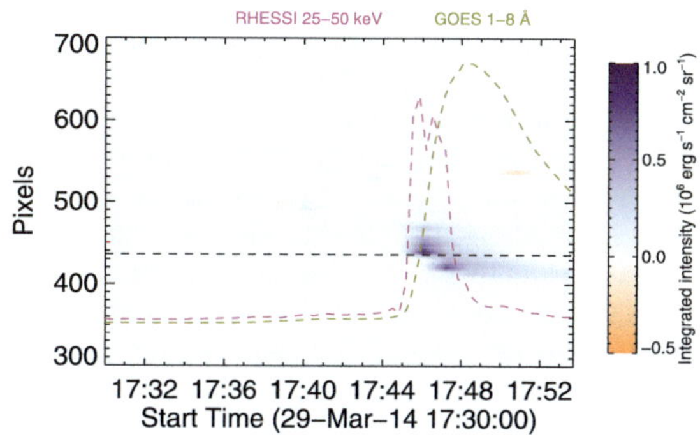

Figure 9 Temporal evolution of the integrated intensities of Fe II at 2794.69 Å over 0.64 Å along slit 5. The dashed horizontal line indicates the position of pixel 436, which is used to plot the temporal evolution of the line profiles in Figure 6. Arbitrarily scaled RHESSI and GOES light curves are also plotted in the figure.

rises impulsively. Then the two ribbons move away from each other: the southern ribbon to the southwest and the northern ribbon to the north. The apparent motion of flare ribbons in the chromosphere may be caused by the flare energy release moving to new coronal loops. Figure 7 shows that the integrated intensities for the Mg II *h* line at pixels 470 – 550 also vary; they are detected before the impulsive phase and continue until the end of the IRIS observations. The penumbra and umbra in slits 7 and 8, respectively, are also clearly seen in Figure 7 with low Mg II *h* line intensities.

Similar to the Mg II *h* line, the temporal evolution of subordinate line intensities is very similar at footpoint pixels along all eight slits. As an example, we plot the temporal evolution of integrated line intensities for the subordinate line at 2791.60 Å over 0.53 Å from pixel 350 to pixel 600 along slit 5 in Figure 8. The Mg II subordinate lines are in absorption in QS regions and non-flaring active regions. Figure 8 shows that the subordinate line at 2791.60 Å goes into emission at the footpoint pixels during the flare, with integrated intensities peaking at 17:46 – 17:47 UT. The integrated intensity of the Mg II subordinate lines at the footpoint pixels rises and peaks as the intensity of the *h* line, but decreases faster than the *h* line.

Figure 6 shows that the metallic lines in the wings of the Mg II *h* and *k* lines at the footpoint pixels also change from absorption to emission during the flare. The temporal evolution of integrated intensities for Fe II at 2794.69 Å over 0.64 Å along slit 5 is plotted in Figure 9 as an example of these metallic lines. Figure 9 shows that this line goes into emission at 17:45 – 17:46 UT and returns to absorption at 17:53:43 UT. Similarly, the integrated intensities for the metallic line also peaks at 17:46 – 17:47 UT, as for the subordinate lines. The metallic line also decreases faster than the Mg II *h* line. This may be related to the fact that the Mg II *h* and *k* lines are formed higher in the chromosphere, while the subordinate lines and the iron line are formed much deeper. The temporal evolution of Fe II (2794.69 Å)

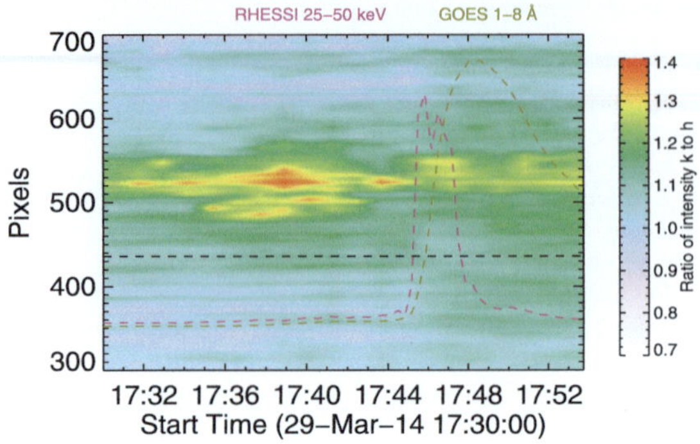

Figure 10 Ratio of the integrated intensity of the k line to that of the h line along slit 5. The dashed horizontal line indicates the position of pixel 436, which is used to plot the temporal evolution of the line profiles in Figure 6. Arbitrarily scaled RHESSI and GOES light curves are also plotted in the figure.

intensities at footpoint pixels along other slits is similar to that along slit 5 (figures are not shown).

We plot in Figure 10 the temporal variations of the k to h line ratio, which is the ratio of integrated intensities along slit 5 from pixels 350 to 600, which covers the ribbons and part of QS regions. Figure 10 shows that the k to h ratio is in the range between 0.9 to 1.4 along slit 5. Along all eight slits, the k to h ratio is between 0.7 to 1.4 for pixels 350 to 600. The ratio at footpoint pixels along all eight slits is around 1.1 (with a standard deviation, $\sigma = 0.068$) and does not change during the flare. This is consistent with previous flare observation by OSO-8/LPSP (Lemaire, Choucq-Bruston, and Vial, 1984) and an M-class flare observed by IRIS and analyzed by Kerr *et al.* (2015), who found the k to h ratio is 1.07 to 1.19 in flaring pixels and does not change much during flare. Figure 10 shows variations of the k to h ratio at pixels 470 – 550 before the impulsive phase of the flare and during the flare.

We used the normalized first-order moment (*e.g.* Druckmüller, Klvaňa, and Druckmüllerová, 2007), which is also known as the center of gravity, to measure the centroid of the line

$$\lambda_g(\lambda_0 - \Delta\lambda, \lambda_0 + \Delta\lambda) = \frac{\int_{\lambda_0-\Delta\lambda}^{\lambda_0+\Delta\lambda} \lambda\, I(\lambda)\, \mathrm{d}\lambda}{\int_{\lambda_0-\Delta\lambda}^{\lambda_0+\Delta\lambda} I(\lambda)\, \mathrm{d}\lambda}, \tag{2}$$

where λ_0 is the theoretical line center, $I(\lambda)$ is the intensity at wavelength λ, $[\lambda_0 - \Delta\lambda, \lambda_0 + \Delta\lambda]$ is the interval of integration, and $\Delta\lambda$ is estimated as twice the distance between the peak intensity and $1/e$ of the peak. To reduce the effect that the location of the integration interval has when the observed line is shifted substantially from its theoretical center, the center of gravity was calculated iteratively (Druckmüller, Klvaňa, and Druckmüllerová, 2007). It is obvious that the center of gravity will move to longer wavelengths if there is a redshift and/or red asymmetry of the line, and it will move to shorter wavelengths as the result of a blueshift and/or blue asymmetry. Therefore, we plot variations of the center of gravity for the Mg II h line along all eight slits in red and blue in Figure 11. We show that the Mg II h line at footpoint pixels is redshifted and/or red asymmetric when these pixels are bright. The center of gravity for pixels at the northern ribbon is less shifted than the pixels in the southern ribbon and is only shifted at the time when the pixel is brightest. For most pixels in the southern ribbon, the center of gravity is also shifted to the red wing after 17:49 UT (peak of the GOES light curve). The profiles of the Mg II h line at these footpoint pixels show that the shift is mainly caused by a red asymmetry, and therefore the line width is greater at the same time. Figure 11 also shows that there are large shifts to the blue at some pixels before the impulsive phase of the flare. Slitjaw images show that the blueshifts and/or

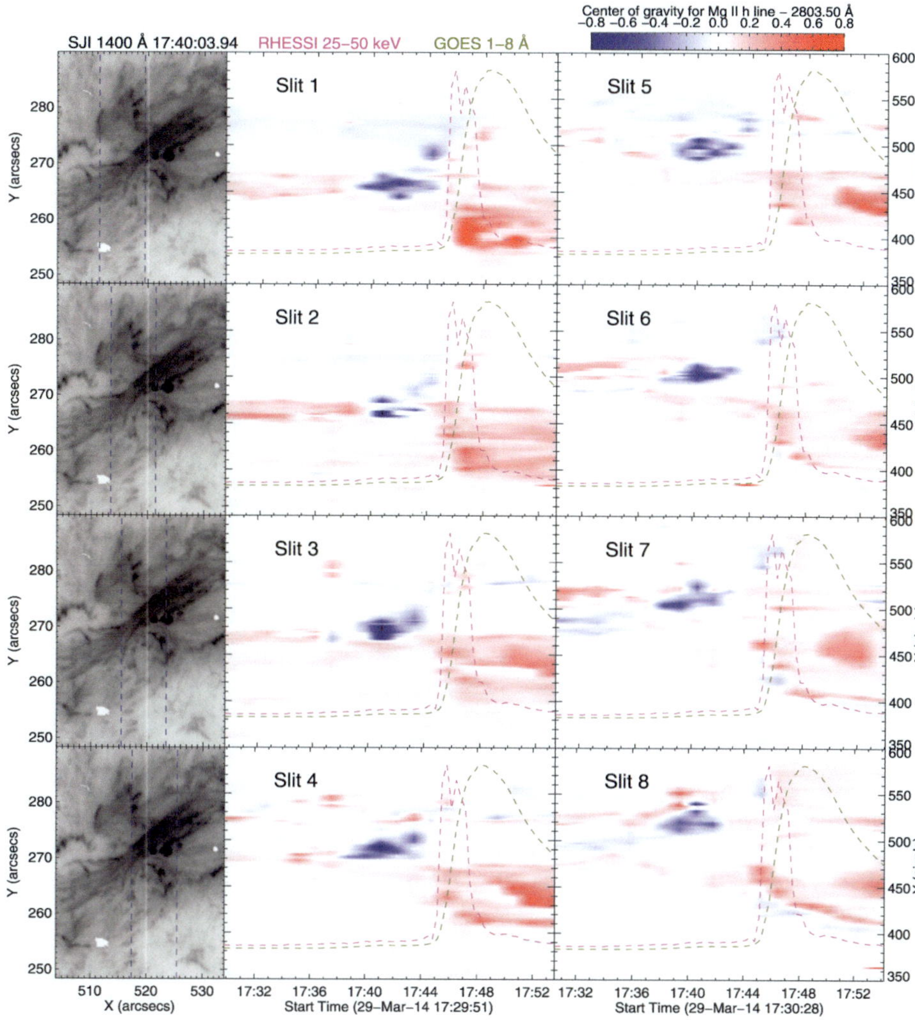

Figure 11 Left: IRIS 1400 Å slitjaw image taken at 17:40:04 UT. The vertical dashed lines indicate the positions of the slits. For these positions the temporal evolutions of the centers of gravity of the Mg II *h* line are plotted in the center and right columns. Arbitrarily scaled RHESSI and GOES light curves are also plotted in the figure. Note that we took 2803.50 Å as the reference wavelength to plot the center of gravity because this wavelength has the largest number of pixels in the histogram of the center of gravity for all pixels along the eight slits from 17:29:22 UT to 17:54:11 UT.

blue asymmetries at these pixels are caused by an erupting filament (Kleint *et al.*, 2015a). There are also some small shifts at pixels unrelated to the filament and flare. Judging from a comparison to the map of the *k* to *h* ratio for slit 5 in Figure 10, these shifts may be related to a higher *k* to *h* ratio.

In Figure 12 we plot the variations of the Mg II *h* intensity at its center of gravity along slit 5. The variations are mainly at footpoint pixels with intensities increasing at 17:45 – 17:46 UT and peaking at 17:46 – 17:47 UT. For other slits, the map of the Mg II *h* intensity

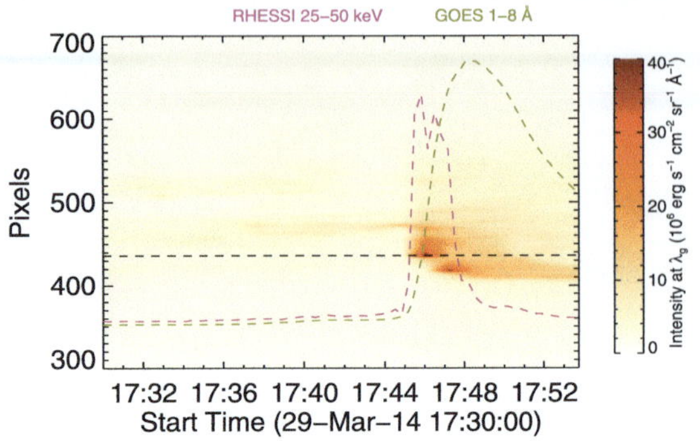

Figure 12 Mg II *h* line intensity at its center of gravity along slit 5. The dashed horizontal line indicates the position of pixel 436, which is used to plot the temporal evolution of line profiles in Figure 6. Arbitrarily scaled RHESSI and GOES light curves are also plotted in the figure.

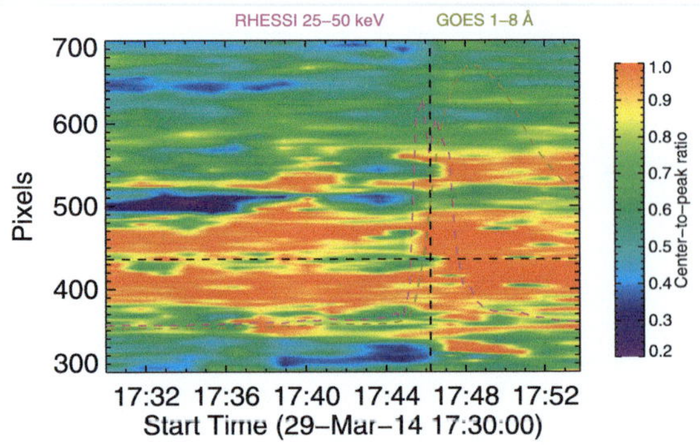

Figure 13 Center-to-peak ratio of the Mg *h* line along slit 5. The dashed horizontal line indicates the position of pixel 436, which is used to plot the temporal evolution of line profiles in Figure 6. The vertical dashed line indicates the time 17:46:13.98 UT when the Mg II bright footpoints along slit 5 are plotted in Figure 5. Arbitrarily scaled RHESSI and GOES light curves are also plotted in the figure.

at its center of gravity also shows clearly when the footpoint pixels start to brighten and become brightest (figures are not shown).

To describe how strongly the cores of the Mg II lines are reversed, we define the center-to-peak ratio as I_c/I_p, where I_c is the intensity at the reversal minimum and I_p is the intensity at the main peak of the line, which is the maximum intensity of the line. This means that the ratio is lower for line profiles with deeper reversal and higher for line profiles that are less strongly reversed. The ratio value is equal to one for a pure emission line. The variations of this ratio for the Mg II *h* line along slit 5 are plotted in Figure 13. The ratio at some footpoint pixels is close to one, while at other pixels it is about 0.7 (for example, pixel 436, whose profile is shown in Figure 6). For the filament, the center-to-peak ratio is very low before its eruption and increases when erupting. However, it is interesting to observe a ratio close to one in some active region pixels also before the flare (for example, pixels 380 to 420 and pixels 450 to 470 along slit 5).

To show variations of the line widths during the flare, we plot in Figure 14 the FWHM of the Mg II *h* line along slit 5 when it appears as an emission line (*i.e.* at the time when its center-to-peak ratio is higher than 0.8). The FWHM of the Mg II *h* line at the footpoint pixels is larger after 17:50 UT when the line shows red asymmetries. For this flare, the maximum FWHM of the Mg II *h* line is 2 Å, which is four times the FWHM at disk center of QS regions, where it is observed to be 0.5 ± 0.02 Å (Kohl and Parkinson, 1976; Staath and Lemaire, 1995). Figure 14 shows that the FWHM of the Mg II *h* line in the filament pixels increases two or three minutes before and during the eruption. The shape of a line is also characterized by its various order moments. The line width is expressed in terms of the second-order central moment, which is the square of the standard deviation. The FWHM of an emission line with a Gaussian profile is proportional to its standard deviation.

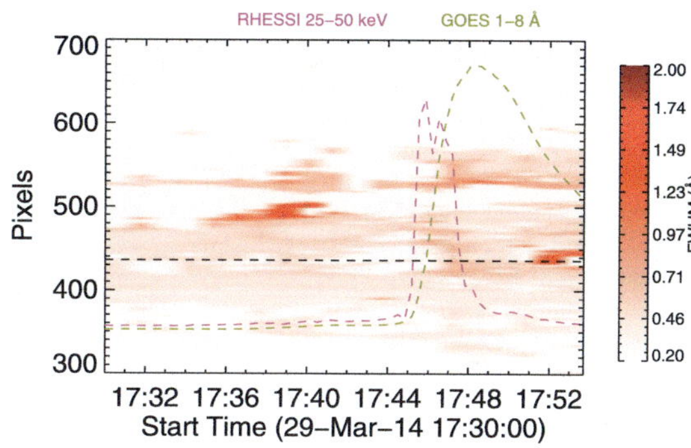

Figure 14 FWHM of the Mg *h* line along slit 5. The dashed horizontal line indicates the position of pixel 436, which is used to plot the temporal evolution of line profiles in Figure 6. Arbitrarily scaled RHESSI and GOES light curves are also plotted in the figure.

Comparing this with the flare reported by Lemaire, Choucq-Bruston, and Vial (1984), we find several similarities. First, there are great emission enhancements in the Mg II *h* and *k* lines and in the subordinate lines. Second, the light curves of the Mg II *h* or *k* lines rise and peak at the same time as the subordinate lines. Finally, the *k* to *h* ratio is 1.1 ± 0.068 at footpoint pixels and does not vary with time during the flare. In addition to these similarities, we observe red asymmetries at the flare peak and the decay phase for the X-class flare. Different from the flare of Lemaire, Choucq-Bruston, and Vial (1984), the light curves of the subordinate lines in our flare decrease faster than the light curves of the *h* and *k* lines after the peak time. It is interesting to note that some Mg II line profiles at some footpoint pixels are reversed, while they are non-reversed at other footpoints during the flare, as we showed before.

4. Synthetic Spectra

In this section, we present preliminary modeling of the Mg II lines in a flaring atmosphere to determine whether existing flare models can quantitatively reproduce the calibrated IRIS spectra. As we have mentioned in the Introduction, the only existing fitting of the Mg II lines in a flare was performed by Lemaire, Choucq-Bruston, and Vial (1984).

We solved the non-LTE radiative-transfer problem to model the Mg II lines using a static semi-empirical flare atmosphere. Our numerical code is based on a 1D plane-parallel geometry and assumes the atmosphere in a hydrostatic equilibrium. The non-LTE problem is treated by the so-called Multilevel Accelerated Lambda Iteration (MALI) technique (Rybicki and Hummer, 1991; Heinzel, 1995; Kašparová and Heinzel, 2002). By taking the temperature and pressure structure given by the flare atmosphere, we obtained hydrogen-level populations and electron densities from a five-level plus continuum atomic model of hydrogen. Hydrogen Ly α and Ly β lines were treated with the partial frequency redistribution (see Hubeny and Mihalas, 2014). Using the fixed electron density structure, the radiative transfer problem was solved for a five-level plus continuum atomic model of Mg II with the two resonance lines *h* and *k* and the $3p-3d$ subordinate triplet around the *k* line; for details of the atomic model see Heinzel, Vial, and Anzer (2014).

After modeling, we compared our synthetic spectra to the observed IRIS spectra. As the flare was of class X1, we chose the classical semi-empirical flare model F2 of Machado *et al.* (1980). From an extended grid of synthetic line profiles as shown in Avrett, Machado, and Kurucz (1986), we noted that the Mg II *k* line profile corresponding to model F2 is non-

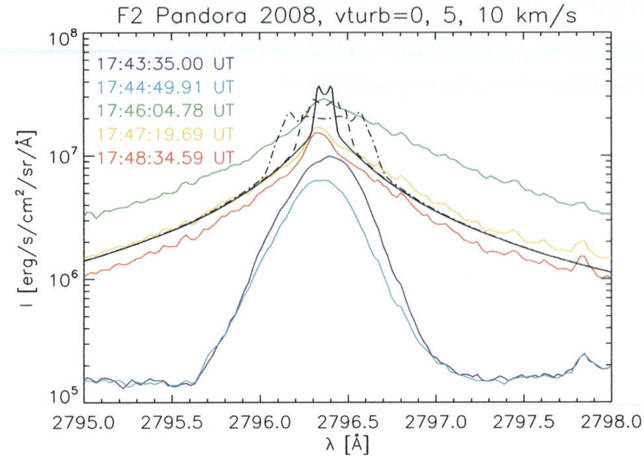

Figure 15 A comparison between the synthetic Mg II *k* line intensities computed from model F2 and IRIS observations for slit 4 in pixel 447. The profiles have been obtained for three microturbulent velocities: 0 $\mathrm{km\,s^{-1}}$ – full line, 5 $\mathrm{km\,s^{-1}}$ – dashed line, and 10 $\mathrm{km\,s^{-1}}$ dash-dotted line. 'Pandora 2008' is the version of the F2 model with a slightly different coronal pressure from the original model of Machado *et al.* (1980). Note the intensity decrease at time 17:44:49.91.

reversed; we observed this at several positions. Its peak intensity is quite comparable to that detected around the maximum of Mg II emission at pixel 447 slit 4, just above the saturation region. Using the above-mentioned non-LTE technique, we computed the Mg II line profiles from the F2 model. The resulting Mg II *k* line is shown in Figure 15, where we present a few examples for various values of the microturbulent velocity ranging from 0 to 10 $\mathrm{km\,s^{-1}}$. We note that this microturbulence is assumed to be uniform in the whole formation region of the Mg II lines, which is certainly a crude approximation. The central line intensities are comparable to the observed ones at about 17:46 UT. The line cores are not much reversed, which is consistent with the observations. We did not perform a convolution with the IRIS instrumental profile with a width of 52 mÅ; doing so, the emission peaks might become slightly depressed toward the observed profile (see *e.g.* Heinzel *et al.*, 2015). The synthetic line profiles are only significantly broadened by the microturbulence in the core, not in the wings. This means that at least with this schematic model the observed bright wings are due to an enhanced line source function and not to additional broadening. This can be seen from the formation heights of the Mg II resonance lines. In Figure 16 we show the contribution functions of the Mg II *k* line where broad line wings corresponding to the F2 atmosphere are formed over an extended region of about 1000 km. In the quiet Sun, *e.g.* VAL C atmosphere (Vernazza, Avrett, and Loeser, 1981), which is shown for a comparison, the line wings are formed much deeper in the atmosphere and over a region that is only some 200 km wide. Finally, we also tested the influence of the coronal pressure on the line spectrum. Model F2 has an upper boundary gas pressure equal to about 95 $\mathrm{dyn\,cm^{-2}}$. The test solution without this pressure leads to a much lower central intensity of the *h* and *k* lines. Therefore, the overlying hot loops seem to have an important influence on the line-core emission. The enhanced pressure is related to the explosive evaporation process, which we describe in the next paragraph.

To demonstrate the dependence of Mg II line intensities and line profile shapes on various model parameters, we also computed a grid of profiles based on models by Ricchiazzi and Canfield (1983) (hereafter called RC models). Although these models have been published more than three decades ago, they still represent the only systematic grid of theoretical flare models. RC models are 1D static atmosphere models in stationary energy balance in which the chromosphere is heated by electron beams. The latter are parameterized by the total energy flux above a low-energy cutoff of 20 keV and a spectral index δ. Another parameter is the coronal pressure at the top of the model atmosphere, which is assumed to be enhanced

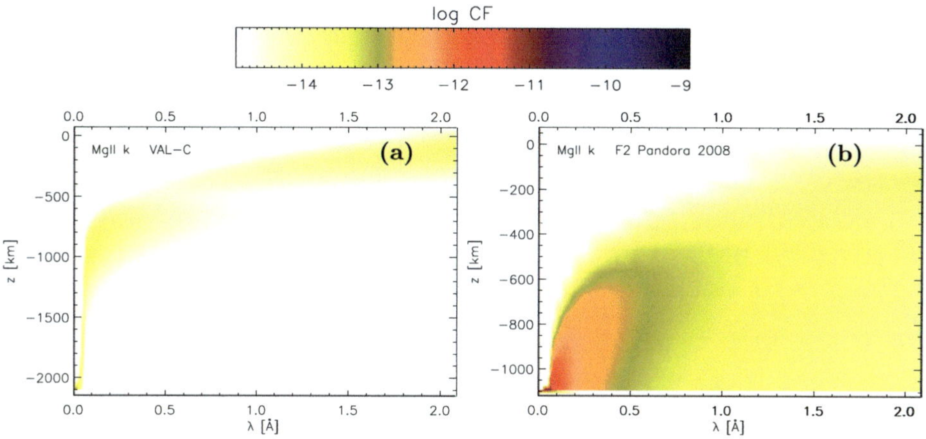

Figure 16 Mg II k line contribution function for (a) VAL C and (b) F2 model atmospheres. Note the height scale z is inverted ($z = 0$ denotes photosphere) and is different for each model. We clearly see how the chromosphere is affected by the heating, contributing significantly to the line wings. The QS wings are formed much deeper.

as a result of explosive evaporation. Using RC models, we computed synthetic Mg II k line profiles that we display in Figure 17. Figure 17a shows the line-profile variations with increasing electron-beam flux from 10^9 to 10^{11} erg s^{-1} cm^{-2}. The spectral index is equal to five and the coronal pressure is 1 dyn cm^{-2}. All profiles are rather strongly reversed, although the averaged line-core intensities for higher fluxes are comparable to the observed ones. The wings are too low, however. The central line reversal is significantly reduced for higher coronal pressures, as shown in Figure 17b. For the highest pressure, the profile is non-reversed, but its central intensity is much higher than in observations. Finally, Figure 17c shows how the wing intensities increase with decreasing δ, while the line core remains practically unchanged. At $\delta = 3$ the wing reaches the observed intensities, which indicates that the beam deeply penetrates the lower atmospheric layers, which are then more heated, in contrast to models with a higher δ or the semi-empirical model F2. From this analysis we conclude that a model with higher flux, between 10^{10} and 10^{11} erg s^{-1} cm^{-2}, a spectral index around three, and enhanced coronal pressure between 10 – 100 dyn cm^{-2} could approximately reproduce the observations. These fluxes and spectral indexes are consistent with those derived for this flare from the detailed analysis of RHESSI spectra obtained at the same time and locations as the Mg II spectra reported in this article (Kleint *et al.*, 2015b).

Still another aspect needs to be taken into account for comparing the enhanced wings of the h and k lines. As found by Heinzel and Kleint (2014), the Balmer-continuum emission during the maximum of this flare detected in the NUV channel of IRIS seems to be fairly significant. Assuming that this emission is optically thin in the chromosphere, it can be subtracted from the NUV spectrum by using the Lockheed Martin Solar and Astrophysics Laboratory (LMSAL) pre-flight calibration for the IRIS NUV spectra as in Heinzel and Kleint (2014). This is shown in Figure 18, where the Balmer continuum is subtracted from the spectrum at 17:46:04.78 UT (depending on the radiometric calibration used (Heinzel and Kleint, 2014), the continuum levels differ by about a factor of three). The effect of the Balmer continuum is marginal in the line core and inner wings where the line intensity is high, but it is significant in the extended wings. Since our preliminary synthesis of the Mg II lines does not account for an enhanced Balmer continuum of hydrogen, we should compare the synthetic profiles from Figure 15 with the reduced ones in Figure 18. This naturally leads

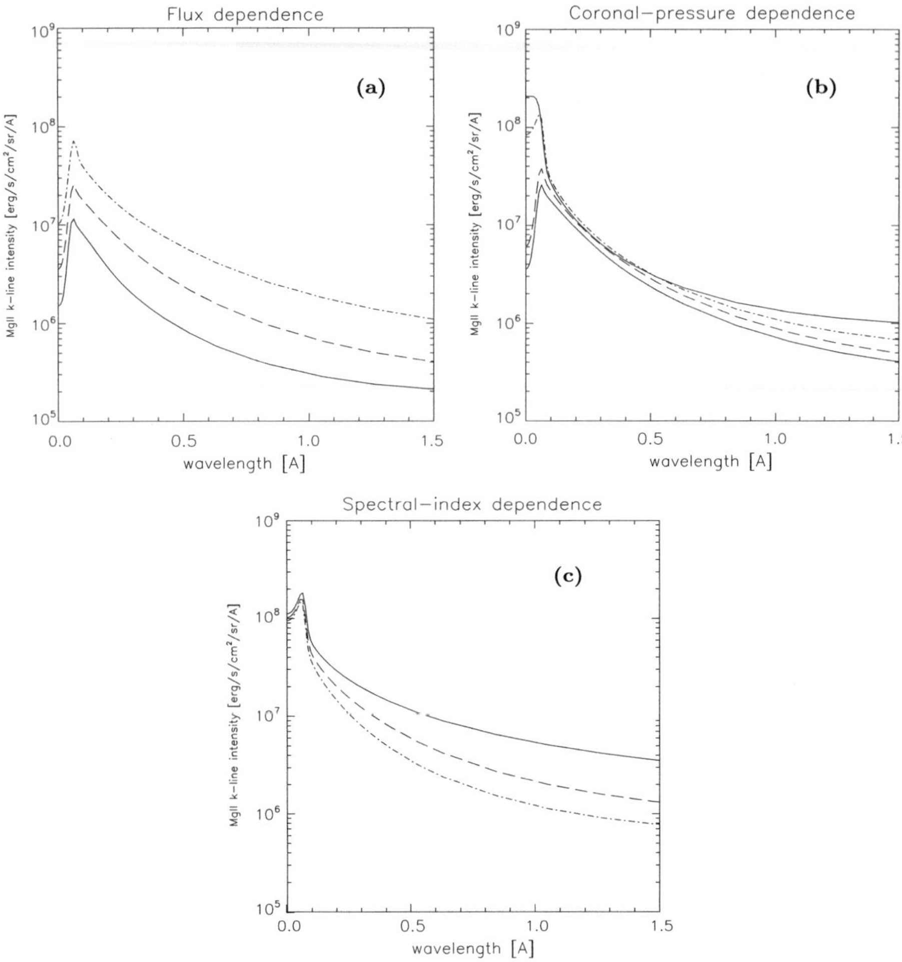

Figure 17 Mg II k line profiles computed from RC models with different parameters. (a) $\delta = 5$, coronal pressure 1 dyn cm^{-2} and three electron-beam fluxes 10^9 (solid), 10^{10} (dashed) and 10^{11} erg s^{-1} cm^{-2} (dash-dotted). (b) $\delta = 5$, flux 10^{10} erg s^{-1} cm^{-2} and four different values of the coronal pressure 1 (solid), 10 (dashed), 100 (dash-dotted) and 1000 dyn cm^{-2} (dash-three-dots). (c) The electron-beam flux 10^{11} erg s^{-1} cm^{-2}, coronal pressure 100 dyn cm^{-2} and three spectral indexes of the electron beam $\delta = 3$ (solid), $\delta = 5$ (dashed) and $\delta = 7$ (dash-dotted).

to a somewhat better agreement. We note that when we calculated the integrated intensities for the subordinate line or metallic line, we subtracted the continuum intensities at the wings near the center of the subordinate line or the metallic line before integration. Therefore the integrated intensities in Figures 8 and 9 are not affected by the Balmer continuum.

5. Conclusions

We studied the 2D spatial and temporal evolution of flare ribbons in Mg II lines in detail. The observations of the Mg II spectrum along all eight slits can be summarized as follows:

Figure 18 Observed Mg II spectrum at 17:46:04.780 UT with (red) and without (black) the Balmer continuum subtracted at pixel 447 slit 4. The pre-flare spectrum at pixel 447 is also plotted as reference with a blue continuous line.

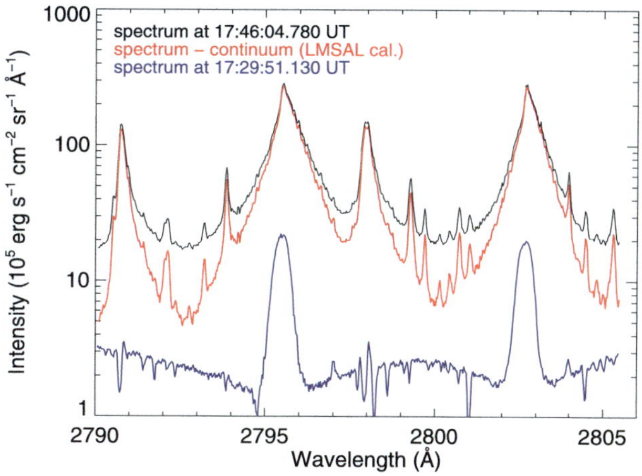

(1) The intensities of the Mg II lines at the footpoints increase during the impulsive phase of the flare and peak at the maximum of the HXR emission. (2) The k to h ratio of the integrated intensities is 1.1 ± 0.068 at the footpoints and does not vary during the flare. (3) Red asymmetries at most footpoints are located in the southern ribbon after the peak. (4) The Mg II h and k lines are non-reversed at some footpoints and reversed at others at the flare peak. Some of the Mg II footpoints with reversed profiles are correlated with HXR sources. After the peak, the Mg II line profiles at most footpoints become non-reversed. (5) The subordinate lines and metallic lines at the footpoints rise from absorption to emission during the flare. (6) The Mg II lines become broader during the flare.

We have compared the observed line intensities with synthetic spectra computed from the semi-empirical model F2 and found several similarities. The temperature structure of the F2 model, which can be regarded as a snapshot of a more realistic model evolution, reproduces some features of the observed profiles, but does not explain the broad line wings well, which will require stronger heating deeper in the chromosphere. Since the original F2 model was constructed without taking into account the non-thermal rates in hydrogen, we also neglected this aspect. For the Mg II lines, the non-thermal processes seem to be of secondary importance, as we found from several tests. Another class of models is represented by a grid of theoretical flare models constructed by RC. For these models we synthesized the Mg II k profiles and showed that for some specific model parameters they roughly fit our observations. However, all used models are static and therefore do not reproduce the line asymmetries observed in some pixels. These mostly red asymmetries are usually associated with the downward moving chromospheric plasma. The velocity field affects the line profiles (asymmetry), but also significantly influences the line intensities *via* the statistical equilibrium for the Mg II atom. Without any systematic modeling, it is rather difficult to interpret the observation that the Mg II line profiles are grouped into reversed (red in Figure 5) and non-reversed profiles (blue in Figure 5). However, based on a limited grid of RC models, we may speculate that reversed profiles will appear at sites of strong heating and moderate coronal pressures where the chromospheric condensation (downflows) is formed as a result of the electron-beam impact, leading to the red asymmetries that are observed. On the other hand, non-reversed and more symmetrical profiles might be related to higher coronal pressures (the loops are filled with the evaporated plasma), and motions inside the ribbons are very small. This might be consistent with Figure 5, where the outer edges of the southern ribbon may represent the footpoints of hot loops in which the electron beams propagate, while the inner edges are places where cooler and denser loops are rooted. From the modeling we found that the line cores are very sensitive to gas pressure in overlying coronal loops. In a future study we plan to perform more realistic radiative–hydrodynamical simulations of the temporal evolution of flare spectra using the code *Flarix* developed at the Ondřejov observatory (Varady *et al.*, 2010). Electron-beam spectra derived from simultaneous RHESSI observations (see Kleint *et al.*, 2015b) will be used. These simulations will also allow us to study the velocity flows that cause the line asymmetries.

Acknowledgements The research leading to these results has received funding from the European Community's Seventh Framework Programme (FP7/2007 – 2013) under grant agreement 606862 (F-CHROMA) and from ASI ASCR project RVO:67985815. L.K. was supported by a Marie-Curie Fellowship. IRIS is a NASA small explorer mission developed and operated by LMSAL with mission operation executed at the NASA Ames Research Center and major contributions to downlink communications funded by the Norwegian Space Center (NSC, Norway) through an ESA PRODEX contract.

Disclosure of Potential Conflicts of Interest The authors declare that they have no conflicts of interest.

References

Asai, A., Ichimoto, K., Kita, R., Kurokawa, H., Shibata, K.: 2012, A study on red asymmetry of Hα flare ribbons using a narrowband filtergram in the 2001 April 10 solar flare. *Publ. Astron. Soc. Japan* **64**, 20. DOI. ADS.

Avrett, E., Landi, E., McKillop, S.: 2013, Calculated resonance line profiles of [Mg II], [C II], and [Si IV] in the solar atmosphere. *Astrophys. J.* **779**, 155. DOI. ADS.

Avrett, E.H., Machado, M.E., Kurucz, R.L.: 1986, Chromospheric flare models. In: Neidig, D.F. (ed.) *The Lower Atmosphere of Solar Flares*, 216. ADS.

Bocchialini, K., Gouttebroze, P.: 1996, Solar chromospheric structures as observed simultaneously in strong UV lines. II. Network and cell modelling. *Astron. Astrophys.* **313**, 949. ADS.

Bonnet, J.: 1968, Recherches sur l'émission continue du Soelil entre 1950 & Aring et 3000 Å. *Ann. Astrophys.* **31**, 597. ADS.

De Pontieu, B., Title, A.M., Lemen, J.R., Kushner, G.D., Akin, D.J., Allard, B., *et al.*: 2014, The Interface Region Imaging Spectrograph (IRIS). *Solar Phys.* **289**, 2733. DOI. ADS.

Deng, N., Tritschler, A., Jing, J., Chen, X., Liu, C., Reardon, K., Denker, C., Xu, Y., Wang, H.: 2013, High-cadence and high-resolution Hα imaging spectroscopy of a circular flare's remote ribbon with IBIS. *Astrophys. J.* **769**, 112. DOI. ADS.

Druckmüller, M., Klvaňa, M., Druckmüllerová, Z.: 2007, Solar spectra analysis based on the statistical moment method. *Cent. Eur. Astrophys. Bull.* **31**, 297. ADS.

Fisher, G.H.: 1989, Dynamics of flare-driven chromospheric condensations. *Astrophys. J.* **346**, 1019. DOI. ADS.

Fisher, G.H., Canfield, R.C., McClymont, A.N.: 1985, Flare loop radiative hydrodynamics – Part seven – Dynamics of the thick target heated chromosphere. *Astrophys. J.* **289**, 434. DOI. ADS.

Freeland, S.L., Handy, B.N.: 1998, Data analysis with the SolarSoft system. *Solar Phys.* **182**, 497. DOI. ADS.

Heinzel, P.: 1995, Multilevel NLTE radiative transfer in isolated atmospheric structures: implementation of the MALI-technique. *Astron. Astrophys.* **299**, 563. ADS.

Heinzel, P., Kleint, L.: 2014, Hydrogen Balmer continuum in solar flares detected by the Interface Region Imaging Spectrograph (IRIS). *Astrophys. J. Lett.* **794**, L23. DOI. ADS.

Heinzel, P., Vial, J.-C., Anzer, U.: 2014, On the formation of Mg II h and k lines in solar prominences. *Astron. Astrophys.* **564**, A132. DOI. ADS.

Heinzel, P., Schmieder, B., Mein, N., Gunár, S.: 2015, Understanding the Mg II and Hα spectra in a highly dynamical solar prominence. *Astrophys. J. Lett.* **800**, L13. DOI. ADS.

Hubeny, I., Mihalas, D.: 2014, *Theory of Stellar Atmospheres*, Princeton University Press, Princeton. ADS.

Ichimoto, K., Kurokawa, H.: 1984, H-alpha red asymmetry of solar flares. *Solar Phys.* **93**, 105. DOI. ADS.

Kašparová, J., Heinzel, P.: 2002, Diagnostics of electron bombardment in solar flares from hydrogen Balmer lines. *Astron. Astrophys.* **382**, 688. DOI. ADS.

Kerr, G.S., Simões, P.J.A., Qiu, J., Fletcher, L.: 2015, IRIS observations of the Mg II h and k lines during a solar flare. *Astron. Astrophys.* **582**, A50. DOI. ADS.

Kleint, L., Battaglia, M., Reardon, K., Sainz Dalda, A., Young, P.R., Krucker, S.: 2015a, The fast filament eruption leading to the X-flare on 2014 March 29. *Astrophys. J.* **806**, 9. DOI. ADS.

Kleint, L., Heinzel, P., Judge, P., Krucker, S.: 2015b, *Astrophys. J.*, in press

Kohl, J.L., Parkinson, W.H.: 1976, The Mg II h and k lines. I – Absolute center and limb measurements of the solar profiles. *Astrophys. J.* **205**, 599. DOI. ADS.

Leenaarts, J., Pereira, T.M.D., Carlsson, M., Uitenbroek, H., De Pontieu, B.: 2013, The formation of IRIS diagnostics. I. A quintessential model atom of Mg II and general formation properties of the Mg II h & k lines. *Astrophys. J.* **772**, 89. DOI. ADS.

Lemaire, P., Choucq-Bruston, M., Vial, J.-C.: 1984, Simultaneous H and K Ca II, h and k Mg II, L-alpha and L-beta H I profiles of the April 15, 1978 solar flare observed with the OSO-8/L.P.S.P. experiment. *Solar Phys.* **90**, 63. DOI. ADS.

Lemaire, P., Gouttebroze, P., Vial, J.C., Artzner, G.E.: 1981, Physical properties of the solar chromosphere deduced from optically thick lines. I – Observations, data reduction, and modelling of an average plage. *Astron. Astrophys.* **103**, 160. ADS.

Lin, R.P., Dennis, B.R., Hurford, G.J., Smith, D.M., Zehnder, A., Harvey, P.R., *et al.*: 2002, The Reuven Ramaty High-Energy Solar Spectroscopic Imager (RHESSI). *Solar Phys.* **210**, 3. DOI. ADS.

Longcope, D.W.: 2014, A simple model of chromospheric evaporation and condensation driven conductively in a solar flare. *Astrophys. J.* **795**, 10. DOI. ADS.

Machado, M.E., Avrett, E.H., Vernazza, J.E., Noyes, R.W.: 1980, Semiempirical models of chromospheric flare regions. *Astrophys. J.* **242**, 336. DOI. ADS.

Milkey, R.W., Mihalas, D.: 1974, Resonance-line transfer with partial redistribution. II – The solar Mg II lines. *Astrophys. J.* **192**, 769. DOI. ADS.

Morrill, J.S., Korendyke, C.M.: 2008, High-resolution center-to-limb variation of the quiet solar spectrum near Mg II. *Astrophys. J.* **687**, 646. DOI. ADS.

Pereira, T.M.D., Leenaarts, J., De Pontieu, B., Carlsson, M., Uitenbroek, H.: 2013, The formation of IRIS diagnostics. III. Near-ultraviolet spectra and images. *Astrophys. J.* **778**, 143. DOI. ADS.

Ricchiazzi, P.J., Canfield, R.C.: 1983, A static model of chromospheric heating in solar flares. *Astrophys. J.* **272**, 739. DOI. ADS.

Rybicki, G.B., Hummer, D.G.: 1991, An accelerated lambda iteration method for multilevel radiative transfer. I – Non-overlapping lines with background continuum. *Astron. Astrophys.* **245**, 171. ADS.

Staath, E., Lemaire, P.: 1995, High resolution profiles of the Mg II h and Mg II k lines. *Astron. Astrophys.* **295**, 517. ADS.

Švestka, Z., Kopecký, M., Blaha, M.: 1962, Qualitative discussion of 244 flare spectra. II. Line asymmetry and helium lines. *Bull. Astron. Inst. Czechoslov.* **13**, 37. ADS.

Tang, F.: 1983, Flare asymmetry as seen in offband H-alpha filtergrams. *Solar Phys.* **83**, 15. DOI. ADS.

Uitenbroek, H.: 1997, The solar Mg II h and k lines – observations and radiative transfer modeling. *Solar Phys.* **172**, 109. DOI. ADS.

Varady, M., Kasparova, J., Moravec, Z., Heinzel, P., Karlicky, M.: 2010, Modeling of solar flare plasma and its radiation. *IEEE Trans. Plasma Sci.* **38**, 2249. DOI. ADS.

Vernazza, J.E., Avrett, E.H., Loeser, R.: 1981, Structure of the solar chromosphere. III – Models of the EUV brightness components of the quiet-sun. *Astrophys. J. Suppl.* **45**, 635. DOI. ADS.

Young, P.R., Tian, H., Jaeggli, S.: 2015, The 2014 March 29 X-flare: subarcsecond resolution observations of Fe XXI λ1354.1. *Astrophys. J.* **799**, 218. DOI. ADS.

Reprinted from *Solar Physics* Journal, DOI 10.1007/s11207-015-0734-8
SOLAR AND STELLAR FLARES

Ionisation Equilibrium for the Non-Maxwellian Electron *n*-Distributions in Solar Flares: Updated Calculations

Elena Dzifčáková[1] · **Jaroslav Dudík**[1]

Received: 10 February 2015 / Accepted: 29 June 2015 / Published online: 4 August 2015
© Springer Science+Business Media Dordrecht 2015

Abstract We use the latest available atomic data to calculate the ionisation and recombination rates for the non-Maxwellian n-distributions, which were shown previously to provide a good fit to the enhanced intensities of dielectronic satellite lines during solar flares. The ionisation and recombination coefficients are subsequently used to derive the ionisation equilibrium. To do so, we consider odd values of n ranging from 1 to 19, *i.e.*, from Maxwellian to strongly non-Maxwellian cases. These calculations involve all elements with proton number up to 30, *i.e.*, H to Zn. The n-distributions modify both the ionisation and the recombination rates. The ionisation rates decrease more steeply at lower pseudo-temperatures, while the radiative recombination rate is reduced due to a lower number of low-energy electrons. The peaks of the dielectronic recombination rates become narrower. These changes are reflected in the ionisation equilibrium. Ion abundance peaks become narrower and can also be shifted, mostly towards higher temperatures. The He-like ions are an important exception, as they are formed in a larger temperature range than that for the Maxwellian distribution. The ions Si XIII – XIV used previously for the diagnostics of the n-distributions are affected only weakly, confirming the determination of n. The ionisation equilibria are available as the electronic supplementary material in a format compatible with the CHIANTI database.

Keywords Energetic particles, acceleration · Energetic particles, electrons · Flares, energetic particles · Flares, spectrum · Spectral line, theory · Spectrum, X-ray

Solar and Stellar Flares: Observations, Simulations, and Synergies
Guest Editors: Lyndsay Fletcher and Petr Heinzel

Electronic supplementary material The online version of this article (doi:10.1007/s11207-015-0734-8) contains supplementary material, which is available to authorized users.

✉ E. Dzifčáková
elena@asu.cas.cz

J. Dudík
dudik@asu.cas.cz

[1] Astronomical Institute, Academy of Sciences of the Czech Republic, Fričova 298, 25165 Ondřejov, Czech Republic

1. Introduction

Solar flares are dynamic phenomena exhibiting a rapid increase of radiation throughout the electromagnetic spectrum. The energy released via the process of magnetic reconnection is converted mostly to kinetic energies of particles, namely the bulk motion of plasma, plasma heating and also acceleration of non-Maxwellian particles. In particular, electrons commonly reach energies of several keV or orders of magnitude higher (*e.g.*, Brown, 1971; Lin and Hudson, 1971; Petkaki and MacKinnon, 2007, 2011; Asai *et al.*, 2009; Veronig *et al.*, 2010; Fletcher *et al.*, 2011; Zharkova *et al.*, 2011; Burge, MacKinnon, and Petkaki, 2014; Bian *et al.*, 2014), and they are observed *e.g.* by the *Reuven Ramaty High-Energy Solar Spectroscopic Imager* (RHESSI; Lin *et al.*, 2002) as the presence of high-energy tails.

However, non-Maxwellian distributions other than the high-energy tails have been detected in solar flares. Seely, Feldman, and Doschek (1987) analyzed Fe XXV line intensities together with the neighbouring Fe XXIV dielectronic satellite lines and found that the dielectronic satellites j and $d13$ were increased during the early and peak phases of the flare and diminished later on. The authors concluded that the observed increase of the $d13$ line cannot be reproduced by multithermal plasma. The observations were interpreted in terms of a non-Maxwellian electron energy distribution more peaked than the Maxwellian one. This is due to the nature of the dielectronic recombination process, which is a resonant one and occurs only at well-defined energies. For the Fe XXIV dielectronic lines j and $d13$, these energies are 4.7 and 5.8 keV, respectively. The occurrence of the more peaked distribution at these energies during the impulsive phase of solar flares coincided with the occurrence of hard X-ray bursts.

Dzifčáková (1998) adapted the analytical formula of the peaked non-Maxwellians distribution used by Seely, Feldman, and Doschek (1987) into the n-distributions used in subsequent works and defined the pseudo-temperature related to the mean energy of the distribution (see Section 2). These authors published also the Fe ionisation equilibrium for the n-distributions, corresponding to the atomic data of Mazzotta *et al.* (1998). Subsequent calculation of synthetic spectra were performed by Dzifčáková and Kulinová (2001), Dzifčáková (2005) and Dudík *et al.* (2011). It was found that both the ionisation equilibrium and subsequently the synthetic spectra are strongly affected by the n-distributions. Individual ionisation peaks and subsequently the line contribution functions become narrower and can be shifted to different temperatures, leading to modifications of the total radiative losses (Dudík *et al.*, 2011).

These effects have been successfully used for diagnostics of n in solar flares despite relative lack of X-ray spectra. Dzifčáková *et al.* (2008) analyzed the spectra at 5 – 6 Å obtained by the RESIK bent-crystal spectrometer (Sylwester *et al.*, 2005) onboard the CORONAS-F satellite. This wavelength range contains allowed lines of the He-like Si XIII together with the neighbouring Si XII dielectronic satellites. During the impulsive phase, the dielectronic satellites were increased up to several times when compared to the Maxwellian prediction constrained by the observed Si XIV / Si XIII ratio of allowed lines. The value of $n = 11$ was diagnosed from the intensities of the Si XII dielectronic lines relative to Si XIII. It was also shown that the spectra cannot be interpreted in terms of a multithermal plasma: in order to fit the observed intensities of the Si XII dielectronic lines, the allowed Si XIII and Si XIV lines were predicted to be significantly increased, up to a factor of two, compared to the observed values.

The analysis of Dzifčáková *et al.* (2008) was extended by Kulinová *et al.* (2011) for three flares as observed by RESIK. These authors confirmed the diagnostics of the n-distributions in all flares studied. Furthermore, the occurrence of the strongest deviation from the Maxwellian distribution in terms of n coincided with the presence of type-III radio bursts observed at 2.7 – 15.4 GHz. Separate diagnostics of n using RHESSI spectra at energies down to 4 keV were also performed. Although such low RHESSI energies are strongly attenuated, the data show presence of n-distributions on top of the thermal component and the high-energy power-law tail (see Figure 11 in Kulinová *et al.*, 2011). The emission measure associated with these n-distribution were several times higher than the thermal component. Finally, the values of n diagnosed separately from RHESSI and RESIK were found to be consistent within their respective errors.

Li-like dielectronic satellites from other elements can be difficult to observe due to low abundances and/or blending with the He-like allowed lines, precluding diagnostics of n-distributions. Nevertheless, Sylwester, Sylwester, and Phillips (2008) reported that at least for some flares studied, the RESIK Ar XVII spectra showed enhancements of line intensities that could occur due to the overabundance of electrons at several keV similar to the n-distribution. Other theoretical diagnostics proposed for the determination of n-distributions in flares include the disappearance of recombination edges in the free-bound continuum with high n (Dudík *et al.*, 2012) and polarisation of the gyroresonance radio emission (Fleishman and Kuznetsov, 2014), which would not arise in case of thermal plasma.

Since the n-distributions are analytic distributions only, their physical significance have been investigated by Karlický (2012) and Karlický, Dzifčáková, and Dudík (2012). Karlický (2012) considered the possibility that the electrons in the plasma have a drift velocity. It was shown that such "moving Maxwellian" indeed has a more mono-energetic shape and that the high-energy slope at few keV matches well that of the n-distribution. However, the drift velocities necessary for production of such distribution are several times the thermal velocity. Namely, such distribution is unstable mainly with respect to the Buneman instability. However, the observed departures from the Maxwellian distribution (Dzifčáková *et al.*, 2008; Kulinová *et al.*, 2011) have lasted for several minutes, implying stable distribution in the flaring plasma. Karlický (2012) searched for such stabilizing processes and found that the formation of double layers along flare loops can produce a stable peak of the distribution, again closely resembling the slope of the n-distribution at several keV. The double layers are narrow structures characterised by the strong gradient of the electric field. The electric potential changes its sign, with the potential being much larger in one portion of the double layer than in the second one with an opposite sign (Figure 6 in Karlický, 2012). This leads to preferential acceleration of electrons in one direction. The double layers considered by Karlický (2012) are formed in a flare loop containing electron beams, background plasma with a return current and reflected beam electrons. The influence of the particle distributions formed in double layers were studied by Dzifčáková, Karlický, and Dudík (2013), who showed that the X-ray spectra arising in such plasma double layers can match the observed ones well, if a background Maxwellian component is also considered.

Diagnostics of the n-distributions or distributions with a similar slope at several keV relies strongly on the underlying atomic data neccessary to synthesise the spectrum. New calculations of the ionisation rates by Dere (2007) and the dielectronic radiative recombination rates by N. Badnell and colleagues (*e.g.*, Badnell *et al.*, 2003; Altun *et al.*, 2006, 2007; Nikolić *et al.*, 2010; Abdel-Naby *et al.*, 2012) lead to significant modifications of the ionisation equilibrium obtained for the Maxwellian distribution (Dere *et al.*, 2009; Bryans, Landi, and Savin, 2009; Landi *et al.*, 2013). These new atomic data calculations have to be reflected in the corresponding ionisation and recombination rates and the subsequent ionisation equilibria also for the non-Maxwellian distributions. This is the purpose of this paper. The calculations obtained here for the n-distributions are made available as the online electronic supplementary material.

This paper is organised as follows. The non-Maxwellian n-distributions and their properties are described in Section 2. Section 3 summarises the atomic data used. The methods for obtaining the individual ionisation and recombination rates are outlined in Section 4. Ionisation equilibria are calculated and described in Section 5. Finally, Section 6 contains a summary and a discussion.

2. The Non-Maxwellian *n*-Distribution

The non-Maxwellian n-distributions (Seely, Feldman, and Doschek, 1987; Dzifčáková, 1998) are a class of analytic, parametric distributions invoked to explain spectra of intense dielectronic satellite lines in the neighbourhood of the allowed lines during solar flares. Compared to the Maxwellian distribution, the n-distributions have higher and narrower peaks (see also Figure 1):

$$f(\mathcal{E}, n)\mathrm{d}\mathcal{E} = \mathcal{B}(n)\frac{2}{k_\mathrm{B}T\sqrt{\pi}}\left(\frac{\mathcal{E}}{k_\mathrm{B}T}\right)^{n/2}\exp\left(-\frac{\mathcal{E}}{k_\mathrm{B}T}\right)\mathrm{d}\mathcal{E}, \tag{1}$$

where $\mathcal{B}(n) = \pi^{1/2}/(2\Gamma(n/2+1))$ is the normalisation constant, $\mathcal{E}$ is the kinetic energy of the free electron, n and T, are free parameters and k_B is the Boltzmann constant. The $n = 1$ distribution is a Maxwellian with temperature T. The deviation from the Maxwellian distribution increases with n. The mean energy $\langle\mathcal{E}(n)\rangle$ of the n-distribution depends on both parameters n and T,

$$\langle\mathcal{E}(n)\rangle = \left(\frac{n}{2}+1\right)k_\mathrm{B}T. \tag{2}$$

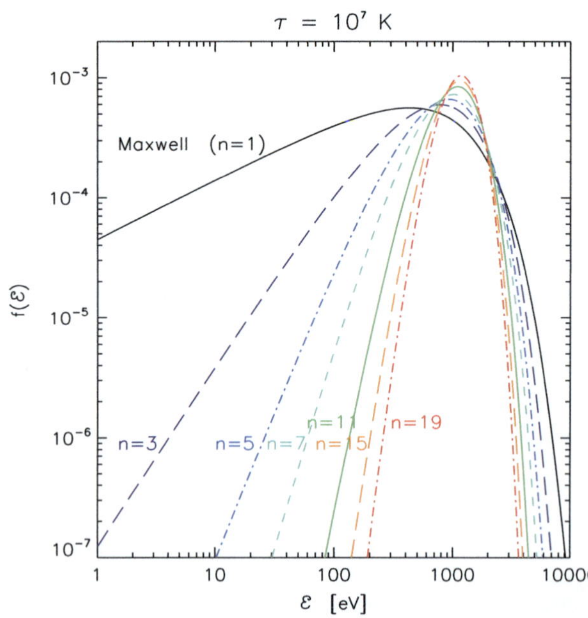

Figure 1 Comparison of the Maxwellian ($n = 1$) and n-distributions with $n = 3, 5, 7, 11, 15$, and 19. These distributions have the same mean energy, corresponding to $\tau = 10^7$ K.

Because of this, a new parameter, $\tau = T(n+2)/3$, called the pseudo-temperature, has been introduced by Dzifčáková (1998). From Equation (2), we get

$$\langle \mathcal{E}(n) \rangle = \frac{3}{2} k_{\mathrm{B}} \tau = \left(\frac{n}{2} + 1 \right) k_{\mathrm{B}} T. \tag{3}$$

Since the pseudo-temperature τ is proportional to the mean energy of the n-distribution, it has an analogous physical meaning to the temperature for Maxwellian distribution. Therefore, τ is used for computation of all dependent quantities, *e.g.* the pressure (Dzifčáková, 1998).

Finally, the most probable energy (peak of the n-distribution) occurs at

$$\mathcal{E}_{\mathrm{p}}(n) = \frac{n}{2} k_{\mathrm{B}} T = \frac{3}{2} \frac{n}{n+2} k_{\mathrm{B}} \tau. \tag{4}$$

Namely, the peak of the n-distributions is shifted to higher energies for higher n and constant $\langle \mathcal{E}(n) \rangle$ (Figure 1).

3. Atomic Data

The CHIANTI atomic database (Dere *et al.*, 1997; Landi *et al.*, 2013) contains continuously updated ionisation and recombination rates for the Maxwellian distribution, as well as the resulting ionisation equilibria. Dere (2007) produced fits to the observed and/or calculated cross sections for direct ionisation and autoionisation and extended these to all astrophysically important elements with $Z \leq 30$, *i.e.*, H to Zn. These approximations to the ionisation cross sections are available in the CHIANTI database since its version 6 (Dere *et al.*, 2009).

Radiative and dielectronic recombination rates for the Maxwellian distribution are also available within the CHIANTI database (see Section 24 of Dere *et al.*, 2009). There, the dielectronic and radiative recombination rates for the H to Al and Ar isoelectronic sequences are taken from Badnell *et al.* (2003), Colgan *et al.* (2003), Colgan, Pindzola, and Badnell (2004), Mitnik and Badnell (2004), Badnell (2006), Altun *et al.* (2005, 2006, 2007), Zatsarinny *et al.* (2005a,b, 2006), Bautista and Badnell (2007), Nikolić *et al.* (2010) and Abdel-Naby *et al.* (2012). For the Si to Mn isoelectronic sequences, the recombination rates are taken from Shull and van Steenberg (1982), Nahar (1996, 1997), Nahar and Bautista (2001), Mazzotta *et al.* (1998) and Mazzitelli and Mattioli (2002). Details can be found in Dere *et al.* (2009) and Landi *et al.* (2013).

To calculate the ionisation and recombination rates, we use the CHIANTI database (Version 7.1; Landi *et al.*, 2013). This approach is the same as in Dzifčáková and Dudík (2013) for obtaining the ionisation and recombination rates and the subsequent equilibria for the non-Maxwellian κ-distributions (a different class of non-Maxwellian distributions exhibiting a high-energy tail). Differences between the currently used datasets and the resulting ionisation equilibrium with respect to the previous ionisation equilibrium calculations of Mazzotta *et al.* (1998) are due to the Maxwellian distribution as discussed *e.g.* in Landi *et al.* (2013) and also in Dzifčáková and Dudík (2013). Here, we note in passing that there are also strong differences in the relative ion abundances of Fe VIII and Fe IX (see, *e.g.*, the top panel of Figure 5 in Dzifčáková and Dudík, 2013).

4. Ionisation and Recombination Rates

Using the atomic data described above, the ionisation and recombination rates for the n-distributions can be obtained as follows. Direct ionisation and autoionisation cross sections $\sigma_{\rm ioniz}$ available in CHIANTI are integrated directly over the distribution to obtain the corresponding ionisation rate $R_{\rm ioniz}$,

$$R_{\rm ioniz}(n) = \langle \sigma_{\rm ioniz} v \rangle = \int_0^\infty \sigma_{\rm ioniz} \left(\frac{2\mathcal{E}}{m_{\rm e}} \right)^{1/2} f(\mathcal{E}, n) {\rm d}\mathcal{E}. \quad (5)$$

The integration is carried out for odd values of n up to $n = 19$. We note that this choice of n is arbitrary, but it covers the parameter space of n values typically derived from observations (Dzifčáková *et al.*, 2008; Kulinová *et al.*, 2011).

The behaviour of the total ionisation rate with n is shown in Figure 2 for C V and Fe XXV. Typically, the total ionisation rate exhibits a steeper increase with τ, with the maxima shifted to a lower τ compared to the Maxwellian case (black and coloured lines in Figure 2).

The dielectronic recombination rate for the n-distribution, $R_{\rm DR}(n)$, can be obtained in the manner derived by Dzifčáková (1998). The Maxwellian dielectronic recombination rate $R_{\rm DR}(n = 1)$ is typically approximated by the expression

$$R_{\rm DR}(n) = \frac{\mathcal{B}(n)}{T^{3/2}} \sum_i A_i \exp\left(-\frac{t_i}{T} \right), \quad (6)$$

where t_i and A_i are parameters available in the CHIANTI database. Dzifčáková (1998) has shown that for the n-distributions, $R_{\rm DR}(n)$ can be expressed as

$$R_{\rm DR}(n) = \frac{\mathcal{B}(n)}{T^{3/2}} \sum_i A_i \left(\frac{t_i}{T} \right)^{(n-1)/2} \exp\left(-\frac{t_i}{T} \right), \quad (7)$$

using the same parameters t_i and A_i as for the Maxwellian distribution.

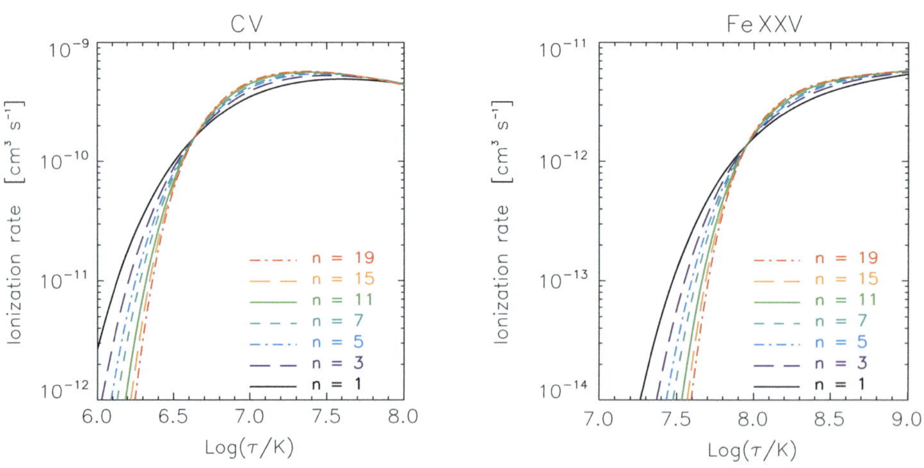

Figure 2 The ionisation rates for the Maxwellian (black lines) and n-distributions with $n = 3$, 5, 7, 11, 15, and 19 for C V (left) and Fe XXV (right). Individual colours and line styles denote different n and are the same as in Figure 1.

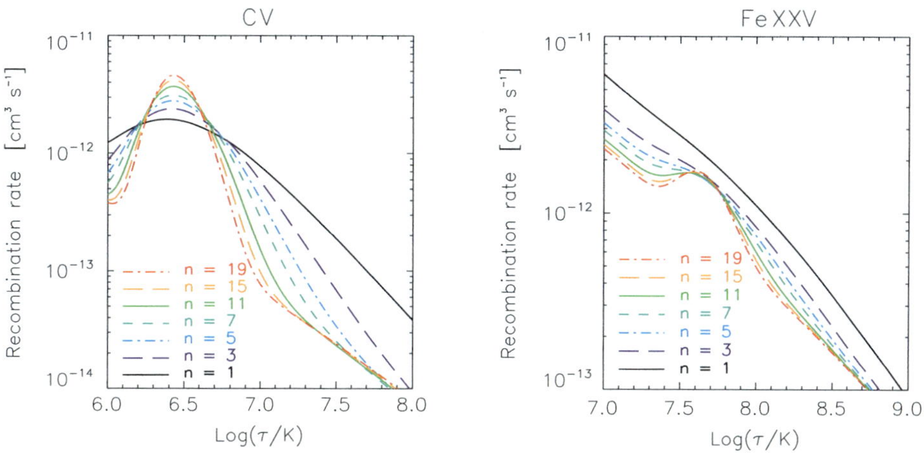

Figure 3 The recombination rates for the Maxwellian (black lines) and n-distributions with $n = 3, 5, 7, 11, 15$, and 19 for C V (left) and Fe XXV (right). Individual colours and line styles denote different n and are the same as in Figure 1.

For the radiative recombination, a power-law energy dependence of the cross section σ_{RR} on energy is assumed (*e.g.*, Osterbrock, 1974),

$$\sigma_{\mathrm{RR}}(\mathcal{E}) = C_{\mathrm{RR}}/\mathcal{E}^{\eta+1/2}, \tag{8}$$

where C_{RR} is a constant and $\eta + 1/2$ is a power-law index. Integrating over the n-distribution, one obtains (Dzifčáková, 1998)

$$R_{\mathrm{RR}}(n) = \mathcal{B}(n)\frac{4C_{\mathrm{RR}}}{(2\pi m_{\mathrm{e}})^{1/2}}\frac{\Gamma(n/2+1-\eta)}{(k_{\mathrm{B}}T)^{\eta}} = R_{\mathrm{RR}}^{\mathrm{Maxw}}\frac{\Gamma(n/2+1-\eta)\Gamma(3/2)}{\Gamma(3/2-\eta)\Gamma(n/2+1)}, \tag{9}$$

where $R_{\mathrm{RR}}^{\mathrm{Maxw}}$ is the radiative recombination rate for the Maxwellian distribution ($n = 1$).

A typical behaviour of the total recombination coefficient $R_{\mathrm{RR}} + R_{\mathrm{DR}}$ with n and τ is shown in Figure 3 for C V and Fe XXV. With the increase in n, the relative number of low-energy electrons decreases (Figure 1). This causes the decrease of the radiative recombination rate. On the other hand, the contribution from the dielectronic recombination can increase with increasing value of n. The peak of the dielectronic recombination rate becomes narrower with higher n, analogously to the shape of the n-distribution (Figure 1). The behaviour of the total recombination coefficient with τ then depends on the relative contributions of the dielectronic and radiative recombination. For some τ, the total recombination rate can become larger for the n-distributions than for the Maxwellian distribution. This happens *e.g.* for C V at $\log(\tau/\mathrm{K}) \approx 6.4$ (Figure 3, left panel). However, for other ions, such as Fe XXV, the total recombination rate for the n-distributions is lower than for the Maxwellian distribution for all τ (Figure 3, right panel).

5. The Ionisation Equilibria

The rates obtained using the method described in Section 4 were subsequently used to calculate the ionisation equilibria for all astrophysically important elements, H to Zn, *i.e.*, elements with $Z \leqq 30$. These ionisation equilibria were calculated for the odd values of n,

$1 \leqq n \leqq 19$. The relative ion abundance of an ion $+k$ of the element Z is obtained by considering the balance between the ionisation and recombination processes to and from the ion $+k$ (see *e.g.* Equation (4.77) in Phillips, Feldman, and Landi, 2008). In doing so, direct ionisation, autoionisation, radiative recombination, and dielectronic recombination are taken into account (see Section 4). Three-body processes and photoionisation are neglected. The relative ion abundances in the ionisation equilibria obtained here are calculated for 101 values of $\log(\tau/\mathrm{K})$, $4 \leqq \log(\tau/\mathrm{K}) \leqq 9$, with the step of 0.05 similarly as in CHIANTI (Dere *et al.*, 2009; Landi *et al.*, 2013), and they are available as electronic supplementary material. The files in there are in the CHIANTI *.ioneq* format. This means that they may be read directly with the CHIANTI *read_ionEquationpro* IDL routine. Details on the format of the *.ioneq* files can be found in Dzifčáková and Dudík (2013, Appendix A therein). We note that previous calculations of the ionisation equilibria for the n-distributions were done only for Fe (Dzifčáková, 1998), Ca and Si (Dzifčáková and Kulinová, 2001; Dzifčáková, 2005), and then for 12 elements necessary for calculation of the radiative losses, *i.e.* C, N, O, Ne, Mg, Al, Si, S, Ar, Ca, Fe, and Ni (Dudík *et al.*, 2011). For the last group of elements, their ionisation equilibria were never published separately.

The behaviour of the relative ion abundances with τ and n are shown for Fe in Figure 4 and for Si in Figure 5. Generally, with increase in n, the ionisation peaks become steeper and narrower (see also Dzifčáková, 1998). We note that this behaviour is in stark contrast to the behaviour of the relative ion abundances for the κ-distributions characterised by the high-energy tail (Dzifčáková and Dudík, 2013). Generally, the shape of the ionisation peak follows that of the distribution; more mono-energetic distributions produce narrower ionisation peaks. In cases where the distribution is strongly non-Maxwellian with high n, such as in the impulsive phase of solar flares (Dzifčáková *et al.*, 2008; Kulinová *et al.*, 2011), this behaviour results in a lower number of ions with very high abundances at a given τ.

However, exceptions to this behaviour can occur. Especially the He-like ions such as Fe XXV and Si XIII (Figures 4 and 5, respectively) have ionisation peaks broader than in the Maxwellian case. This is the result chiefly of the suppressed dielectronic recombination with n (Section 4 and Figure 3, right panel), especially at higher τ. The broadness of these He-like ion abundance peaks means that such ions are formed at wider ranges of τ, and this will have strong presence in multi-τ flare spectra.

The ion abundance maxima for the n-distributions can also be shifted to different τ compared to the Maxwellian distribution. Typically, these shifts are towards higher τ. This is the result of the strong decrease of the total ionisation rate with τ (Figure 2). The shift of the maximum of the ion abundance can sometimes be towards lower τ, especially if $\tau > 10^7$ K. Such situations occur *e.g.* for Fe XXIV (Figure 4).

Finally, an example comparison with the older calculations first performed by Dzifčáková (1998) is shown in Figure 6. There, the ionisation equilibria are shown for $n = 11$ and Fe and Si ion species at $\tau = 10^6 - 10^8$ K. We note that $n = 11$ is the largest departure from the Maxwellian distribution typically diagnosed within the observational uncertainties (Dzifčáková *et al.*, 2008; Kulinová *et al.*, 2011). We see that the new calculations significantly change the relative ion abundances of Si ions up to Si XI. Similarly, the relative ion abundances of various Fe ions up to Fe XXV are different from those calculated by Dzifčáková (1998). However, the relative ion abundances of Si XIII and Si XIV do not change significantly, despite the new atomic data used (Section 3). We note that these ions were the basis of the diagnostics of Dzifčáková *et al.* (2008) and Kulinová *et al.* (2011) and that the weak changes in the ionisation equilibrium derived here increase our confidence in the results of the non-Maxwellian diagnostics presented therein.

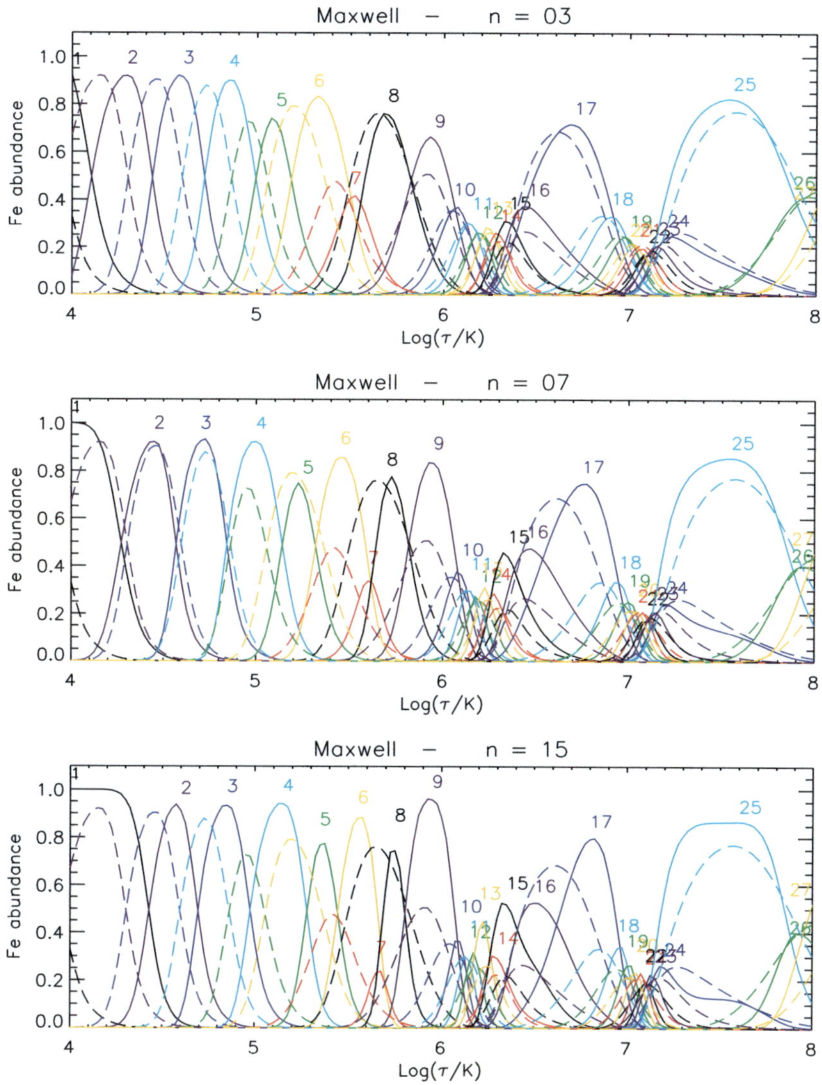

Figure 4 Comparison of the ionisation equilibrium for the Maxwellian (dashed lines) and n-distributions with different n (full lines) for Fe at $\tau = 10^4 - 10^8$ K. Colours represent individual ions.

6. Summary and Discussion

We calculated the ionisation equilibria for the non-Maxwellian n-distributions using the latest atomic datasets available within CHIANTI. Direct ionisation and autoionisation rates are obtained numerically from the corresponding cross sections. The radiative and dielectronic recombination rates are obtained using analytical approximations. While the radiative recombination rates are decreased for the n-distributions due to the lower number of low-energy electrons, the peak of the dielectronic recombination rate becomes narrower with

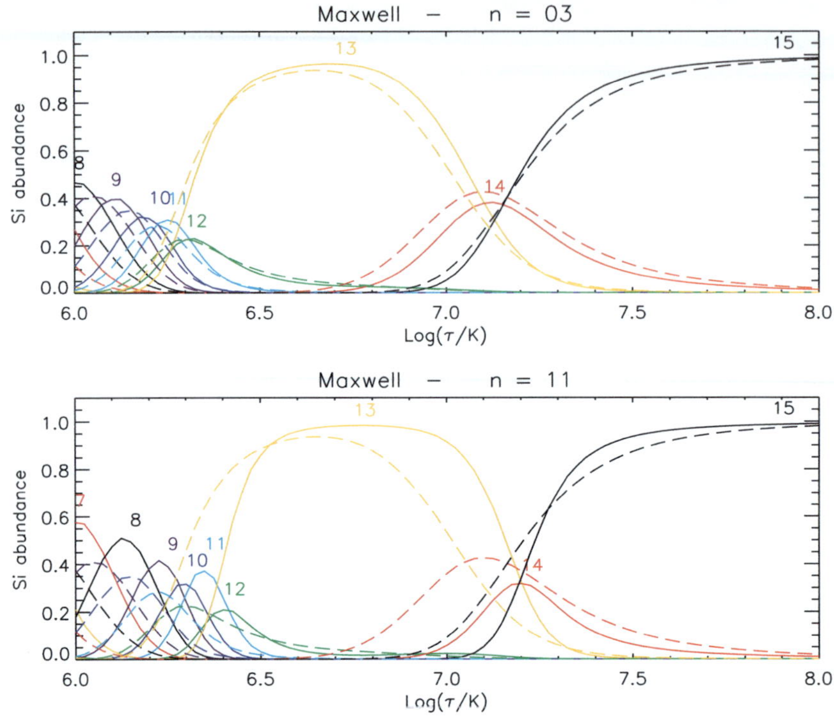

Figure 5 Comparison of the ionisation equilibrium for the Maxwellian (dashed lines) and n-distributions with different n (full lines) for Si at $\tau = 10^6 - 10^8$ K. Colours represent individual ions.

increase in n, following the shape of the n-distribution. The total ionisation rate increases more steeply at lower τ than the Maxwellian one, and its peak is shifted towards lower τ.

This behaviour is reflected in the resulting ionisation equilibrium. Relative ion abundance of most of the ions exhibit steeper and narrower peaks. Namely, the number of ions with high relative abundances at a given τ is decreased to the Maxwellian distribution. However, He-like ions such as Si XIII or Fe XXV have broader ion abundance peaks than for the Maxwellian distribution, a consequence mainly due to decreased total recombination rates at all τ. These ions will be prominent emitters in a flare plasma exhibiting n-distributions. The ion abundance peaks for the n-distributions can also be shifted towards different τ.

We note here that the n-distribution is not the only non-Maxwellian distribution observed during flares. High-energy power-law tails are ubiquitous (*e.g.*, Veronig *et al.*, 2010; Fletcher *et al.*, 2011; Zharkova *et al.*, 2011; Battaglia and Kontar, 2013; Simões and Kontar, 2013; Milligan *et al.*, 2014; Oka *et al.*, 2013, 2015). The n-distribution is typically prominent at the energies of only a few keV (Kulinová *et al.*, 2011), where the RHESSI observations are strongly attenuated. Nevertheless, the emission measure of the n-distributions as derived from RHESSI can be several times higher than the emission measure of the thermal component and orders of magnitude higher than the emission measure of the power-law tail (see Table 2 in Kulinová *et al.*, 2011). However, if the number of electrons in the high-energy tail is large in some spatial locations, as suggested by Oka *et al.* (2013, 2015), these will affect the ionisation equilibrium (Dzifčáková, Homola, and Dudík, 2011) and also the line intensities (*e.g.*, Gabriel and Phillips, 1979; Dzifčáková, Homola, and Dudík, 2011; Ferouani, Inal, and Csanak, 2013). Typically, the presence of a high-energy power-law tail together with a

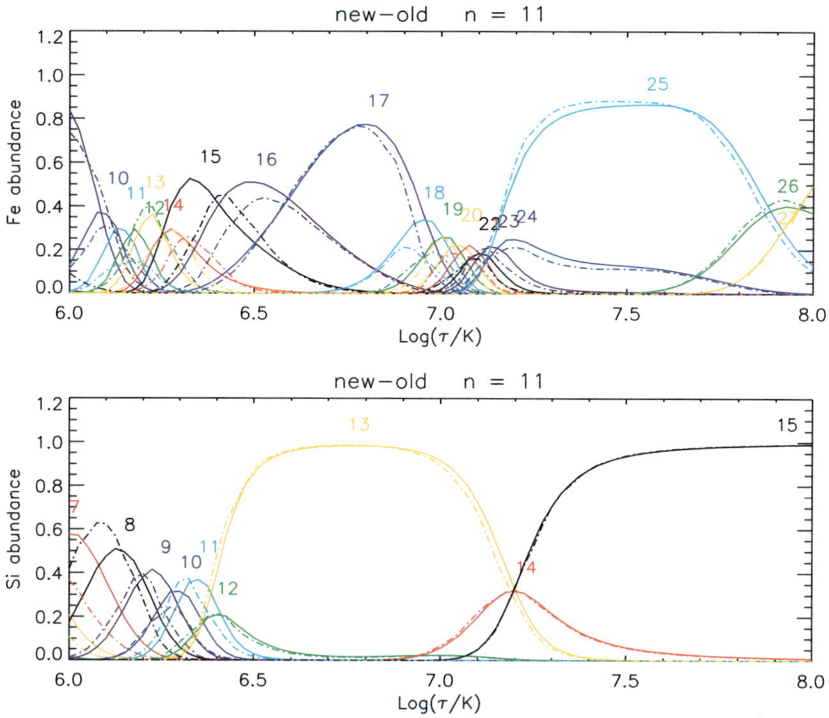

Figure 6 Comparison of the Fe (top) and Si (bottom) ionisation equilibrium for the present atomic data (full lines) and the earlier calculation with the atomic data of Mazzotta *et al.* (1998). The comparison is shown at $n = 11$ and $\tau = 10^6 - 10^8$ K. Colours represent individual ions.

Maxwellian core significantly increases the ionisation rates, especially at low τ, but leaves the recombination rates unaffected, since these are much more sensitive to low-energy electrons. Instead, if the bulk (core) of the distribution corresponds to the n-distribution, arising *e.g.* in the double layers (Dzifčáková, Karlický, and Dudík, 2013), the resulting behaviour of the ionisation equilibrium becomes complex (Dzifčáková, Homola, and Dudík, 2011), since the bulk and tail shifts the peaks of the ionisation equilibrium in the opposite directions. The behaviour of the ionisation equilibrium is then strongly sensitive to the nature of the bulk of the distribution, as well as the high-energy tail, and also to the bulk/tail particle ratio. In such complex situations, this work represents a contribution towards the description of the ionic composition, with the effect of the high-energy tails, described *e.g.* by the κ-distributions (Oka *et al.*, 2013, 2015) that have already been described elsewhere (Dzifčáková and Dudík, 2013).

Finally, we note that the bulk and tail of the distribution will affect different parts of the spectrum, with the n-distributions strongly modifying the intensities of dielectronic satellite lines in soft X-rays, the high-energy tail strongly modulating the bremsstrahlung emission observed by RHESSI (see also Dudík *et al.*, 2012). This only underlines the need for future combined observations involving both imaging and spectroscopy in both hard X-rays and soft X-rays in order to elucidate all non-Maxwellian and possibly non-equilibrium processes occurring in solar flares simultaneously.

References

Abdel-Naby, S.A., Nikolić, D., Gorczyca, T.W., Korista, K.T., Badnell, N.R.: 2012, Dielectronic recombination data for dynamic finite-density plasmas. XIV. The aluminum isoelectronic sequence. *Astron. Astrophys.* **537**, A40. DOI. ADS.

Altun, Z., Yumak, A., Badnell, N.R., Colgan, J., Pindzola, M.S.: 2005, Erratum: Dielectronic recombination data for dynamic finite-density plasmas. VI. The boron isoelectronic sequence. *Astron. Astrophys.* **433**, 395. DOI. ADS.

Altun, Z., Yumak, A., Badnell, N.R., Loch, S.D., Pindzola, M.S.: 2006, Dielectronic recombination data for dynamic finite-density plasmas. XI. The sodium isoelectronic sequence. *Astron. Astrophys.* **447**, 1165. DOI. ADS.

Altun, Z., Yumak, A., Yavuz, I., Badnell, N.R., Loch, S.D., Pindzola, M.S.: 2007, Dielectronic recombination data for dynamic finite-density plasmas. XIII. The magnesium isoelectronic sequence. *Astron. Astrophys.* **474**, 1051. DOI. ADS.

Asai, A., Nakajima, H., Shimojo, M., Yokoyama, T., Masuda, S., Krucker, S.: 2009, Imaging spectroscopy on preflare coronal nonthermal sources associated with the 2002 July 23 flare. *Astrophys. J.* **695**, 1623. DOI. ADS.

Badnell, N.R.: 2006, Dielectronic recombination data for dynamic finite-density plasmas. X. The hydrogen isoelectronic sequence. *Astron. Astrophys.* **447**, 389. DOI. ADS.

Badnell, N.R., O'Mullane, M.G., Summers, H.P., Altun, Z., Bautista, M.A., Colgan, J., Gorczyca, T.W., Mitnik, D.M., Pindzola, M.S., Zatsarinny, O.: 2003, Dielectronic recombination data for dynamic finite-density plasmas. I. Goals and methodology. *Astron. Astrophys.* **406**, 1151. DOI. ADS.

Battaglia, M., Kontar, E.P.: 2013, Electron distribution functions in solar flares from combined X-ray and extreme-ultraviolet observations. *Astrophys. J.* **779**, 107. DOI. ADS.

Bautista, M.A., Badnell, N.R.: 2007, Dielectronic recombination data for dynamic finite-density plasmas. XII. The helium isoelectronic sequence. *Astron. Astrophys.* **466**, 755. DOI. ADS.

Bian, N.H., Emslie, A.G., Stackhouse, D.J., Kontar, E.P.: 2014, The formation of kappa-distribution accelerated electron populations in solar flares. *Astrophys. J.* **796**, 142. DOI. ADS.

Brown, J.C.: 1971, The deduction of energy spectra of non-thermal electrons in flares from the observed dynamic spectra of hard X-ray bursts. *Solar Phys.* **18**, 489. DOI. ADS.

Bryans, P., Landi, E., Savin, D.W.: 2009, A new approach to analyzing solar coronal spectra and updated collisional ionization equilibrium calculations. II. Updated ionization rate coefficients. *Astrophys. J.* **691**, 1540. DOI. ADS.

Burge, C.A., MacKinnon, A.L., Petkaki, P.: 2014, Effect of binary collisions on electron acceleration in magnetic reconnection. *Astron. Astrophys.* **561**, A107. DOI. ADS.

Colgan, J., Pindzola, M.S., Badnell, N.R.: 2004, Dielectronic recombination data for dynamic finite-density plasmas. V: The lithium isoelectronic sequence. *Astron. Astrophys.* **417**, 1183. DOI. ADS.

Colgan, J., Pindzola, M.S., Whiteford, A.D., Badnell, N.R.: 2003, Dielectronic recombination data for dynamic finite-density plasmas. III. The beryllium isoelectronic sequence. *Astron. Astrophys.* **412**, 597. DOI. ADS.

Dere, K.P.: 2007, Ionization rate coefficients for the elements hydrogen through zinc. *Astron. Astrophys.* **466**, 771. DOI. ADS.

Dere, K.P., Landi, E., Mason, H.E., Monsignori Fossi, B.C., Young, P.R.: 1997, CHIANTI – An atomic database for emission lines. *Astron. Astrophys. Suppl.* **125**, 149. DOI. ADS.

Dere, K.P., Landi, E., Young, P.R., Del Zanna, G., Landini, M., Mason, H.E.: 2009, CHIANTI – An atomic database for emission lines. IX. Ionization rates, recombination rates, ionization equilibria for the elements hydrogen through zinc and updated atomic data. *Astron. Astrophys.* **498**, 915. DOI. ADS.

Dudík, J., Dzifčáková, E., Karlický, M., Kulinová, A.: 2011, The bound-bound and free–free radiative losses for the nonthermal distributions in solar and stellar coronae. *Astron. Astrophys.* **529**, A103. DOI. ADS.

Dudík, J., Kašparová, J., Dzifčáková, E., Karlický, M., Mackovjak, Š.: 2012, The non-Maxwellian continuum in the X-ray, UV, and radio range. *Astron. Astrophys.* **539**, A107. DOI. ADS.

Dzifčáková, E.: 1998, The ionization equilibrium in the solar corona for the electron power distribution. *Solar Phys.* **178**, 317. ADS.

Dzifčáková, E.: 2005, The excitation equilibrium of Si XII and Ca XIX in the solar corona for electron power distribution. In: Danesy, D., Poedts, S., De Groof, A., Andries, J. (eds.) *Proceedings of the 11th European Solar Physics Meeting – The Dynamic Sun: Challenges for Theory and Observations. ESA* **SP-600**, 120. ADS.

Dzifčáková, E., Dudík, J.: 2013, H to Zn ionization equilibrium for the non-Maxwellian electron κ-distributions: Updated calculations. *Astrophys. J. Suppl.* **206**, 6. DOI. ADS.

Dzifčáková, E., Kulinová, A.: 2001, The Fe XXV excitation equilibrium in the solar corona for the electron power distributions using a pseudo-temperature. *Solar Phys.* **203**, 53. DOI. ADS.

Dzifčáková, E., Homola, M., Dudík, J.: 2011, The ionization equilibrium and flare line spectra for the electron distribution with a power-law tail. *Astron. Astrophys.* **531**, A111. DOI. ADS.

Dzifčáková, E., Karlický, M., Dudík, J.: 2013, Dielectronic satellite lines and double layers in solar flares. *Astron. Astrophys.* **550**, A60. DOI. ADS.

Dzifčáková, E., Kulinová, A., Chifor, C., Mason, H.E., Del Zanna, G., Sylwester, J., Sylwester, B.: 2008, Nonthermal and thermal diagnostics of a solar flare observed with RESIK and RHESSI. *Astron. Astrophys.* **488**, 311. DOI. ADS.

Ferouani, A.K., Inal, M.K., Csanak, G.: 2013, Ne IX line G-ratio in a non-Maxwellian and anisotropic plasma. *J. Phys. B* **46**, 085701. DOI. ADS.

Fleishman, G.D., Kuznetsov, A.A.: 2014, Theory of gyroresonance and free-free emissions from non-Maxwellian quasi-steady-state electron distributions. *Astrophys. J.* **781**, 77. DOI. ADS.

Fletcher, L., Dennis, B.R., Hudson, H.S., Krucker, S., Phillips, K., Veronig, A., Battaglia, M., Bone, L., Caspi, A., Chen, Q., Gallagher, P., Grigis, P.T., Ji, H., Liu, W., Milligan, R.O., Temmer, M.: 2011, An observational overview of solar flares. *Space Sci. Rev.* **159**, 19. DOI. ADS.

Gabriel, A.H., Phillips, K.J.H.: 1979, Dielectronic satellite spectra for highly charged helium-like ions. IV. Iron satellite lines as a measure of non-thermal electron energy distributions. *Mon. Not. R. Astron. Soc.* **189**, 319. ADS.

Karlický, M.: 2012, The n-distribution of electrons and double layers in the electron-beam-return-current system of solar flares. *Astrophys. J.* **750**, 49. DOI. ADS.

Karlický, M., Dzifčáková, E., Dudík, J.: 2012, On the physical meaning of n-distributions in solar flares. *Astron. Astrophys.* **537**, A36. DOI. ADS.

Kulinová, A., Kašparová, J., Dzifčáková, E., Sylwester, J., Sylwester, B., Karlický, M.: 2011, Diagnostics of non-thermal distributions in solar flare spectra observed by RESIK and RHESSI. *Astron. Astrophys.* **533**, A81. DOI. ADS.

Landi, E., Young, P.R., Dere, K.P., Del Zanna, G., Mason, H.E.: 2013, CHIANTI – An atomic database for emission lines. XIII. Soft X-ray improvements and other changes. *Astrophys. J.* **763**, 86. DOI. ADS.

Lin, R.P., Hudson, H.S.: 1971, 10 – 100 keV electron acceleration and emission from solar flares. *Solar Phys.* **17**, 412. DOI. ADS.

Lin, R.P., Dennis, B.R., Hurford, G.J., Smith, D.M., Zehnder, A., Harvey, P.R., Curtis, D.W., Pankow, D., Turin, P., Bester, M., Csillaghy, A., Lewis, M., Madden, N., van Beek, H.F., Appleby, M., Raudorf, T., McTiernan, J., Ramaty, R., Schmahl, E., Schwartz, R., Krucker, S., Abiad, R., Quinn, T., Berg, P., Hashii, M., Sterling, R., Jackson, R., Pratt, R., Campbell, R.D., Malone, D., Landis, D., Barrington-Leigh, C.P., Slassi-Sennou, S., Cork, C., Clark, D., Amato, D., Orwig, L., Boyle, R., Banks, I.S., Shirey, K., Tolbert, A.K., Zarro, D., Snow, F., Thomsen, K., Henneck, R., McHedlishvili, A., Ming, P., Fivian, M., Jordan, J., Wanner, R., Crubb, J., Preble, J., Matranga, M., Benz, A., Hudson, H., Canfield, R.C., Holman, G.D., Crannell, C., Kosugi, T., Emslie, A.G., Vilmer, N., Brown, J.C., Johns-Krull, C., Aschwanden, M., Metcalf, T., Conway, A.: 2002, The Reuven Ramaty High-Energy Solar Spectroscopic Imager (RHESSI). *Solar Phys.* **210**, 3. DOI. ADS.

Mazzitelli, G., Mattioli, M.: 2002, Ionization balance for optically thin plasmas: Rate coefficients for Cu, Zn, Ga, and Ge ions. *At. Data Nucl. Data Tables* **82**, 313. DOI. ADS.

Mazzotta, P., Mazzitelli, G., Colafrancesco, S., Vittorio, N.: 1998, Ionization balance for optically thin plasmas: Rate coefficients for all atoms and ions of the elements H to NI. *Astron. Astrophys. Suppl.* **133**, 403. DOI. ADS.

Milligan, R.O., Kerr, G.S., Dennis, B.R., Hudson, H.S., Fletcher, L., Allred, J.C., Chamberlin, P.C., Ireland, J., Mathioudakis, M., Keenan, F.P.: 2014, The radiated energy budget of chromospheric plasma in a major solar flare deduced from multi-wavelength observations. *Astrophys. J.* **793**, 70. DOI. ADS.

Mitnik, D.M., Badnell, N.R.: 2004, Dielectronic recombination data for dynamic finite-density plasmas. VIII. The nitrogen isoelectronic sequence. *Astron. Astrophys.* **425**, 1153. DOI. ADS.

Nahar, S.N.: 1996, Total electron–ion recombination of Fe III. *Phys. Rev. A* **53**, 2417. DOI. ADS.

Nahar, S.N.: 1997, Electron–ion recombination of Fe II. *Phys. Rev. A* **55**, 1980. DOI. ADS.

Nahar, S.N., Bautista, M.A.: 2001, Electron–ion recombination rate coefficients and photoionization cross sections for astrophysically abundant elements. VI. Ni II. *Astrophys. J. Suppl.* **137**, 201. DOI. ADS.

Nikolić, D., Gorczyca, T.W., Korista, K.T., Badnell, N.R.: 2010, Dielectronic recombination of argon-like ions. *Astron. Astrophys.* **516**, A97. DOI. ADS.

Oka, M., Ishikawa, S., Saint-Hilaire, P., Krucker, S., Lin, R.P.: 2013, Kappa distribution model for hard X-ray coronal sources of solar flares. *Astrophys. J.* **764**, 6. DOI. ADS.

Oka, M., Krucker, S., Hudson, H.S., Saint-Hilaire, P.: 2015, Electron energy partition in the above-the-looptop solar hard X-ray sources. *Astrophys. J.* **799**, 129. DOI. ADS.

Osterbrock, D.E.: 1974, *Astrophysics of Gaseous Nebulae*. ADS.

Petkaki, P., MacKinnon, A.L.: 2007, Particle acceleration by fluctuating electric fields at a magnetic field null point. *Astron. Astrophys.* **472**, 623. DOI. ADS.

Petkaki, P., MacKinnon, A.L.: 2011, Acceleration of charged particles by fluctuating and steady electric fields in a X-type magnetic field. *Adv. Space Res.* **48**, 884. DOI. ADS.
Phillips, K.J.H., Feldman, U., Landi, E.: 2008, *Ultraviolet and X-ray Spectroscopy of the Solar Atmosphere*, Cambridge University Press, Cambridge. ADS.
Seely, J.F., Feldman, U., Doschek, G.A.: 1987, Observation of nonthermal energy distributions during the impulsive phase of solar flares. *Astrophys. J.* **319**, 541. DOI. ADS.
Shull, J.M., van Steenberg, M.: 1982, The ionization equilibrium of astrophysically abundant elements. *Astrophys. J. Suppl.* **48**, 95. DOI. ADS.
Simões, P.J.A., Kontar, E.P.: 2013, Implications for electron acceleration and transport from non-thermal electron rates at looptop and footpoint sources in solar flares. *Astron. Astrophys.* **551**, A135. DOI. ADS.
Sylwester, J., Sylwester, B., Phillips, K.J.H.: 2008, RESIK observations of helium-like argon X-ray line emission in solar flares. *Astrophys. J. Lett.* **681**, L117. DOI. ADS.
Sylwester, J., Gaicki, I., Kordylewski, Z., Kowaliński, M., Nowak, S., Płocieniak, S., Siarkowski, M., Sylwester, B., Trzebiński, W., Bakała, J., Culhane, J.L., Whyndham, M., Bentley, R.D., Guttridge, P.R., Phillips, K.J.H., Lang, J., Brown, C.M., Doschek, G.A., Kuznetsov, V.D., Oraevsky, V.N., Stepanov, A.I., Lisin, D.V.: 2005, Resik: A bent crystal X-ray spectrometer for studies of solar coronal plasma composition. *Solar Phys.* **226**, 45. DOI. ADS.
Veronig, A.M., Rybák, J., Gömöry, P., Berkebile-Stoiser, S., Temmer, M., Otruba, W., Vršnak, B., Pötzi, W., Baumgartner, D.: 2010, Multiwavelength imaging and spectroscopy of chromospheric evaporation in an M-class solar flare. *Astrophys. J.* **719**, 655. DOI. ADS.
Zatsarinny, O., Gorczyca, T.W., Korista, K.T., Fu, J., Badnell, N.R., Mitthumsiri, W., Savin, D.W.: 2005a, Dielectronic recombination data for dynamic finite-density plasmas. II. The oxygen isoelectronic sequence. *Astron. Astrophys.* **438**, 743. DOI. ADS.
Zatsarinny, O., Gorczyca, T.W., Korista, K.T., Fu, J., Badnell, N.R., Mitthumsiri, W., Savin, D.W.: 2005b, Dielectronic recombination data for dynamic finite-density plasmas. IV. The carbon isoelectronic sequence. *Astron. Astrophys.* **440**, 1203. DOI. ADS.
Zatsarinny, O., Gorczyca, T.W., Fu, J., Korista, K.T., Badnell, N.R., Savin, D.W.: 2006, Dielectronic recombination data for dynamic finite-density plasmas. IX. The fluorine isoelectronic sequence. *Astron. Astrophys.* **447**, 379. DOI. ADS.
Zharkova, V.V., Arzner, K., Benz, A.O., Browning, P., Dauphin, C., Emslie, A.G., Fletcher, L., Kontar, E.P., Mann, G., Onofri, M., Petrosian, V., Turkmani, R., Vilmer, N., Vlahos, L.: 2011, Recent advances in understanding particle acceleration processes in solar flares. *Space Sci. Rev.* **159**, 357. ADS. DOI.

Reprinted from *Solar Physics* Journal, DOI 10.1007/s11207-015-0731-y

Particle Acceleration and Plasma Heating in the Chromosphere

V.V. Zaitsev[1] · **A.V. Stepanov**[2]

Received: 17 November 2014 / Accepted: 23 June 2015 / Published online: 16 July 2015
© Springer Science+Business Media Dordrecht 2015

Abstract We propose a new mechanism of electron acceleration and plasma heating in the solar chromosphere, based on the magnetic Rayleigh–Taylor instability. The instability develops at the chromospheric footpoints of a flare loop and deforms the local magnetic field. As a result, the electric current in the loop varies, and a resulting inductive electric field appears. A pulse of the induced electric field, together with the pulse of the electric current, propagates along the loop with the Alfvén velocity and begins to accelerate electrons up to an energy of about 1 MeV. Accelerated particles are thermalized in the dense layers of the chromosphere with the plasma density $n \approx 10^{14}-10^{15}\ \mathrm{cm}^{-3}$, heating them to a temperature of about several million degrees. Joule dissipation of the electric current pulse heats the chromosphere at heights that correspond to densities $n \le 10^{11}-10^{13}\ \mathrm{cm}^{-3}$. Observations with the *New Solar Telescope* at Big Bear Solar Observatory indicate that chromospheric footpoints of coronal loops might be heated to coronal temperatures and that hot plasma might be injected upwards, which brightens ultra-fine loops from the photosphere to the base of the corona. Thereby, recent observations of the Sun and the model we propose stimulate a *déjà vu* – they are reminiscent of the concept of the chromospheric flare.

Keywords Chromosphere Rayleigh–Taylor instability · Particle acceleration · Plasma heating

---

Solar and Stellar Flares: Observations, Simulations, and Synergies
Guest Editors: Lyndsay Fletcher and Petr Heinzel

✉ A.V. Stepanov
stepanov@gao.spb.ru

V.V. Zaitsev
za130@appl.sci-nnov.ru

[1] Institute of Applied Physics, Ulianova str. 46, Nizhny Novgorod, 603950 Russia

[2] Pulkovo Observatory, Pulkovo chaussee 65, Saint Petersburg, 196140 Russia

1. Introduction

At the dawn of the studies of solar activity, it was suggested that solar flares occur in sufficiently dense layers of the solar atmosphere, where the electron number density is $(0.2-6.4) \times 10^{13}$ cm^{-3} (Švestka and Fritzová-Švestková, 1967; Fritzová-Švestková and Švestka, 1967). This notion led to the concept of a 'chromospheric flare' (Giovanelli, 1946). At present, the standard (thick-target) flare model is quite popular, which assumes that the primary energy release occurs due to magnetic reconnection high in the corona and that the heating of the chromosphere is caused by accelerated particles that penetrate from the corona into the chromosphere (Priest and Forbes, 2000). Obviously, this case requires a suitable configuration of the magnetic field that triggers the reconnection. For coronal magnetic loops it is usually assumed that this configuration occurs during the formation of the cusp near the top of a loop – the area where the magnetic field is stretched by the plasma flow and a neutral current sheet is formed. However, flares rather frequently develop without forming a cusp. Moreover, sometimes the cusp is formed after the impulsive phase of the flare, *i.e.* it is produced by the flare. These facts apparently suggest that the magnetic reconnection is not the only possible mechanism of the origin of flares and that there are other causes of the flare energy release.

Recent simultaneous observations of the active region NOAA 11259 with *the Solar Dynamic Observatory* (AIA/*SDO*) and the 1.6 m *New Solar Telescope* at Big Bear Solar Observatory on June 22, 2011 unexpectedly revealed aggregates of ultrafine hot magnetic loops extending from the photosphere to the base of the corona (Ji, Cao, and Goode, 2012). Most of these loops display a surprisingly small diameter of about 100 km and originate on the solar surface from intense, compact elements of the magnetic field. The NST observations also detected evidence for upward injections of hot ($\sim 10^6$ K) plasma that were excited by ultrafine loops from the photosphere to the corona. Sharykin and Kosovichev (2014) presented both ground-based (NST) and space-based (GOES, SDO, FERMI) observations of the 2013 August 15 flare, which revealed a subarcsecond structure of the flare ribbons consisting of numerous small-scale bright knots. They also suggested that these Hα knots may be energized by Joule heating.

The hot sources in the footpoints of magnetic loops are also located in *Yohkoh* and TRACE data (Aschwanden *et al.*, 2007); in particular, the updrafts of hot plasma flows in coronal magnetic loops, propagating upward as wave disturbances increases the density in the loops compared with that of the surrounding plasma. These examples clearly indicate that plasma might be heated and particles might be accelerated directly in the chromosphere (see also Hudson *et al.*, 1994; Mrozek, Tomczak, and Grubek, 2007; Fletcher *et al.*, 2011, 2013).

The detection of sub-THz radiation from active regions (Kaufmann *et al.*, 2004) presents an additional factor in the occurrence of particle acceleration and plasma heating in the chromosphere. The plasma mechanism, along with others, may be responsible for the sub-THz emission in compact and hot sources located in the heated chromosphere with the electron number density $\geq 10^{14}$ cm^{-3} (Sakai *et al.*, 2006; Zaitsev, Stepanov, and Melnikov, 2013).

These circumstances largely determine the interest in heating mechanisms of the chromosphere, which is inherent in the implementation of the international ALMA project. Below, we consider a possible mechanism of sporadic particle acceleration and direct plasma heating in the chromosphere. We analyze how strong electric fields induced by magnetic Rayleigh–Taylor (R–T) instability in the chromospheric footpoints of current-carrying magnetic loops are generated and how charged particles are accelerated by these electric fields.

2. Rayleigh–Taylor Instability in Current-Carrying Magnetic Loops

The magnetic version of the Rayleigh–Taylor instability (Kruskal and Schwarzschild, 1954; Priest, 1982) leads to penetration of external chromospheric plasma with $\beta = 8\pi n k_{\mathrm{B}} T/B^2 \leq 1$ into a magnetic loop, where $\beta \ll 1$. As a result, the magnetic field of the loop is deformed and causes an increase in electrical current. Thus, an electric field is induced and effectively accelerates electrons, which results in heating of the chromosphere *in situ*. It is essential that conditions for the R–T instability are realized near chromospheric footpoints of magnetic loops, where, because of a sharp decrease in the plasma pressure with height and an unfavorable curvature of the magnetic field, the effective gravity is directed inward of the loop. This means that the region in which the inductive electric field is generated also produces efficient particle acceleration and plasma heating.

We consider a thin vertical magnetic flux tube in an inhomogeneous chromosphere. For simplicity, we assume that the magnetic field inside the tube is independent of the radial coordinate r and only depends on the vertical coordinate z. We assume that there is no magnetic field outside the tube and the radial component of the magnetic field B_{r} is smaller than the vertical component B_{z}. This approximation is valid in the vicinity of a magnetic loop footpoint. We assume that the internal pressure in the flux tube $p_{\mathrm{i}}(z)$ is lower than the pressure at the interface of the tube $p_{\mathrm{e}}(z)$. Then we can write the equilibrium condition in the radial coordinate

$$B_{\mathrm{z}}^2(z) = 8\pi p_{\mathrm{e}}(z). \tag{1}$$

In addition, we write the magnetic flux conservation condition through the cross section of the flux tube

$$\pi a^2(z) B_{\mathrm{z}}(z) = const, \tag{2}$$

where $a(z)$ is the tube radius. External pressure decreases with height as

$$p_{\mathrm{e}}(z) = p_0 \exp\left(-\int_0^z \frac{\mathrm{d}z'}{H(z')}\right), \qquad H = \frac{k_{\mathrm{B}} T_{\mathrm{e}}(z)}{m_{\mathrm{i}} g}. \tag{3}$$

From Equations (1) and (2) it follows that the decrease in pressure with height results in the increase of the magnetic tube radius, which develops a curvature of the magnetic field. The radius of curvature is (Figure 1)

$$R_{\mathrm{c}} = \frac{[1 + (a')^2]^{3/2}}{a''}, \tag{4}$$

$$a(z) = a(0) \exp\left(\int_0^z \frac{\mathrm{d}z'}{4H(z')}\right). \tag{5}$$

Given that the derivatives a', a'' are determined by formulas $a' = a/4H$, $a'' = a/16H^2$, we obtain the following expression for the curvature radius of the magnetic field:

$$R_{\mathrm{c}} = \frac{16H^2}{a}\left[1 + \frac{a^2}{16H^2}\right]^{3/2}. \tag{6}$$

The minimum value of the radius of curvature is obtained when $a(z) = 2\sqrt{2}H(z)$. In this case

$$R_{\mathrm{c\,min}} = 6\sqrt{3}H. \tag{7}$$

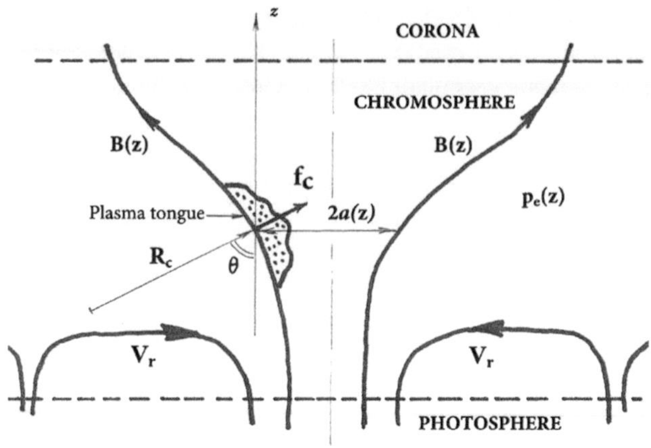

Figure 1 Sketch of the footpoint of a flaring loop. A tongue of external plasma (dotted) penetrates the loop due to the R–T instability and deforms the magnetic field of the loop.

As a result of the curvature of the magnetic field, the centrifugal force appears:

$$\mathbf{f}_{\mathrm{c}} = \frac{2nk_{\mathrm{B}}T}{R_{\mathrm{c}}^2}\mathbf{R}_{\mathrm{c}}, \tag{8}$$

which acts per cubic centimeter of the plasma at the outer boundary of the tube with the density $\rho \approx (n + n_{\mathrm{a}})m_{\mathrm{i}}$. Therefore, the effective centrifugal acceleration acting on chromospheric plasma surrounding a loop footpoint is

$$\mathbf{g}_{\mathrm{c}} = (\mathbf{f}_{\mathrm{c}}/\rho) = \frac{2k_{\mathrm{B}}T}{m_{\mathrm{i}}R_{\mathrm{c}}^2}\frac{n}{n + n_{\mathrm{a}}}\mathbf{R}_{\mathrm{c}}, \tag{9}$$

where n, n_{a} are densities of electrons and neutral atoms, respectively, and T is the plasma temperature at the outer boundary of the tube, which, in general, is different from the temperature of the ambient chromosphere T_{e} because of additional heating by electric currents in the tube. Furthermore, the outer boundary of the tube forms the gradient of the dynamic pressure of the convective flow, $m_{\mathrm{i}}(n + n_{\mathrm{a}})\partial V_{\mathrm{r}}^2/\partial r$ (Stepanov, Zaitsev, and Nakariakov, 2012).

For the development of a local disturbance (ballooning mode of the R–T instability) with the scale of a plasma tongue invading the magnetic loop $l \approx H \approx (1-5) \times 10^7$ cm, it is necessary that the projection of forces on the direction of the curvature of the magnetic field lines must be positive. In the case of a convergent photospheric plasma flow inside the flux tube, we assume the radial dependence of the flow velocity in the form $V_{\mathrm{r}}(r,t) = -V_0(t)r/a$ when $r \le a$, and thereby obtain the instability condition against the ballooning mode:

$$g_{\mathrm{c}} + \frac{2V_0^2}{a}\sin\theta - g\cos\theta > 0, \tag{10}$$

where θ is the angle between the direction of the radius of curvature R_{c} and the vertical:

$$\cos\theta = \frac{a'}{\sqrt{1 + (a')^2}}, \qquad \sin\theta = \frac{1}{\sqrt{1 + (a')^2}}. \tag{11}$$

As a result, the instability criterion takes the form

$$\frac{nT}{2(n + n_{\mathrm{a}})T_{\mathrm{e}}}\left(1 + \frac{a^2}{16H^2}\right)^{-1} + \frac{8HV_0^2}{a^2 g} > 1. \tag{12}$$

If the radius of the tube a is smaller than the scale height H, the instability criterion is simplified:

$$\frac{nT}{2(n+n_\mathrm{a})T_\mathrm{e}} + \frac{8HV_0^2(t)}{a^2 g} > 1. \tag{13}$$

From Equation (13) it follows that the R–T instability develops when the external shell of the flux tube is heated (thereby, the density of neutral atoms reduces and the ratio n/n_a increases) and when the velocity of the convective plasma flow increases sharply. To determine the temperature to which the chromosphere should be heated, we used a modified Saha formula, which is applicable in the conditions of the lower layers of the chromosphere (Brown, 1973):

$$\frac{(n+n_\mathrm{a})x^2}{1-x} = 7.2\times 10^{18}\sqrt{T}\exp\left(-6.583 - \frac{1.185\times 10^5}{T}\right), \tag{14}$$

where $x = n/(n+n_\mathrm{a})$. From Equation (14) it follows that for a chromospheric layer with the density, for example, $n_\mathrm{tot} = n + n_\mathrm{a} = 10^{15}\ \mathrm{cm}^{-3}$, the instability criterion (Equation (13)) is satisfied for relatively low values of the plasma flow velocity $V_0(t)$, provided the tube shell is heated to a temperature of $T \approx 2.5\times 10^4$ K. At this temperature, the degree of ionization x reaches 90 %, and the instability criterion takes the form $xT/2T_\mathrm{e} \approx 2.8 > 1$. In this case, the characteristic time of the instability is on the order of seconds, $\tau_\mathrm{R-T} \approx (3\sqrt{3}T_\mathrm{e}a/4\pi Tg)^{1/2} \approx 5$ s. This is the order of the time for which an external plasma with the ion thermal velocity penetrates a flux tube of radius a. Heating may be associated with electrical currents in the magnetic flux tube. If the flux tube is formed by converging photospheric plasma flows with velocity $V_\mathrm{r}(r)$, electric currents appear, and the current dissipation leads to additional heating. The main contribution to the current dissipation is provided by the Cowling resistivity, which is related to electron-atom collisions.

The dissipation rate is determined by the formula (Sen and White, 1972)

$$q_\mathrm{J} = \frac{\alpha B^2}{(1+\alpha B^2)^2}\frac{\sigma B^2 V_\mathrm{r}^2}{c^2}, \quad \alpha B^2 = \frac{\sigma(1-x)^2 B^2}{c^2 n m_\mathrm{i}\nu'_\mathrm{ia}(1+x)}, \quad \sigma = \frac{e^2 n}{m_\mathrm{e}(\nu'_\mathrm{ei}+\nu'_\mathrm{ea})}. \tag{15}$$

Here, B is the magnetic field inside a loop, σ is classical (Spitzer) conductivity, m_e, m_i are the electron and ion mass, respectively, ν'_ei, ν'_ea, ν'_ia are effective electron–ion, electron–atom, and ion–atom collision frequencies, respectively, $\nu'_\mathrm{ij} = \nu_\mathrm{ij}m_\mathrm{j}/(m_\mathrm{i}+m_\mathrm{j})$, e is the electron charge, and c is the light velocity. In case of $\alpha B^2 \gg 1$, the formula for the Joule dissipation rate does not depend on the magnetic field

$$q_\mathrm{J} = \frac{n m_\mathrm{i}\nu'_\mathrm{ia}V_\mathrm{r}^2(1+x)}{(1-x)^2}. \tag{16}$$

The main contribution to the effective frequency of ion–atom collisions under the conditions of the chromosphere is made by the charge exchange process, for which, within the temperature range $10^4 < T < 10^5$ K, we can write

$$\nu'_\mathrm{ia} \approx 2.25\times 10^{-11} n_\mathrm{a}\sqrt{T}. \tag{17}$$

As a result, Equation (16) is transformed into

$$q_\mathrm{J} = 3.76\times 10^{-35}\frac{1+x}{1-x}(n+n_\mathrm{a})n\sqrt{T}V_\mathrm{r}^2(r)\ \mathrm{erg\,cm^{-3}\,s^{-1}}. \tag{18}$$

For the heating of the chromosphere, the rate of Joule heating should exceed the optical radiation losses, whose rate in the temperature range $10^{4.3}$ K $< T < 10^{4.6}$ K is determined by the formula (Peres *et al.*, 1982)

$$q_r = 10^{-21.85}(n + n_a)n. \tag{19}$$

From the condition $q_r \leq q_J$, with regard to Equation (18), we obtain a lower value for the rate of photospheric convection, which provides an additional heating and ionization degree for the flute instability, $V_r \geq 3.5 \times 10^4\ \mathrm{cm\,s^{-1}}$. This is a sufficient condition, because Equation (18) only gives a lower limit for Joule dissipation of the electric current. As we show below, under optimal conditions, heating by electric currents can be significantly more effective. For the velocity $V_r > 10^5\ \mathrm{cm\,s^{-1}}$, the second term on the left side of Equation (13) is on the order of the first one, and the flute instability becomes more likely. The horizontal velocity of convective flows near the boundaries of supergranulation cells, where current-carrying magnetic loops are commonly formed, is about $(3-5) \times 10^4\ \mathrm{cm\,s^{-1}}$. Thereby, photospheric convection penetrates the chromosphere to the heights of several hundred kilometers (Kostyk, 2010). In five-minute photospheric oscillations, the velocity of the plasma flow can be as high as $(1-2) \times 10^5\ \mathrm{cm\,s^{-1}}$. Therefore, we can expect the occurrence of the R–T instability in the lower chromosphere.

3. Electric Field Excitation

Using the model of a vertical cylindrically symmetric magnetic flux tube, we consider an example of how an induced electric field is generated near the footpoint of a magnetic loop, driven by the R–T instability. We assume that before the development of the R–T instability, the components of the tube magnetic field have the form $B_{z0}(r) = const$ (*i.e.* the vertical component is uniform over the cross section of the tube), and $B_{\varphi 0}(r) = B_{\varphi 0} r/a$ (the electric current is uniform over the cross section of the loop). We also assume that at time $t = 0$ the surrounding chromospheric matter begins to penetrate into a flux tube with the velocity $V_r(r,t) = -V_r(t) r/a$. Then, from the equation

$$\frac{\partial \vec{B}}{\partial t} = \mathrm{curl}[\vec{V} \times \vec{B}] \tag{20}$$

we obtain time evolution of the magnetic field components:

$$B_\varphi(r,t) = B_{\varphi 0}\frac{r}{a}\exp\left(-\frac{2}{a}\int_0^t V_0(t')\,\mathrm{d}t'\right), \qquad B_z(r,t) = B_{z0}\exp\left(-\frac{2}{a}\int_0^t V_0(t')\,\mathrm{d}t'\right). \tag{21}$$

It is easy to verify from Equation (21) and from the second electrodynamics equation

$$\mathrm{curl}\,\vec{E} = -\frac{1}{c}\frac{\partial \vec{B}}{\partial t} \tag{22}$$

that in the vicinity of the magnetic loop footpoint, where the flute instability develops, the induced electric field is perpendicular to the magnetic field $\vec{E} = -(1/c)[\vec{V} \times \vec{B}]$. This means that the induced field is not the reason for direct particle acceleration. However, for the time of about $\tau_A \approx l/V_A \approx 5 \div 25$ s, where $l = (1-5) \times 10^7$ cm is the scale of a plasma tongue invading the magnetic loop, a pulse of the magnetic field B_φ escapes from the R–T instability region along the flux tube in the form of a nonlinear Alfvén wave. In the Appendix we show that if the electric current pulse is sufficiently strong, $p \ll B_{\varphi}^2/8\pi$, and if it propagates along the magnetic loop, an accelerating electric field E_z appears. From Equation (53) it follows that the maximum electric field with $B_{\varphi} \approx B_{z0}$ is approximately $E_{zm} \approx (V_A/2c^2)(dI_0/d\xi)$. Assuming $I_0 \approx 10^9$ A, $V_A \approx 10^7\ \mathrm{cm\,s^{-1}}$, $\Delta\xi \approx 5 \times 10^7$ cm, we obtain an estimate of the maximum electric field $E_{zm} \approx 0.1\ \mathrm{V\,cm^{-1}}$. The electron acceleration rate depends on the ratio of the Dreicer field E_D to the accelerating field E_{zm} (Knoepfel and Spong, 1979):

$$\frac{dn_s}{dt} \approx 0.35 n \nu_{ei} \left(\frac{E_D}{E_{zm}}\right)^{3/8} \exp\left(-\sqrt{\frac{2E_D}{E_{zm}}} - \frac{E_D}{4E_{zm}}\right), \quad E_D \approx 6 \times 10^{-8} \frac{n}{T}\ \mathrm{V\,cm^{-1}}. \tag{23}$$

Thus the acceleration rate increases as the electric current pulse increases, *i.e.* towards decreasing plasma density. For a plasma electron number density of $10^{11}\ \mathrm{cm^{-3}}$ and temperature $T = 10^4$ K (Avrett and Loeser, 2008), the ratio $E_D/E_{zm} \approx 6$. With $E_{zm} \approx 0.1\ \mathrm{V\,cm^{-1}}$ electrons can be accelerated at a distance of $\approx 10^7$ cm to an energy of $\varepsilon_{sm} = \int E_{zm}\, d\xi \approx 1$ MeV. As was shown before by Zaitsev, Stepanov, and Melnikov (2013), the density of the accelerated ~ 500 keV electrons reaches a value $n_s \approx 10^9 - 10^{11}\ \mathrm{cm^{-1}}$, which is sufficient to heat the lower layers of the chromosphere to a temperature of about a million degrees by thermalizing these electrons. The longitudinal electric field grows from the tube axis and reaches its maximum on the tube surface. It also increases with the increase in the electric current, and for $B_{\varphi}^{max} \approx \sqrt{4\pi\rho} V_A$ reaches the values (51) that were obtained by Zaitsev, Stepanov, and Melnikov (2013).

Here we consider the case when the pulse of the longitudinal current leaves the instability region and begins to spread along the flux tube with approximately the Alfvén speed. In this case, formula (23) relates to the electric field that occurs at the front of the current pulse. This longitudinal electric field also provides the electromotive force (emf) in the flaring magnetic loop as a global electric circuit. The magnitude of this emf is approximately $\Xi \approx \bar{E}_z \Delta\xi$ and its duration is $\mathcal{L}/V_A$, where $\mathcal{L}$ is the length of the loop. If this time is shorter than the settling time of the current in the circuit, *i.e.* $\tau_{\Xi} \ll L/c^2 R$, then the equation for the current variation in the global electric chain presented as an equivalent electric circuit will be

$$\frac{L}{c^2}\frac{dI_z}{dt} + R I_z = \Xi, \tag{24}$$

where R is the resistance and L the inductance of the circuit, respectively, and the resistance can be neglected. Then the equation for the circulation of the electric field in a closed loop

$$\oint E_z\, dz = -\frac{L}{c^2}\frac{dI_z}{dt} \tag{25}$$

gives an average induction field in the magnetic loop associated with the propagation of the current pulse along the loop

$$\bar{E}_{ind} \approx (\bar{E}_z \Delta\xi)/\mathcal{L} \ll \bar{E}_z. \tag{26}$$

In Equation (25), we integrate over the coronal part of the loop and over the closed current domain under the photosphere $\mathcal{L}_1$. This field is much smaller than the electric field that arises at the front of the current pulse as $\Delta\xi \ll \mathcal{L} + \mathcal{L}_1$.

4. Heating of Chromospheric Plasma

Joule pre-heating of the chromosphere by steady-state electric current in a loop required for the R–T instability was considered in Section 2. Here we discuss *in situ* chromospheric heating by accelerated electrons, as well as Joule heating by the electric current pulse arising due to the R–T instability.

The mean free path of accelerated electrons with the energy $\varepsilon_s \approx 500$ keV in a plasma with the number density $n_a \approx 10^{15}$ cm^{-3} under the assumption of complete ionization is close to the size of the acceleration region:

$$\lambda_{ee} \approx \lambda_{ei} \approx \lambda_{ii} \approx \frac{\varepsilon_s^2}{2\pi n e^4 \Lambda} \approx 10^8 \text{ cm}. \tag{27}$$

Here, Λ is the Coulomb logarithm. This means that the accelerated electrons mainly remain within the acceleration region and their total energy is transformed both into the heating of the chromosphere *in situ* and into radiation.

Heating of the chromosphere caused by fast electrons, which are generated by the development of the R–T instability, was considered by Zaitsev, Stepanov, and Melnikov (2013). It was shown that for currents $I_z \geq 10^{10}$ A, the number of fast electrons accelerated in electric fields of the flare region is sufficient to heat the chromospheric part of a coronal magnetic loop with a plasma density of $n \approx 10^{15}$ cm^{-3} to a temperature of about several million Kelvin.

We consider another source of chromospheric heating, related to Joule dissipation of the electric current pulse that propagates with the Alfvén velocity along a magnetic loop from the region of the development of the R–T instability. The rate of plasma heating due to dissipation of electric currents is specified by the equation

$$q_j(r) = \left(\vec{E} + \frac{1}{c}\vec{V} \times \vec{B}\right)\vec{j} = \frac{j^2}{\sigma} + \frac{F^2}{(2-F)c^2 n m_i \nu'_{ia}}(\vec{j} \times \vec{B})^2. \tag{28}$$

In the right-hand part of Equation (28), the first term describes the current dissipation in the magnetic flux tube due to classical conductivity σ, caused by collisions of the electrons with ions and atoms. The second term in the right-hand part only appears when neutral atoms are present in the plasma; then, the equation describes the current dissipation due to ion–atom collisions, and Cowling conductivity is to be taken into account. Here $F = n_a/(n + n_a)$, where n and n_a are number densities of electrons and neutral atoms, respectively, ν'_{ia} the effective frequency of collisions between ions and neutral atoms (see Equation (17)). In our case, the current flows along the axis of the tube and is uniformly distributed within the cylinder of radius a, whose axis coincides with the direction of the outer magnetic field B_{z0}. Then, taking into account that $\vec{j} \times \vec{B} = -jB_\varphi$, where $B_\varphi = (2\pi jr/c)$, $r \leq a$ is the azimuthal component of the magnetic field formed by the current, and that the total current through the cross-section of the cylinder is equal to $I = \pi a^2 j$, for the Joule heating, we obtain

$$q = \frac{j^2}{\sigma} + \frac{F^2 B_\varphi^2 j^2}{(2-F)c^2 n m_i \nu'_{ia}}. \tag{29}$$

It can be shown that for a current $I \geq 10^{10}$ A and temperature $T \geq 10^4$ K, the second term in the right-hand part of Equation (29), related to Cowling conductivity, exceeds the first. To determine the degree of ionization in the chromosphere $x = n/(n + n_a)$, we applied the modified Saha formula (Equation (14)). As a result, for the rate of plasma heating by a current pulse, we obtain

$$q = 1.5 \times 10^{-24} \frac{I^4(z - V_A t)}{a^6(n + n_a)T} \exp\left(\frac{1.185 \times 10^5}{T}\right) \text{erg}\,\text{cm}^{-3}\,\text{s}^{-1}. \tag{30}$$

The temperature of the heated plasma may be estimated from the condition $q \approx q_r$ for the function of radiative losses (Equation (19)). For example, a current pulse with an amplitude of $I \approx 3 \times 10^{10}$ A $\approx 9 \times 10^{19}$ cgs begins to heat plasma within a magnetic flux tube starting from levels with the density $n_{tot} = (n + n_a) \approx 10^{11}$ cm^{-3}. At these levels, the temperature increases to 1.5×10^4 K, which, according to the standard model of the atmosphere (Avrett and Loeser, 2008), is more than twice the non-perturbed temperature. In the course of further propagation upwards, the heating rate increases, while the rate of radiative losses decreases, and at levels with the density $n \approx 10^{10}$ cm^{-3}, the temperature reaches $T \approx 1.5 \times 10^6$ K, exceeding the temperature of the non-perturbed atmosphere at the levels with the same plasma density by about an order of magnitude.

In the case of large currents, the heating begins from denser layers in the chromosphere. If, for example, $I \approx 3 \times 10^{11}$A, then already at levels with $n_{tot} \approx 2 \times 10^{13}$ cm^{-3} the temperature increases to 1.5×10^4 K and later reaches $T \approx 1.5 \times 10^6$ K at levels with density $n \approx 2 \times 10^{11}$ cm^{-3}. In this case, the characteristic time of plasma heating by a current pulse $\tau_H \approx n k_B T / q$ in dense layers of the chromosphere is substantially shorter than the pulse duration $\tau_A \approx \Delta\xi / V_A$. In less dense layers (for example, for a density $n \approx 10^{11}$ cm^{-3}), the condition $\tau_H < \tau_A$ may be fulfilled only for moderate Alfvén velocities, or for a sufficiently large size of the current pulse $\Delta\xi$.

Thereby, the Joule dissipation that arises in the course of the propagation of a pulse of an electric current in a magnetic flux tube may heat the chromosphere at the heights that correspond to densities $n \leq 10^{11} - 10^{13}$ cm^{-3}. Denser layers of the chromosphere are likely to be heated by fast particles accelerated by the electric field on the fronts of a current pulse.

5. UV Radiation from Heated Loop Footpoints

It should be noted that the heating of the chromospheric footpoints of a coronal magnetic loop is not accompanied by excessively high flux of ultraviolet radiation. Under the assumption that the optical thickness of the chromosphere and corona is low, we can find the total intensity of the radiation at Earth in the frequency interval $\nu_1 < \nu < \nu_2$ (Shklovskii, 1965)

$$\begin{aligned} \Im &= 1.55 \times 10^{-52} \left(\frac{10^6}{T}\right)^{1/2} \\ &\times \left[e^{\frac{-h(\nu_1 - \nu_0)}{k_B T}} - e^{\frac{-h(\nu_2 - \nu_0)}{k_B T}} + \frac{k_B T}{2h\nu_0} \left(e^{\frac{-h\nu_1}{k_B T}} - e^{\frac{-h\nu_2}{k_B T}}\right) \right] EM \, \text{erg}\,\text{cm}^{-2}\,\text{s}^{-1}. \end{aligned} \tag{31}$$

Here, $EM = \iint n^2 \, dl \, dS \approx n^2 V$ is the emission measure of a source of UV-radiation with the volume V, $\nu_0 = 3.28 \times 10^{15}$ Hz is the frequency that corresponds to the ionization potential of the hydrogen atom. This interval corresponds to a wavelength of 225 Å $< \lambda <$ 900 Å, *i.e.*, the range of extreme ultraviolet radiation. Assuming a plasma density in the chromospheric source of $n = 10^{15}$ cm^{-3}, the volume of the source $V = (1-5) \times 10^{21}$ cm^3, we obtain for the interval of 225 Å $< \lambda <$ 900 Å the value of the flux $\Im \approx (0.05-0.25)$ erg cm^{-2} s^{-1} for the temperature $T = 10^6$ K and $\Im \approx (0.08-0.4)$ erg cm^{-2} s^{-1} for $T = 10^7$ K.

The overall maximum flux of extreme ultraviolet radiation during a flare ranges from 10^{-2} to 10 erg cm^{-2} s^{-1}, in accordance with ground-based measurements (Bruzek and Durrant, 1977). Thus, the flux of the ultraviolet radiation from the heated chromospheric footpoints of coronal magnetic loops is on the order of magnitude corresponding to the UV emission of conventional flares. The flux of ultraviolet radiation from the coronal part of a magnetic loop filled with hot plasma with a density of $n \approx 10^{11} \div 10^{12}$ cm^{-3} and a temperature of $T \approx 10^7$ K, for a flare loop volume of $V \approx 6 \times 10^{25}$ cm^3, is about $\Im \approx 4 \times (10^{-3} - 10^{-5})$ erg cm^{-2} s^{-1}, which is significantly lower than the UV flux from the heated chromosphere.

6. Discussion and Conclusions

The Sun is an excellent natural physical laboratory in which to study fundamental instabilities. For example, Carlyle *et al.* (2014) presented a good illustration of the Rayleigh–Taylor instability when CME matter fell back to the solar surface. There are indications of both Rayleigh–Taylor and Kelvin–Helmholtz instabilities in the solar prominence evolution (Ryutova *et al.*, 2010; Hillier *et al.*, 2011). *SDO*/AIA observations of the CME event on 2010 November 3 revealed perturbations along the flank of the ejecta that may have resulted from the Kelvin–Helmholtz instability (Foullon *et al.*, 2011).

Here, we studied the role of the magnetic Rayleigh–Taylor instability in accelerating charged particles in the solar chromosphere and *in situ* heating of chromospheric plasma. Our analysis showed that under certain conditions, an R–T instability may develop in chromospheric footpoints of coronal magnetic loops. The distorsion of the magnetic flux tube in chromospheric loop footpoints is necessary for the instability. This distortion produces the curvature of the magnetic field that is directed into the loop. The Joule dissipation of the electric currents within the magnetic flux tube increases the degree of ionization of chromospheric plasma. The R–T instability has a threshold against the velocity of photospheric convection. As a result of the instability, the external plasma of the chromosphere penetrates the loop. This leads to a deformation of the magnetic field of the loop and to generation of the induced electric field, which arises when the magnetic field B_φ of the electric current pulse is sufficiently large ($p \ll B_\varphi^2/8\pi$). The characteristic scale of the instability is approximately $(1-5) \times 10^7$ cm, the density range about $10^{14} - 10^{15}$ cm^{-3} (Avrett and Loeser, 2008). A pulse of the induced electric field propagates along the loop with the Alfvén velocity (along with the pulse of the longitudinal electric current), and, as the electric field emerges from the instability region, it begins to accelerate particles because it is projected in the direction of the magnetic field of the loop. The electrons are accelerated to energies of about 1 MeV and are thermalized in the dense layers of the chromosphere, with the plasma density $n \approx 10^{15}$ cm^{-3} heating them to a temperature on the order of several million degrees. The density of the accelerated electrons reaches $10^{10} - 10^{11}$ cm^{-3} and may greatly exceed the threshold of the collisional damping of plasma waves. Along with the heating, this suggests that plasma turbulence is excited in the chromospheric footpoints of coronal magnetic loops (Zaitsev, Stepanov, and Melnikov, 2013).

The Joule dissipation of a pulse of electric current in a magnetic flux tube may heat the chromosphere at heights that correspond to densities of $n \leq 10^{11} - 10^{13}$ cm^{-3}.

The flux of ultraviolet radiation from the heated chromospheric loop footpoints by an order of magnitude corresponds to the UV emission from conventional solar flares. The heated plasma from the region of the R–T instability evaporates toward the upper chromosphere with a velocity of about 10^7 cm s^{-1} and heats it through Coulomb collisions.

We note that the proposed mechanism of particle acceleration and plasma heating in the chromosphere can also be applied to the prominence matter located near the top of a flaring loop arcade (Zaitsev and Stepanov, 1992).

The results presented here may be seen as a kind of *déjà vu*: they recall the notion of a chromospheric flare (Giovanelli, 1946; Švestka, 1976).

Acknowledgements This work is supported partially by RFBR grants No 14-02-00133 and 15-02-08028, by the Programs of Presidium of RAS No 9 and 41, grant of Leading Scientific School NSh 1041.2014.2, as well as by grant of Ministry of Education and Sciences 14.Z50.31.0007. The authors thank D. Melrose and one of the Guest Editors for valuable remarks.

Appendix: Generation of Accelerating Electric Field

We consider a cylindrical magnetic flux tube of radius a with the longitudinal magnetic field B_z, uniform over the flux tube cross section, and with the electric current density $j_z(z,t)$, homogeneous over the tube cross section. To analyze the magnetic and electric fields arising in the course of the propagation of the current pulse, we start from the equations for the plasma

$$\frac{\partial \rho}{\partial t} + \mathrm{div}(\rho \vec{V}) = 0, \tag{32}$$

$$\frac{\partial \vec{V}}{\partial t} = \frac{1}{4\pi\rho} \,\mathrm{curl}\, \vec{B} \times \vec{B} - \frac{\nabla p}{\rho}, \tag{33}$$

Equations (20) and (22) for the fields, and Ohm's law

$$\vec{E} = -\frac{1}{c}[\vec{V} \times \vec{B}]. \tag{34}$$

In a cylindrical coordinate system, we assume axial symmetry, *i.e.* $\frac{\partial}{\partial \varphi} = 0$. We also assume that the plasma is incompressible, *i.e.* $\rho = const$. Then from Equation (32) we obtain

$$\frac{1}{r}\frac{\partial}{\partial r}(r V_r) + \frac{\partial V_z}{\partial z} = 0. \tag{35}$$

We assume that in the first approximation for the amplitude of the electric current (B_φ), the radial velocity component $V_r = 0$ (provided this approximation is valid). In this case, it follows from Equation (35) that $V_z = const = 0$, as the velocity is zero at infinity. Then, from Equations (33) and (20) it follows

$$B_r = 0, \qquad B_z = const = B_{z0}, \tag{36}$$

$$\frac{\partial B_\varphi}{\partial t} = B_{z0}\frac{\partial V_\varphi}{\partial z}, \tag{37}$$

$$\frac{\partial V_\varphi}{\partial t} = \frac{B_{z0}}{4\pi\rho}\frac{\partial B_\varphi}{\partial z}, \tag{38}$$

$$\frac{\partial V_r}{\partial t} = -\frac{1}{\rho}\left(\frac{\partial p}{\partial r} + \frac{1}{4\pi}\frac{B_\varphi}{r}\frac{\partial (r B_\varphi)}{\partial r}\right) = 0, \tag{39}$$

$$\frac{\partial V_z}{\partial t} = -\frac{1}{\rho}\left(\frac{\partial p}{\partial z} + \frac{B_\varphi}{4\pi}\frac{\partial B_\varphi}{\partial z}\right) = 0. \tag{40}$$

From Equations (37) and (38), we obtain

$$\frac{\partial^2 B_\varphi}{\partial t^2} = \frac{B_{z0}^2}{4\pi\rho}\frac{\partial^2 B_\varphi}{\partial z^2}, \tag{41}$$

i.e., the pulse of the electric current propagates along the magnetic flux tube with the Alfvén velocity $V_A = B_{z0}/\sqrt{4\pi\rho}$. If the amplitude of the electric current is sufficiently small, then in the classical Alfvén wave the pressure gradients by the coordinates r, z are balanced by the corresponding magnetic field gradients, and it follows from Equations (39) and (40) that the approximation of $V_r = 0$, $V_z = 0$ is valid. From Equations (20), (22), and (37), we obtain

$$\frac{\partial E_z}{\partial r} = \frac{\partial E_r}{\partial z} + \frac{1}{c}\frac{\partial B_\varphi}{\partial t} = -\frac{1}{c}B_{z0}\frac{\partial V_\varphi}{\partial z} + \frac{1}{c}\frac{\partial B_\varphi}{\partial t} = 0, \tag{42}$$

i.e., in this case, there is no longitudinal electric field (Pikel'ner, 1964). However, if an electric current is sufficiently large, so that the condition $p \ll B_\varphi^2/8\pi$ is fulfilled, the velocity components V_φ, V_z are nonzero, and for them the following approximate equations can be written

$$\frac{\partial V_r}{\partial t} = -\frac{1}{4\pi\rho}\frac{B_\varphi}{r}\frac{\partial (rB_\varphi)}{\partial r}, \tag{43}$$

$$\frac{\partial V_z}{\partial t} = -\frac{B_\varphi}{4\pi\rho}\frac{\partial B_\varphi}{\partial z}. \tag{44}$$

From the conditions $\partial/\partial\varphi = 0$ and $E_\varphi = 0$, it follows that $B_r = 0$, $B_z = B_{z0}$,

$$\frac{\partial B_\varphi}{\partial t} = \frac{\partial}{\partial z}(V_\varphi B_{z0} - V_z B_\varphi) - \frac{\partial}{\partial r}(V_r B_\varphi). \tag{45}$$

Considering now

$$E_r = -\frac{1}{c}(V_\varphi B_{z0} - V_z B_\varphi), \tag{46}$$

we obtain

$$\frac{\partial E_z}{\partial r} = \frac{\partial E_r}{\partial z} + \frac{1}{c}\frac{\partial B_\varphi}{\partial t} = -\frac{1}{c}\frac{\partial}{\partial r}(V_r B_\varphi) = -\frac{1}{c}\left(B_\varphi\frac{\partial V_r}{\partial r} + V_r\frac{\partial B_\varphi}{\partial r}\right). \tag{47}$$

Using Equation (34), we find

$$\frac{\partial E_z}{\partial r} = \frac{1}{c}B_\varphi\frac{\partial V_z}{\partial z} + \frac{1}{c}\frac{V_r}{r}B_\varphi - \frac{1}{c}V_r\frac{\partial B_\varphi}{\partial r}. \tag{48}$$

The sum of the last two terms in the right-hand side of Equation (48) for the uniform current distribution over the cross section of a flux tube is zero; therefore,

$$\frac{\partial E_z}{\partial r} = \frac{1}{c}B_\varphi\frac{\partial V_z}{\partial z}. \tag{49}$$

Assuming $\frac{\partial}{\partial z} \approx -\frac{1}{V_A}\frac{\partial}{\partial t}$, and using Equation (44), we obtain

$$\frac{\partial E_z}{\partial r} = -\frac{1}{c}\frac{B_\varphi^2}{4\pi\rho V_A^2}\frac{\partial B_\varphi}{\partial t}. \tag{50}$$

The longitudinal electric field increases from the flux tube axis and reaches its maximum on the surface. It also increases with the increase in the electric current, and for $B_\varphi^{\max} \approx \sqrt{4\pi\rho}V_A$ reaches the values

$$\frac{\partial E_{zm}}{\partial r} \approx -\frac{1}{c}\frac{\partial B_\varphi}{\partial t}, \tag{51}$$

which were used in Zaitsev, Stepanov, and Melnikov (2013).

As a result of the plasma effect, the field changes its sign relative to that in vacuum. At the leading edge of the current pulse, the derivative $\partial B_\varphi/\partial t$ is positive and the electric field accelerates electrons upwards in the direction of the current pulse. In the tail of the pulse, B_φ decreases with time and the electric field accelerates electrons downwards to dense layers of the chromosphere. Given that $B_\varphi = 2rI_0(z - V_A t)/ca^2$, the formula (50) can be written as

$$E_z = -\frac{2I_0^2 r^4}{c^4 a^6 B_{z0}^2}\frac{\partial I_0}{\partial t}. \tag{52}$$

The maximum field corresponds to $r = a$:

$$E_{zm} = \frac{2I_0^2 V_A}{c^4 a^2 B_{z0}^2}\frac{\partial I_0}{\partial \xi}, \quad \xi = z - V_A t. \tag{53}$$

The electric field averaged along the tube radius is

$$\bar{E}_z = \frac{1}{5}E_{zm}. \tag{54}$$

Equation (50) for the parallel component of the electric field can be written as

$$E_z \approx \frac{k_\parallel}{k_\perp}\frac{B_\varphi^2}{B_{z0}^2}E_r, \tag{55}$$

where $k_\parallel \approx 1/\Delta\xi$, $k_\perp \approx 1/a$ are characteristic parallel and perpendicular wavenumbers of the electric current pulse, respectively, $E_r = -\frac{1}{c}(V_\varphi B_{z0} - V_z B_\varphi) = \frac{V_A}{c}B_\varphi(1 + \frac{1}{2}\frac{B_\varphi^2}{B_{z0}^2}) \approx \frac{V_A}{c}B_\varphi$ (if $B_\varphi^2 \ll B_{z0}^2$) is the perpendicular electric field in the current pulse. Because the current pulse has both longitudinal and transverse scales, *i.e.* there is an effective dispersion, this pulse corresponds to the *nonlinear dispersive Alfvén wave*. In the case of *linear dispersive Alfvén waves* (both inertial and kinetic) there is a parallel electric field as well. If the condition $8\pi p/B^2 \le m_e/m_i$ is fulfilled, the Alfvén waves are called inertial (Stasiewicz *et al.*, 2000; Cramer, 2001). In this case, the parallel electric field is determined by the formula (Stasiewicz *et al.*, 2000)

$$E_z = \frac{k_\parallel}{k_\perp}\frac{(k_\perp\lambda_e)^2}{1 + (k_\perp\lambda_e)^2}E_r, \tag{56}$$

where $\lambda_e = c/\omega_{pe}$ is the electron skin depth. If $m_e/m_i \le 8\pi p/B^2 \le 1$ the Alfvén waves are called kinetic. In this case (Malovichko, 2013),

$$E_z = \frac{k_\parallel}{k_\perp}\frac{k_\perp^2 V_{Ti}^2}{\omega_{Bi}^2}E_r, \tag{57}$$

where V_{Ti}, ω_{Bi} are ion thermal velocity and ion gyrofrequency, respectively. A parallel electric field (Equation (55)) in a nonlinear dispersive Alfvén wave can be higher by many orders of magnitude than the corresponding value in linear Alfvén waves. For example, for the case under consideration, $\Delta\xi \approx a \approx 10^7$ cm, we have $k_{\parallel} \approx k_{\perp} \approx 10^{-7}\ \mathrm{cm}^{-1}$, and, as is easily seen from Equations (56) and (57), $E_z \approx 2.5 \times 10^{-19} E_r$ in the case of inertial Alfvén waves, and $E_z \approx 10^{-14} E_r$ in the case of kinetic Alfvén waves. For nonlinear dispersive Alfvén waves (Equation (55)), the parallel electric field can be on the order of the perpendicular one if the wave amplitude $B_\varphi \sim B_{z0}$. The approximation in which the electric field (Equation (55)) was obtained ($m_e/m_i \leq 8\pi p/B^2 \leq 1$) formally corresponds to the nonlinear kinetic Alfvén wave. But as the appropriate solution is obtained within an MHD approach, it is more correctly called a nonlinear dispersive Alfvén wave.

References

Aschwanden, M.J., Winebarger, A., Tsiklauri, D., Peter, H.: 2007, *Astrophys. J.* **659**, 1673.
Avrett, E.H., Loeser, R.: 2008, *Astrophys. J. Suppl.* **175**, 229.
Brown, J.C.: 1973, *Solar Phys.* **29**, 421.
Bruzek, A., Durrant, C.J. (eds.): 1977, *Illustrated Glossary for Solar and Solar-Terrestrial Physics*, Dordrecht, Reidel.
Carlyle, J., Williams, D.R., van Driel-Gesztelyi, L., Innes, D., Hillier, A., Matthews, S.: 2014, *Astrophys. J.* **782**, 87.
Cramer, N.F.: 2001, *The Physics of Alfvén Waves*, Wiley, New York, 312 p.
Fletcher, L., Dennis, B.R., Hudson, H.S., Krucker, S., Phillips, K., Veronig, A., *et al.*: 2011, *Space Sci. Rev.* **159**, 19.
Fletcher, L., Hannan, I.G., Hudson, H.S., Innes, D.E.: 2013, *Astrophys. J.* **771**, 104.
Foullon, C., Verwichte, E., Nakariakov, V.M., Nykyri, K., Farrugia, C.J.: 2011, *Astrophys. J. Lett.* **729**, L8.
Fritzová-Švestková, L., Švestka, Z.: 1967, *Solar Phys.* **2**, 87.
Giovanelli, R.G.: 1946, *Nature* **158**, 81.
Hillier, A., Isobe, H., Shibata, K., Berger, T.: 2011, *Astrophys. J. Lett.* **736**, L1.
Hudson, H.S., Strong, K.T., Dennis, B.R., Zarro, D., Inda, M., Kosugi, T., Sakao, T.: 1994, *Astrophys. J. Lett.* **422**, L25.
Ji, H., Cao, W., Goode, P.R.: 2012, *Astrophys. J. Lett.* **750**, L25.
Kaufmann, P., Raulin, J.-P., de Castro, C.G.G., Levato, H., Gary, D.J., Costa, J.E.R., *et al.*: 2004, *Astrophys. J. Lett.* **603**, L121.
Knoepfel, H., Spong, D.A.: 1979, *Nucl. Fusion* **19**, 785.
Kostyk, R.I.: 2010, *Kinemat. Phys. Celest. Bodies* **26**, 233.
Kruskal, M., Schwarzschild, M.: 1954, *Proc. Roy. Soc. London A* **223**, 348.
Malovichko, P.P.: 2013, *Kinemat. Phys. Celest. Bodies* **29**, 269.
Mrozek, T., Tomczak, M., Gburek, S.: 2007, *Astron. Astrophys.* **472**, 945.
Peres, G., Rosner, R., Serio, S., Vaiana, G.S.: 1982, *Astrophys. J.* **252**, 791.
Pikel'ner, S.B.: 1964, *Fundamentals of Cosmic Electrodynamics*, NASA, Washington.
Priest, E.R.: 1982, *Solar Magnetohydrodynamics*, Dordrecht, Reidel.
Priest, E.R., Forbes, T.: 2000, *Magnetic Reconnection. MHD Theory and Applications*, Cambridge University Press, Cambridge.
Ryutova, M., Berger, T., Frank, Z., Tarbell, T., Title, A.: 2010, *Solar Phys.* **267**, 75.
Sakai, J.I., Nagasugi, Y., Saito, S., Kaufmann, P.: 2006, *Astron. Astrophys.* **457**, 313.
Sen, N.K., White, M.L.: 1972, *Solar Phys.* **23**, 146.
Sharykin, I.N., Kosovichev, A.G.: 2014, *Astrophys. J. Lett.* **788**, L18.
Shklovskii, I.S.: 1965, *Physics of Solar Corona*, Pergamon Press, Oxford.
Stasiewicz, K., Bellan, P., Chaston, C., Kletzing, C., Lysak, R., Maggs, J., *et al.*: 2000, *Space Sci. Rev.* **92**, 423.
Stepanov, A.V., Zaitsev, V.V., Nakariakov, V.M.: 2012, *Phys. Usp.* **55**, 929.
Švestka, Z.: 1976, *Solar Flares*, Dordrecht, Reidel.
Švestka, Z., Fritzová-Švestková, L.: 1967, *Solar Phys.* **2**, 75.
Zaitsev, V.V., Stepanov, A.V.: 1992, *Solar Phys.* **139**, 343.
Zaitsev, V.V., Stepanov, A.V., Melnikov, V.F.: 2013, *Astron. Lett.* **39**, 650.

Reprinted from *Solar Physics* Journal, DOI 10.1007/s11207-015-0709-9

Impulsive Heating of Solar Flare Ribbons Above 10 MK

P.J.A. Simões[1] · D.R. Graham[1,2] · L. Fletcher[1]

Received: 31 January 2015 / Accepted: 18 May 2015 / Published online: 3 June 2015
© Springer Science+Business Media Dordrecht 2015

Abstract The chromospheric response to the input of flare energy is marked by extended extreme ultraviolet (EUV) ribbons and hard X-ray (HXR) footpoints. These are usually explained as the result of heating and bremsstrahlung emission from accelerated electrons colliding in the dense chromospheric plasma. We present evidence of impulsive heating of flare ribbons above 10 MK in a two-ribbon flare. We analyse the impulsive phase of SOL2013-11-09T06:38, a C2.6 class event using data from *Atmospheric Imaging Assembly* (AIA) on board of the *Solar Dynamics Observatory* (SDO) and the *Reuven Ramaty High Energy Solar Spectroscopic Imager* (RHESSI) to derive the temperature, emission measure and differential emission measure of the flaring regions and investigate the evolution of the plasma in the flaring ribbons. The ribbons were visible at all SDO/AIA EUV/UV wavelengths, in particular, at 94 and 131 Å filters, sensitive to temperatures of 8 MK and 12 MK. The time evolution of the emission measure of the plasma above 10 MK at the ribbons has a peak near the HXR peak time. The presence of hot plasma in the lower atmosphere is further confirmed by a RHESSI imaging spectroscopy analysis, which shows resolved sources at 11 – 13 MK that are associated with at least one ribbon. We found that collisional beam-heating can only marginally explain the power necessary to heat the 10 MK plasma at the ribbons.

Keywords Sun: flares · Sun: particle emission · Sun: X-rays, gamma rays

Solar and Stellar Flares: Observations, Simulations, and Synergies
Guest Editors: Lyndsay Fletcher and Petr Heinzel

✉ P.J.A. Simões
paulo.simoes@glasgow.ac.uk

D.R. Graham
dgraham@arcetri.astro.it

L. Fletcher
lyndsay.fletcher@glasgow.ac.uk

1 SUPA School of Physics and Astronomy, University of Glasgow, Glasgow G12 8QQ, UK

2 Present address: INAF-Osservatorio Astrofisico di Arcetri, 50125 Firenze, Italy

1. Introduction

The impulsive phase of solar flares is characterised by intense emission of microwaves and hard X-rays (HXR), showing the presence of non-thermal electrons in the corona and chromosphere, respectively. These electrons are believed to be accelerated by the magnetic energy release mechanisms in the corona (*e.g.* Zharkova *et al.*, 2011), although this interpretation has been contested recently (Fletcher and Hudson, 2008; Brown *et al.*, 2009; Varady *et al.*, 2014). Strong ultraviolet (UV) and extreme ultraviolet (EUV) emission is also commonly observed in association with the non-thermal emission (*e.g.* Hinteregger and Hall, 1969; Emslie and Noyes, 1978; Horan, Kreplin, and Fritz, 1982; Alexander and Coyner, 2006; Coyner and Alexander, 2009), indicating fast heating at the transition region and chromosphere. This heating is usually explained by the energy deposition of the accelerated electrons hitting the high-density plasma at these locations (Brown, 1973; Hudson, 1972; Brown *et al.*, 1990). The increase of pressure due to heating drives the plasma upwards into the coronal loops, a process termed chromospheric evaporation. The coronal loops filled with hot plasma will be bright in soft X-rays (SXR) and EUV, peaking at temperatures between 8 and 40 MK (*e.g.* Ryan *et al.*, 2012). The maximum temperature is usually reached after the impulsive phase, *i.e.* after most of the released energy is deposited in the ambient plasma. The hot coronal loops eventually cool by conduction, sending energy to the lower atmosphere, followed by a long radiative cooling phase (*e.g.* Švestka, 1987).

McTiernan *et al.* (1993), using *Yohkoh Soft X-ray Telescope* (SXT) images, identified impulsive SXR emission associated with an HXR footpoint. Its duration was shorter than one minute, reaching its maximum about 20 seconds before an associated HXR sub-burst was detected in the low-energy channel, and 40 seconds before the main HXR burst at the same location. This SXR burst corresponded to a sharp temperature–time profile, rising from ≈ 7 MK to about ≈ 10 MK in ≈ 30 seconds, followed by a fast cooling. Hudson *et al.* (1994) reported similar observations for the limb event SOL1992-01-26, where the footpoint SXR peak time matched the HXR peak time within the instrument's temporal resolution. Mrozek and Tomczak (2004) studied SXR impulsive brightenings in footpoints in 46 *Yohkoh* events and concluded that these can be explained as result of collisional heating mainly by non-thermal electrons with relatively low energies.

More recently, Graham *et al.* (2013) presented the emission measure distributions (EMD) as a function of temperature for the footpoints of six flares, derived from EUV spectroscopic observations, and showed that the footpoint EMDs had a peak around 8 MK, indicating the presence of high-temperature plasma within the footpoints. Evidence of hot ($T > 1$ MK) and dense ($n_e > 10^{10}$ cm^{-3}) plasma at flare footpoints during the impulsive phase was also presented by several other authors (Milligan and Dennis, 2009; Watanabe *et al.*, 2010; Del Zanna *et al.*, 2011; Milligan, 2011; Graham, Fletcher, and Hannah, 2011; Fletcher *et al.*, 2013). A comprehensive review of EUV spectroscopic observations of hot footpoints has been presented by Milligan (2015).

While the *Hinode/Extreme ultraviolet Imaging Spectrometer* (EIS: Culhane *et al.*, 2007) provides great spectroscopic information on the plasma across a wide temperature range, the slit-raster observational technique does not allow evaluation of the plasma evolution at the same spatial location with high temporal cadence. Imaging from the *Atmospheric Imager Assembly* (AIA: Lemen *et al.*, 2012), on board the *Solar Dynamics Observatory* (SDO), and methods to recover the EMDs allow us to track the plasma evolution at different regions of flares simultaneously. We present the analysis of the SOL2013-11-09 flare ribbons, recovering the temperature and emission measure, and show evidence for a sudden heating to ≈ 10 MK temperatures of the ribbon plasma during the impulsive phase.

2. Observational Data of SOL2013-11-09

The SOL2013-11-09 flare occurred in Active Region NOAA 11890 near disc centre and was characterised by two flaring ribbons plus a bright compact source between them during the impulsive phase. For this study we used data from SDO/AIA. AIA has nine filters, or passbands, in the EUV and UV range: 94, 131, 171, 193, 211, 304, 335, 1600, and 1700 Å filters, taking full-Sun images with a pixel resolution of 0.6 arcsec and 12 seconds cadence for the EUV passbands and 24 seconds for the UV passbands. We also used data from the *Reuven Ramaty High-Energy Solar Spectroscopic Imager* (RHESSI: Lin *et al.*, 2002). The RHESSI imager consists of nine bi-grid rotating modulation collimators (RMCs), with a rotation period of about 4 seconds. The data from individual RMC are combined by the imaging algorithms to cover different spatial scales, from 2.3 arcsec (RMC 1) up to 180 arcsec (RMC 9). To avoid saturation of the detectors at times with a high rate of incident photons, two-stage (thin and thick) attenuators can be automatically employed. No attenuators (state A0) were in during this event, a rare occurrence that gives us the opportunity to investigate the evolution of the low-energy X-rays during the impulsive phase without gaps and with full count rates. Complementary SXR data from the *Soft X-ray Sensor* on board of the *Geostationary Operational Environmental Satellite* (GOES) were also employed. GOES observes the Sun as a star in two broadband SXR channels, 1 – 8 Å and 0.5 – 4 Å, which can be used to estimate the temperature, T, and emission measure, EM, of the flaring plasma (Thomas, Crannell, and Starr, 1985; White, Thomas, and Schwartz, 2005).

In Figure 1 we summarise some aspects of the event. Figure 1a shows RHESSI HXR count rates at several energy bands, where the high-energy bands (9 – 12, 12 – 25 and 25 – 50 keV) show a clear impulsive peak with a maximum at 06:25:46 UT, and are associated with bremsstrahlung emission from non-thermal electrons, as we discuss in Section 4. A second, weaker and less impulsive peak is seen around 06:27:20 UT. The lower energy bands (below 9 keV), associated with thermal bremsstrahlung, show a gradual rise starting slightly after 06:22 UT, and also a pronounced peak simultaneous with the high-energy channels. Both GOES channels, 1 – 8 Å and 0.5 – 4 Å (Figure 1b), have a steep rise, with a shoulder at the same time as the HXR peak. Spatially integrated EUV emission was obtained by summing the pixels of AIA images taking the whole flare region; they are shown in Figure 1c. The 171 Å emission is clearly associated with the high-energy HXR, with a main impulsive peak around 06:25:46 UT and a secondary peak around 06:27:20 UT. Similar characteristics are observed at 193, 211, 304, 1600, and 1700 Å. Emission in the 94 and 131 Å passbands is similar to the GOES SXR channels: a sharp rise matching the HXR peak, followed by a more gradual rise. These AIA passbands are sensitive to emission from hot plasma: the main contribution to the 94 Å passband is the Fe XVIII 93.93 Å emission line, formed at $\log T = 6.8$, while the dominant contribution to the AIA 131 Å passband during flares comes from Fe XXI 128.75 Å ($\log T = 7.1$) (O'Dwyer *et al.*, 2010; Boerner *et al.*, 2014). These main contributions were verified using data from the *EUV Variability Experiment* (EVE: Woods *et al.*, 2012) for this event reported by Simões, Graham, and Fletcher (2015).

3. SDO/AIA Data Analysis

3.1. Impulsive EUV Emission from Flaring Ribbons

We now focus our analysis on the impulsive phase, 06:22 to 06:30 UT, when the impulsive ribbon emission was observed. AIA images (see Figures 2 and 3) reveal two ribbons as the brightest features in all AIA EUV/UV passbands during the impulsive phase

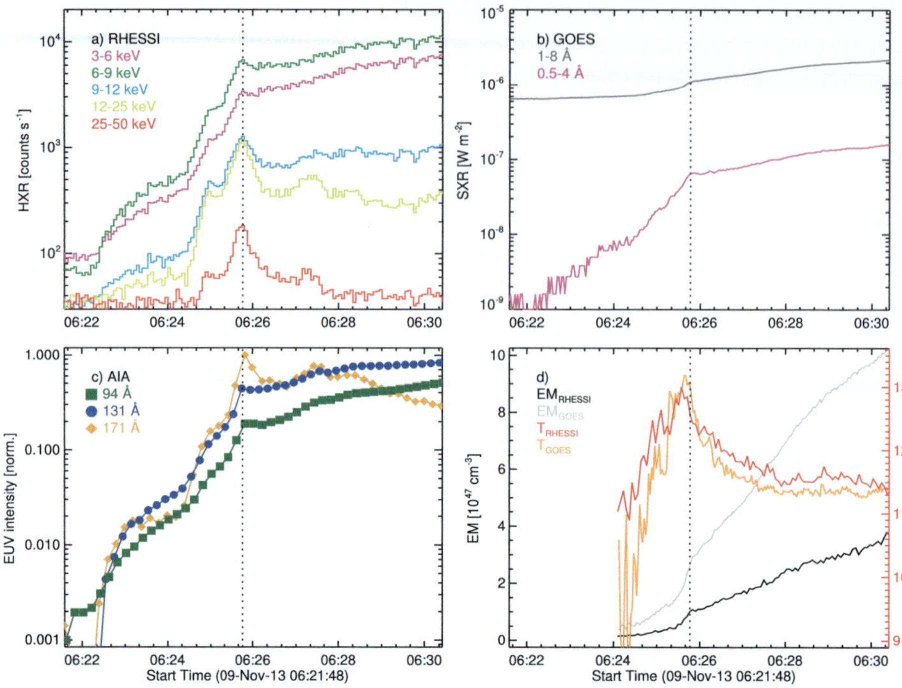

Figure 1 SOL2013-11-09 flare overview. (a) RHESSI HXR count rates (energy bands indicated in the figure). (b) GOES flux at 1 – 8 and 0.5 – 4 Å channels. (c) Spatially integrated emission of AIA passbands 94, 131, and 171 Å. (d) Temperature, T, and emission measure, EM, determined using from GOES and RHESSI. In all frames, the vertical dotted line indicates the HXR peak.

of the event. The ribbons are located in regions with opposite magnetic field polarity, as can be seen in Figure 4, which shows the line-of-sight (LOS) photospheric magnetic field obtained by SDO/*Helioseismic and Magnetic Imager* (HMI: Scherrer *et al.*, 2012). Coronal loops become the dominant features at 94, 131, and 335 Å after the main impulsive peak of the flare. The bright source located between the ribbons, visible at the impulsive phase in all AIA wavelengths and also seen in HXR (Section 4), was analysed in detail by Simões, Graham, and Fletcher (2015). They characterised this source as the main energy release site of this event. We summarise their findings as follows: 1) The source is located in the corona (possibly low down), it has a filamentary shape along the loops seen in EUV/UV images, the lack of hot loops connecting the region after the impulsive phase, and the weak and featureless photospheric magnetic field at the same location (see Figure 4). 2) It has intense and impulsive EUV and HXR emission. 3) Consistently, the source is found to be dense ($\log n = 11.50 \pm 0.82$) and hot ($T \approx 12 \sim 16$ MK). 4) Strong red-shifts observed in many EUV emission lines observed by *Hinode*/EIS, including Fe XII and Fe XXIV, indicate plasma downflows of $40-250\ \mathrm{km\,s^{-1}}$, which, along with plasma outflows observed in AIA images, are interpreted as plasma outflows along the magnetic loops.

We investigated the emission from the ribbons and the coronal source (for context) separately by marking regions of interest (ROI) associated with each of the three main sources as shown in Figure 5: eastern (orange) and western (green) ribbons, and coronal source (blue). We obtained light curves for all AIA passbands by summing the pixels for each ROI, which are shown in Figure 6. We also show RHESSI HXR 15 – 25 keV counts for comparison.

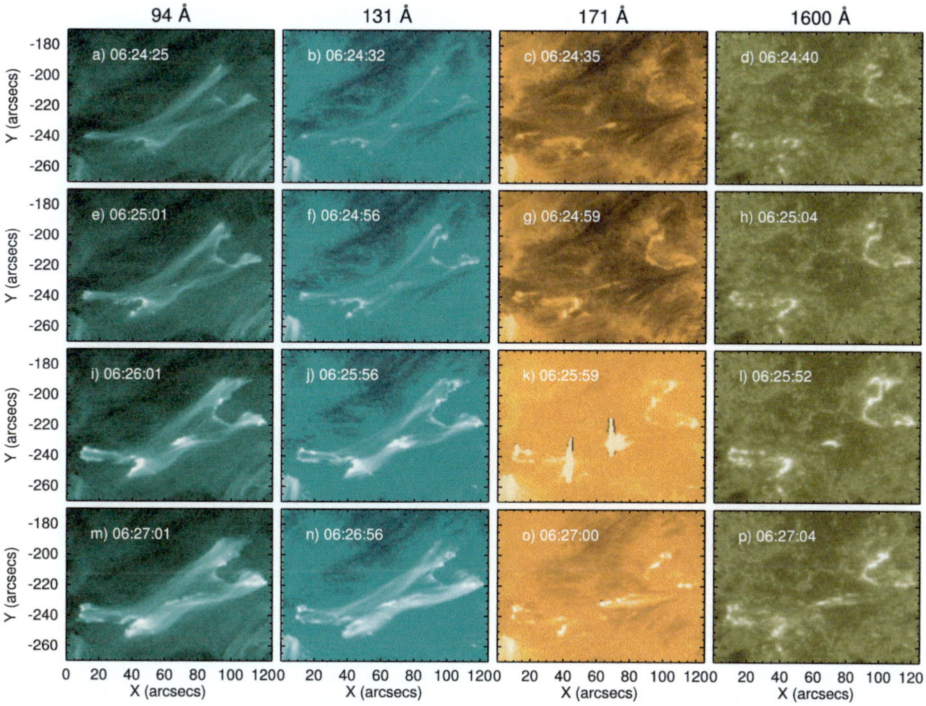

Figure 2 SDO/AIA images at 94, 131, 171, and 1600 Å filters of SOL2013-11-09, at four times during the impulsive phase.

The UV (1600 and 1700 Å), 304 Å, and the warm EUV (171, 193, and 211 Å) channels have similar enhancements during the impulsive phase, showing impulsive emission from both ribbons and the coronal source. The bursts are coincident with the HXR main peak (06:25:46 UT), within the instrumental cadence. A secondary, less impulsive HXR peak at 06:27:14 UT is also well associated with most AIA channels. The peaks are simultaneous at the three ROIs in all nine AIA filters, within the instrumental cadence. After about 06:30 UT, the coronal source fades out at all wavelengths. At 94 and 131 Å the coronal source ROI is dominated by the emission from the coronal loops.

3.2. Emission Measure Distributions

To investigate the evolution of the plasma at the flaring ribbons, we applied a method of regularised inversion to the AIA data developed by Hannah and Kontar (2012) to obtain the *differential emission measure* (DEM), $\xi(T) = n_e^2\,\mathrm{d}h/\mathrm{d}T$ [$\mathrm{cm}^{-5}\,\mathrm{K}^{-1}$], defining the quantity of emitting material as a function of temperature, T, along the given LOS, h, with an average density n_e. We employed up-to-date AIA temperature response functions (Boerner *et al.*, 2012, 2014) that have empirical corrections for the missing emission lines from the CHIANTI v7.1.3 database (Dere *et al.*, 1997; Landi *et al.*, 2013), time-dependent response corrections for each channel due to degradation of the filters, and normalisation to ensure agreement with SDO/EVE spectroscopic full-disk observations. The regularisation method also provides the DEM uncertainty and effective temperature resolution. To obtain the DEM it must be assumed that the emitting plasma is optically thin, in local thermodynamic equilibrium, and in ionisation equilibrium. In the case of a flare ribbon, where the emitting

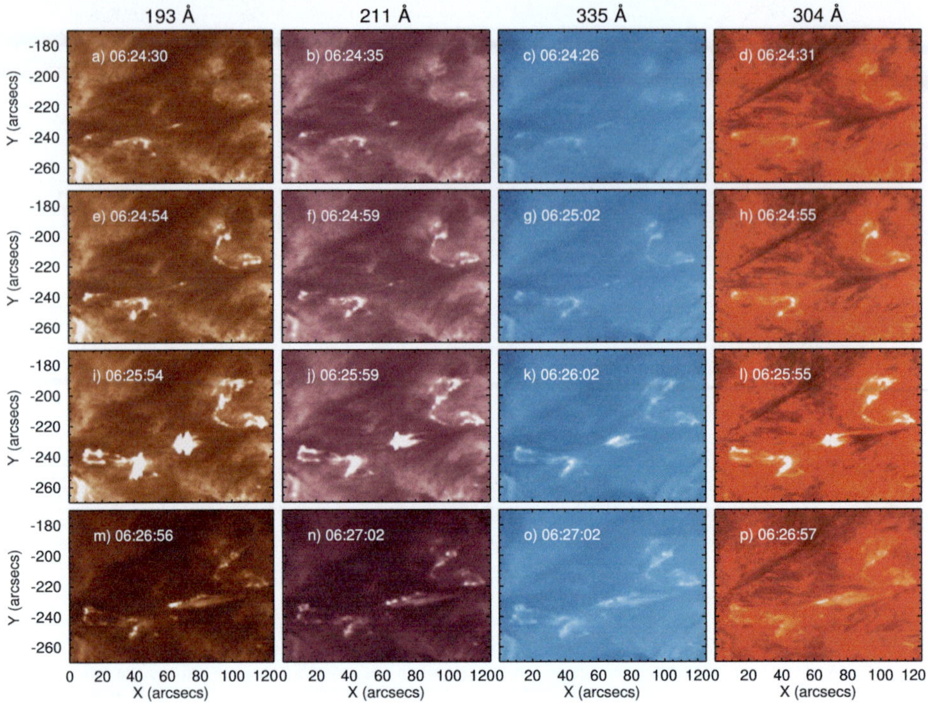

Figure 3 SDO/AIA images at 193, 211, 335 and 304 Å filters of SOL2013-11-09, at four times during the impulsive phase.

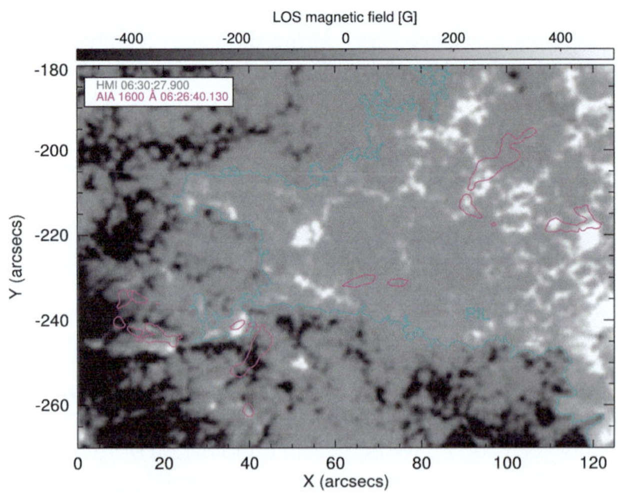

Figure 4 SDO/HMI line-of-sight (LOS) magnetogram, overlaid by SDO/AIA 1600 Å contours at 500 DN s^{-1}, showing the position of the ribbons and the coronal source, analysed by Simões, Graham, and Fletcher (2015).

plasma is both hot and dense, this may hold, although departures from equilibrium and optical depth effects should never be completely discounted (see Graham *et al.*, 2013, for further discussion). Continuum emission may also contribute to the AIA passbands (O'Dwyer *et al.*, 2010), but this contribution is likely to be small for C class flares and can probably

Figure 5 Regions of interest (ROI) defined for the AIA images: eastern (orange) and western (green) ribbons, and coronal source (blue).

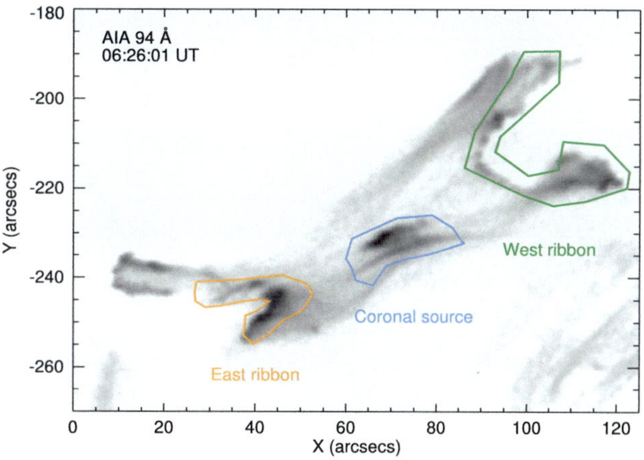

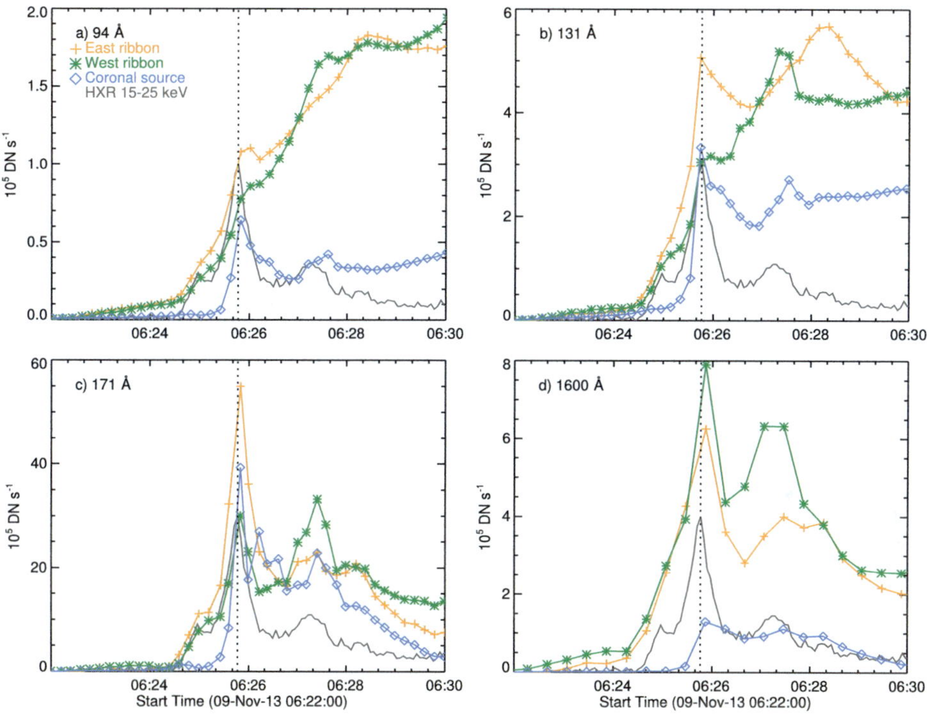

Figure 6 SDO/AIA light curves for EUV/UV filters, (a) 94 Å, (b) 131 Å, (c) 171 Å, and (d) 1600 Å, for each region of interest (ROI), as defined in Figure 5: eastern (orange) and western (green) ribbons, and coronal source (blue). RHESSI 15 – 25 keV counts (grey) are shown as a reference of the impulsive non-thermal emission, with the time of its maximum shown by the vertical dotted line.

be neglected (Milligan and McElroy, 2013). We refer to Hannah and Kontar (2012) for an extended explanation of the method and comparison with other available methods.

We applied the method to the average emission inside each ROI indicated in Figure 5 throughout the impulsive phase. Integrating the DEM over fixed temperature intervals gives

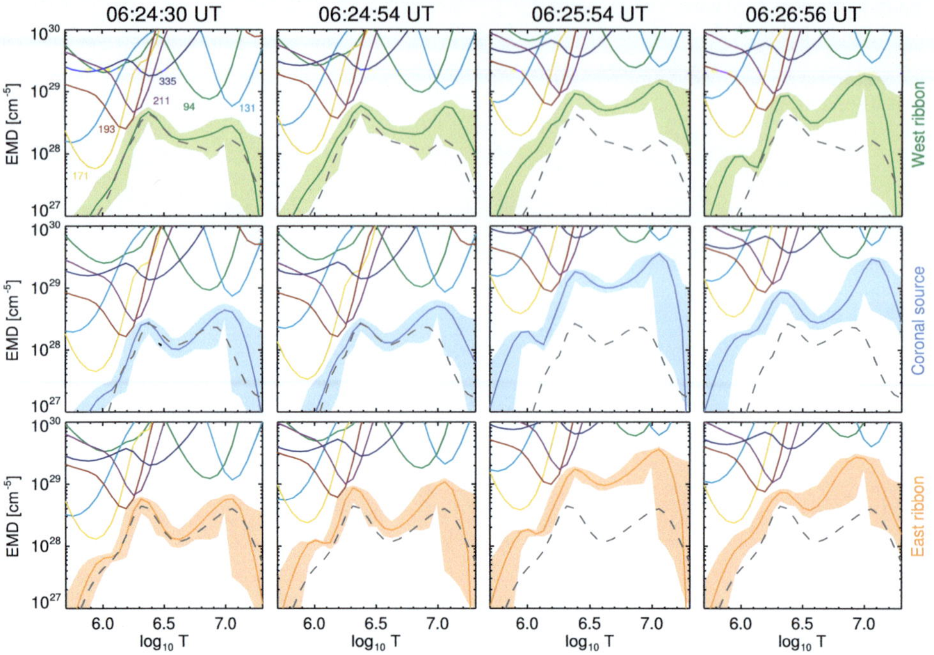

Figure 7 Emission measure distribution EMD (cm^{-5}) for the eastern (orange) and western (green) ribbons and coronal source (blue) at selected times from the early phase into the impulsive phase. Pre-flare (06:20:19 UT) EMD is shown by the grey dashed line for comparison. The EM-*loci* curves for each filter are indicated by different colours and identified in the top-left panel.

the emission measure distribution, $\text{EMD} = n^2 h$, in the more practical units of cm^{-5} (*e.g.* Graham *et al.*, 2013; Hannah and Kontar, 2013). The resulting EMD for the eastern and western ribbons and coronal source for selected times are shown in Figure 7 along with the EM-*loci* curves for each filter, these represent the EMD for an isothermal plasma, *i.e.* the maximum theoretical EM.

The pre-flare EMD is shown by the dashed line with a peak around $\log T = 6.3 \sim 6.4$ and falling off towards higher temperatures. A second peak is visible around $\log T = 6.9 \sim 7.1$, most notable in the eastern ribbon, but the 94 and 131 Å AIA channels both contain contributions from lines at approximately 1 MK and 10 MK, therefore it is likely that this apparent hot emission is in fact much cooler. The EMD for the three sources share a similar development; a pre-flare EMD that evolves with an overall increase of the emission measure at all temperatures in the AIA passband response range (roughly $5.7 < \log T < 7.3$), a stronger increase in high-temperature emission at $\log T \simeq 7.0$, and the formation of a low temperature shoulder just below $\log T = 6.0$.

We then integrated the EMD at temperature ranges 0.5 – 1.5 MK, 1.8 – 3.2 MK and 7.9 – 12.6 MK, which show the most prominent peaks, to obtain the column emission measure, $\text{EM}_c = n^2 h$ [cm^{-5}], at these temperature ranges. We did not attempt this at the temperatures around $\log T = 6.6$ because this range is not very well constrained by the AIA filters. To compare these results with (volume), $\text{EM} = n^2 V$ [cm^{-3}], values from GOES and RHESSI, we converted the EM_c to EM by multiplying the values by the projected area, A, of the sources, *i.e.* making $V = hA$. The projected area A of the emitting plasma was estimated by finding the number of pixels inside each ROI with a value above a determined

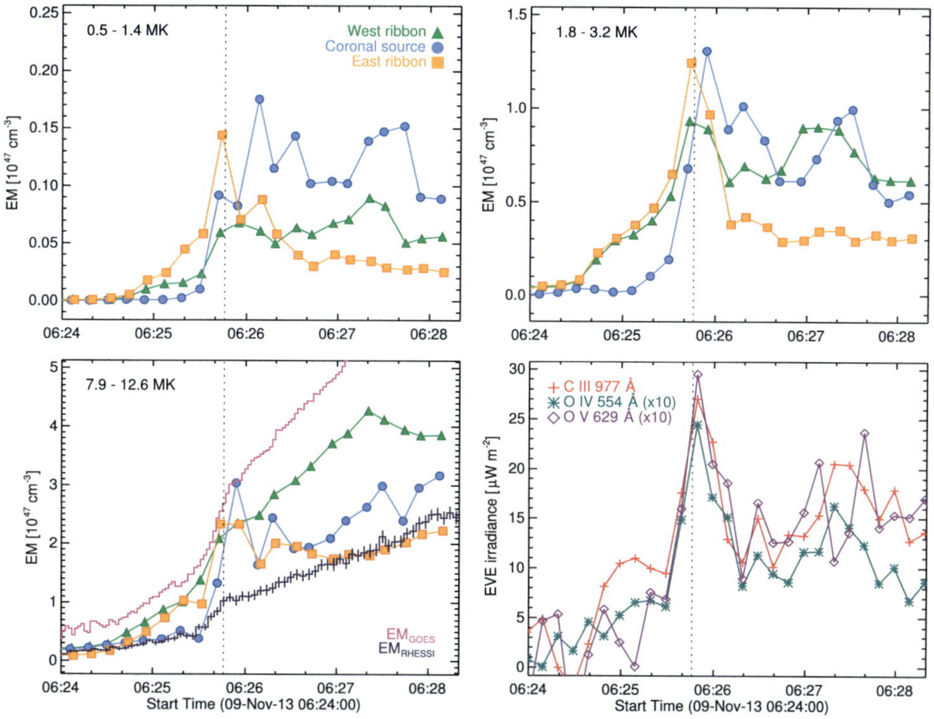

Figure 8 Time evolution of the EM (converted to cm^{-3} for comparison with GOES and RHESSI results) at three temperature ranges: (a) 0.5 – 1.4, (b) 1.8 – 3.2, and (c) 7.9 – 12.6 MK, for the eastern ribbon, western ribbon, and the coronal source from AIA from pre-flare into the impulsive phase. (d) SDO/EVE light curves of selected emission lines formed at transition region temperature ($\log T \approx 5$). The vertical dotted line indicates the time of the HXR peak.

threshold. Here we used 94 Å images and chose a threshold value of 35 $\mathrm{DN\,s^{-1}\,pixel^{-1}}$ that captures the weaker but relevant emission at the early phase, and also the stronger emission at the impulsive phase. We verified if the emitting pixels at the other AIA filters sensitive to cooler plasma temperatures co-aligned well with the emitting area at 94 Å, generally finding a good agreement within a couple of pixels. The time evolution of the EM (now in cm^{-3}) at the three temperature ranges for the three sources is shown in Figure 8, along with the values from GOES and RHESSI obtained for full-Sun observations. We subtracted the EM pre-flare values to consider the flare excess alone, taking the pre-flare values as the mean in the interval 6:00 – 6:20 UT. Figures 8a–c in the early phase (6:22:00 – 6:25:20 UT) show that the EM for the two ribbons in all three temperature bands starts to rise very slowly from about 6:22 UT (clearly visible on a logarithmic scale, not shown here), with a faster enhancement after 6:24:40 UT. During the main impulsive phase (6:25:20 – 6:26:40 UT), the EM in the three temperature ranges for all the three sources rises impulsively, peaking around 6:25:40 – 6:26:20 UT. The GOES and RHESSI emission measures also show a steeper enhancement. This phase coincides with the main HXR peak. The rapid increase, followed by a decrease of the EM, implies fast heating and cooling (or removal) of the plasma, and it is more pronounced at the eastern ribbon. After 6:26:40 UT, at 7.9 – 12.6 MK, the EM keeps rising, which is consistent with loop filling by chromospheric evaporation. AIA images at

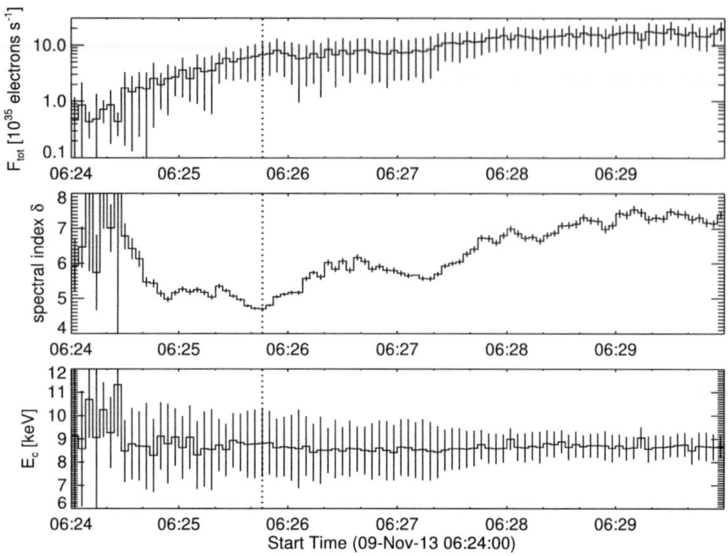

Figure 9 Spatially unresolved RHESSI spectral analysis results for the non-thermal electrons: F_{tot}, the total electron rate above the low-energy cutoff E_c, and the spectral index δ.

94, 131, and 335 Å show the enhancement of bright loops in this phase, and the increase is also supported by the GOES and RHESSI EM values.

At lower temperature ranges, the EM peak is more pronounced, showing a fast increase and decrease in the amount of material below ≈ 3 MK. This behaviour is similar to emission lines observed by SDO/EVE formed at chromospheric and transition region temperatures, $4.2 < \log T < 5.7$ K, which show impulsive emission in time with HXR. In Figure 8d we show the light curves of C III 977 Å, O IV 554 Å and O V 629 Å, noting that other lines formed at similar temperatures display a similar behaviour, namely Ne VII 465 Å, O III 526 Å, He I 584 Å, O III 599 Å, O IV 790 Å, and O VI 1032 Å, with some being noisier and less impulsive. Simulations by Fisher, Canfield, and McClymont (1985) showed that the entire flare chromosphere rapidly cools on a radiative timescale after the heating is turned off. However, after the main HXR peak there is still HXR and EUV emission in this flare, indicating that the energy deposition still continues.

4. RHESSI Hard X-ray Observations

We now employ RHESSI HXR observations to characterise the hot plasma and the non-thermal electrons. First, we evaluate the HXR spectra during the impulsive phase, with a four-second time resolution. RHESSI spectra were fitted with an isothermal plus a single power-law thick-target model, using *Object Spectral Executive* (OSPEX) software (Schwartz *et al.*, 2002). No attenuator was active during the course of the event (state A0). The non-thermal power-law is assumed to originate from a cold, collisional thick-target (Brown, 1971). The fitting parameters for the non-thermal electrons derived from RHESSI are shown in Figure 9. We use these parameters to derive the collisional beam heating in Section 5.2. The thermal plasma parameters (EM and T) are shown in Figure 1 along with the same parameters estimated from GOES data.

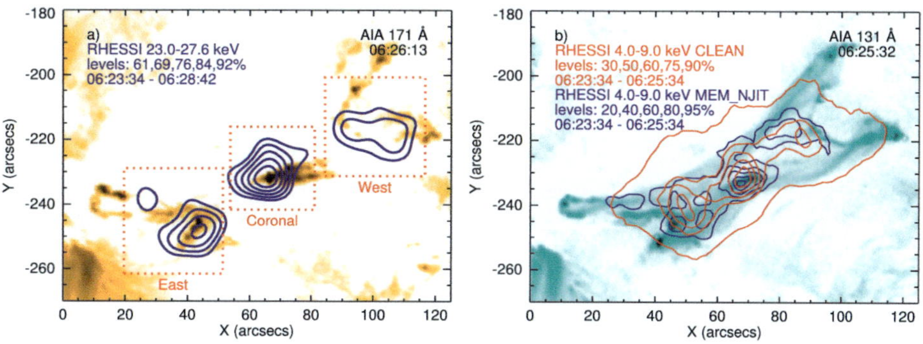

Figure 10 (a) RHESSI HXR 23 – 27 keV contours overlaid on an SDO/AIA 171 Å image (inverted colours) with (red) dashed boxes indicating the regions for imaging spectroscopy. (b) RHESSI HXR 4 – 9 keV contours obtained with Clean (red) and MEM NJIT (dark blue) overlaid on an SDO/AIA 131 Å image (inverted colours).

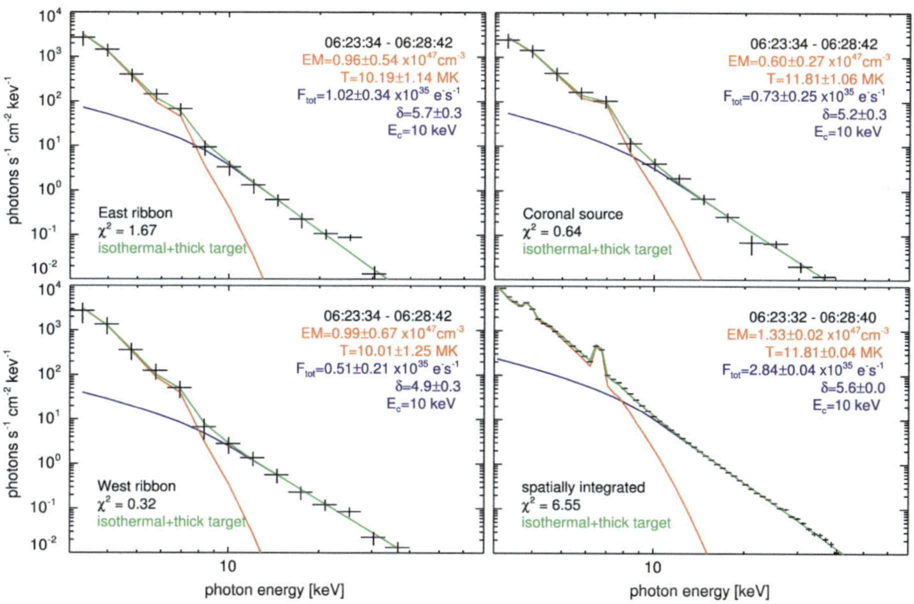

Figure 11 RHESSI photon spectra for the three main HXR sources indicated in Figure 10 and spatially integrated, for the interval 06:23:34 – 06:26:42 UT.

We applied the imaging spectroscopy technique for the three main HXR sources observed to be associated with the EUV ribbons and coronal source, indicated in Figure 10. Because of the dynamic range of RHESSI, it was not possible to obtain imaging spectroscopy for the western ribbon. We constructed RHESSI Clean images for 14 energy bands, logarithmically spaced between 3 and 40 keV, using detectors 3 to 8, integrated during the main impulsive phase 06:23:34 – 06:26:42 UT. The spectra were fitted with an isothermal plus thick-target model, shown in Figure 11, and the resulting parameters are listed in Table 1. For comparison, we also fitted the spatially unresolved full-Sun spectrum integrated for the same time interval. The thermal and non-thermal properties of the three HXR sources are very similar,

Table 1 RHESSI imaging spectroscopy results.

	EM $\times 10^{47}$ cm^{-3}	T [MK]	$F_{35} \times 10^{35}$ s^{-1}	δ
Eastern ribbon	1.0 ± 0.5	10 ± 1	1.0 ± 0.3	5.7 ± 0.3
Coronal	0.6 ± 0.3	12 ± 1	0.7 ± 0.2	5.2 ± 0.3
Western ribbon	1.0 ± 0.7	10 ± 1	0.5 ± 0.2	4.9 ± 0.3
Full Sun	1.33 ± 0.02	11.81 ± 0.05	2.84 ± 0.04	5.61 ± 0.02

confirming that T of the ribbon plasma is higher than 10 MK during the impulsive phase, as indicated by our AIA analysis in Section 3.1. The non-thermal properties also agree with the findings of Simões, Graham, and Fletcher (2015), who showed that the coronal source is $9-18$ arcsec long, with a plasma density of $\log n = 11.5$, which means that it is collisionally thick for electrons with energies of up to $45-65$ keV.

5. Discussion

5.1. Impulsive Heating of Ribbons

We now investigate the evolution of the properties of the thermal plasma at the eastern ribbon and coronal source compared to the full-Sun spectral results. The presence of low-energy HXR from thermal bremsstrahlung from the ribbon sources is shown by imaging spectroscopy results and can also be seen directly in the reconstructed images. HXR $4-9$ keV emission associated with both ribbons (and coronal source) is shown in Figure 10b. We obtained images using two different imaging algorithms, Clean and the Maximum Entropy Method of the New Jersey Institute of Technology (MEM-NJIT: Schmahl *et al.*, 2007), to confirm that the sources were not spurious artefacts of the algorithms. MEM-NJIT has a tendency to over-resolve the sources, but in this case, it works to confirm the overall spatial distribution of the low-energy HXR emission over part of the ribbons. We then constructed Clean maps (with `Clean_beam_width`[1] $= 1.5$; detectors $3-8$) for nine time bins in the interval 06:22:22 and 06:27:51 UT. The spectra were fitted with an isothermal plus thick-target component. The latter was included to achieve a better fitting for the temperature, as fluxes around $8-10$ keV and above could not be fitted with a thermal component alone, requiring a non-thermal tail. In Figure 12 we show the EM and T for the eastern ribbon (orange), coronal source (blue), and the full Sun (grey). Although the uncertainties in the imaging spectroscopy method are larger due to the smaller time intervals considered, the temperature peak around 06:25:40 UT is evident for both the eastern ribbon and the coronal source, in good agreement with the full-Sun values from both RHESSI and GOES (see Figure 1d). The evolution of the plasma temperature in the impulsive phase of this event is in fact quite unusual. The plasma temperature from GOES and RHESSI (Figure 14) shows that the temperature peak (06:25:38 UT) occurs about 8 to 10 seconds[2] before the HXR peak (06:25:46 UT). Even considering the different instrumental responses from GOES and RHESSI, both sample the hottest plasma present in flares, weighted by the EM (Ryan *et al.*,

[1] Arbitrary factor applied to the beam convolved with Clean sources. Values between $1.5-2.0$ tend to give better results when compared to different imaging algorithms (see Simões and Kontar, 2013, and references therein.)

[2] Given the time resolution of GOES (≈ 2 s) and RHESSI (≈ 4 s)

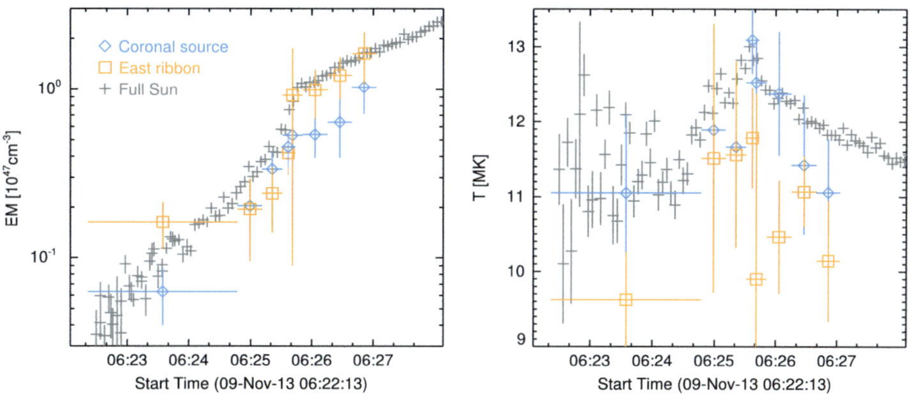

Figure 12 EM and T from RHESSI imaging spectroscopy for the coronal source (blue crosses) and East ribbon source (orange). The EM and T from the unresolved spectral fitting are also shown (grey).

2014), indicating that this temperature peak reveals an impulsive heating followed by a fast cooling process. The temperature decrease from 13 MK to 11 MK in about two minutes is consistent with conduction losses, with a timescale of about 14 to 38 seconds (for the coronal source, Simões, Graham, and Fletcher, 2015). The e-folding cooling time for the (full-Sun) temperatures in Figure 14 is $\tau = 35 - 55$ seconds.

Since the spatial resolution of RHESSI is lower than that of AIA, we confirmed the location of the highest EM by applying the AIA-DEM method for each pixel in AIA images, following Hannah and Kontar (2013). In Figure 13 we show EM maps from the AIA-DEM method, where the highest concentrations ($\mathrm{EM}_c > 10^{29.5}\ \mathrm{cm}^{-5}$) of hot plasma ($T > 8$ MK) are near the flare ribbons and the coronal source.

EM maps (Figure 13) obtained with the AIA-DEM method also suggest the presence of a short-lived, high-EM ribbon ($\mathrm{EM}_c \approx 10^{30}\ \mathrm{cm}^{-5}$) at $11 - 14$ MK near the HXR peak time at 06:25:46 UT.

The gradual rise of the EM values from the AIA-DEM (Figure 8) starting about 06:24 UT shows the filling of the loops with hot material, but the sudden enhancement between 06:25:46 and 06:25:55 UT comes from this impulsively heated material, with the two components then contributing to the overall EM. The fast cooling of the 13 MK source can only be attributed to conduction to the lower atmosphere because radiative cooling is much slower than the observed e-folding time, of about $35 - 55$ seconds. The energy source that produces the temperature peak is unclear, but we envisage a scenario where a rise in the plasma temperature leads to the acceleration of particles.

5.2. Energy Budget at the 10 MK Ribbon

We now examine the energy budget of the 10 MK plasma in the eastern ribbon. For that, we considered a slab of plasma at $T = 10$ MK in the lower atmosphere (*i.e.* at transition region heights), with a thickness L, being collisionally heated by the non-thermal electrons, balanced by radiative and conductive losses.

Following Fletcher *et al.* (2013), we calculated the energy budget of the 10 MK plasma, considering radiative and conductive losses and the collisional energy loss of the non-thermal electrons as the energy input mechanism. The total collisional power input

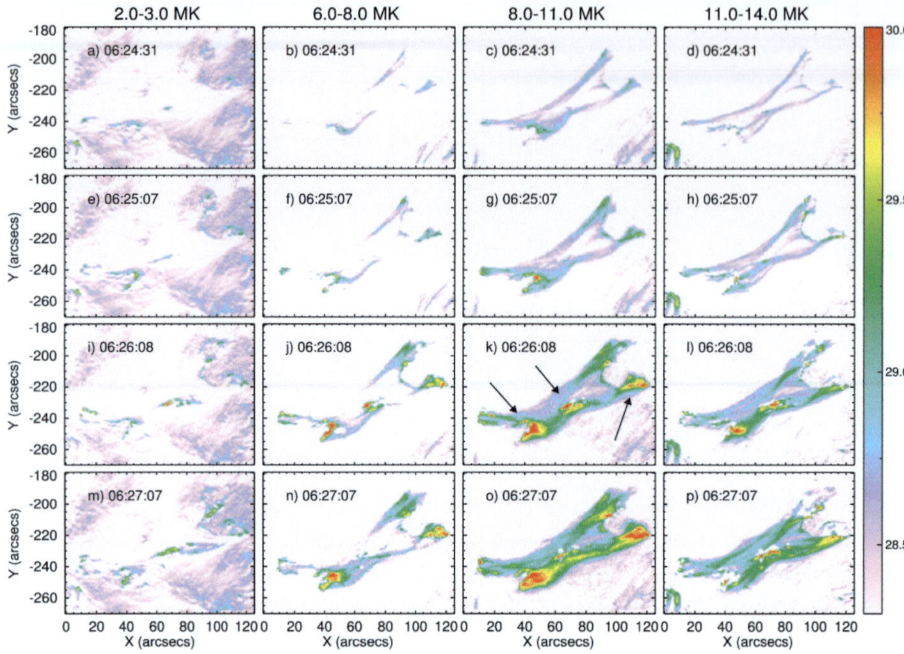

Figure 13 EM maps obtained with the AIA-DEM method (Hannah and Kontar, 2012), for temperature ranges 2 – 3, 6 – 8, 8 – 11, and 11 – 14 MK at four times during the impulsive phase. The arrows (from left to right) in panel *k* indicate the positions of the eastern ribbon, the coronal source, and the western ribbon.

Figure 14 Plasma temperature inferred from RHESSI (blue) and GOES (orange) data, peaking about 10 seconds before the RHESSI HXR counts at 3 – 6 keV (green) and 25 – 45 keV (violet). The values of the *e*-folding cooling time, τ (in seconds), are indicated in the figure.

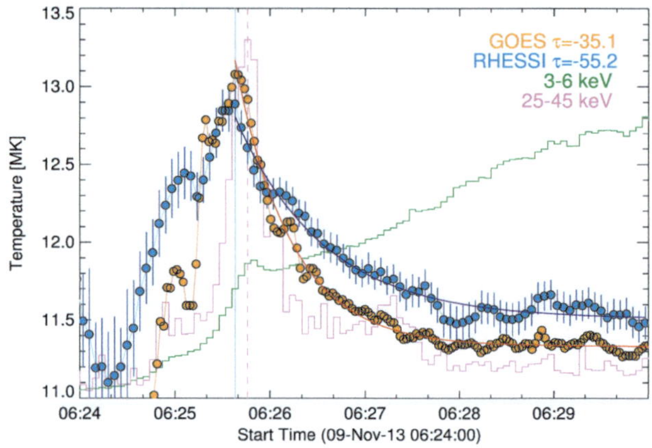

P_{coll} [erg s^{-1}] to the plasma can be inferred from the collisional thick-target model by integrating the energy loss (Emslie, 1978; Fletcher *et al.*, 2013)

$$P_{\text{coll}} = F_{\text{tot}} E_c \frac{\delta - 1}{\delta - 2}\left[1 - \left(\frac{\delta}{2} - 1\right) x_c^{1-\delta/2} B\left(x_c; \frac{\delta}{2} - 1; \frac{3}{2}\right)\right], \tag{1}$$

where B is the incomplete beta function and $x_c = 3KN/E_c^2$, where $K = 2\pi e^4$, and also

$$F_{\text{tot}} = \int_{E_c}^{\infty} F_0 E_0^{-\delta} \, \mathrm{d}E_0, \tag{2}$$

where $F_{\rm tot}$ is the total number of electrons per second above E_c injected into the source.

The values for $F_{\rm tot}$, δ and E_c are taken from RHESSI spectral analysis (Figure 9). To estimate the time evolution of these parameters for the eastern ribbon source, we used the results from the full-Sun spectroscopy and assumed that about a third of the total number of electrons are associated with each of the three HXR sources, based on the HXR imaging spectroscopy results in Table 1.

The column depth, N, can be estimated from the column, EM_c, inferred with the AIA-DEM analysis, as $N = n_e L \simeq (\mathrm{EM}\ L)^{1/2}$, for a uniform source of thickness L.

The hot plasma will lose energy by conduction, along the field direction z, to cooler layers of the atmosphere at a rate $L_{\rm cond}$ [$\mathrm{erg\,s^{-1}}$]. We estimated the conductive losses following Battaglia, Fletcher, and Benz (2009), considering the case of flux-limited conduction:

$$L_{\rm cond} = \varrho(x)\kappa_0 T^{5/2}\frac{\mathrm{d}T}{\mathrm{d}z}A, \tag{3}$$

where the factor $\varrho(x)$, a function of $x = \log(l_{\rm mfp}/L)$, reduces the classical (Spitzer, 1965) conduction coefficient $\kappa_0 = 10^{-6}\ \mathrm{erg\,cm^{-1}\,s^{-1}\,K^{-7/2}}$ to the flux-limited conduction (Campbell, 1984). Battaglia, Fletcher, and Benz (2009) fitted the values of ϱ published by Campbell (1984) as $\varrho(x) = 1.01\mathrm{e}^{-0.05(x+6.63)^2}$. This condition applies because the chromospheric scale length is small and the electron collisional mean free path, $l_{\rm mfp} = 5.21 \times 10^3 T^2/n_e$, is significant compared to the temperature scale length. Here, we approximate $\mathrm{d}T/\mathrm{d}z$ by T/L and set $T = 10$ MK and $L = 2000$ km.

The optically thin hot plasma radiates energy at a rate $L_{\rm rad}$ that can be estimated by (Rosner, Tucker, and Vaiana, 1978)

$$L_{\rm rad} = 10^{-17.73}\ \mathrm{EM}\ T^{-2/3} \quad \left(10^{6.3} < T < 10^7\ \mathrm{K}\right). \tag{4}$$

Using the EM values derived from the AIA-DEM analysis (Figure 8c) and the area A of the 10 MK source estimated as defined in Section 3.1, we found that the collisional heating is not sufficient to balance the estimated conductive losses, as shown in Figure 15.

Fletcher *et al.* (2013) found a substantial amount of plasma at 10 MK in the flare ribbons during the early impulsive phase of the flare SOL2010-08-07T18:24, with an average column EM of a few times $10^{28}\ \mathrm{cm^{-5}}$. They found that the energy carried by an electron beam is not sufficient to heat the ribbon plasma when radiative and conductive losses are taken into account, unless a low-energy cutoff of $E_c \approx 5$ keV is considered. The total electron energy inferred from the HXR spectrum is mostly defined by the low-energy cutoff E_c (*e.g.* Saint-Hilaire and Benz, 2005), which is sometimes chosen arbitrarily to balance the energy deposited by the non-thermal electrons and the value of the maximum thermal energy (*e.g.* Mrozek and Tomczak, 2004). The E_c values we found from the HXR spectral fitting are already quite low, approximately 8 keV during the impulsive phase (see the bottom panel in Figure 9), only a few times the average thermal energy kT of electrons in the 10 MK plasma. It is of course possible to reduce the value of the low-energy cutoff to cause the energy losses and gains to balance, and a lower value of the low-energy cutoff is permitted by the HXR spectrum. A lower E_c means an increased total non-thermal power and an increased electron number flux. For this event, a value of $E_c \approx 4$ keV is necessary to equate energy gains and losses. On the other hand, using the E_c values obtained from the HXR spectral analysis, a slab thickness L of ≈ 30 Mm is required to balance the energy budget, which would comprise a large portion of the loops, a picture not supported by the AIA images, which show very compact footpoint sources.

As shown by Simões, Graham, and Fletcher (2015), the coronal source is both dense ($n \approx 10^{11}\ \mathrm{cm^{-3}}$) and hot ($T \approx 13$ MK), a downward thermal conduction from the coronal

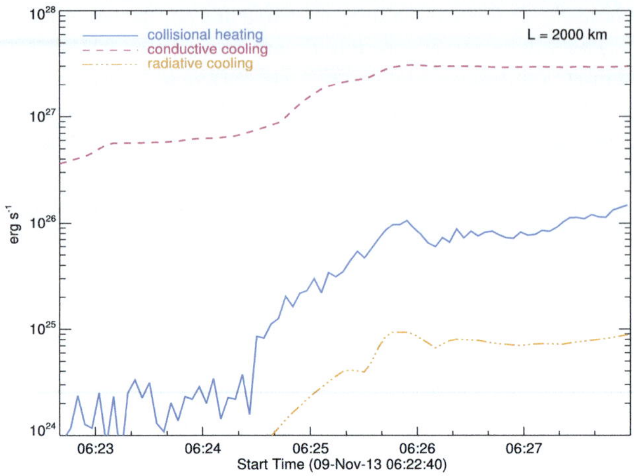

Figure 15 Power budget of the eastern ribbon source, considering beam collisional heating (blue), conductive (magenta) and radiative (gold) cooling in a $L_T = 2000$ km thick slab of plasma at a temperature of 10 MK.

source to the ribbons may contribute to the observed heating at these regions (*e.g.* Hori *et al.*, 1998; Qiu *et al.*, 2013; Battaglia, Fletcher, and Benz, 2009).

If the low-energy cutoff is indeed as low as 4 keV to balance energy gains and losses by the chromospheric plasma, then E_c approaches the mean thermal energy of the 10 MK plasma, and it becomes difficult to separate the beam from the thermal core. We may find it profitable in the future to think about the electrons in the chromosphere as forming a single distribution, such as a κ-distribution, which describes a single population transitioning smoothly from a power-law-like tail to a thermal-like core. This distribution is being discussed in connection with dense coronal HXR sources, where populations of heated electrons and accelerated electrons are present in the same volume (Oka *et al.*, 2013, 2015). Bian *et al.* (2014) have demonstrated how a κ-distribution arises naturally in a volume where diffusive acceleration and collisional losses operate simultaneously and co-spatially. If this volume is coronal and a beam is formed by a high energy tail that 'leaks out', then that beam looks very much like a standard power-law beam, with a low-energy cutoff determined by the particle velocity at which the escape time is shorter than the acceleration or diffusion timescales; then, we are still in the standard collisional thick-target scenario, but with a slightly differently shaped coronal beam. However, if we move the site of both heating and acceleration to the chromosphere, which is a very different interpretation, then the properties of the κ-distribution extending across all of the electron energy space can be investigated to evaluate the total required power and number flux needed to account for all radiation signatures. This is an interesting possibility; it does require that some other agent, such as wave turbulence, is present in the chromosphere to locally heat and accelerate electrons, however. A full investigation is beyond the scope of this article.

6. Summary

We have presented an analysis of the plasma in the flare ribbons of the event SOL2013-11-09, a C2.6 class non-eruptive, two-ribbon flare using SDO/AIA EUV and RHESSI HXR observations. The ribbons have impulsive EUV/UV emission seen in all SDO/AIA filters, clearly associated with non-thermal HXR emission observed by RHESSI. Using the method of regularised inversion of SDO/AIA data (Hannah and Kontar, 2012), we obtained the differential emission measure (DEM) of the two flare ribbons and investigated the time evolution of the emission measure (EM) in three temperature ranges (0.5 – 1.4, 1.8 – 3.2 and 7.9 – 12.6 MK). From these, we have shown that the plasma heats rapidly to 12 – 13 MK during the impulsive phase of the event, marked by the HXR peak. The EM temporal evolution shows a peak near the time of maximum HXR emission, indicating fast heating and cooling, with the hottest plasma ($T = 7.9 - 12.6$ MK) reaching EM values of $1 - 3 \times 10^{47}$ cm^{-3}; these values agree with those obtained from GOES and RHESSI. The rapid evolution of the ribbon plasma temperature and its high peak temperature were confirmed by a RHESSI imaging spectroscopic analysis and also agree with the temperature derived from GOES (although without spatial resolution). We also note that the evolution of the ribbon plasma characteristics are very similar to those found in the intense and compact coronal source, as studied in detail by Simões, Graham, and Fletcher (2015). Performing RHESSI HXR imaging spectroscopy, we obtained the parameters to describe the distribution of non-thermal electrons at each source (both ribbons and coronal source). With the information about the plasma and non-thermal electrons at the eastern ribbon, we deduced the energy balance of the plasma, considering collisional beam-heating (Emslie, 1978; Fletcher *et al.*, 2013) against conductive and radiative losses. We found that beam heating alone is not sufficient to heat and maintain the ribbon plasma at $T = 10$ MK, even with a low-energy cutoff of $E_c \approx 8$ keV, and speculated whether the dense and hot coronal source can provide a heat source for the ribbons by conduction (*e.g.* Hori *et al.*, 1998; Qiu *et al.*, 2013).

Acknowledgements The authors would like to thank Paul Boerner for providing updated SDO/AIA temperature response functions and Iain Hannah for making the DEM regularised inversion software freely available. The research leading to these results has received funding from the European Community's Seventh Framework Programme (FP7/2007-2013) under grant agreement no. 606862 (F-CHROMA), from STFC grant ST/I001808/1 (PJAS, LF) and ST/L000741/1 (LF), and from an STFC 'STEP' award to the University of Glasgow (DRG).

References

Alexander, D., Coyner, A.J.: 2006, Temporal and spatial relationships between ultraviolet and hard X-ray emission in solar flares. *Astrophys. J.* **640**, 505. DOI. ADS.

Battaglia, M., Fletcher, L., Benz, A.O.: 2009, Observations of conduction driven evaporation in the early rise phase of solar flares. *Astron. Astrophys.* **498**, 891. DOI. ADS.

Bian, N.H., Emslie, A.G., Stackhouse, D.J., Kontar, E.P.: 2014, The formation of kappa-distribution accelerated electron populations in solar flares. *Astrophys. J.* **796**, 142. DOI. ADS.

Boerner, P., Edwards, C., Lemen, J., Rausch, A., Schrijver, C., Shine, R., Shing, L., Stern, R., Tarbell, T., Title, A., Wolfson, C.J., Soufli, R., Spiller, E., Gullikson, E., McKenzie, D., Windt, D., Golub, L., Podgorski, W., Testa, P., Weber, M.: 2012, Initial calibration of the Atmospheric Imaging Assembly (AIA) on the Solar Dynamics Observatory (SDO). *Solar Phys.* **275**, 41. DOI. ADS.

Boerner, P.F., Testa, P., Warren, H., Weber, M.A., Schrijver, C.J.: 2014, Photometric and thermal cross-calibration of solar EUV instruments. *Solar Phys.* **289**, 2377. DOI. ADS.

Brown, J.C.: 1971, The deduction of energy spectra of non-thermal electrons in flares from the observed dynamic spectra of hard X-ray bursts. *Solar Phys.* **18**, 489. DOI. ADS.

Brown, J.C.: 1973, The temperature structure of chromospheric flares heated by non-thermal electrons. *Solar Phys.* **31**, 143. DOI. ADS.

Brown, J.C., Karlický, M., MacKinnon, A.L., van den Oord, G.H.J.: 1990, Beam heating in solar flares – electrons or protons? *Astrophys. J. Suppl.* **73**, 343. DOI. ADS.

Brown, J.C., Turkmani, R., Kontar, E.P., MacKinnon, A.L., Vlahos, L.: 2009, Local re-acceleration and a modified thick target model of solar flare electrons. *Astron. Astrophys.* **508**, 993. DOI. ADS.

Campbell, P.M.: 1984, Transport phenomena in a completely ionized gas with large temperature gradients. *Phys. Rev. A* **30**, 365. DOI. ADS.

Coyner, A.J., Alexander, D.: 2009, Implications of temporal development of localized ultraviolet and hard X-ray emission for large solar flares. *Astrophys. J.* **705**, 554. DOI. ADS.

Culhane, J.L., Harra, L.K., James, A.M., Al-Janabi, K., Bradley, L.J., Chaudry, R.A., Rees, K., Tandy, J.A., Thomas, P., Whillock, M.C.R., Winter, B., Doschek, G.A., Korendyke, C.M., Brown, C.M., Myers, S., Mariska, J., Seely, J., Lang, J., Kent, B.J., Shaughnessy, B.M., Young, P.R., Simnett, G.M., Castelli, C.M., Mahmoud, S., Mapson-Menard, H., Probyn, B.J., Thomas, R.J., Davila, J., Dere, K., Windt, D., Shea, J., Hagood, R., Moye, R., Hara, H., Watanabe, T., Matsuzaki, K., Kosugi, T., Hansteen, V., Wikstol, O.: 2007, The EUV imaging spectrometer for Hinode. *Solar Phys.* **243**, 19. DOI. ADS.

Del Zanna, G., Mitra-Kraev, U., Bradshaw, S.J., Mason, H.E., Asai, A.: 2011, The 22 May 2007 B-class flare: new insights from Hinode observations. *Astron. Astrophys.* **526**, A1. DOI. ADS.

Dere, K.P., Landi, E., Mason, H.E., Monsignori Fossi, B.C., Young, P.R.: 1997, CHIANTI – an atomic database for emission lines. *Astron. Astrophys. Suppl.* **125**, 149. DOI. ADS.

Emslie, A.G.: 1978, The collisional interaction of a beam of charged particles with a hydrogen target of arbitrary ionization level. *Astrophys. J.* **224**, 241. DOI. ADS.

Emslie, A.G., Noyes, R.W.: 1978, The characteristics of impulsive solar EUV bursts. *Solar Phys.* **57**, 373. ADS.

Fisher, G.H., Canfield, R.C., McClymont, A.N.: 1985, Flare loop radiative hydrodynamics. V – Response to thick-target heating. VI – Chromospheric evaporation due to heating by nonthermal electrons. VII – Dynamics of the thick-target heated chromosphere. *Astrophys. J.* **289**, 414. DOI. ADS.

Fletcher, L., Hudson, H.S.: 2008, Impulsive phase flare energy transport by large-scale Alfvén waves and the electron acceleration problem. *Astrophys. J.* **675**, 1645. DOI. ADS.

Fletcher, L., Hannah, I.G., Hudson, H.S., Innes, D.E.: 2013, Flare ribbon energetics in the early phase of an SDO flare. *Astrophys. J.* **771**, 104. DOI. ADS.

Graham, D.R., Fletcher, L., Hannah, I.G.: 2011, Hinode/EIS plasma diagnostics in the flaring solar chromosphere. *Astron. Astrophys.* **532**, A27. DOI. ADS.

Graham, D.R., Hannah, I.G., Fletcher, L., Milligan, R.O.: 2013, The emission measure distribution of impulsive phase flare footpoints. *Astrophys. J.* **767**, 83. DOI. ADS.

Hannah, I.G., Kontar, E.P.: 2012, Differential emission measures from the regularized inversion of Hinode and SDO data. *Astron. Astrophys.* **539**, A146. DOI. ADS.

Hannah, I.G., Kontar, E.P.: 2013, Multi-thermal dynamics and energetics of a coronal mass ejection in the low solar atmosphere. *Astron. Astrophys.* **553**, A10. DOI. ADS.

Hinteregger, H.E., Hall, L.A.: 1969, Solar extreme ultraviolet emissions in the range 260 1300 Å observed from OSO-III. *Solar Phys.* **6**, 175. DOI. ADS.

Horan, D.M., Kreplin, R.W., Fritz, G.G.: 1982, Direct measurements of impulsive extreme ultraviolet and hard X-ray solar flare emission. *Astrophys. J.* **255**, 797. DOI. ADS.

Hori, K., Yokoyama, T., Kosugi, T., Shibata, K.: 1998, Single and multiple solar flare loops: hydrodynamics and Ca XIX resonance line emission. *Astrophys. J.* **500**, 492. DOI. ADS.

Hudson, H.S.: 1972, Thick-target processes and white-light flares. *Solar Phys.* **24**, 414. DOI. ADS.

Hudson, H.S., Strong, K.T., Dennis, B.R., Zarro, D., Inda, M., Kosugi, T., Sakao, T.: 1994, Impulsive behavior in solar soft X-radiation. *Astrophys. J. Lett.* **422**, L25. DOI. ADS.

Landi, E., Young, P.R., Dere, K.P., Del Zanna, G., Mason, H.E.: 2013, CHIANTI – an atomic database for emission lines. XIII. Soft X-ray improvements and other changes. *Astrophys. J.* **763**, 86. DOI. ADS.

Lemen, J.R., Title, A.M., Akin, D.J., Boerner, P.F., Chou, C., Drake, J.F., Duncan, D.W., Edwards, C.G., Friedlaender, F.M., Heyman, G.F., Hurlburt, N.E., Katz, N.L., Kushner, G.D., Levay, M., Lindgren, R.W., Mathur, D.P., McFeaters, E.L., Mitchell, S., Rehse, R.A., Schrijver, C.J., Springer, L.A., Stern, R.A., Tarbell, T.D., Wuelser, J.-P., Wolfson, C.J., Yanari, C., Bookbinder, J.A., Cheimets, P.N., Caldwell, D., Deluca, E.E., Gates, R., Golub, L., Park, S., Podgorski, W.A., Bush, R.I., Scherrer, P.H., Gummin, M.A., Smith, P., Auker, G., Jerram, P., Pool, P., Soufli, R., Windt, D.L., Beardsley, S., Clapp, M., Lang, J., Waltham, N.: 2012, The Atmospheric Imaging Assembly (AIA) on the Solar Dynamics Observatory (SDO). *Solar Phys.* **275**, 17. DOI. ADS.

Lin, R.P., Dennis, B.R., Hurford, G.J., Smith, D.M., Zehnder, A., Harvey, P.R., Curtis, D.W., Pankow, D., Turin, P., Bester, M., Csillaghy, A., Lewis, M., Madden, N., van Beek, H.F., Appleby, M., Raudorf, T., McTiernan, J., Ramaty, R., Schmahl, E., Schwartz, R., Krucker, S., Abiad, R., Quinn, T., Berg, P., Hashii, M., Sterling, R., Jackson, R., Pratt, R., Campbell, R.D., Malone, D., Landis, D., Barrington-Leigh, C.P., Slassi-Sennou, S., Cork, C., Clark, D., Amato, D., Orwig, L., Boyle, R., Banks, I.S., Shirey, K., Tolbert, A.K., Zarro, D., Snow, F., Thomsen, K., Henneck, R., McHedlishvili, A., Ming, P., Fivian, M., Jordan, J., Wanner, R., Crubb, J., Preble, J., Matranga, M., Benz, A., Hudson, H., Canfield, R.C., Holman, G.D., Crannell, C., Kosugi, T., Emslie, A.G., Vilmer, N., Brown, J.C., Johns-Krull, C., Aschwanden, M., Metcalf, T., Conway, A.: 2002, The Reuven Ramaty High-Energy Solar Spectroscopic Imager (RHESSI). *Solar Phys.* **210**, 3. DOI. ADS.

McTiernan, J.M., Kane, S.R., Loran, J.M., Lemen, J.R., Acton, L.W., Hara, H., Tsuneta, S., Kosugi, T.: 1993, Temperature and density structure of the November 2 1991 flare observed by the YOHKOH soft X-ray telescope and hard X-ray telescope. *Astrophys. J. Lett.* **416**, L91. DOI. ADS.

Milligan, R.O.: 2011, Spatially resolved nonthermal line broadening during the impulsive phase of a solar flare. *Astrophys. J.* **740**, 70. DOI. ADS.

Milligan, R.O.: 2015, Extreme ultra-violet spectroscopy of the flaring solar chromosphere. *Solar Phys.*, submitted. ADS.

Milligan, R.O., Dennis, B.R.: 2009, Velocity characteristics of evaporated plasma using Hinode/EUV imaging spectrometer. *Astrophys. J.* **699**, 968. DOI. ADS.

Milligan, R.O., McElroy, S.A.: 2013, Continuum contributions to the SDO/AIA passbands during solar flares. *Astrophys. J.* **777**, 12. DOI. ADS.

Mrozek, T., Tomczak, M.: 2004, Solar impulsive soft X-ray brightenings and their connection with footpoint hard X-ray emission sources. *Astron. Astrophys.* **415**, 377. DOI. ADS.

O'Dwyer, B., Del Zanna, G., Mason, H.E., Weber, M.A., Tripathi, D.: 2010, SDO/AIA response to coronal hole, quiet Sun, active region, and flare plasma. *Astron. Astrophys.* **521**, A21. DOI. ADS.

Oka, M., Ishikawa, S., Saint-Hilaire, P., Krucker, S., Lin, R.P.: 2013, Kappa distribution model for hard X-ray coronal sources of solar flares. *Astrophys. J.* **764**, 6. DOI. ADS.

Oka, M., Krucker, S., Hudson, H.S., Saint-Hilaire, P.: 2015, Electron energy partition in the above-the-looptop solar hard X-ray sources. *Astrophys. J.* **799**, 129. DOI. ADS.

Qiu, J., Sturrock, Z., Longcope, D.W., Klimchuk, J.A., Liu, W.-J.: 2013, Ultraviolet and extreme-ultraviolet emissions at the flare footpoints observed by atmosphere imaging assembly. *Astrophys. J.* **774**, 14. DOI. ADS.

Rosner, R., Tucker, W.H., Vaiana, G.S.: 1978, Dynamics of the quiescent solar corona. *Astrophys. J.* **220**, 643. DOI. ADS.

Ryan, D.F., Milligan, R.O., Gallagher, P.T., Dennis, B.R., Tolbert, A.K., Schwartz, R.A., Young, C.A.: 2012, The thermal properties of solar flares over three solar cycles using GOES X-ray observations. *Astrophys. J. Suppl.* **202**, 11. DOI. ADS.

Ryan, D.F., O'Flannagain, A.M., Aschwanden, M.J., Gallagher, P.T.: 2014, The compatibility of flare temperatures observed with AIA, GOES, and RHESSI. *Solar Phys.* **289**, 2547. DOI. ADS.

Saint-Hilaire, P., Benz, A.O.: 2005, Thermal and non-thermal energies of solar flares. *Astron. Astrophys.* **435**, 743. DOI. ADS.

Scherrer, P.H., Schou, J., Bush, R.I., Kosovichev, A.G., Bogart, R.S., Hoeksema, J.T., Liu, Y., Duvall, T.L., Zhao, J., Title, A.M., Schrijver, C.J., Tarbell, T.D., Tomczyk, S.: 2012, The Helioseismic and Magnetic Imager (HMI) investigation for the Solar Dynamics Observatory (SDO). *Solar Phys.* **275**, 207. DOI. ADS.

Schmahl, E.J., Pernak, R.L., Hurford, G.J., Lee, J., Bong, S.: 2007, Analysis of RHESSI flares using a radio astronomical technique. *Solar Phys.* **240**, 241. DOI. ADS.

Schwartz, R.A., Csillaghy, A., Tolbert, A.K., Hurford, G.J., McTiernan, J., Zarro, D.: 2002, RHESSI data analysis software: rationale and methods. *Solar Phys.* **210**, 165. DOI. ADS.

Simões, P.J.A., Kontar, E.P.: 2013, Implications for electron acceleration and transport from non-thermal electron rates at looptop and footpoint sources in solar flares. *Astron. Astrophys.* **551**, A135. DOI. ADS.

Simões, P.J.A., Graham, D.R., Fletcher, L.: 2015, Direct observation of the energy release site in a solar flare by SDO/AIA, Hinode/EIS, and RHESSI. *Astron. Astrophys.* **577**, A68. DOI. ADS.

Spitzer, L.: 1965, *Physics of Fully Ionized Gases*. ADS.

Švestka, Z.: 1987, Cooling of a coronal flare loop through radiation and conduction. *Solar Phys.* **108**, 411. DOI. ADS.

Thomas, R.J., Crannell, C.J., Starr, R.: 1985, Expressions to determine temperatures and emission measures for solar X-ray events from GOES measurements. *Solar Phys.* **95**, 323. DOI. ADS.

Varady, M., Karlický, M., Moravec, Z., Kašparová, J.: 2014, Modifications of thick-target model: re-acceleration of electron beams by static and stochastic electric fields. *Astron. Astrophys.* **563**, A51. DOI. ADS.

Watanabe, T., Hara, H., Sterling, A.C., Harra, L.K.: 2010, Production of high-temperature plasmas during the early phases of a C9.7 flare. *Astrophys. J.* **719**, 213. DOI. ADS.

White, S.M., Thomas, R.J., Schwartz, R.A.: 2005, Updated expressions for determining temperatures and emission measures from Goes soft X-ray measurements. *Solar Phys.* **227**, 231. DOI. ADS.

Woods, T.N., Eparvier, F.G., Hock, R., Jones, A.R., Woodraska, D., Judge, D., Didkovsky, L., Lean, J., Mariska, J., Warren, H., McMullin, D., Chamberlin, P., Berthiaume, G., Bailey, S., Fuller-Rowell, T., Sojka, J., Tobiska, W.K., Viereck, R.: 2012, Extreme ultraviolet variability experiment (EVE) on the solar dynamics observatory (SDO): Overview of science objectives, instrument design, data products, and model developments. *Solar Phys.* **275**, 115. DOI. ADS.

Zharkova, V.V., Arzner, K., Benz, A.O., Browning, P., Dauphin, C., Emslie, A.G., Fletcher, L., Kontar, E.P., Mann, G., Onofri, M., Petrosian, V., Turkmani, R., Vilmer, N., Vlahos, L.: 2011, Recent advances in understanding particle acceleration processes in solar flares. *Space Sci. Rev.* **159**, 357. DOI. ADS.

Reprinted from *Solar Physics* Journal, DOI 10.1007/s11207-015-0783-z
SOLAR AND STELLAR FLARES

Temperature Dependence of the Flare Fluence Scaling Exponent

M. Kretzschmar[1]

Received: 14 February 2015 / Accepted: 16 September 2015 / Published online: 27 October 2015
© Springer Science+Business Media Dordrecht 2015

Abstract Solar flares result in an increase of the solar irradiance at all wavelengths. While the distribution of the flare fluence observed in coronal emission has been widely studied and found to scale as $f(E) \sim E^{-\alpha}$, with α slightly below 2, the distribution of the flare fluence in chromospheric lines is poorly known. We used the solar irradiance measurements observed by the SDO/EVE instrument at a 10 s cadence to investigate the dependency of the scaling exponent on the formation region of the lines (or temperature). We analyzed all flares above the C1 level since the start of the EVE observations (May 2010) to determine the flare fluence distribution in 16 lines covering a wide range of temperatures, several of which were not studied before. Our results show a weak downward trend with temperature of the scaling exponent of the PDF that reaches from above 2 at lower temperature (a few 10^4 K) to ~ 1.8 for hot coronal emission (several 10^6 K). However, because colder lines also have fainter contrast, we cannot exclude that this behavior is caused by including more noise for smaller flares for these lines. We discuss the method and its limitations and tentatively associate this possible trend with the different mechanisms responsible for the heating of the chromosphere and corona during flares.

Keywords Flares · Solar irradiance · Heating, in flares

1. Introduction

Solar flares are huge releases of energy in the atmosphere of the Sun that are frequently observed through bursts of electromagnetic radiation throughout the spectrum (Hudson, 2011; Fletcher *et al.*, 2011). Because flares affect large parts of the solar atmosphere with various phenomena, the emission comes from plasma in different states and therefore arises

Solar and Stellar Flares: Observations, Simulations, and Synergies
Guest Editors: Lyndsay Fletcher and Petr Heinzel

✉ M. Kretzschmar
matthieu.kretzschmar@cnrs-orleans.fr

[1] LPC2E, UMR 7328 CNRS, University of Orléans, 3a av. de la recherche scientifique, 45071 Orléans, France

at different wavelengths with different contrasts. Two phenomena are usually distinguished during solar flares: 1) the deposition of energy by accelerated particles in the chromosphere, and 2) the chromospheric evaporation that leads to strong emission of coronal loops in the extreme-ultraviolet (EUV) and soft X-ray (SXR) wavelength range. The EUV and SXR coronal emission is composed of spectral lines and free–free continuum emitted by a very hot plasma with highly ionized elements. Chromospheric emission arises at different wavelengths. The most highly contrasted flare signals in the chromosphere are probably the hard X-rays (HXR) (see *e.g.* Veronig *et al.*, 2002a). They are caused by the braking of accelerated electrons in the dense chromosphere. Emission with high contrast originating from the chromosphere is also observed at radio frequencies, but most of the radiative energy from the chromosphere is emitted at shorter wavelengths, in the ultraviolet (UV) and even extreme-ultraviolet wavelength range. Part of the energy initially received by the chromosphere early on during the flare is quickly used to form bright coronal loops filled with hot plasma, but another part is also radiated at these wavelengths by the chromospheric plasma itself during the impulsive phase. Furthermore, the chromosphere also receives energy by conduction as well as radiation from the surrounding coronal loops (*e.g.*, Berlicki and Heinzel, 2004; Longcope, 2014). All these energy transfers bring the chromosphere into a very complex state, and modeling the chromospheric emission in the visible and UV during flare remains a very difficult task (Allred *et al.*, 2005; Heinzel and Avrett, 2012; Berlicki, 2007). During flares, the whole atmosphere evolves very rapidly, and multiple interactions take place between plasma and photons, which often obscure the reasons for the phenomena detected in the observations (see *e.g.* Veronig *et al.*, 2010 for *e.g.* one – relatively – clear case and another, more complex case). In this study, we concentrate on the flare emission by chromospheric EUV lines and sometimes refer to it simply as chromospheric emission.

Chromospheric evaporation causes coronal loops to be filled by very hot plasma. This leads to strong contrasts in the SXR and short EUV wavelength ranges, even when the light is integrated over the solar disk, *i.e.* in solar irradiance measurements. These SXR and EUV flare signatures are therefore easier to observe than the less strongly contrasted chromospheric emission. When observed at these coronal wavelengths, the flare fluence is well known to be distributed according to a power law,

$$f(E) \sim E^{-\alpha}, \tag{1}$$

with E the fluence, $f(E)$ the probability distribution function (PDF) of the flare fluence, and α the scaling exponent that is usually observed to be slightly below 2 (Hudson, 1991; Crosby, Aschwanden, and Dennis, 1993; Hannah *et al.*, 2011). For example, Veronig *et al.* (2002b) made an in-depth analysis of soft X-ray flares observed by GOES and found a power-law frequency distribution for SXR flare fluences with an exponent of about 1.8 ± 0.1 (background subtracted). In this study, we find a value of 1.86 ± 0.02 for a different period of time (see Figure 3). Although small flares are more difficult to observe and measure, there are indications that small (nano) and large flares belong to the same distribution (*e.g.*, Hannah *et al.*, 2011; Schrijver *et al.*, 2012). HXR emission is produced in the chromosphere, but with different processes than EUV lines. The statistics of flare emission in the HXR has also been often studied and is reviewed by Hannah *et al.* (2011). The peak flux values (Dennis, 1985; Crosby, Aschwanden, and Dennis, 1993) is found to be distributed according to a power law with a scaling exponent of between 1.5 and 2.

The value of the scaling exponent α is important since only a value $\alpha > 2$ allows a significant contribution of the smallest flares to the total energy released by all flares (Hudson, 1991); nanoflares have been invoked as one way to heat the solar corona, but their ubiquitous nature also gives them the potential to contribute to the variability of the solar radiative flux received at Earth, *i.e.* the solar spectral irradiance. This latest scenario requires a mechanism to cause the (nano) flaring rate to vary, however, which has not yet been observed (but cannot be ruled out because of the in-phase variability of most of the solar activity manifestations with the sunspot cycle).

The potential dependency of the scaling exponent α on wavelengths is also important because it would constrain the flare models. It also reflects the modification of the solar spectrum shape with the flare size. Indeed, if the flare fluences E_{λ_1} and E_{λ_2} measured at two wavelengths λ_1 and λ_2 have different scaling exponents, the energy radiated at these two wavelengths will not change at the same rate with the flare size (we detail this in Section 2.2). This parameter α is therefore also important for investigating the effect of flares on the Earth's upper atmosphere, whose response is also dependent on wavelengths.

Because of the weaker contrast and the difficulty of observing the UV chromosphere with space-based instruments (many UV chromospheric lines can be measured from the ground), previous studies of the flare energy distribution have focused on observations at coronal wavelengths or in the HXR and radio domains, which have different physical mechanisms than the chromospheric lines. There has been no study to our knowledge that dealt with the statistics of flare fluences in E/UV chromospheric lines. Measuring the flare fluence for an ensemble of flares requires a certain level of calibration that is difficult to obtain from the ground. Temmer *et al.* (2001), for instance, presented a statistical analysis of Hα flare, characterizing various parameters, but without reporting on the flare fluence for this line. However, chromospheric emission can represent a large part of the total energy radiated by flare (Woods, Kopp, and Chamberlin, 2006; Kretzschmar *et al.*, 2010; Kretzschmar, 2011; Fletcher *et al.*, 2011; Milligan *et al.*, 2012, 2014), and determining α for these emissions is therefore important.

The main aim of this article is to investigate the scaling of the flare fluence at different wavelengths, searching in particular for potential differences for EUV spectral lines that are predominantly of coronal or chromospheric origin. Recent observations by the *Extreme Ultraviolet Variability Experiment* (EVE; Woods *et al.*, 2012) onboard the *Solar Dynamic Observatory* (SDO) gives us an excellent opportunity to investigate the flare distribution in EUV chromospheric lines. SDO/EVE is observing the Sun quasi-permanently since May 2010, providing high-resolution irradiance spectrum with a 10 s cadence and at EUV wavelengths representing both the (hot) corona and the (warm) upper chromosphere and lower transition region. In this article we use four years of flare observations by SDO/EVE to investigate if and how the flare fluence depends on wavelength and temperature.

Section 2 presents the SDO/EVE data and the spectral lines that are used in the analysis. We also describe the computation of the flare fluence and the determination of the scaling exponent of the fluence distribution. Section 3 shows and discusses the results, and we conclude in Section 4.

2. Data Analysis

2.1. EVE Flare Spectrum and Line Fluxes

We used version 4 of the EVE data product, which contains observations from the two spectrometers MEGS-A (5 – 37 nm) and MEGS-B (35 – 105 nm) with a spectral resolution of 0.02 nm and from several spectrophotometers (ESP) with broad passbands between 0.1 nm and 39 nm.

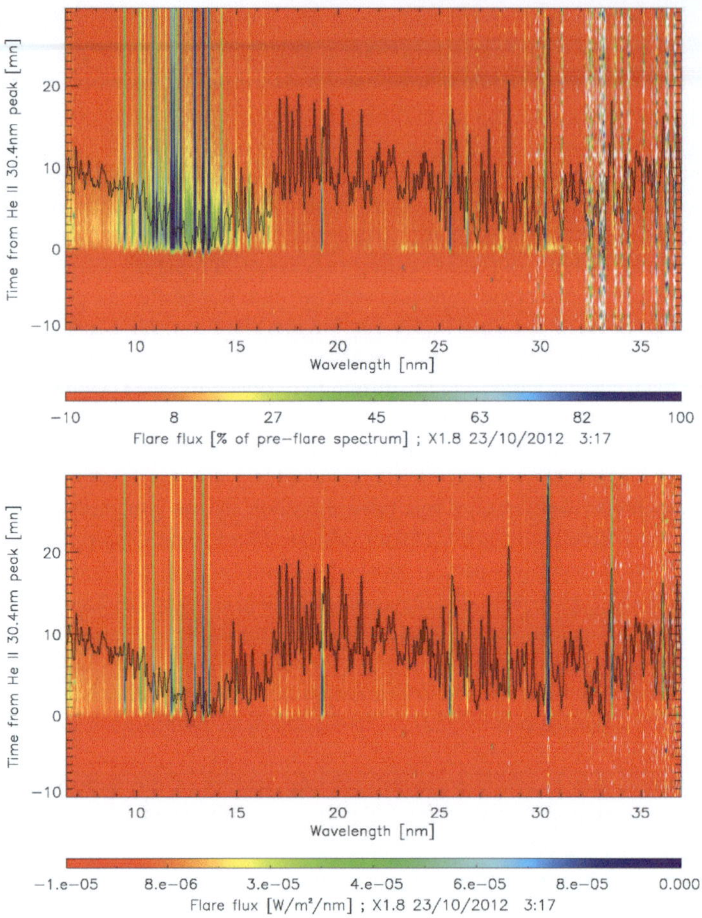

Figure 1 Flare flux observed by SDO/EVE for the X1.8 flare on 23 October 2012 as a function of time and wavelength. Time is given in minutes from the peak of the flare in the He II 30.4 nm line. The black line shows the spectrum (no units) for reference. The color represents the intensity in the following way. Top panel: the flare flux is expressed in percent of the pre-flare flux, which is equivalent to the irradiance contrast of the flare. Bottom panel: the flare flux is expressed in physical units $\mathrm{W/m^2/nm}$.

An example of a flare observation by EVE is presented in Figures 1 and 2. Figure 1 shows the EVE flare spectrum at the full wavelength resolution in two forms: the upper panel represents the flare (irradiance) contrast in percent, while the bottom panel shows the flare flux in physical units. These two representations allow us to realize that the flare energy can reside at wavelengths where the contrast is moderate, as is the case for the Fe XVI line at 36.07 nm, for example. We can also note by looking at the top panel that noise will disturb the flare signal when the contrast becomes lower than a few percent. Figure 2 shows time series for spectral lines available in level 2 of the EVE data product; these line fluxes are integrated with fixed spectral integration limits. As expected, the hottest lines have the highest contrast. When moving to colder (chromospheric and transition region) lines, *e.g.* the O V line and below in the figure, the flare profile appears to be more impulsive. For this M1-class flare, the contrast of the colder lines is on the order of 5 % in these one-minute smoothed data. This is comparable to the noise level, and for these low-contrast lines only

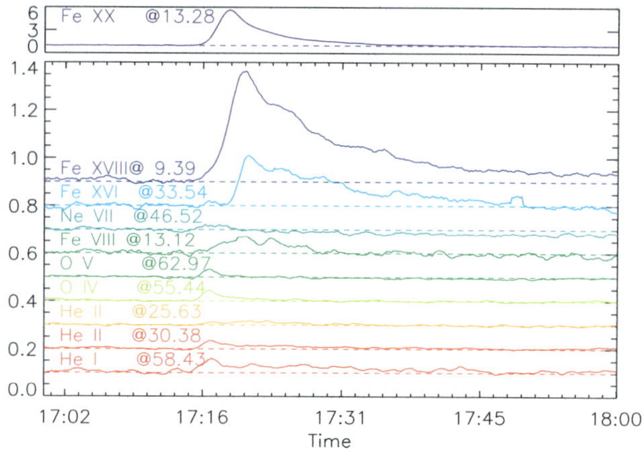

Figure 2 Light curves of the relative increase (or irradiance contrast) for different lines for the M1.2 flare on 5 May 2010. The Fe XX line at 13.28 nm is isolated in the upper panel because of its high contrast. In the lower panel, all curves are shifted vertically by 0.1 for clarity, but an increase of 0.1 corresponds to an increase of 10 %.

flares above the M1 level can be processed meaningfully through our analysis. Flares were identified using the GOES flare catalog, and we first considered the 4664 flares above the C level that were observed by EVE from May 2010 to mid-2014. As explained below, the low contrast of some EUV lines forced us to use only the 435 M- and X-flares during this period.

In level 2 of the EVE data products, the fluxes of several spectral lines are already provided together with the full spectrum. These lines are integrated with a fixed spectral width on each spectrum, however. We used these line-flux values, but we also directly computed the fluxes by integrating the lines observed in the spectrum to include possible modifications of the spectrum and line shapes during the flare. This was tested in two different ways; first, by automatically determining the spectral integration limits as follows: we started from the peak of the line and defined the blue and red limits of the integration by identifying the first wavelength at which the intensity stops to decrease when moving away from the peak. Second, we attempted to fit the lines with a continuum and a Gaussian profile. This second method gave valuable but also very intermittent results: several spectral lines needed to be fitted at each time step of a flare, and the fit failed several times during this procedure, which makes the final statistics difficult to interpret. In the end, we used the line intensities provided in the EVE data product and those that we computed by direct integration (with automatic determination of the spectral limits). We required the line-flux values retrieved by these two methods to agree to keep them in the statistics.

Several of the lines available in the EVE level 2 data product have very low flare contrast and/or do not show a clear scaling of the flare signal with the SXR class of the flare (see also Figure 7 and Section 2.3). Moreover, because of the strong degradation of the MEGS-B spectrometer, observations above 35 nm have been reduced to three hours per day, and consequently, fewer flares are observed at these wavelengths. We kept only two of these lines (the O IV line at 55.44 nm and the Si XII line at 49.94 nm). Furthermore, because the dominant spectral lines of the spectrum can change during flares with respect to those of a quieter Sun, we analyzed a few flare spectra to determine other lines of interest, with the objective to cover the widest possible range of temperatures. These lines were identified with the use of the CHIANTI database (Dere *et al.*, 1997; Landi *et al.*, 2006) by simulating two spectra for a flare and an active region to ensure that the dominant line in the integrated spectral bins is the same for large and small flares. The flare spectrum simulated with the CHIANTI database is based on a differential emission measure (DEM) computed from a flaring region of the Sun (Dere and Cook, 1979) and not an integrated Sun; therefore we do not expect to

Table 1 List of spectral lines that we used. The asterisk indicates lines that are not already provided in the EVE data product.

Ions	Wavelength (nm)	Log(T)
He II	30.38	4.7
O IV	55.44	5.19
O VI*	15.01	5.5
Fe VIII*	10.37	5.6
Fe VIII	13.12	5.57
Fe IX*	10.52	5.8
Fe XI*	10.06	6.1
Ni XI*	14.89	6.1
Si XII	49.94	6.29
Fe XV	28.42	6.3
Fe XVI	33.54	6.43
Fe XVII	25.49	6.6
Fe XVIII	9.39	6.81
Fe XX	13.28	6.97
Fe XXI*	12.87	7.0
Fe XXII*	13.57	7.1

observe the same lines in the EVE irradiance spectrum, but we used the ratio of the flare intensity to the active region intensity as a criterium to determine which line is dominant. Table 1 shows the lines that we finally used, with the lines that are not present in the EVE data product marked with an asterisk.

2.2. Determination of the Power-Law Exponent

We start this section by explaining the relations between the scaling exponents α_x and α_y of two power-law distributed variables x and y. We first assume that x and y are the flare fluences at two different wavelengths (SXR and EUV *e.g.*), and that:

- x is distributed according to a power law with scaling exponents α_x: $P(x) = C_1 x^{-\alpha_x}$
- x and y are related through $y \sim x^\gamma$.

As $P(y)\,\mathrm{d}y = P(x)\,\mathrm{d}x$, it is straightforward to verify that

$$P(y) = C_2 y^{-\alpha_y}$$

with

$$\alpha_y = \frac{\alpha_x + \gamma - 1}{\gamma}. \tag{2}$$

The uncertainty $U(\alpha_y)$ on α_y can be computed with

$$U(\alpha_y) = \frac{U(\alpha_x)}{\gamma} + \left| 1 - \frac{\alpha_x}{\gamma} - \frac{\gamma - 1}{\gamma} \right| \frac{U(\gamma)}{\gamma}. \tag{3}$$

We therefore have two ways to estimate the exponent of the distribution function for the flare fluence of one line. The first one is to compute the probability distribution function (PDF) and to fit it with a power law. The second one, which is particularly relevant when only very

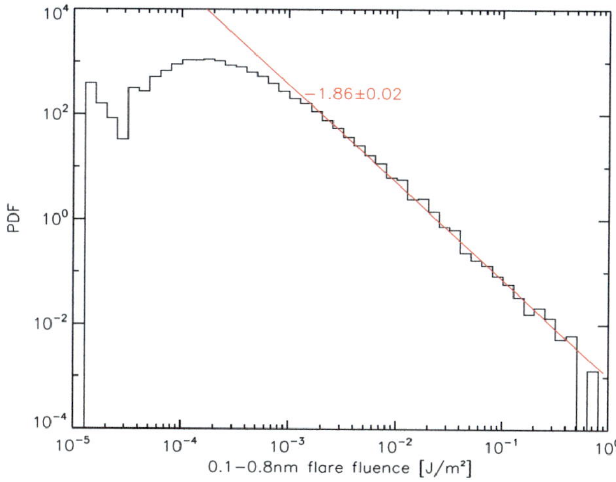

Figure 3 Probability distribution function (PDF) of the flare fluence in the 0.1 – 0.8 nm range.

few flares have a high enough contrast to build a reliable PDF, is to plot the flare fluence of the line *versus* the flare fluence observed in the 0.1 nm – 0.8 nm by GOES and then apply Equation (2). The GOES SXR fluence is well known to have a power-law distribution with an exponent slightly below 2 (see Veronig *et al.*, 2002b and reference therein). For the flares that we consider here, we found $\alpha_{\mathrm{GOES}} = 1.86$ (see Figure 3), which was determined by using the one-minute data of the GOES satellite.

This procedure is illustrated in Figures 3, 4, and 5. The power-law distribution of flare fluence in the 0.1 – 0.8 nm range has an exponent of 1.86 ± 0.02, while for the flare fluence observed by the EVE/ESP diode in a wavelength range around the He II 30.4 nm line, we found an exponent of 2.16 ± 0.07 (top panel of Figure 4). By scaling the He II flare fluence to the SXR flare fluence, we found the relation $E_{\mathrm{HeII}} \sim E_{\mathrm{SXR}}^{\gamma}$ with $\gamma = 0.75 \pm 0.03$ (bottom panel of Figure 4), when we only considered flares above the M1 level. By applying the relation of Equation (2) between the exponents above, this value of γ leads to $\alpha_{304} = 2.15 \pm 0.07$ for the value of the power-law exponent of the PDF of the flare fluence at He II 30.4 nm. This compares very well to the value of 2.16 ± 0.07 found by fitting the PDF. We also note that γ decreases when we include C-class flares, or equivalently, the scaling exponent increases. We attribute this to the fact that for these smaller flares, part of the flares observed at 30.4 nm are below the noise level between the GOES start- and end-times; the flare fluence is then artificially increased by the noise.

The same procedure is illustrated for the hot coronal emission produced by the Fe XX ion at 13.28 nm in Figure 5. By again applying Equation (2), the scaling of the Fe XX line fluence to the SXR fluence leads to a value of 1.83 ± 0.04 for the scaling exponent of the PDF, which agrees well with the value 1.88 ± 0.03 found by fitting the PDF. Here, the α value found with the C-class flares included is very similar (1.87 ± 0.04), suggesting that this line has good contrast throughout the flare duration as defined by GOES.

Computing a robust power-law distribution requires many flare observations, but only hot coronal lines have a high enough contrast to compute their flare fluence over several flare magnitudes. Therefore, we first determined γ by scaling the flare fluence of a spectral line $E(\lambda)$ to the flare fluence in the SXR E_{SXR} and next applied Equation (2) to compute the scaling exponent α of the fluence distribution.

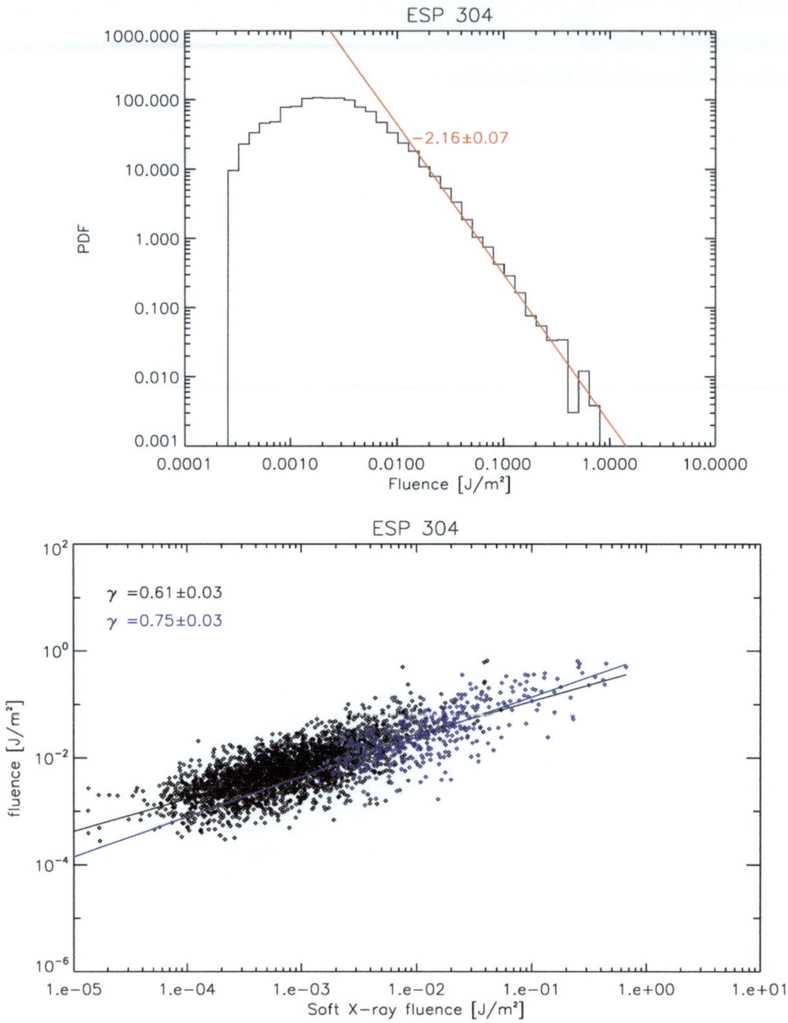

Figure 4 Top panel: probability distribution function (PDF) of the flare fluence for the EVE MEGS 304 diode. In red, the best fit to the power law gives an exponent of 2.16. Bottom panel: scaling of EVE MEGS 304 diode flare fluence *versus* 0.1 – 0.8 nm. Only flares with a signal-to-noise ratio higher than two sigma are plotted. Flares above the M1 level are shown in blue. The two fits are for flares above the C1 (black) and M1 (blue) levels.

2.3. Flare Fluence Computation

The flare fluence is defined as follows:

$$E(\lambda) = \int_{t_{st}}^{t_{end}} \big(I(\lambda, t) - I_{bg}(\lambda)\big)\, dt, \tag{4}$$

where $I(\lambda, t)$ is the irradiance at time t during the flare, $I_{bg}(\lambda)$ is the background value, and t_{st} and t_{end} are the start- and end-time of the flare. $I_{bg}(\lambda)$ was determined as the median of the solar irradiance during ten minutes before the start of the flare.

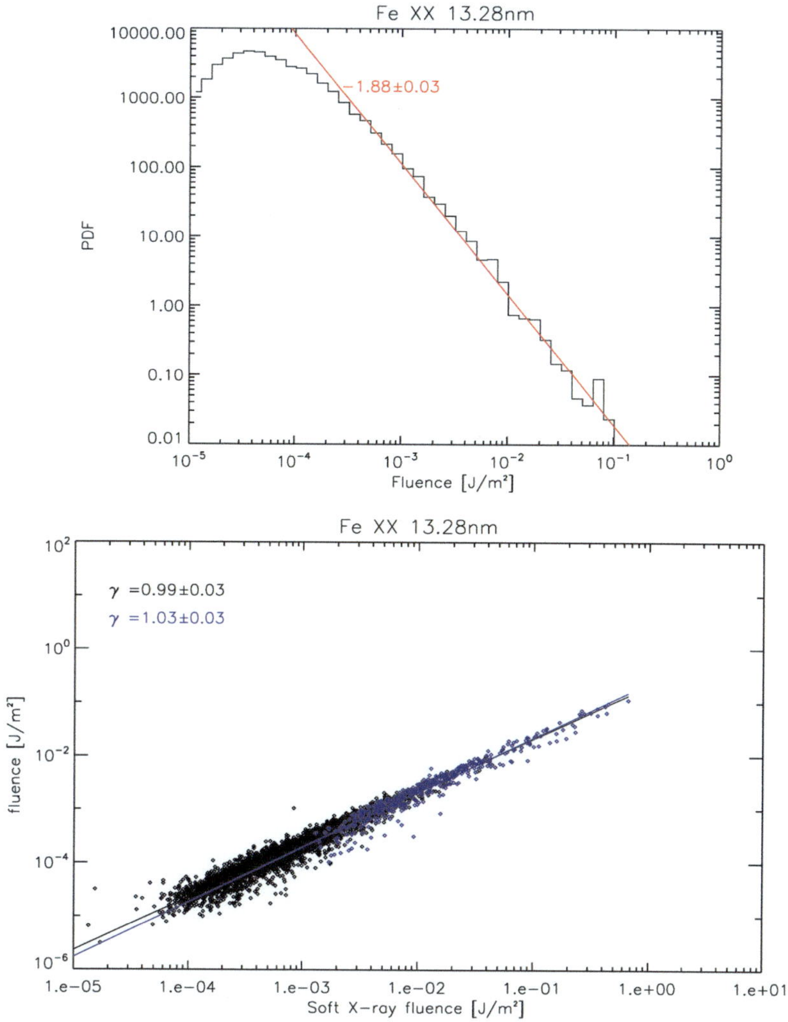

Figure 5 Top panel: probability distribution function (PDF) of the flare fluence for the EVE Fe XX line at 13.28 nm. In red, the best fit to the power-law gives an exponent of 1.88. Bottom panel: scaling of the EVE Fe XX line at 13.28 nm flare fluence *versus* 0.1 – 0.8 nm flare fluence. Only flares with a signal-to-noise ratio higher than two sigma are plotted. Flares above the M1 level are shown in blue. The two fits are for flares above the C1 (black) and M1 (blue) levels.

The integration time is an important parameter for computing the fluence. The end-time of the flare indicated by the GOES flare catalog is the time at which the 0.1 – 0.8 nm flux returns to less than half of its value at peak time. This usually underestimates the length of the flare, at least for the hot lines. The starting time of the flare is less difficult to define, but we must be sure to include the emission during the impulsive phase. Here again, we considered several definitions of the start- and end-time for our study: 1) we attempted to determine the start- and end-time for each line and each flare in a similar manner to the procedures employed for GOES; 2) the integration was made from five minutes before the

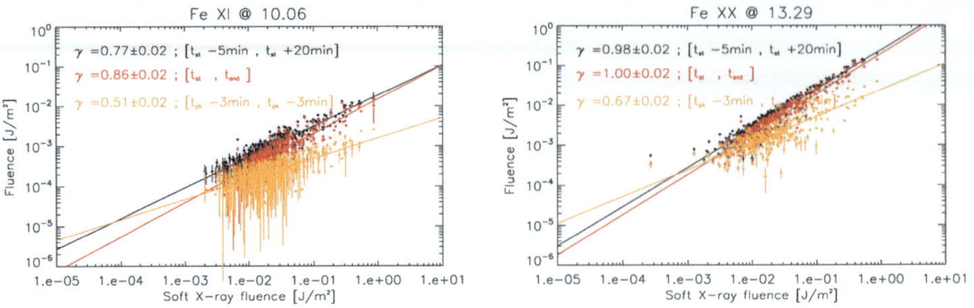

Figure 6 Influence of the integration time on the computation of fluence and on the scaling parameter for Fe XI at 10.06 nm (left panel) and Fe XX at 13.28 nm (right panel). Fluences are computed by integration between 1) five minutes before the GOES start-time and ten minutes after the GOES end-time (black diamonds), 2) the GOES start-time and the GOES end-time (red diamonds), and 3) six minutes centered around the GOES peak time.

GOES start-time to 20 minutes after the GOES end-time, and 3) the integration was made between the GOES start- and end-time.

Methods 2) and 3) are similar because they assume the same start- and end-time for all lines, but method 2) ensures that no flare emission is missed during the integration. For lines with shorter duration (*e.g.* lines with lower contrast or "impulsive" profile), we expected that the integration over a longer time period would not change the fluence values because the background is removed before the integration. However, this was not the case, as illustrated in Figure 6, which shows that integrating over both a too long or too short time with respect to the flare duration leads to an underestimation of γ and therefore to an overestimation of the scaling exponent. For a too long integration time, noise will artificially increase the fluence for the smallest flares, as explained previously. It is more difficult to explain why this is also the case for a very short interval centered around the peak time, where noise should not play a role; one possible explanation is that using a short and fixed integration time leads to an underestimation of the fluence for the large flares because in proportion, less and less of the flare emission is taken into account. Therefore, method 3) was found to be more reliable. For method 1), the determination of the start time was found to be not robust enough to be used. However, we were able to determine the end-time of the flare as the time at which the background-removed solar flux reaches 25 % of its peak value or one standard deviation of the time series outside of the flare. The main effect is to increase the flare duration for most wavelengths; this method is more reliable at wavelengths where the flare contrast is high, however. Finally, we have compared the values of the scaling exponent using both method 1) and 3), and the values usually agree within the error bars (see Figure 8 and and next section).

We checked that no other flares occurred during the integration time and used only line fluences with a signal-to-noise ratio (SNR) higher than two sigma. The SNR $S(E)$ of the fluence E for each flare and line is defined as

$$S(E) = \frac{E}{\int_{t_{\mathrm{st}}}^{t_{\mathrm{end}}} \sigma_{\mathrm{preflare}}\,\mathrm{d}t},$$

where $\sigma_{\mathrm{preflare}}$ is the standard deviation of the time series measured before the flare. We required that $S(E) > 2$ to consider the flare fluence in the statistics. Figure 7 shows the SNR for different lines. In general, only flares higher than M-class can be used. While the fluence

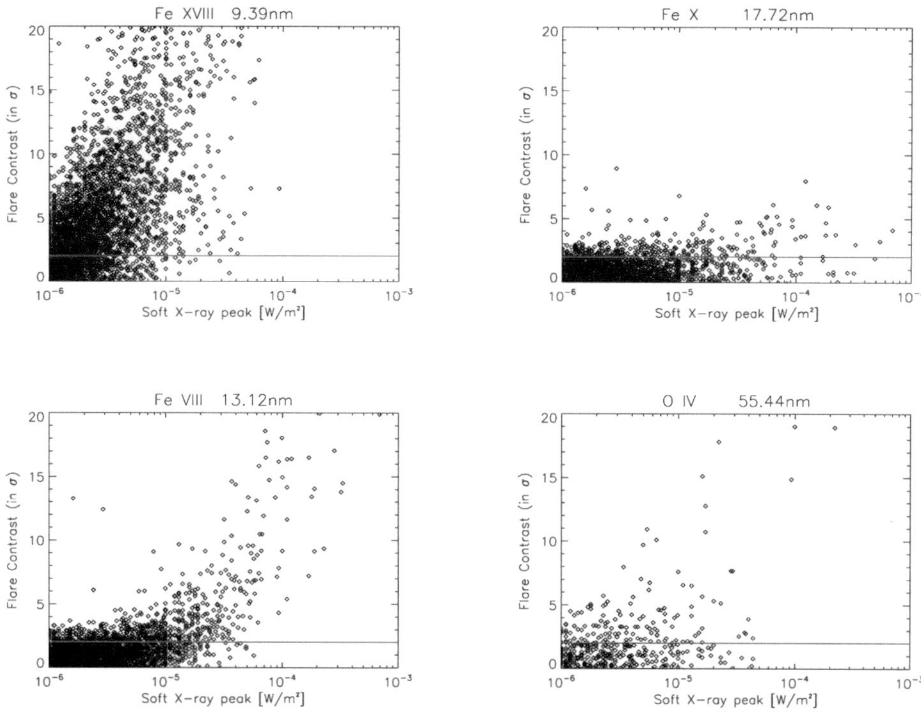

Figure 7 Signal-to-noise ratio of the flare fluence (denoted flare contrast in the figure) for four lines. In each panel, a red line is drawn at the two-sigma level, and only flares above this level were considered.

of the Fe XVIII line at 9.39 nm generally has a good SNR, the Fe X line at 17.22 nm cannot be used at all. The set of lines that we used is shown in Table 1 and is based on the criteria ($S(E) > 2$) and on the requirement that the fluxes computed with fixed and varying spectral limits agree well. We are interested in chromospheric (low-temperature) UV lines, but they usually have a relatively low contrast and mainly show up clearly only for M-flares and higher. To meaningfully compare different lines, all the scalings were made on flares higher than the M1 level in the following, although C-class flares are also shown in the plots.

3. Results and Discussion

Figure 8 shows the scaling of the flare fluence for 16 spectral lines and the determination of the γ exponent, such that $E(\lambda) \simeq E_{\mathrm{SXR}}^{\gamma}$. The black diamonds represent the flare fluence computed using the EVE spectral line product with fixed spectral integration limits, and all flares with $S(E) > 2$ are plotted. Red diamonds are the fluences that we computed by determining for each spectrum the blue and red limits of the integration for a spectral line. Only flares with $S(E) > 2$ and that have an SXR class higher than M1.0 are plotted. In both cases, the fit is only performed on flares higher than the M1 level to avoid taking into account flares with too low contrast that do not appear to belong to the distribution.

Some lines have larger scatter than others; this is the case of the Fe XV line at 28.42 nm and the Fe XVI line at 33.54 nm. Some lines also have fewer valid flares: O IV at 55.44 nm, Si XII at 49.94 nm, both observed by MEGS-B. We note that because several wavelengths

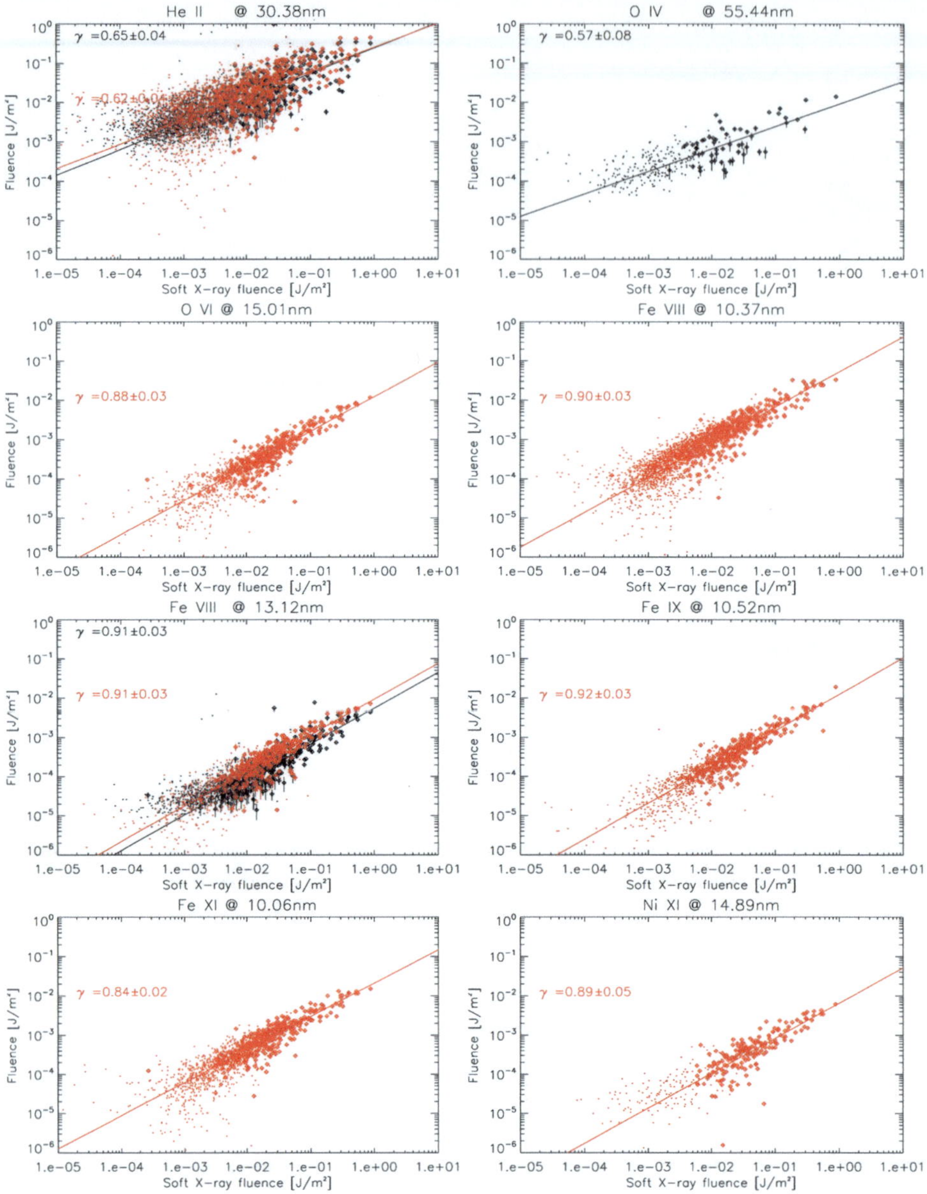

Figure 8 Scaling of the flare fluence for different lines *versus* the SXR GOES flare fluence. Black corresponds to fluences computed with the GOES start- and end-time and with the EVE lines product ("fixed spectral integration"). Red corresponds to fluences computed by automatically determining the line spectral width and the end of the flare. The fits are performed on the flares above the M1 level, which are represented with larger diamonds. For clarity, error bars are drawn only for flares above the M1 level.

are missing in the MEGS B spectrum product, we were unable to validate the fluences computed from the EVE line product for these lines by reproducing them with the spectrum product. The scattering and small number of points for the regression are reflected in the

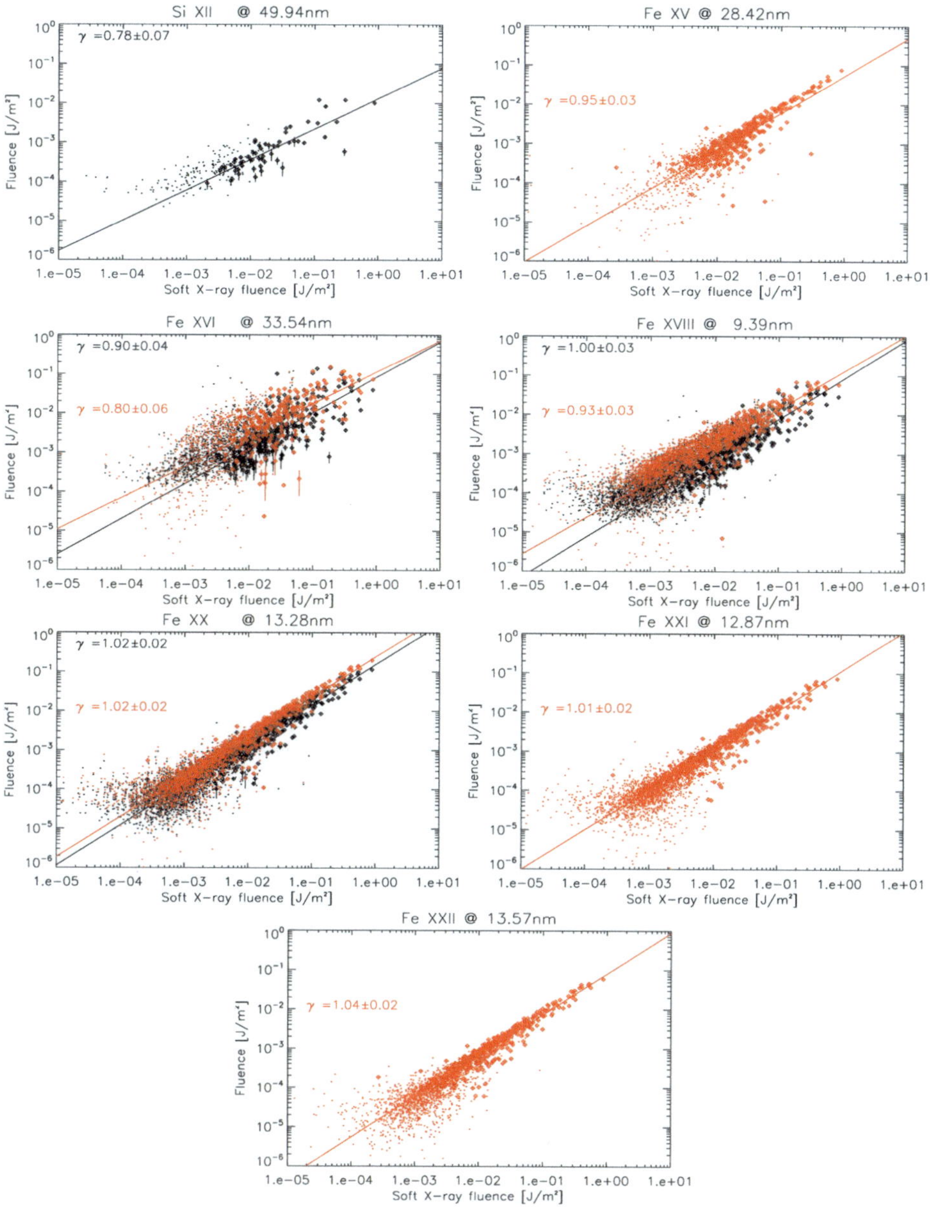

Figure 8 (*Continued*)

1-σ uncertainty of γ, which is then used to compute the uncertainty on α represented in Figure 10. From the top left panel to the bottom panel of Figure 8 there is a weak tendency for the scaling to steepen; in other words, γ slightly increases or the scaling exponent of the PDF α slightly decreases with the formation temperature of the lines.

Figure 9 shows a similar result for the H I Lyman-α line. We used the same methods as for the other lines and applied them to the one-minute observations of the *Extreme Ul-*

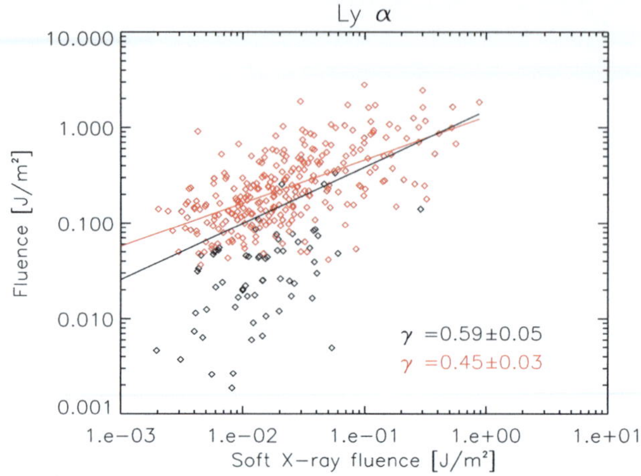

Figure 9 Scaling of the Ly-α fluence with the GOES SXR fluence. The straight lines show the fit of all flares higher than the M1 level (in black) and flares with SNR $> 2\sigma$ (in red).

traviolet Sensor (EUVS)[1] onboard the GOES 15 satellite (Viereck *et al.*, 2007). Channel E of the GOES/EUVS instrument is a broad passband centered around Ly-α. Because of strong degradation, the absolute value of this channel is scaled to the value of a 1 nm bin around Ly-α as observed by SORCE/SOLSTICE. The scaling is made on the basis of a quiet Sun reference spectrum, which may affect the flare fluence. This channel also suffers from geocoronal hydrogen absorption a few hours per day. These effects may partly explain the scatter observed in Figure 9, but the scatter could also reflect the complexity of the formation of the Ly-α line, which is the most intense line of the solar spectrum. More detailed analyses of Ly-α flare profiles are worthwhile, but Figure 9 already shows a reasonable scaling of its fluence with the GOES SXR fluence. The values of γ indicated in the figure lead to a value of α between 2.3 and 2.9.

Figure 10 shows the evolution of the scaling exponent with the formation temperature of the lines (Lyman-α is not represented in this figure). Error bars represent the 1-σ uncertainties on α values and are computed using Equation (3) and the 1-σ uncertainties of γ values coming from the fitting procedure. The O IV line at 55.44 nm ($\text{Log}(T) = 5.5$) and Si XII ($\log(T) = 6.3$) have larger error bars, in agreement with the small number of flares and the large scatter observed in Figure 8. More precise results are obtained as expected for the hottest and more contrasted lines, but also for the He II and the Fe VIII lines ($\log(T) = 5.6$).

Figure 10 shows that the He II and O IV lines, which form at a low temperature, have scaling exponents higher than two and are higher than the other lines. While the result for the O IV lines is more uncertain since we were unable to reproduce it directly from the spectrum, the α value for the He II line is more robust; even if it is higher than the value of 2.16 found for the ESP diode centered around 30.4 nm (see Figure 4), this could be explained by the fact that the ESP 304 bandpass includes more emission from hot material. It also agrees with the value found for Ly-α. Both the He II and H Ly-α lines are strongly optically thick and are emitted from the most abundant elements in the Sun.

The flare fluences computed for lines forming between about $\text{Log}(T) = 5.5$ and $\text{Log}(T) = 6.5$ have similar scaling exponents between 1.9 and 2.1. Lines forming at very high temperature, around 10^7 K, look to have scaling exponents slightly lower, around 1.8. These differences are small, however, and might be explained by uncontrolled biases in

[1] http://ngdc.noaa.gov/stp/satellite/goes/doc/GOES_NOP_EUV_readme.pdf.

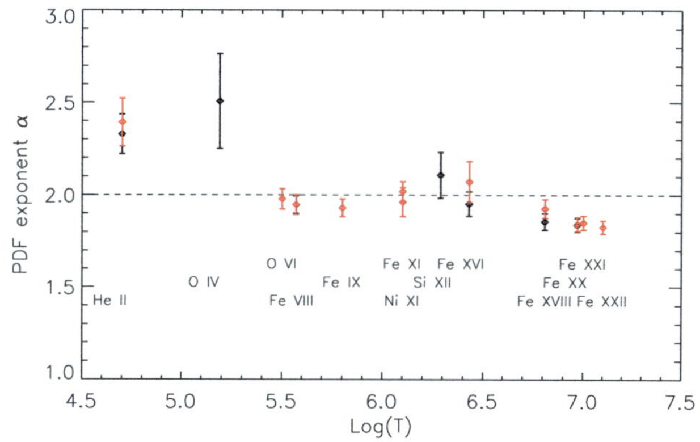

Figure 10 Scaling exponent of the fluence PDF *versus* temperature.

our analysis; in particular, we are forced to recognize that the lines with the lower scaling exponent, which are hotter, also have the higher contrast.

This study suggests, but doest not clearly show, that EUV chromospheric lines might have a higher scaling exponent than the coronal emission. We now discuss possible reasons. The hypothesis that different spectral lines might have different scaling exponents can be understood by geometrical consideration: coronal lines are mostly formed in the hot thermal plasma that is contained in the $V \sim sL$ volume of a loop (with s being the loop footprint area and L the length of the loop), while during flare the increase of chromospheric lines relies on the energy received by the loop footpoint that is first provided by the energetic particles (during the impulsive phase) and later on and for longer by conduction. Even if schematic, these different physical processes can account for the fact that chromospheric lines do not scale linearly with coronal lines. This different geometry was also well shown by Warmuth and Mann (2013a,b), who found from RHESSI observations of hard X-rays emission for 24 flares that the thermal source volume of the flares is well correlated with the SXR GOES class while the nonthermal footpoint areas are not.

The fact that $\alpha_{\text{chromosphere}} > \alpha_{\text{corona}}$ would also mean that when considering smaller flares, in the sense of the X-ray classification, the chromospheric emission becomes increasingly stronger than coronal emission. This can hardly be explained if the chromospheric flare signal depends mostly on the energy provided by conduction from the hot coronal loop. The finding of Warmuth and Mann (2013b) that the nonthermal footpoints area, and therefore one of the parameters governing the energy received by the chromosphere, is not correlated with the soft X-ray class (while thermal volumes are) may explain this behavior. Longcope (2014) recently studied flares for which no HXR signatures are observed and where conduction is thought to be the only energy driver between the corona and the chromosphere. In his model, the flare energy is deposited at the loop apex and is transported from the corona to the chromosphere by conduction alone. Simulating a number of flares with different energies, Longcope found that both the evaporation and condensation velocity scale with the flare energy, but that the condensation velocity has a steeper slope. If we associate the chromospheric emission with the energy brought by condensation, this result contradicts the conclusion suggested here. We also note that there is no evidence in the literature that the flare coronal emission, as observed in SXRs, for example, would scale differently depending on the presence or absence of HXR signatures of the flare.

Finally, if the scaling between flare chromospheric and coronal emissions would hold down to ever smaller flares (of class A and B, for example), a relatively strong increase of chromospheric emission would be expected for these flares, which is not *a priori* observed;

one possible explanation would be that the chromospheric contrast, even in solar images, is too weak to be detected. Another possibility is that the observed scaling in this study is specific to large (*e.g.* higher than C1 level) flares and that the exponent of the power law, or, less probably, the shape of the PDF, changes for smaller flares. This will need to be investigated in the future.

4. Conclusion

We have investigated the scaling of the flare fluence for different lines formed slightly above the chromosphere to the corona. Line fluxes already available in the SDO/EVE data were recomputed, and we also identified other lines of interest for this study. To properly account for weak contrasts, we investigated the probability distribution function of the flare fluence for these lines by examining the scaling of the line flare fluence with the SXR fluence as observed by GOES in the $0.1-0.8$ nm. We found a weak tendency for lines forming at very high temperatures to have a lower exponent (about 1.8) than lines forming between about $\mathrm{Log}(T) = 5.5$ and $\mathrm{Log}(T) = 6.5$ (around 2). The Lyman-α lines of He II and H seem to have a still higher value of the scaling exponent, above 2. However, colder lines also have lower contrast, and despite our efforts, we cannot exclude that some biases are responsible for the described behavior. More work is needed to confirm or reject the hypothesis of different scalings for emissions forming at different layers, where the energy is transferred in different ways.

Acknowledgements The author acknowledges the SDO/EVE and GOES teams (in particular Janet Machol for the Ly-α data) for providing the data, as well as the organizers and attendants of the "Solar and Stellar Flares" meeting in Prague, 2014, for a very successful and interesting meeting. The author also acknowledges the referee for comments that significantly helped to improve the article. Part of this research has been funded by the European Community's Seventh Framework Programme (FP7 2012) under grant agreement no 313188 (SOLID project).

References

Allred, J.C., Hawley, S.L., Abbett, W.P., Carlsson, M.: 2005, Radiative hydrodynamic models of the optical and ultraviolet emission from solar flares. *Astrophys. J.* **630**, 573. DOI. ADS.

Berlicki, A.: 2007, Observations and modeling of line asymmetries in chromospheric flares. In: Heinzel, P., Dorotovič, I., Rutten, R.J. (eds.) *The Physics of Chromospheric Plasmas*, *Astron. Soc. Pac. Conf. Ser.* **368**, 387. ADS.

Berlicki, A., Heinzel, P.: 2004, Soft X-ray heating of the solar chromosphere during the gradual phase of two solar flares. *Astron. Astrophys.* **420**, 319. DOI. ADS.

Crosby, N.B., Aschwanden, M.J., Dennis, B.R.: 1993, Frequency distributions and correlations of solar x-ray flare parameters. *Solar Phys.* **143**, 275. ADS.

Dennis, B.R.: 1985, Solar hard X-ray bursts. *Solar Phys.* **100**, 465. DOI. ADS.

Dere, K.P., Cook, J.W.: 1979, The decay of the 1973 August 9 flare. *Astrophys. J.* **229**, 772. DOI. ADS.

Dere, K.P., Landi, E., Mason, H.E., Fossi, B.C.M., Young, P.R.: 1997, Chianti – An atomic database for emission lines. *Astron. Astrophys.* **125**, 149. ADS.

Fletcher, L., Dennis, B.R., Hudson, H.S., Krucker, S., Phillips, K., Veronig, A., Battaglia, M., Bone, L., Caspi, A., Chen, Q., Gallagher, P., Grigis, P.T., Ji, H., Liu, W., Milligan, R.O., Temmer, M.: 2011, An observational overview of solar flares. *Space Sci. Rev.* **159**, 19. DOI. ADS.

Hannah, I.G., Hudson, H.S., Battaglia, M., Christe, S., Kašparová, J., Krucker, S., Kundu, M.R., Veronig, A.: 2011, Microflares and the statistics of X-ray flares. *Space Sci. Rev.* **159**, 263. DOI. ADS.

Heinzel, P., Avrett, E.H.: 2012, Optical-to-radio continua in solar flares. *Solar Phys.* **277**, 31. DOI. ADS.

Hudson, H.S.: 1991, Solar flares, microflares, nanoflares, and coronal heating. *Solar Phys.* **133**, 357. ADS.

Hudson, H.S.: 2011, Global properties of solar flares. *Space Sci. Rev.* **7**. DOI. ADS.

Kretzschmar, M.: 2011, The Sun as a star: Observations of white-light flares. *Astron. Astrophys.* **530**, A84. DOI. ADS.

Kretzschmar, M., Dudok de Wit, T., Schmutz, W., Mekaoui, S., Hochedez, J., Dewitte, S.: 2010, The effect of flares on total solar irradiance. *Nat. Phys.* **6**, 690. DOI. ADS.

Landi, E., Zanna, G.D., Young, P.R., Dere, K.P., Mason, H.E., Landini, M.: 2006, Chianti-an atomic database for emission lines. vii. New data for x-rays and other improvements. *Astrophys. J.* **162**, 261. DOI. ADS.

Longcope, D.W.: 2014, A simple model of chromospheric evaporation and condensation driven conductively in a solar flare. *Astrophys. J.* **795**, 10. DOI. ADS.

Milligan, R.O., Chamberlin, P.C., Hudson, H.S., Woods, T.N., Mathioudakis, M., Fletcher, L., Kowalski, A.F., Keenan, F.P.: 2012, Observations of enhanced extreme ultraviolet continua during an X-class solar flare using SDO/EVE. *Astrophys. J. Lett.* **748**, L14. DOI. ADS.

Milligan, R.O., Kerr, G.S., Dennis, B.R., Hudson, H.S., Fletcher, L., Allred, J.C., Chamberlin, P.C., Ireland, J., Mathioudakis, M., Keenan, F.P.: 2014, The radiated energy budget of chromospheric plasma in a major solar flare deduced from multi-wavelength observations. *Astrophys. J.* **793**, 70. DOI. ADS.

Schrijver, C.J., Beer, J., Baltensperger, U., Cliver, E.W., Güdel, M., Hudson, H.S., McCracken, K.G., Osten, R.A., Peter, T., Soderblom, D.R., Usoskin, I.G., Wolff, E.W.: 2012, Estimating the frequency of extremely energetic solar events, based on solar, stellar, lunar, and terrestrial records. *J. Geophys. Res.* **117**, 8103. DOI. ADS.

Temmer, M., Veronig, A., Hanslmeier, A., Otruba, W., Messerotti, M.: 2001, Statistical analysis of solar Hα flares. *Astron. Astrophys.* **375**, 1049. DOI. ADS.

Veronig, A., Vršnak, B., Dennis, B.R., Temmer, M., Hanslmeier, A., Magdalenić, J.: 2002a, Investigation of the Neupert effect in solar flares. I. Statistical properties and the evaporation model. *Astron. Astrophys.* **392**, 699. DOI. ADS.

Veronig, A., Temmer, M., Hanslmeier, A., Otruba, W., Messerotti, M.: 2002b, Temporal aspects and frequency distributions of solar soft X-ray flares. *Astron. Astrophys.* **382**, 1070. DOI. ADS.

Veronig, A.M., Rybák, J., Gömöry, P., Berkebile-Stoiser, S., Temmer, M., Otruba, W., Vršnak, B., Pötzi, W., Baumgartner, D.: 2010, Multiwavelength imaging and spectroscopy of chromospheric evaporation in an M-class solar flare. *Astrophys. J.* **719**, 655. DOI. ADS.

Viereck, R., Hanser, F., Wise, J., Guha, S., Jones, A., McMullin, D., Plunket, S., Strickland, D., Evans, S.: 2007, Solar extreme ultraviolet irradiance observations from GOES: Design characteristics and initial performance. In: *Soc. Photo-Optical Eng. (SPIE) Conf. Ser.*, **66890K**.

Warmuth, A., Mann, G.: 2013a, Thermal and nonthermal hard X-ray source sizes in solar flares obtained from RHESSI observations. I. Observations and evaluation of methods. *Astron. Astrophys.* **552**, A86. DOI. ADS.

Warmuth, A., Mann, G.: 2013b, Thermal and nonthermal hard X-ray source sizes in solar flares obtained from RHESSI observations. II. Scaling relations and temporal evolution. *Astron. Astrophys.* **552**, A87. DOI. ADS.

Woods, T.N., Kopp, G., Chamberlin, P.C.: 2006, Contributions of the solar ultraviolet irradiance to the total solar irradiance during large flares. *J. Geophys. Res.* **111**, A10S14. DOI.

Woods, T.N., Eparvier, F.G., Hock, R., Jones, A.R., Woodraska, D., Judge, D., Didkovsky, L., Lean, J., Mariska, J., Warren, H., McMullin, D., Chamberlin, P., Berthiaume, G., Bailey, S., Fuller-Rowell, T., Sojka, J., Tobiska, W.K., Viereck, R.: 2012, Extreme Ultraviolet Variability Experiment (EVE) on the Solar Dynamics Observatory (SDO): Overview of science objectives, instrument design, data products, and model developments. *Solar Phys.* **275**, 115. DOI. ADS.

Reprinted from *Solar Physics* Journal, DOI 10.1007/s11207-015-0688-x
SOLAR AND STELLAR FLARES

Flare Hybrids

M. Tomczak[1] · P. Dubieniecki[1]

Received: 9 December 2014 / Accepted: 13 April 2015 / Published online: 29 April 2015
© The Author(s) 2015. This article is published with open access at Springerlink.com

Abstract On the basis of the *Solar Maximum Mission* observations, Švestka (*Solar Phys.* **121**, 399, 1989) introduced a new class of flares, the so-called *flare hybrids*. When they start, they look like typical compact flares (phase 1), but later on, they look like flares with arcades of magnetic loops (phase 2). We summarize the characteristic features of flare hybrids in soft and hard X-rays as well as in the extreme ultraviolet; these features allow us to distinguish flare hybrids from other flares. In this article, additional energy release or long plasma cooling timescales are suggested as possible causes of phase 2. We estimate the frequency of flare hybrids, and study the magnetic configurations favorable for flare hybrid occurrence. Flare hybrids appear to be quite frequent, and the difference between the lengths of magnetic loops in the two interacting loop systems seem to be a crucial parameter for determining their characteristics.

Keywords Flares · X-Rays · Magnetic fields

1. Introduction

There are no two identical flares, but it is useful to classify flares following some schemes. The most commonly accepted classification was introduced by Pallavicini, Serio, and Vaiana (1977) based on soft X-ray images obtained by the S-054 experiment onboard *Skylab*. The authors proposed two separate classes of events, namely compact flares (class 1) and flares occurring in large and diffuse systems of loops (class 2). They found that the separation is supported by the different values of several physical parameters such as height, volume, energy density, and characteristic times of rise, decay, and duration. They also perceived that flares of class 1 are located very low in active regions and, in contrast to flares of class 2, do not appear to be associated with coronal mass ejections (CMEs) and prominence eruptions or activations.

Solar and Stellar Flares: Observations, Simulations, and Synergies
Guest Editors: Lyndsay Fletcher and Petr Heinzel

✉ M. Tomczak
tomczak@astro.uni.wroc.pl

[1] Astronomical Institute, University of Wrocław, ul. Kopernika 11, 51-622 Wrocław, Poland

The division into two classes opposed to each other by contradiction is called dichotomy, therefore we call the classification of Pallavicini, Serio, and Vaiana (1977) flare dichotomy for short. The flare dichotomy has been supported by several classifications, *e.g.* impulsive *vs.* long-duration flares, single-loop *vs.* arcade flares, confined *vs.* eruptive flares, or two-footpoint *vs.* two-ribbon flares.

Beyond any doubt, the division into two classes is very rough; therefore, it is possible that observed flares can share features of both class 1 and class 2. Švestka (1989) introduced for them the term *flare hybrids*. What does a flare hybrid look like? Its evolution can be divided into two phases: during phase 1 it looks like a flare of class 1, and during phase 2 it looks like a flare of class 2. Recalling a private communication with Cornelius de Jager, Švestka (1989), discussed that a flare of class 1 may trigger a flare of class 2. He also investigated the process that causes the magnetic field to open and thus start a flare of class 2.

Further observations made with many instruments at different wavelengths have provided a more complete picture of flare hybrids. In Section 2 we present characteristic features of flare hybrids in soft X-ray (SXR), hard X-ray (HXR), and extreme-ultraviolet (EUV) ranges. In Section 3 we lay down some rules for the frequency of occurrence. The magnetic configuration suggested for flare hybrids is described in Section 4. In Section 5 we propose the most likely scenario for a flare hybrid.

2. Observations

2.1. Soft X-Rays

In Figure 1 we present an example of the flare hybrid, SOL1992-11-05T06:22 (M2.0), observed by the *Soft X-ray Telescope* (SXT: Tsuneta *et al.*, 1991) onboard the *Yohkoh* satellite. We show the three SXR images made with the AlMg filter during phase 1 (Figure 1a), during phase 2 (Figure 1c), and in the intermediate time (Figure 1b). During phase 1 the SXR emission of the flare is clearly dominated by a small ($h \approx 10^4$ km) system of bright loops. Later on, a higher magnetic arcade ($h \approx 5 \times 10^4$ km) is seen in SXRs during phase 2. In each image we mark the borders of two areas, 1 and 2, within which the SXR signal was measured. We plot the light curves for these areas and the total signal from all the images in Figure 1d. The time gaps in the light curves are mainly caused by the satellite night. The light curves for areas 1 and 2 are different, but together they have a double-peaked shape. The same double-peaked light curve was recorded by the GOES satellite (Figure 1e), where the time interval of the satellite night is also marked.

However, a double-peaked GOES light curve cannot be considered as a typical signature of a flare hybrid. Another example of a flare hybrid, SOL1992-09-09T18:03 (M1.9), observed by the SXT, is shown in Figure 2. The panels in the figure are organized in the same way as in Figure 1. The difference is due to the choice of SXT filter, Al12. The evolution of the flare presented in Figure 2 is very similar to that presented in Figure 1. During phase 1 the SXR emission of the flare is dominated by a smaller area 1 around a system of lower loops, and during phase 2 an emission from a larger area 2 around a higher magnetic arcade dominates. The light curves for areas 1 and 2 (Figure 2d) have their maxima shifted in time as in Figure 1d, but this time they only compose a one-peak light curve, as also seen in the GOES record (Figure 2e).

We investigated nine flare hybrids that were well observed by *Yohkoh* (Table 1). In each case the evolution in the SXT images was very similar, namely during phase 1 the emission was concentrated in a system of rather small loops, and during phase 2 the emission came

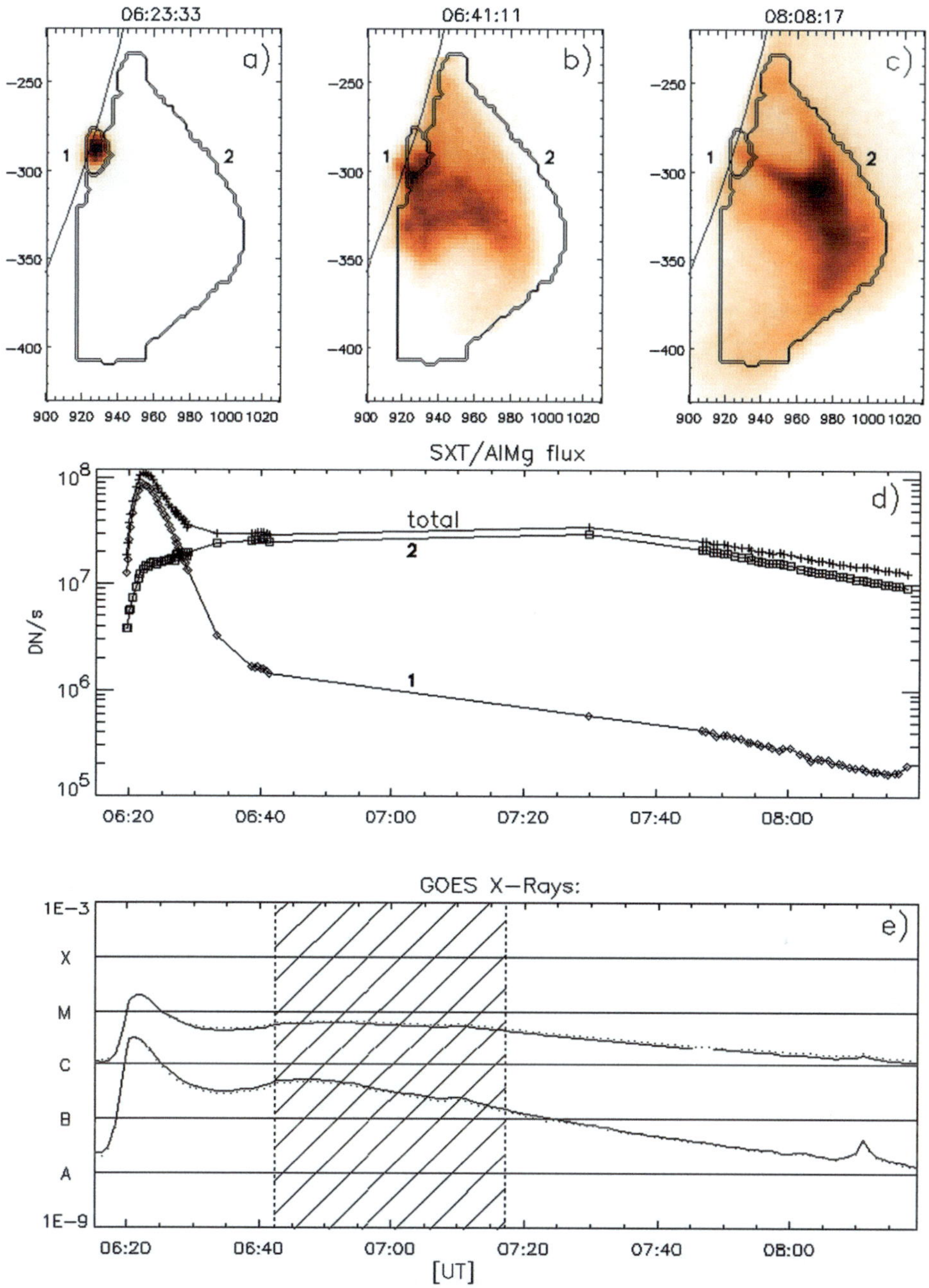

Figure 1 a) – c) The SXT/AlMg images of the flare hybrid SOL1992-11-05T06:22 (M2.0). The intensity scale is reversed. The solid line shows the solar limb, the double solid lines show the border of areas 1 and 2 which brightened in phase 1 and 2, respectively. d) The SXT/AlMg light curves for areas 1 (diamonds), 2 (squares), and the total signal (crosses). e) The GOES light curves (upper curve – 1 – 8 Å range, lower curve – 0.5 – 4 Å range). The hatched areas show the *Yohkoh* satellite nights.

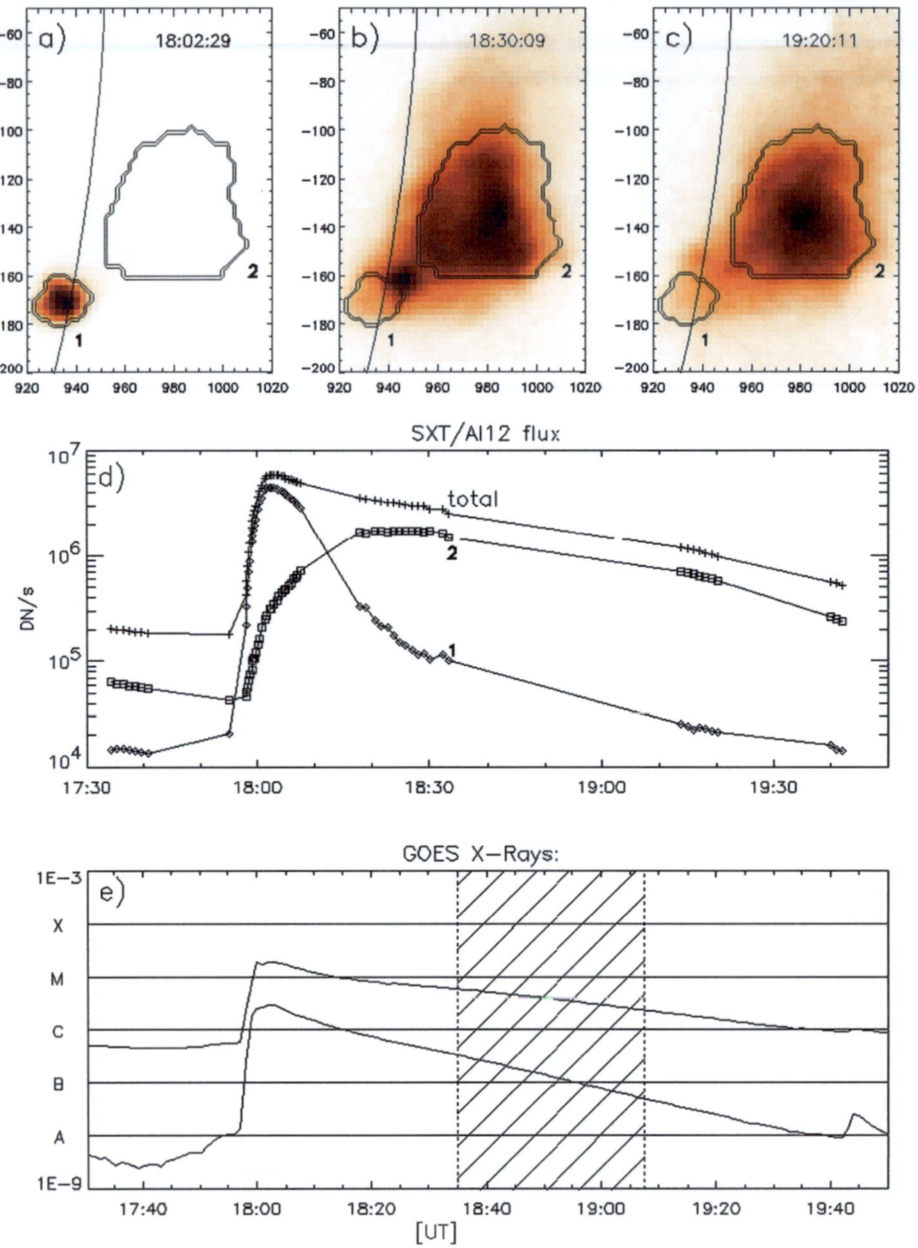

Figure 2 a) – c) The SXT/Al12 images of the flare hybrid SOL1992-09-09T18:03 (M1.9). d) The SXT/Al12 light curves for areas 1, 2, and the total signal. e) The GOES light curves. For detailed explanations, see caption to Figure 1.

from a larger arcade of loops. However, only the flare hybrid from Figure 1 had a double-peaked GOES light curve, while for the other events, GOES recorded light curves with a single peak. Thus, we conclude that an intrinsic feature of a flare hybrid GOES light curve

Table 1 List of investigated flare hybrids.

No.	Date	Max. time [UT]	GOES class	Coordinates	NOAA AR
1	30-Jan-92	17:15	M1.6	S12 E84	7042
2	8-Jul-92	09:50	X1.2	S11 E46	7220
3	11-Aug-92	22:28	M1.4	N16 E90+	7260
4	21-Aug-92	11:10	M1.0	N14 W40	7260
5	6-Sep-92	09:07	M3.3	S11 W38	7270
6	9-Sep-92	18:03	M1.9	S11 W78	7270
7	5-Nov-92	06:22	M2.0	S18 W90+	7323
8	9-Oct-93	08:11	M1.1	N11 W78	7590
9	22-Sep-97	14:16	C4.7	S28 E43	8088

is a rather strong asymmetry consisting of by a short rise typical for flares of class 1 followed by a slow decay typical for flares of class 2. For events 1 to 6, 8, and 9 from Table 1 we found that the rise phase was 5 to 20 times shorter than the decay phase.

We used SXT data to calculate values of some parameters averaged over two systems of loops that form the investigated flare hybrids. For this aim we used the filter ratio method (Hara *et al.*, 1992) employing the Be119 and Al12 image pairs as the first choice or the Al12 and Al.1 image pairs when the Be119 images were not available. In this way, we obtained a set of values characterizing the evolution of the temperature, T, and emission measure, EM. We estimated the total volume, V, of the two components, hence values of the electron density number, $N_{\mathrm{e}} = \sqrt{EM/V}$, became available. Next, the total thermal energy, $E_{\mathrm{th}} = 3N_{\mathrm{e}}k_{\mathrm{B}}TV$ was calculated, where k_{B} is the Boltzmann constant. Finally, the heating rate *per* unit volume, $E_{\mathrm{H}} = (\mathrm{d}E_{\mathrm{th}}/\mathrm{d}t) + E_{\mathrm{C}} + E_{\mathrm{R}}$, was calculated, where E_{C} are the conductive losses and E_{R} are the radiative losses.

In Table 2 the highest values for these parameters are presented. Moreover, the Be119 light-curve parameters are summarized in Columns 2 to 4, where the time of maximum, the strongest signal, and its full width half maximum (FWHM) are given, respectively. We estimate that the typical values of the relative errors are lower than $2-3$ % for the intensity, temperature, and emission measure. For the volume and the related parameters the relative errors are definitely larger, about 15 to 25 %. In each case, large amounts of energy were released in the two flare hybrid components. The relations between parameters characterizing phase 1 and phase 2 of particular flare hybrids can be different, but some trends are seen. Phase 1 always occurs in a smaller magnetic structure than phase 2 and lasts not as long. The smaller volume involved in phase 1 leads to its larger electron number density, higher heating rate, and lower total thermal energy than phase 2. Compare the mean values and their standard deviations given in the bottom rows of Table 2.

Figures 1a–c and 2a–c show that the areas labeled 1 are situated near footpoints of arcades 2. Moreover, the light curves in Figures 1d and 2d show that the signal started to rise in arcades 2 at the beginning of phase 1, not seen in Figures 1a and 2a, which are scaled to the brightest pixel. These facts strongly support the scenario that the reported flares were not coincidences of two independent events; one of class 1, another of class 2, but were instead the result of an interaction between the loop systems 1 and 2. We have observed similar behaviors for other investigated flare hybrids.

A fundamental question arises. Is phase 2 caused by an additional energy release in the arcade, or is it an effect of the long plasma cooling time within the larger structure? There is no doubt that magnetic reconnection between loop systems 1 and 2 can heat both

Table 2 Highest values of some parameters obtained for events from Table 1.

No./phase	Max. time [UT]	I_{max} [10^6 DN s^{-1}]	$\Delta t_{1/2}$ [min]	V [10^{28} cm^3]	T [MK]	EM [10^{49} cm^{-3}]	N_e [10^{10} cm^{-3}]	E_{th} [10^{30} ergs]	E_H [ergs cm^{-3} s^{-1}]
1/1	17:14	1.2	15.5	0.4	10.1	1.6	6.0	1.0	1.1
1/2	17:20	1.1	25.2	10.4	10.2	N/A	1.1	4.6	0.3
2/1	09:49.5	14.3	5.3	0.3	13.5	13.0	20.5	3.2	14.0
2/2	10:04	2.0	19.0	10.9	16.4	3.2	1.8	10.8	1.2
3/1	22:27.5	1.4	3.5	0.15	12.5	1.4	9.6	0.6	5.8
3/2	22:34.3	1.4	17.7	5.5	13.4	1.8	1.8	4.2	1.1
4/1	11:06	0.5	11.5	0.3	9.5	0.6	4.9	0.6	0.8
4/2	11:14.5	1.0	32.3	5.3	12.3	1.1	1.6	3.5	0.2
5/1	09:06.5	3.8	5.3	0.3	13.3	4.5	13.0	1.5	10.2
5/2	09:14	1.0	16.3	3.6	13.2	1.4	2.2	3.5	0.9
6/1	18:02.3	2.7	8.2	0.7	11.0	3.2	6.8	2.0	1.9
6/2	18:21	2.6	56.5	5.3	9.5	1.8	0.6	9.6	0.05
7/1	06:21.5	2.8	4.5	0.45	12.3	3.3	8.5	1.6	4.8
7/2	06:41	1.4	91.0	5.0	9.2	N/A	0.5	7.4	0.07
8/1	08:11.7	1.4	5.8	0.45	12.3	1.7	6.2	1.2	3.6
8/2	08:19	0.6	19.8	12.8	11.6	0.8	0.8	4.1	0.25
9/1	14:16.3	0.5	5.7	0.15	10.8	0.6	6.4	0.4	2.8
9/2	14:20	0.4	20.5	3.8	11.5	0.5	1.2	2.0	0.55
phase 1 (mean)		3.2	7.3	0.36	11.7	3.3	9.1	1.3	5.0
phase 1 (st. dev.)		4.3	3.9	0.17	1.4	3.9	4.9	0.9	4.4
phase 2 (mean)		1.3	33	7.0	11.9	1.5	1.3	5.5	0.5
phase 2 (st. dev.)		0.7	25	3.4	2.3	0.9	0.6	3.0	0.4

systems. The occurrence of magnetic of reconnection is supported by intense HXR emission occurring during phase 1, when SXR emission strongly rises in both systems. There is no doubt either that conductive and radiative losses are higher in system 1 than in system 2 as a result of smaller sizes and higher electron number densities, respectively (see Table 2). The higher energy losses should justify the shorter evolution timescales for system 1 in comparison with system 2, $\tau_1 \ll \tau_2$. However, the real evolution of flare loops is a complex interplay between heating and cooling processes.

Jakimiec *et al.* (1992) introduced the density–temperature ($N_e - T$) diagram as a very useful diagnostic tool to determine heating process in a single flaring loop based on SXR observations. They showed that flare evolutionary paths during the decay phase in this diagram strongly depend on the duration of the energy release. When the heating is switched off abruptly, a cooling due to conductive and radiative losses quickly decreases the temperature and causes a steep slope of the path of ~ 2. When the decay of the heating rate is rather slow, the cooling is much slower and the slope of the path is ~ 0.5.

We investigated evolutionary paths of the analyzed flare hybrids on the $\sqrt{EM} - T$ diagram. When we built the paths with the data from the whole flare area, the paths were more complex than those obtained for simple hydrodynamic flare models (Jakimiec *et al.*, 1992). Sylwester *et al.* (1993) interpreted the complex evolutionary paths as a consequence of involving a set of distinct loops within the same flaring structure. The separate evolutionary paths on the $\sqrt{EM} - T$ diagram of the two components of flare hybrids resemble a path modeled hydrodynamically for a single loop. Moreover, the decay slopes for components 1 and 2 usually are very similar, suggesting a slow decay of the heating rate. We stress that the same phases in the evolutionary paths are shifted in time, for example, when we observe signatures of a prolonged energy release in system 2 (a slope of the path ~ 0.5), the evolutionary path for system 1 is completed.

Unfortunately, phase 2 in the investigated flare hybrids lasted long enough to be interrupted by the satellite night. Further observations made during the next orbit in the late decay phase of flares suffered from a lack of Be119 images, which prevented a good temperature and emission measure diagnostics. For these reasons we cannot be sure that phase 2 of the investigated flare hybrids is caused by an additional energy release in system 2. Moreover, the available images do not allow us to identify the place of an additional reconnection.

2.2. Hard X-Rays

The HXR light curves in four energy channels for the flare in Figure 1 (5 November 1992) are given in Figure 3. They were recorded by the *Hard X-ray Telescope* (HXT: Kosugi *et al.*, 1991) onboard the *Yohkoh* satellite. In all the channels a sequence of almost equally spaced pulses, $P \approx 13$ s, is seen. At lower energies (below 33 keV), this sequence lasts for about three minutes, between 06:19 and 06:22 UT, while at higher energies (above 33 keV), the pulses can only be detected above the background between 06:19 and 06:20 UT because of lower count statistics. Similar pulsations, called quasi-periodic pulsations (QPP), are observed in many solar flares, and it is commonly accepted that they are caused by MHD oscillations excited in flaring magnetic structures (Nakariakov and Melnikov, 2009, and references therein).

The pulses are modulated by an additional gradual component, which is well observed below 33 keV and absent above 53 keV. It consists of two broad enhancements lasting at least two minutes each, the maxima of which are separated in time by about 90 s (06:19:50 and 06:21:20 UT, respectively). The hardness ratio of a net signal above the background for two successive channels proves that the photon energy spectra of pulses are harder than the spectra of the gradual component.

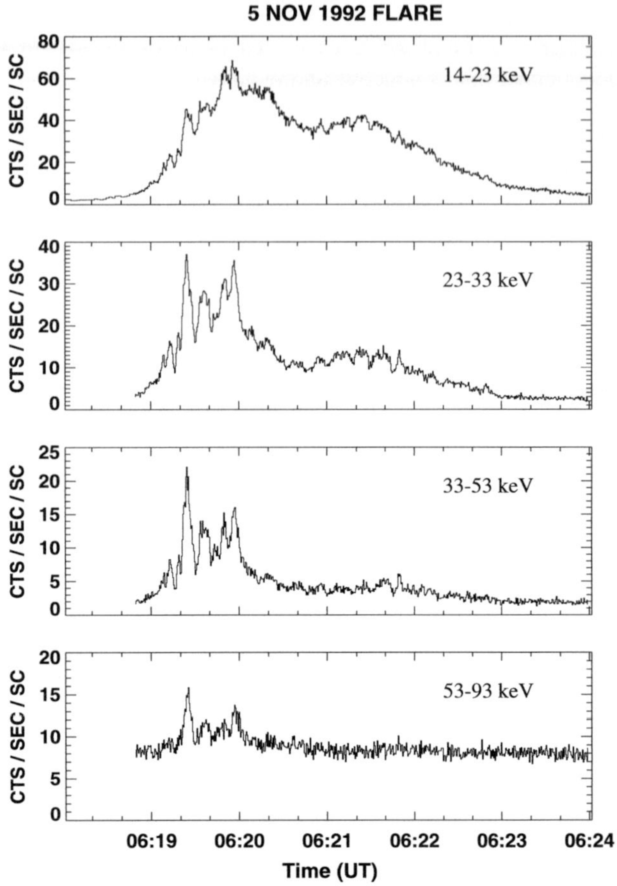

Figure 3 The *Yohkoh* hard X-ray (HXR) light curves in four energy ranges. The number of counts are averaged *per* second and *per* subcollimator.

Recently, Tomczak and Szaforz (2014) analyzed the similar solar flare of 2 October 2001 (SOL2001-10-02T17:31, C4.7), in which more energetic pulses with a period $P_1 = 26-31$ s were modulated by three more gradual enhancements with a period $P_2 = 110$ s. The authors found that these periods were excited simultaneously in a flare hybrid by an interaction between a system of small loops and a high arcade of loops. They also proved that the shorter period was excited in the small loops and the longer period in the arcade.

The observations available for the flare of 5 November 1992 do not allow us to spatially separate the two periods, but the close similarity of the overall magnetic configuration with the flare of 2 October 2001 suggests an interaction between small loops and a high arcade excited simultaneously by MHD oscillations in both magnetic structures. We would like to stress that in some articles reporting the presence of QPP with two distinctly different periods excited simultaneously in HXRs, an interaction between two magnetic structures of different sizes is always mentioned, see *e.g.* Asai *et al.* (2001), and Nakariakov *et al.* (2006).

We investigated HXR light curves of all the flare hybrids from Table 1 looking for similarities with the flare of 5 November 1992, *i.e.* showing two distinctly different periods excited simultaneously. Unfortunately, some light curves were not complete due to the passage of the satellite through the South Atlantic Anomaly, therefore we incorporated the available light curves derived by the *Burst and Transient Source Experiment* (BATSE: Fishman *et al.*, 1992) on board the *Compton Gamma Ray Observatory*. In summary, we confirm the presence of QPPs for seven flare hybrids, but only for three events two distinctly different periods were excited simultaneously. Apart from the flare of 5 November 1992, these were flares of 8 July 1992, 21 August 1992, and 9 September 1992. Four events from nine are for sure too few to claim that QPPs with two distinctly different periods excited simultaneously are intrinsic features of flare hybrids in HXRs. Nevertheless, this characteristic HXR pattern suggests an influence of the magnetic configuration of flare hybrids, in which two interacting magnetic structures have distinctly different sizes.

2.3. Extreme-Ultraviolet

Taking advantage of the new instruments the *Atmospheric Imaging Assembly* (AIA: Lemen *et al.*, 2012), and the *Extreme-ultraviolet Variability Experiment* (EVE: Woods *et al.*, 2012) onboard the NASA *Solar Dynamics Observatory* (SDO), Woods *et al.* (2011) introduced a new class of flares, called EUV late-phase (ELP) flares. Their defining feature is an additional peak of the coronal emission seen in the spectral line of Fe XVI 335 Å occurring half an hour to two hours after the GOES SXR peak. This line is an indicator of plasma with a temperature of about 3 MK. For other EUV spectral lines use for detecting the warm plasma, *e.g.* Fe XVIII 94 Å ($\sim$6 MK) or Fe XV 284 Å ($\sim$2 MK), the late peaks also occur (Sun *et al.*, 2013). Woods (2014) specified other features of ELP flares: (1) no significant counterpart of the late peak in the hot plasma (the GOES SXR or Fe XX/Fe XXIII 133 Å), (2) an eruption followed by a coronal dimming seen in the cooler plasma, *e.g.* Fe IX 171 Å, preceding the late peak, and (3) a different system of longer loops visible above the place where the peak of the hot plasma was emitted.

These features of ELP flares strongly resemble those presented in Section 2.1 for flare hybrids observed in SXRs. The evolution of ELP flares consists of the two following phases: the first, when the emission is concentrated in a small system of loops (system 1), and the second, when an additional system of longer loops (system 2) occurs and its emission dominates. Moreover, an interaction between the two systems undoubtedly exists. Therefore we consider the ELP flares simply as a new face of flare hybrids that we can investigate thanks to the new EUV instruments onboard the SDO. A broad wavelength range and high temporal, spatial, and spectral resolutions allow us to measure changes in many EUV spectral lines that are indicators of plasma in a wide range of temperatures. This gives us the opportunity to decide whether additional energy release or a long timescale of plasma cooling is responsible for the prolonged evolution.

In several case studies of ELP flares (Hock *et al.*, 2012; Liu *et al.*, 2013; Dai, Ding, and Guo, 2013; Sun *et al.*, 2013), the authors investigated sequences of AIA light curves ordered with decreasing temperature of the filters from Fe XX/Fe XXIII 133 Å ($\sim$10 MK) to Fe IX 171 Å ($\sim$0.7 MK). The signal was formed by selected fragments of ELP flares. The authors agree that during phase 1 in system 1 of the investigated flares the maximum is shifted in time, first occurring for the hot plasma, later on for the warm plasma, and finally for the cold plasma. This appearance is interpreted as a consequence of plasma cooling due to radiative and conductive losses. Liu *et al.* (2013) reported a similar delay during phase 2 in system 2. They obtained timescales definitely longer than for phase 1 as a result of fewer conductive (larger loops) and radiative (lower electron density number) losses. Sun *et al.* (2013) reported similar results, but only for the hot and warm plasma. The evolution of the cold plasma was more complex.

On the other hand, Dai, Ding, and Guo (2013) obtained a more complicated picture, in which each light curve for system 2 during phase 2 shows several maxima. They interpreted this observations as a proof of several episodes of energy release in this system. Hock *et al.*

(2012) noted a lack of hot plasma in system 2 during phases 1 and 2 and interpreted this as a proof of additional energy release in system 2 causing only a modest increase of the temperature. The same interpretation can be used to explain the double maxima visible in light curves of cold filters in the flare analyzed by Sun *et al.* (2013).

Recently, Li *et al.* (2014) have published additional arguments supporting a twofold explanation of phase 2 in ELP flares. They modeled the EUV emission from sets of loops having different lengths and different heating rates using the enthalpy-based thermal evolution of loops (EBTEL) model (Klimchuk, Patsourakos, and Cargill, 2008). They found that two separate maxima in the Fe XVI 335 Å can be modeled by simultaneous heating during phase 1 in two distinct loops of different lengths and by repeating the heating during phase 2 in system 2. Li *et al.* (2014) pointed out the importance of the AIA UV 1600 Å channel to distinguish between cooling and heating for the ELP flares. Some contribution to the emission in this channel comes from the C IV line formed in the upper chromosphere. Therefore the AIA UV 1600 Å channel is very sensitive to flare energy release. Indeed, the light curves presented by Li *et al.* (2014) support the interpretation of additional energy release for the flares investigated by Dai, Ding, and Guo (2013) and Sun *et al.* (2013) as well as a lack of energy release in phase 2 of flares investigated by Liu *et al.* (2013).

3. How Common Are Flare Hybrids?

We have investigated the occurrence of flare hybrids between September 1991 and April 1999. For this aim we analyzed SXR GOES light curves and qualified as flare hybrids (1) those showing a double maximum, under the condition that both maxima occurred in the same active region, or (2) those having a strongly asymmetric light curve, *i.e.* a fast rise and a slow decay. If possible, doubtful events were verified on the basis of SXT images. In summary, we identified 577 flare hybrids. Altogether, 15178 flares occurred in the investigated time interval. This translates into a 3.8 % contribution of flare hybrids.

Figure 4 presents the number of flare hybrids for each trimester in the investigated time interval. The total number of all flares is also given for comparison. The number of flare hybrids roughly mimics the solar cycle phase. However, the contribution of this class in the full population was the highest around the activity minimum (1995 – 1997) ≈9 – 13 %, whereas during enhanced activity (1991, 1993, 1998 – 1999) it decreased to ≈2 – 4 %.

Recently, a more comprehensive statistical research of ELP flares has been reported by Woods (2014). He investigated SXR GOES light curves from 1974 to 2013 and searched

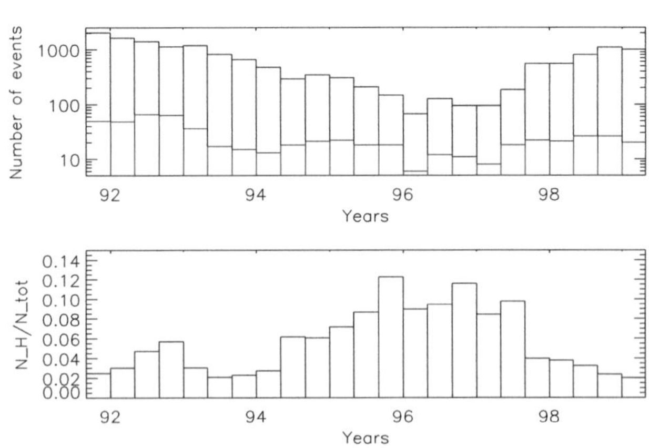

Figure 4 a) Number of flare hybrids (red bins) and all flares (black bins) that occurred between September 1991 and April 1999. The size of a bin is four months. b) Ratio of flare hybrids for each trimester bin.

for a dual-decay behavior, *i.e.* a steep slope followed by a moderate slope. He argued that the first slope represents cooling of shorter loops (system 1), whereas the second slope represents the cooling of longer loops (system 2). Woods (2014) obtained that the contribution of flares showing dual-decay is 7.9 %, in particular, from 2010 to 2013 it was 10.5 %. He found that the contribution of ELP flares is the highest (20 – 30 %) around a solar-cycle minimum. He also found that the higher the flare class, the higher the contribution of the dual-decay flares. The very important work of Woods (2014) aimed at validating the results obtained with SDO/EVE and AIA data. He reported that 36 % and 57 % of flares showing double-decay from 2010 to 2011 and from 2011 to 2013, respectively, do not show features that would allow us to classify them as ELP flares.

In spite of fundamental differences between the methodologies used by Woods and us, there are also some similarities. The most striking is that in both studies the highest frequency occurs during low solar activity. The reason for this seems to be rather trivial. During low activity, flares are not as frequent, thus their GOES light curves do not overlap, and every selection criterion works perfectly. Keeping in mind that during higher activity the probability of occurrence of active regions with complex magnetic structure with loops of different lengths is even higher, the contribution of 9 – 13 % can be treated as a lower limit of flare hybrids. It is interesting that the frequency values of 20 – 30 % given by Woods (2014) for the low activity phase, after adopting his 43 %-validation obtained for events from 2011 to 2013, decrease to 9 – 13 %.

Woods *et al.* (2011) noted that ELP flares show a tendency of occurring within the same active regions. For example, their Table 2 shows that of the 22 ELP flares, six events occurred in NOAA AR 11069 and another six in NOAA AR 11121. We did not complete the list of flare hybrids including the active region identification, but during the investigation of some active regions within which a flare hybrid occurred, we found that the same magnetic configuration was flaring several times. For example, the flare hybrid described by Tomczak and Szaforz (2014) was preceded by three other flare hybrids that occurred in the same active region NOAA AR 9628.

4. Magnetic Configuration

All the available observations of flare hybrids strongly suggest the existence of two sets of magnetically related loop systems. This means a multipolar magnetic configuration, in which magnetic reconnection plays a crucial role in energy release and in shaping a new configuration that becomes more potential. In previous studies the following particular configurations have been proposed: a classical quadrupolar topology based on breakout reconnection (Hock *et al.*, 2012), an asymmetric quadrupolar topology with a sigmoidal core (Liu *et al.*, 2013), a multistep reconnection in a multipolar topology (Dai, Ding, and Guo, 2013), and a fan-spine topology (Sun *et al.*, 2013).

We usually conclude about the reconnection indirectly by observing the emission of non-thermal electrons in HXRs and the emission of the multithermal plasma in SXRs and EUV. Sometimes eruptions can be treated as a signature of the reconnection, and expanding loops can even initiate a following reconnection with a higher magnetic system (Su *et al.*, 2012; Dai, Ding, and Guo, 2013). Occasionally, it is possible to identify the loops that are the product of reconnection, *e.g.* see Figure 4 in Sun *et al.* (2013). Tomczak (2013) reported a unique flare hybrid (SOL2001-10-02T17:31, C4.7), in which the reconnection occurred between a new emerging flux and an overlying coronal field. In SXR images recorded by the SXT the whole process is seen very clearly, starting from a fast expansion of emerging loops and their evident deformation in the vicinity of the reconnection site, followed by vigorous plasma motions inside the reconfiguring loops and the formation of a new system of loops.

The emerging flux model (Heyvaerts, Priest, and Rust, 1977), in which subphotospheric magnetic fields emerge as a result of buoyancy within an already existing active region and meet overlying coronal magnetic fields easily explains the main characteristics of flare hybrids. Continuous emergence of a new flux under a stable magnetic environment can produce homologous flares occurring in the same location with similar morphologies. The examples mentioned in Section 3 are most likely the illustration of this process, explaining the tendency of flare hybrids to occur in the same active region.

5. Conclusions and Prospects

Our intention is to recall the forgotten term introduced by the late Zdeněk Švestka many years ago. His experience and intuition suggested the importance of flare hybrids that show a complex evolution in which their appearance changes completely. Events like these warn us against general conclusions drawn from a limited set of observations. The conclusion can be incorrect even though it agrees with the available data. A closer insight sometimes requires a detailed study of the evolution of the active region in which the investigated event occurred.

For the past 25 years, an abundance of new data has been obtained by successive solar satellites. On this basis, the following typical observational features of flare hybrids are easily recognized: (1) separate systems of loops seen in EUV and SXR images, (2) double-peaked or strongly asymmetrical light curves in these wavelengths, (3) multiperiodicity of pulses recorded in HXRs. Now it is possible to give comprehensive answers to the general questions asked by Švestka (1989).

The defining condition for the occurrence of a flare hybrid seems to be the reconnection between two systems of magnetic loops, a system 1 and a system 2, one shorter, the other longer. This means that the condition necessary for the occurrence is a multipolar magnetic configuration. The process can be initiated, for example, by a new magnetic flux (system 1) emerging within the already existing active region (system 2). The reconnection triggers energy release and the chromospheric evaporation, which fills both systems with plasma and initiates intense SXR and EUV emissions. System 1 is quickly cooled by strong radiative and conductive losses, completing phase 1. Further evolution of flare hybrids (phase 2) is connected with the evolution of system 2. The prolonged SXR and EUV emissions in this system might be the effect of the long timescale of plasma cooling due to fewer radiative and conductive losses. However, in some events proofs of additional energy release in system 2 are also given. A new reconnection site can be somehow connected with eruptions observed in phase 1. The amount of energy released in system 2 during phase 2 establishes the highest temperature of the plasma and in this way, the accessibility of observations in different SXR and EUV filters.

It is easier to recognize typical features of flare hybrids when the differences between the lengths of the interacting systems of loops are greater. However, even very different systems do not always cause the clear features of these events. The frequency estimates of flare hybrids presented in Section 3 should therefore be treated as a lower limit of the actual value.

Flares are important for space weather because they produce photons that are energetic enough to quickly enhance the ionization in Earth's upper atmosphere. Integration of emitted flux over wavelengths shows the importance of the EUV flux. The prolonged EUV emission makes the flare hybrids extremely geoeffective through its prolonged impact on the Earth's atmosphere. For this reason, new methods introducing an early warning of an occurrence of these events would be useful. Future complex investigation including X-ray and EUV observations should verify, for example, the usefulness of the multiperiodicity in HXRs for predicting the EUV late maximum.

Acknowledgements The *Yohkoh* satellite is a project of the Institute of Space and Astronautical Science of Japan. We are very thankful to the referee for important comments that helped to improve this article. We acknowledge financial support from the Polish National Science Centre grant 2011/03/B/ST9/00104.

Open Access This article is distributed under the terms of the Creative Commons Attribution 4.0 International License (http://creativecommons.org/licenses/by/4.0/), which permits unrestricted use, distribution, and reproduction in any medium, provided you give appropriate credit to the original author(s) and the source, provide a link to the Creative Commons license, and indicate if changes were made.

References

Asai, A., Shimojo, M., Isobe, H., Morimoto, M., Yokoyama, T., Shibasaki, K., *et al.*: 2001, *Astrophys. J.* **562**, L103. DOI.

Dai, Y., Ding, M.D., Guo, Y.: 2013, *Astrophys. J.* **773**, L21. DOI.

Fishman, G.J., Meegan, C.A., Wilson, R.B., Paciesas, W.S., Pendleton, G.N.: 1992, In: Shrader, C.R., Gehrels, N., Dennis, B. (eds.) *The Compton Observatory Science Workshop*, *NASA CP* **3137**, 26.

Hara, H., Tsuneta, S., Lemen, J.R., Acton, L.W., McTiernan, J.M.: 1992, *Publ. Astron. Soc. Japan* **44**, L135.

Heyvaerts, J., Priest, E.R., Rust, D.M.: 1977, *Astrophys. J.* **216**, 123. DOI.

Hock, R.A., Woods, T.N., Klimchuk, J.A., Eparvier, F.G., Jones, A.R.: 2012. arXiv.

Jakimiec, J., Sylwester, B., Sylwester, J., Serio, S., Peres, G., Reale, F.: 1992, *Astron. Astrophys.* **253**, 269.

Klimchuk, J.A., Patsourakos, S., Cargill, P.J.: 2008, *Astrophys. J.* **682**, 1351. DOI.

Kosugi, T., Makishima, K., Murakami, T., Sakao, T., Dotani, T., Inda, M., *et al.*: 1991, *Solar Phys.* **136**, 17. DOI.

Lemen, J.R., Title, A.M., Akin, D.J., Boerner, P.F., Chou, C., Drake, J.F., *et al.*: 2012, *Solar Phys.* **275**, 17. DOI.

Li, Y., Ding, M.D., Guo, Y., Dai, Y.: 2014, *Astrophys. J.* **793**, 85. DOI.

Liu, K., Zhang, J., Wang, Y., Cheng, X.: 2013, *Astrophys. J.* **768**, 150. DOI.

Nakariakov, V.M., Melnikov, V.F.: 2009, *Space Sci. Rev.* **149**, 119. DOI.

Nakariakov, V.M., Foullon, C., Verwichte, E., Young, N.P.: 2006, *Astron. Astrophys.* **452**, 343. DOI.

Pallavicini, R., Serio, S., Vaiana, G.: 1977, *Astrophys. J.* **216**, 108. DOI.

Švestka, Z.: 1989, *Solar Phys.* **121**, 399. DOI.

Su, Y., Dennis, B.R., Holman, G.D., Wang, T., Chamberlin, P.C., Savage, S., *et al.*: 2012, *Astrophys. J.* **746**, L5. DOI.

Sun, X., Hoeksema, J.T., Liu, Y., Aulanier, G., Su, Y., Hannah, I.G., *et al.*: 2013, *Astrophys. J.* **778**, 139. DOI.

Sylwester, B., Sylwester, J., Serio, S., Reale, F., Bentley, R.D., Fludra, A.: 1993, *Astron. Astrophys.* **267**, 586.

Tomczak, M.: 2013, *Cent. Eur. Astrophys. Bull.* **37**, 585.

Tomczak, M., Szaforz, Ż: 2014, *Cent. Eur. Astrophys. Bull.* **38**, 111.

Tsuneta, S., Acton, L., Bruner, M., Lemen, J., Brown, W., Caravalho, R., *et al.*: 1991, *Solar Phys.* **136**, 37. DOI.

Woods, T.N.: 2014, *Solar Phys.* **289**, 3391. DOI.

Woods, T.N., Hock, R., Eparvier, F., Jones, A.R., Chamberlin, P.C., Klimchuk, J.A., *et al.*: 2011, *Astrophys. J.* **739**, 59. DOI.

Woods, T.N., Eparvier, F.G., Hock, R., Jones, A.R., Woodraska, D., Judge, D., *et al.*: 2012, *Solar Phys.* **275**, 115. DOI.

Reprinted from *Solar Physics* Journal, DOI 10.1007/s11207-015-0691-2

Soft X-Ray Pulsations in Solar Flares

P.J.A. Simões[1] · H.S. Hudson[1,2] · L. Fletcher[1]

Received: 9 December 2014 / Accepted: 20 April 2015 / Published online: 8 May 2015
© Springer Science+Business Media Dordrecht 2015

Abstract The soft X-ray emissions ($h\nu > 1.5$ keV) of solar flares mainly come from the bright coronal loops at the highest temperatures normally achieved in the flare process. Their ubiquity has led to their use as a standard measure of flare occurrence and energy, although the overwhelming bulk of the total flare energy goes elsewhere. Recently Dolla *et al.* (*Astrophys. J. Lett.* **749**, L16, 2012) noted quasi-periodic pulsations (QPP) in the soft X-ray signature of the X-class flare SOL2011-02-15, as observed by the standard photometric data from the GOES (*Geostationary Operational Environmental Satellite*) spacecraft. In this article we analyse the suitability of the GOES data for this type of analysis and find them to be generally valuable after September, 2010 (GOES-15). We then extend the result of Dolla *et al.* to a complete list of X-class flares from Cycle 24 and show that most of them (80 %) display QPPs in the impulsive phase. The pulsations show up cleanly in both channels of the GOES data, making use of time-series of irradiance differences (the digital time derivative on the 2-s sampling). We deploy different techniques to characterise the periodicity of GOES pulsations, considering the red-noise properties of the flare signals, finding a range of characteristic time scales of the QPPs for each event, but usually with no strong signature of a single period dominating in the power spectrum. The QPP may also appear on somewhat longer time scales during the later gradual phase, possibly with a greater tendency towards coherence, but the sampling noise in GOES difference data for high irradiance values (X-class flares) makes these more uncertain. We show that there is minimal phase difference between the differenced GOES energy channels, or between them and the

Solar and Stellar Flares: Observations, Simulations, and Synergies
Guest Editors: Lyndsay Fletcher and Petr Heinzel

✉ P.J.A. Simões
paulo.simoes@glasgow.ac.uk

H.S. Hudson
hhudson@ssl.berkeley.edu

L. Fletcher
lyndsay.fletcher@glasgow.ac.uk

1 SUPA, School of Physics and Astronomy, University of Glasgow, Glasgow G12 8QQ, UK

2 SSL, UC Berkeley, Berkeley, CA 94720, USA

hard X-ray variations on short time scales. During the impulsive phase, the footpoints of the newly forming flare loops may also contribute to the observed soft X-ray variations.

Keywords Flares, impulsive phase · Oscillations, solar · X-ray bursts, hard · X-ray bursts, soft

1. Introduction

The soft X-ray (SXR) emission of a solar flare (its GOES class) has practically supplanted the Hα brightening as the defining property of a solar flare (*e.g.* Fletcher *et al.*, 2011). The *Geostationary Operational Environmental Satellite* (GOES) is a satellite series initially deployed in 1974, with GOES-15 being the latest in the line. The GOES "Sun-as-a-star" soft X-ray observations, standardised in essentially the same form since the 1970s, come from two photometers that sample $1-8$ and $0.5-4$ Å bands.

We understand the soft X-ray sources to consist of plasma ablated (evaporated) from the lower solar atmosphere into coronal magnetic structures, and that the residence time scale (for cooling and draining) substantially exceeds the injection time scale. The time of injection corresponds to the impulsive phase of the flare, although it can be prolonged in some cases to tens of minutes; the subsequent gradual phase sees continued X-ray emission for many hours in extreme cases.

The $1-8$ Å and $0.5-4$ Å bands reflect a complicated mixture of line and continuum contributions (*e.g.* White, Thomas, and Schwartz, 2005) and respond to temperatures above about 4 MK. The temperature sensitivity of this filter pair strongly favours flare emissions, but there is sufficient sensitivity in the long-wavelength band to detect quiescent active regions as well, at least during active times. The two broad passbands overlap the thermal/non-thermal regions of solar X-ray emission; they generally reflect bound-bound transitions at lower temperatures and continuum components (thermal free-free and free-bound) at higher temperatures. Although the spectral response extends above 10 keV, the GOES time histories of solar flares typically do not show an impulsive-phase signature and invariably lag the hard X-ray peak emission time, a manifestation of the so-called Neupert effect (Neupert, 1968). We attribute this to the rapid filling of the coronal magnetic reservoir with hot plasma, which gradually obscures the hot footpoint sources detectable in soft X-ray images (McTiernan *et al.*, 1993; Hudson *et al.*, 1994; Mrozek and Tomczak, 2004; Simões, Graham, and Fletcher, 2015). By comparison with the extreme ultraviolet (EUV) passbands (*e.g.* O'Dwyer *et al.*, 2010), the GOES photometry isolates the hottest part of the coronal plasma.

Pulsations in the X-ray emission of solar flares were discovered in hard X-rays (above 20 keV) by Parks and Winckler (1969) in the impulsive phase of a major flare (SOL1968-08-08T18) and had a timescale of about 16 s. Simultaneous and closely synchronous peaks appeared at 15.4 GHz, identifiable as gyrosynchrotron radiation from MeV-range electrons, and Janssens and White (1969) noted earlier pulsations at longer radio wavelengths with somewhat longer (24 s) spacings. These longer-wavelength microwaves normally would be interpreted as thermal free-free radiation, and so in this early example we could already see pulsation signatures in very different physical domains. In our current interpretation of flare hard X-ray emission, this behaviour reflects time variations in an ill-understood mechanism for particle acceleration, by which the flare magnetic energy release predominantly goes into fast electrons at energies above about 20 keV.

The term "quasi-periodic pulsations" (QPP) is usually employed to refer generally to this kind of behaviour – short trains of roughly periodic emissions. Throughout the literature, the QPPs are usually identified by periodic signatures in the Fourier or wavelet power spectrum of the flare signal.

Since these early observations, a great deal of progress has been made in coronal seismology, in which one attempts to identify the QPP with resonant MHD wave properties and thereby learn something about the physical conditions in the corona (*e.g.* Nakariakov and Verwichte, 2005). Nakariakov and Melnikov (2009) specifically reviewed the now extensive observations of flare-related QPP mainly in this context. The identification of discrete eigenmodes during the early stages of a solar flare may not make much sense, however, since the flaring structure must evolve rapidly during this time and thus substantially change its geometry until the energy release subsides. On the other hand, structures external to the flare site clearly react to the disturbance in an oscillatory fashion (Aschwanden *et al.*, 1999), which may also involve a relaxation to a new equilibrium state (*e.g.* Simões *et al.*, 2013). Far-reaching disturbances have long been known to emanate from flares (*e.g.* Švestka, 1976) and can frequently be identified with the shock-wave exciters of type II radio bursts (Hudson and Warmuth, 2004). Today, the EUV observations from the TRACE and SDO satellites in particular have provided a wealth of new observations through high-resolution movies of coronal structures.

Soft X-ray data (and EUV imaging at wavelengths corresponding to higher temperatures) in principle probe a frontier area between the dynamical process of energy release and the possibly oscillatory relaxation into the new equilibrium of the flaring active region. We note early observations of X-ray emission lines by the soft X-ray spectrometer onboard the *Yohkoh* satellite (Mariska, 2005, 2006), making use mainly of S XV and Ca XIX lines. These observations reveal average oscillation periods of 5.5 ± 2.7 min, which is shorter than the typical periods associated with the large-scale structures seen in the EUV. The flare QPP phenomena discovered by Parks and Winckler (1969) typically have time scales more than one order of magnitude shorter, generally suggesting smaller scale lengths, stronger magnetic fields, and lower altitudes.

In an important recent comment, Gruber *et al.* (2011) (see also Vaughan, 2005, 2010) pointed out that an underlying power-law distribution of variability could mimic the appearance of a resonant process for a short time-series as an artefact of band-limited noise. In this case, the apparent discrete frequencies of power-spectrum analyses such as that of Parks and Winckler (1969) and many subsequent examples might readily be misinterpreted (Ireland, McAteer, and Inglis, 2015; Inglis, Ireland, and Dominique, 2015).

In this article we characterise the time variability of the GOES soft X-ray fluxes, following the discovery by Dolla *et al.* (2012) of QPP in the 1 – 30 s period range in the first X-class flare of Cycle 24, the well-studied SOL2011-02-15. Such QPPs were also noted by Simões *et al.* (2013) in the M-class flare SOL2012-03-09, with a similar behaviour observed in SDO/AIA EUV hot channels. We analyse the whole set of X-class flares from this cycle to date, a total of 35 at the time of writing, in an effort to understand the relationship between the relatively fast impulsive-phase QPP seen in the non-thermal signatures, with the longer periods with possible resonant properties seen in thermal signatures formed in coronal magnetic structures with relatively low plasma temperatures.

2. GOES Data Analysis

To sharpen the variability, we took advantage of the excellent signal-to-noise ratio of the current GOES soft X-ray irradiance data and mainly studied the *digital time-derivative* of

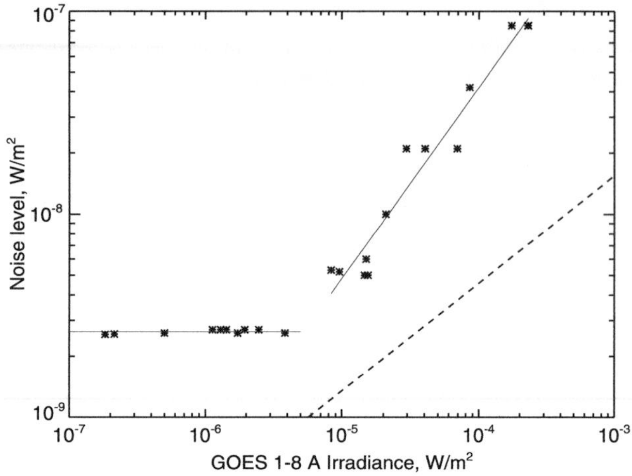

Figure 1 Empirical noise levels for GOES irradiance measurements in the 1 – 8 Å band. The dashed line shows the estimated Poisson noise from photon-counting statistics, the data points show the irradiance steps resulting from the digitization of the data. We derived these levels from a series of flares covering the range of irradiance values shown in the plot.

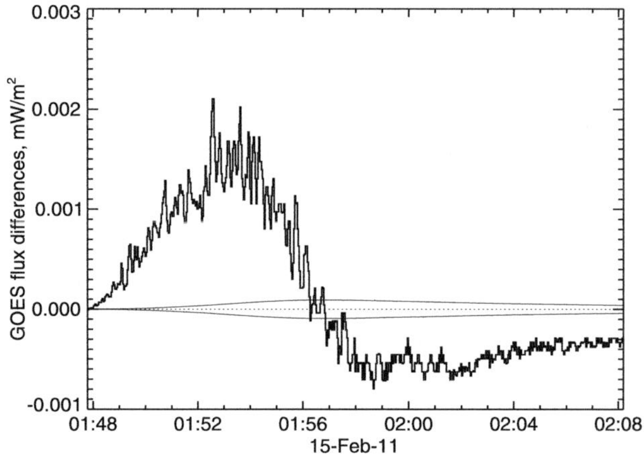

Figure 2 The time derivative of 1 – 8 Å GOES data for SOL2011-02-15, with the solid lines showing the digitization errors ± from the fit in Figure 1.

the original high-resolution (2 s sampling) time-series. The data before GOES-13 had poorer sampling and thus are not so useful for this type of analysis. Since the time derivative of the SXRs is known to track the hard X-rays (HXR), peaks in the derivative time series should correspond well with HXR spikes, if both reflect the impulsive energy release (the Neupert effect). However, differencing data can degrade their precision, therefore we first studied this problem. A casual inspection of the GOES irradiance data shows discrete steps in the noise level imposed by the digitization, and Figure 1 characterises this as a function of irradiance level for the 1 – 8 Å band. We have taken observed irradiance digital steps from several flares, ranging from C to X class, in the time frame 2012 – 2014, and the solid lines show a fit to the points. We estimated the Poisson noise from counting statistics and also show this, noting that it lies below the digital step size. Because of this, the precision of the GOES data does not increase with irradiance level, as should be the case for finer sampling (see also Chamberlin *et al.*, 2009).

The iconic flare SOL2011-02-15 illustrates some of the properties of the differenced time-series; see Figure 2. For the purposes of this article we note a working definition of the two phases of this flare: we identify the impulsive phase as between GOES onset and the first zero crossing of the 1 – 8 Å differences, with the gradual phase the time between that

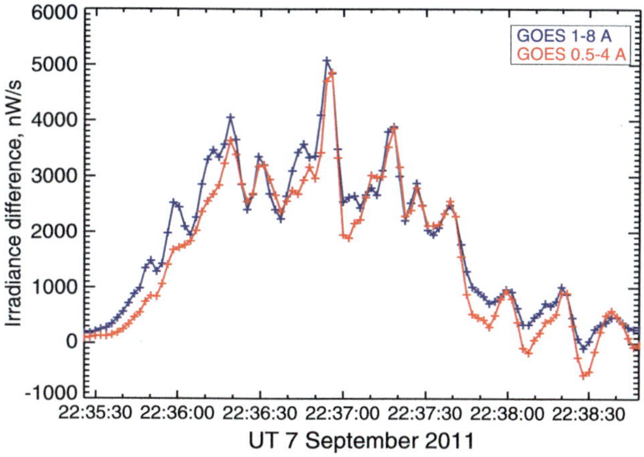

Figure 3 Correlation of the time-derivative of GOES 1 – 8 Å with GOES 0.5 – 4 Å (scaled up by a factor 2.35) for SOL2011-09-07.

zero crossing and the GOES end time. In this flare the zero-crossing time is 01:56:49 UT, whereas the listed time of GOES maximum is 01:56 UT, and this distinction will normally not matter. The strong variability seen in the impulsive phase of this flare has a rough time scale of about 12 s, estimated either by peak-counting or a power spectrum, which is similar to the original observation of hard X-ray pulsations (about 16 s) by Parks and Winckler. The variability in the gradual phase is much less pronounced and indeed has a magnitude generally comparable to the digitization error (Figure 1).

We conclude from this representative event that the differencing technique properly reveals the QPP of the impulsive phase, but that the variability in the gradual phase competes in a complicated way with the digitization levels of the GOES data; we do not believe that this can be characterised quantitatively because the digital levels do not adequately sample the noise, meaning that successive points may have identical values reported. This limitation should be borne in mind for the survey material presented below, which deals with X-class flares of comparable magnitudes to SOL2011-02-15.

As further evidence of the fidelity of the QPP signature in the time derivative of GOES data for the impulsive phase, we show another example (Flare 4 of the survey; see below) in Figure 3. This compares the two independent GOES bands (0.5 – 4 Å and 1 – 8 Å), whose ratio characterises the spectral hardness. The variations match well at the shorter time scales (the QPP), with some soft-hard-soft pattern indicated by the hardness ratio (*e.g.* Fletcher *et al.*, 2011) on the longer time scales as well.

3. Survey of X-Class Flares

We have used the time derivative of GOES time series with all of the X-class flares in Cycle 24, up to the time of writing and commencing with Event 0 (SOL2011-02-15), the subject of the article of Dolla *et al.* (2012) and Figure 2. This sample contains 35 events, the strongest of which was Flare 2, SOL2011-03-09 (X6.9). The basic analysis tool was wavelet transforms, as described below (Section 3.1), with additional checks. All of these events showed QPP in the impulsive phase.

3.1. Wavelet Analysis

To characterise the variability in the SXR signal, we employed the wavelet analysis using the methods and routines provided by Torrence and Compo (1998). To evaluate the red-noise spectrum, we used the method described in their Section 4 and provided by their wavelet software. The power-law slope that resulted from this was -1.9 for all the events in our sample. As described, we used the time derivative of the GOES channels instead of the more usual approach of detrending the signal (*e.g.* Dolla *et al.*, 2012), *i.e.* subtracting a smoothed curve, which depends on an arbitrarily chosen time scale. The time derivative maintains the fluctuations in the signal, retaining the original periodicity (if present). For completeness, however, we also applied the detrending approach and found similar results. We investigated the presence of periodic signals in the local wavelet power spectrum (WPS) and in the WPS averaged over the impulsive and gradual phases of each event in our sample. The significance of the WPS was tested against a red-noise model at 95 % and 99.7 % confidence levels; see Gruber *et al.* (2011), Inglis, Ireland, and Dominique (2015), Ireland, McAteer, and Inglis (2015) for a discussion of this type of analysis. We used the Morlet mother wavelet (parameter $w = 6$), but note that similar results were found for varying w and also for the Paul mother wavelet (see Torrence and Compo, 1998, for details). In short, we defined the presence of QPP as the identification of power in the wavelet power spectrum (averaged over the impulsive phase) above the red-noise model, with a confidence level of 99.7 %.

3.2. Survey Results

For our sample of 35 X-class flares, we characterised the time variability in several basic ways, as summarised in Table 1. For each event, we found the period range where the wavelet power spectrum, averaged for the impulsive and gradual phases, was above the 95 % confidence level of the red-noise model. We also verified the periods of peaks in the wavelet power spectrum.

As an example, in Figure 4 we show the results of the wavelet analysis for our first event, SOL2011-02-15, also analysed by Dolla *et al.* (2012). The four panels of Figure 4 show a) the wavelet power spectrum, b) and c) the time-averaged power spectrum for the two GOES spectral bands, d) the GOES timeseries, e) the time derivative of the time series of both channels, and f) the cross-correlation between the two channels, confirming the oscillatory non-dispersive nature of the signals (0.94 correlation coefficient with effectively zero lag). The wavelet power spectrum shows a periodic signal above the red-noise 99.7 % confidence level during the impulsive phase in the range 12 – 24 seconds, in agreement with the findings of Dolla *et al.* (2012). In the gradual phase, the periodic power starts to vanish, but still appears to last for about 150 seconds, with a period range 20 – 40 seconds (see Figure 2 for the caveat about sampling precision).

We show time derivative of the timeseries plots for each of the 35 events in Figure 5, normalised to the irradiance values. From the wavelet analysis, we found an average characteristic time scale of 16 – 53 seconds (at 95 % confidence level) and 22 – 42 seconds (at 99.7 %) (Table 1). Overall, periodic signals in the range 8 – 112 seconds (at a confidence level of 99.7 %) appear across our flare sample. For most of the events, we detected similar characteristic time scales in both channels during the rise impulsive phase. During the gradual phase, only seven flares (20 % of the sample) had QPPs detected in the average wavelet power spectrum, although some events gave the appearance of weak broad-band power in the accessible time range. Moreover, during the impulsive phase of some events (indicated in Table 1 by †), the average WPS level of the low channel was below the red-noise 95 % confidence level, yet there are significant regions clustered together in time and period, indicating processes with a less random nature. For these cases, the high channel average WPS confirmed the presence of quasi-periodic power at the confidence level of 95 %; but we found very low persistent quasi-periodic power for these 7 events at 99.7 %.

Table 1 X-class flares of Cycle 24 along with the results of the wavelet analysis, where we show the characteristic time scale of the WPS above the red-noise level for confidence levels of 95 % and 99.7 %. † indicates events with quasi-periodic power found locally, but not evident in the average WPS.

No.	Flare ID	Class	Characteristic time scale		QPP gradual
			95 %	99.7 %	
0	SOL2011-02-15	X2.2	10 – 30	12 – 24	no
1	SOL2011-03-09	X1.5	10 – 30†	–	no
2	SOL2011-08-09	X6.9	10 – 55	14 – 46	yes
3	SOL2011-09-06	X2.1	10 – 25	12 – 23	no
4	SOL2011-09-07	X1.8	10 – 45	10 – 37	no
5	SOL2011-09-22	X1.4	25 – 80†	–	no
6	SOL2011-09-24	X1.9	10 – 40	10 – 34	no
7	SOL2011-11-03	X1.9	10 – 25	–	no
8	SOL2012-01-27	X1.7	50 – 200	65 – 99	yes
9	SOL2012-03-05	X1.1	25 – 70	30 – 54	yes
10	SOL2012-03-07	X5.4	15 – 50	20 – 26	no
11	SOL2012-03-07	X1.3	30 – 55†	–	no
12	SOL2012-07-06	X1.1	10 – 30	12 – 25	no
13	SOL2012-07-12	X1.4	9 – 20, 50 – 200	70 – 106	yes
14	SOL2012-10-23	X1.8	10 – 20†	10 – 14	no
15	SOL2013-05-13	X1.7	10 – 60	17 – 45	no
16	SOL2013-05-13	X2.8	15 – 40	19 – 31	yes
17	SOL2013-05-14	X3.2	10 – 90	16 – 29	yes
18	SOL2013-05-15	X1.2	30 – 70	35 – 57	no
19	SOL2013-10-25	X1.7	10 – 45	11 – 25	no
20	SOL2013-10-25	X2.1	10 – 25	10 – 18	no
21	SOL2013-10-28	X1.0	15 – 65	17 – 24	no
22	SOL2013-10-29	X2.3	20 – 65	24 – 54	no
23	SOL2013-11-05	X3.3	9 – 12†	–	no
24	SOL2013-11-08	X1.1	8 – 20	10 – 18	no
25	SOL2013-11-10	X1.1	9 – 25	18 – 22	no
26	SOL2013-11-19	X1.0	10 – 30	12 – 26	no
27	SOL2014-01-07	X1.2	9 – 18†	–	no
28	SOL2014-02-25	X4.9	8 – 45	12 – 40	no
29	SOL2014-03-29	X1.0	8 – 60	8 – 56	no
30	SOL2014-04-25	X1.3	15 – 54	17 – 48	yes
31	SOL2014-06-10	X2.2	24 – 45†	–	no
32	SOL2014-06-10	X1.5	22 – 82	26 – 35	no
33	SOL2014-06-11	X1.0	10 – 16	11 – 14	no
34	SOL2014-09-10	X1.6	28 – 126	80 – 112	no

3.3. Association with Hard X-Rays

For a subset of six of our sample, we compared the SXR time series with hard X-rays (HXR) data from *Fermi*/GBM (Meegan *et al.*, 2009; Paciesas, 2011; Schwartz *et al.*, 2010). We de-trended the intensity signals by subtracting their smoothed component using a boxcar of 30 s

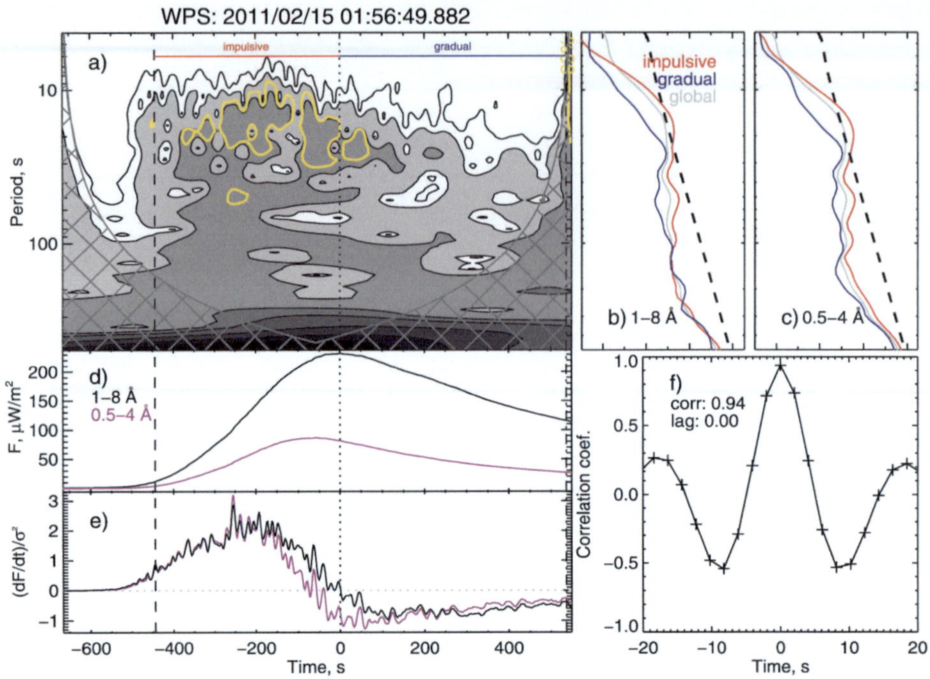

Figure 4 Results of the wavelet analysis of SOL2011-02-15. Reference times are indicated by the vertical lines: onset (dashed), peak (dotted), and end (dash-dotted). a) Wavelet power spectrum of the time derivative of the signal of the $1-8$ Å channel, with the power above the red-noise model (99.7 % confidence level) shown in yellow, b) the time-averaged power spectrum for the low channel ($1-8$ Å) for the impulsive phase (red) and gradual phase (blue) compared to the 99.7 % confidence level of the red-noise spectrum model (thick dashed curve), and c) same for the high channel ($0.5-4$ Å), d) time series of both GOES channels, e) time derivative of both channels, and f) lag cross-correlation between the time derivative of the signal of both channels.

(note that Dolla *et al.* (2012) used 20 s). The time derivative of SXR and HXR light curves and the detrended components are shown in Figure 6. The SXR and HXR detrended signals were cross-correlated, and the results found are listed in Table 2. The SXR and HXR pulsations are positively correlated in general, with the cross-correlation coefficient r frequently greater than 0.5. For these cases of $r > 0.5$, no delays were found between the signals. For events 18, 21, 25, and 30 (see Table 2), we found a good correlation for many HXR energy bands; for the event 17, a good correlation appeared for the higher energy channels (> 25 keV), and finally, for event 28, we derived $r \approx 0.4$ at all energy channels, with delays of $12-20$ sec. This particular example, SOL2014-02-25, exhibited periods with an anti-correlation between detrended > 25 keV and the GOES. This may result from a well-known artefact present in many scintillation-counter spectrometers that operate at excessively high count rates, as explained below. We note that this particular flare was an unusual and powerful γ-ray event that showed sustained emission of > 100 MeV photons (*e.g.* Ackermann *et al.*, 2014). The correlations, although not perfect, indicate that this is a typical situation in which the QPP excitations detected in the GOES time series closely match those seen in hard X-rays.

We note that applying a filter (*e.g.* Fourier) to remove the slowly varying component would encounter the same problems as the detrending technique, *i.e.* the possibility of

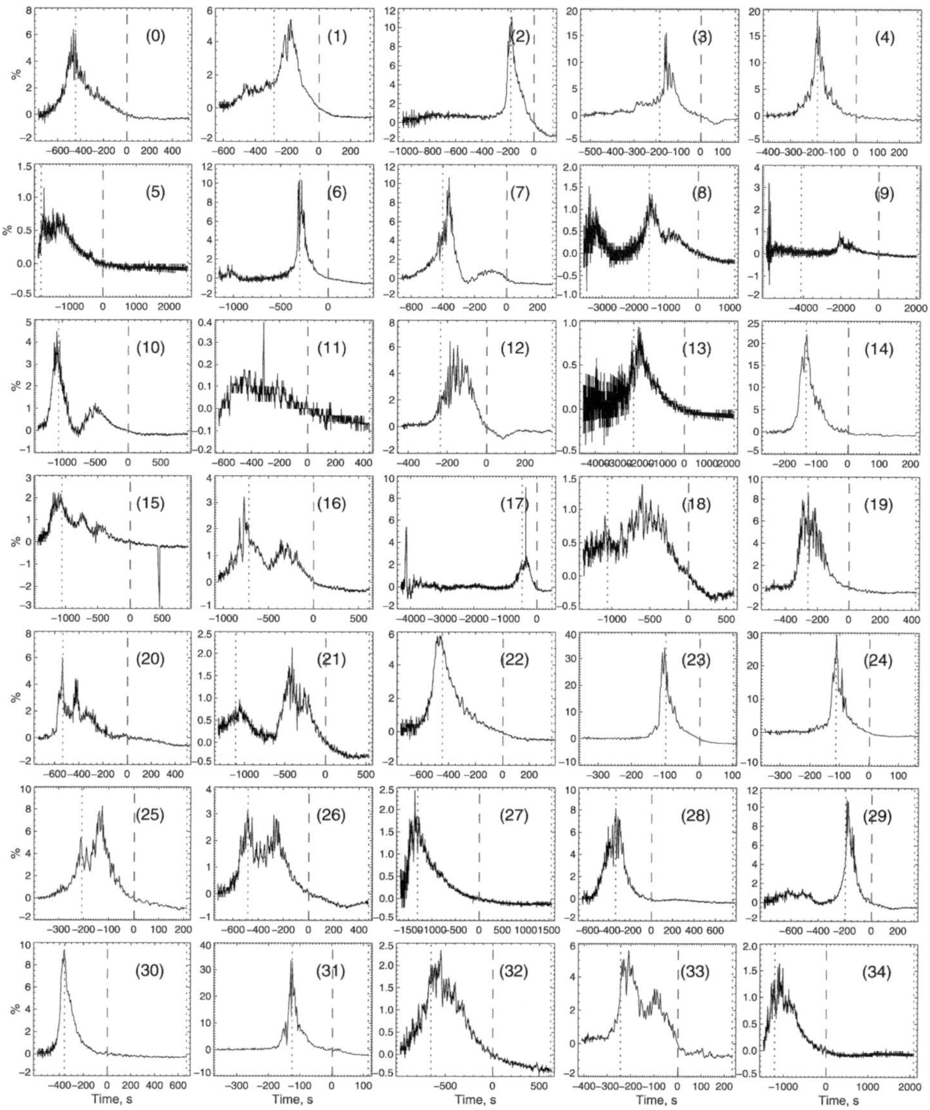

Figure 5 Time derivative of the GOES 1 – 8 Å time series normalised to the irradiance values, showing a wide variety of patterns, but the general presence of the QPP in the impulsive phase. Vertical dashed lines indicate the 1 – 8 Å peak, and dotted lines indicate the arbitrarily defined flare onset (5 % of the maximum flux density) and end times.

introducing spurious signals. We chose the latter because 1) it is easily reproducible and 2) it is the preferred method of the majority of authors investigating QPPs, and in particular, Dolla *et al.* (2012), making it easy to compare the results of different authors. Moreover, our point here is to show the correlation of the variability at small amplitudes of soft and hard X-ray data and the implications of this, and not to perform a strict periodicity analysis. Periodic or not, this variability is real (as noted by *e.g.* Gruber *et al.* 2011; Inglis, Ireland, and Dominique 2015) and possibly affecting a wide range of the flare emission spectrum, as shown here and by Dolla *et al.* (2012).

Table 2 Cross-correlation results between GOES SXR 0.5 – 4 Å and *Fermi*/GBM HXR channels (in keV).

No.	Flare ID	NaI det.	4 – 9	9 – 12	12 – 15	15 – 25	25 – 50	50 – 100
17	SOL2013-05-14	n1	0.25	0.19	0.23	0.22	0.42	0.75
18	SOL2013-05-15	n4	0.30	0.73	0.75	0.71	0.65	0.54
21	SOL2013-10-28	n1	0.67	0.29	0.31	0.67	0.61	0.46
25	SOL2013-11-10	n4	0.59	0.62	0.59	0.67	0.65	0.33
28	SOL2014-02-25	n4	0.34	0.38	0.37	0.40	0.44	0.40
30	SOL2014-04-25	n4	0.38	0.72	0.75	0.67	0.37	0.22

As an important caveat regarding the hard X-ray data from *Fermi*, we note that in many cases the archived light curves clearly exhibit nonlinear responses as a result of high counting rates (note that our sample specifically contains only the most energetic events). We have not tried to understand these effects in detail and are unaware of any published analysis of such effects in the *Fermi* data, but past experience with scintillation counters at high rates suggests that both the spectrum and the time series can suffer. At high rates the photomultiplier gain may change, for example, affecting the energy calibration of the detectors. We have not used the most sunward NaI detectors to avoid these effects (see Table 2 for detector used in this analysis); however, taking data from detectors with larger offset angles reduces the count rates at high energies and also affects the detection of the small-scale pulsations. We note, however, that during some parts of SOL2014-02-25's time history, the impulsive energy signatures appear to change phase by 180°, becoming minima (at some energies) rather than maxima; this would have a natural explanation in terms of gain changes. Nevertheless, for this event we found similar results for different *Fermi*/GBM NaI detectors, suggesting that the lack of correlation between the SXR and HXR may not be an instrumental effect.

4. Discussion

We have found that most of the X-class flares in our sample show clear pulsation signatures in their time series, as would be expected from the presence of spikes in hard X-ray emission (*e.g.* Dennis and Zarro, 1993) and the operation of the Neupert effect on these time scales. These signatures appear strongly in the impulsive phase of the flare, but much less prominently in the gradual phase.

Our analysis typically covers a frequency range of about four octaves below the Nyquist frequency of the sampling, in the band 0.01 – 0.25 Hz. Although the GOES peaks may have time scales down to our limit of 4 s, most of the variability is slower. The wavelet analysis for the most part does not show narrow features in spite of the presence of multiple peaks in the time series, and therefore need not be interpreted in terms of a resonant process. This seems reasonable, given the rapid structural changes of the emitting regions during the impulsive phase; one would not expect stable eigenfrequencies to dominate the time signatures. Later on and remote from the site of energy release, distinct signatures of damped oscillations appear commonly (Aschwanden *et al.*, 1999).

The QPP variability found here extends into the 0.1 Hz band, as discovered by Parks and Winckler (1969), but involves longer time scales (8 – 112 seconds) and agrees overall with the periods found in many QPP studies in different spectral ranges in the literature (*e.g.* Nakariakov, Melnikov, and Reznikova, 2003; Melnikov *et al.*, 2005; Reznikova *et al.*,

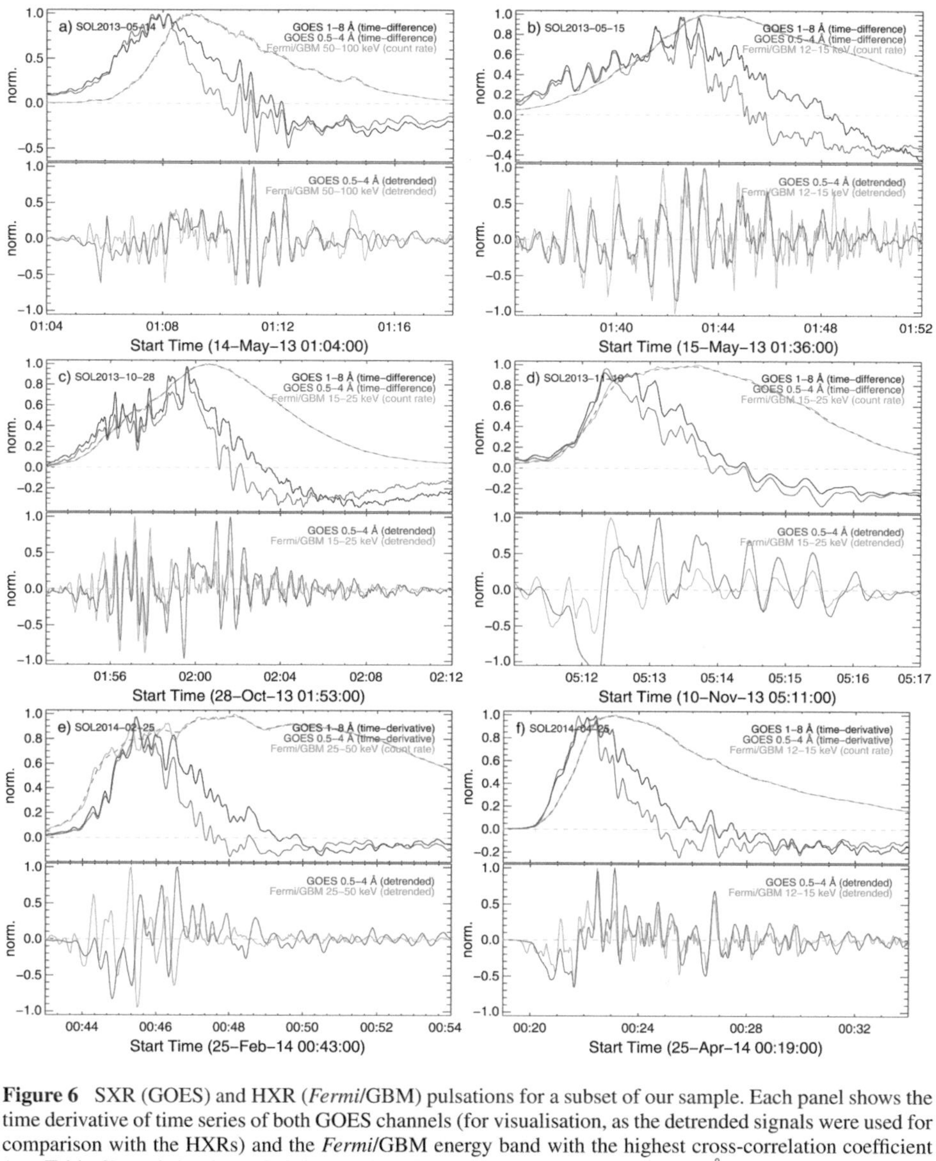

Figure 6 SXR (GOES) and HXR (*Fermi*/GBM) pulsations for a subset of our sample. Each panel shows the time derivative of time series of both GOES channels (for visualisation, as the detrended signals were used for comparison with the HXRs) and the *Fermi*/GBM energy band with the highest cross-correlation coefficient (see Table 2), and the detrended time series of *Fermi*/GBM and GOES 0.5 – 4 Å. The upper right panel (SOL2013-05-15) shows an excellent pulse-by-pulse match between GOES and the hard X-rays, limited by the decreasing S/N at late times. The only poor example is SOL2014-02-25 (lower left) in this set.

2007; Inglis, Nakariakov, and Melnikov, 2008; Inglis and Nakariakov, 2009; Kallunki and Pohjolainen, 2012). Parks and Winckler identified a specific time scale (16 s, or roughly 60 mHz), and we found a few cases in which there is a spectral peak in the range above about 80 mHz (or about 12 seconds). Curiously, one of the best-studied recent QPP events, SOL1998-05-08, shows almost exactly the same time scale (Asai *et al.*, 2001; Stepanov *et al.*, 2004; Inglis, Nakariakov, and Melnikov, 2008; Inglis, Ireland, and Dominique, 2015), and is one of the best examples of QPP that could arise from a true oscillation of some

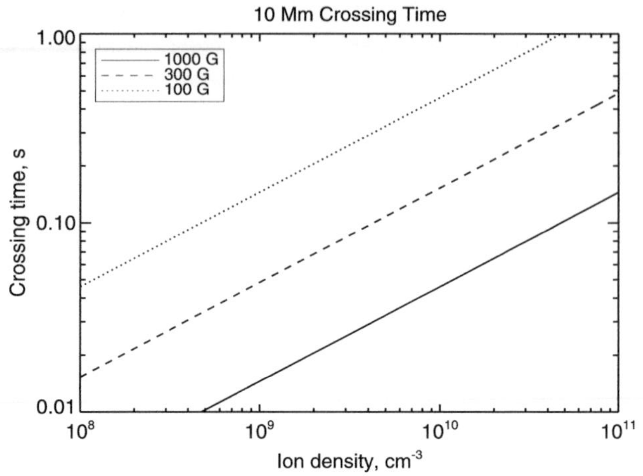

Figure 7 Estimates of the Alfvén time scale for coronal or chromospheric structures for ion densities in the range 10^{8-11} cm^{-3} and for three plausible values of the magnetic field intensity, for a 10 Mm scale.

sort. Kupriyanova *et al.* (2010) investigated QPPs at 17 GHz using data from the *Nobeyama Radio Heliograph* and found periods in the range 5 – 60 seconds for ten events. In particular, in our analysis of SOL2011-02-15 (see Figure 4), we found a characteristic time scale of 12 – 24 seconds, in agreement with the results of Dolla *et al.* (2012). We note that the periodicity analysis of this event reported by Inglis, Ireland, and Dominique (2015), which used several HXR energy bands from *Fermi*/GBM and extreme ultraviolet (EUV) data from PROBA2/LYRA, did not find a periodic signal. We argue that analysing the global wavelet power or Fourier power (*e.g.* Inglis, Ireland, and Dominique, 2015) seems to reduce the chances of detecting any periodic power because the signal is averaged over the entire duration of the events. Any periodic signal would have to be persistently strong through the entire flare to appear. Moreover, the variability in the GOES SXR time series has a much smaller amplitude than the slowly varying component, thus a filter must be applied to retrieve it.

The observed time scales are therefore inconsistent with Alfvénic processes in the core of an active region, as reflected in Figure 7 (see also Fletcher and Hudson, 2008). In the impulsive phase of a flare, we expect low plasma densities since evaporation is just beginning to occur, and high magnetic fields since the events typically occur in active regions with sunspots. Thus the preferred domain in Figure 7 would be the lower right quadrant. This suggests that the chromospheric or photospheric regions of the flares may help to define the time scales of the flare development, or the travel times of the QPP disturbances, since the deep atmosphere has slower characteristic time scales.

The soft X-ray variability is highest at the very beginning (Figure 5), which we readily explain by the accumulation of hot material in the corona, which systematically tends to obscure new contributions.

5. Conclusions

We have surveyed the X-class flares of the current solar cycle (35 events from SOL2011-02-15 to SOL2014-09-10), as viewed by the standard soft X-ray irradiance measurements from the GOES satellites. Over this time interval, the GOES-15 detectors provided the data, and (as described in Section 2) its sampling precision greatly exceeds that of earlier members of the GOES series. This and the excellent signal-to-noise ratio of the observations allowed us to use the digital time derivative of the time series to study the variability in detail, up to the Nyquist frequency of the sampling at 0.25 Hz. Our main conclusion is, therefore, to confirm the result reported by Dolla *et al.* (2012): the recent GOES measurements work very well in studies of QPP. By "recent" we mean GOES 13 – 15, with coverage from September 2010, and therefore the source of all of the data on X-class flares in Cycle 24 thus far. These data have a better precision than that of earlier GOES instruments, but still do not adequately sample the detector noise (Section 2).

Using the time derivative of time series, we found that 28 out of 35 X-class flares show QPP to some degree, especially during the impulsive phase of the event development. Here we defined the presence of QPP as the identification of power in the wavelet power spectrum (averaged over the impulsive phase) above the red-noise model, with a confidence level of 99.7 %. Overall, the characteristic time scale found is in the range 8 – 112 seconds (see Table 1). Because of the sampling problem we have discussed, the data are unable clearly to show QPP during the gradual phase of the flares. We find that most of the spectral power typically is at lower frequencies (between roughly 8 – 112 seconds) and is explicitly inconsistent with signal propagation at reasonable values of the Alfvén times for relevant scales in the active-region corona and upper chromosphere.

The QPP phenomenon in solar flares could be interpreted in several ways (*e.g.* Nakariakov and Melnikov, 2009). Nakariakov and Zimovets (2011), for example, described a scenario in which slow-mode MHD waves trigger episodes of magnetic reconnection at successive locations. We think this unlikely for the impulsive phase of a flare; the presence of non-steady evolution of the magnetic field, as required to release magnetic energy rapidly, would tend to destroy the identities of eigenmodes as the impulsive phase develops. Such an evolution is not typically considered in theoretical work in the area of coronal seismology, which is largely based on the structure of a fixed cylinder in the absence of non-steady flows (Nakariakov and Verwichte, 2005). The problem of time scales also remains: Nakariakov and Zimovets hypothesised that there might be slow coupling between arcade loops through perpendicular transport in the footpoint regions, but there would be no obvious source of resonant behaviour or periodicity in such a process.

Acknowledgements We would like to thank the anonymous referee for the comments and suggestions that helped to improve the paper. We would like to thank Marie Dominique and Rodney Viereck for helpful discussions. HSH thanks NASA for support under contract NAS 5-98033 for RHESSI, and acknowledges warm hospitality at the University of Glasgow. The research leading to these results has received funding from the European Community's Seventh Framework Programme (FP7/2007-2013) under grant agreement no. 606862 (F-CHROMA) (PJAS, LF), from STFC grant ST/I001808/1 (PJAS, LF) and ST/L000741/1 (LF).

References

Ackermann, M., Ajello, M., Albert, A., Allafort, A., Baldini, L., Barbiellini, G., Bastieri, D., Bechtol, K., Bellazzini, R., Bissaldi, E., Bonamente, E., Bottacini, E., Bouvier, A., Brandt, T.J., Bregeon, J., Brigida, M., Bruel, P., Buehler, R., Buson, S., Caliandro, G.A., Cameron, R.A., Caraveo, P.A., Cecchi, C., Charles, E., Chekhtman, A., Chen, Q., Chiang, J., Chiaro, G., Ciprini, S., Claus, R., Cohen-Tanugi, J., Conrad, J., Cutini, S., D'Ammando, F., de Angelis, A., de Palma, F., Dermer, C.D., Desiante, R., Digel, S.W., Di Venere, L., Silva, E.d.C.e., Drell, P.S., Drlica-Wagner, A., Favuzzi, C., Fegan, S.J., Focke, W.B., Franckowiak, A., Fukazawa, Y., Funk, S., Fusco, P., Gargano, F., Gasparrini, D., Germani, S., Giglietto, N., Giordano, F., Giroletti, M., Glanzman, T., Godfrey, G., Grenier, I.A., Grove, J.E., Guiriec, S., Hadasch, D., Hayashida, M., Hays, E., Horan, D., Hughes, R.E., Inoue, Y., Jackson, M.S., Jogler, T., Jóhannesson, G., Johnson, W.N., Kamae, T., Kawano, T., Knödlseder, J., Kuss, M., Lande, J., Larsson, S., Latronico, L., Lemoine-Goumard, M., Longo, F., Loparco, F., Lott, B., Lovellette, M.N., Lubrano, P., Mayer, M., Mazziotta, M.N., McEnery, J.E., Michelson, P.F., Mizuno, T., Moiseev, A.A., Monte, C., Monzani, M.E., Moretti, E., Morselli, A., Moskalenko, I.V., Murgia, S., Murphy, R., Nemmen, R., Nuss, E., Ohno, M., Ohsugi, T., Okumura, A., Omodei, N., Orienti, M., Orlando, E.,

Ormes, J.F., Paneque, D., Panetta, J.H., Perkins, J.S., Pesce-Rollins, M., Petrosian, V., Piron, F., Pivato, G., Porter, T.A., Rainò, S., Rando, R., Razzano, M., Reimer, A., Reimer, O., Ritz, S., Schulz, A., Sgrò, C., Siskind, E.J., Spandre, G., Spinelli, P., Takahashi, H., Takeuchi, Y., Tanaka, Y., Thayer, J.G., Thayer, J.B., Thompson, D.J., Tibaldo, L., Tinivella, M., Tosti, G., Troja, E., Tronconi, V., Usher, T.L., Vandenbroucke, J., Vasileiou, V., Vianello, G., Vitale, V., Werner, M., Winer, B.L., Wood, D.L., Wood, K.S., Wood, M., Yang, Z.: 2014, High-energy gamma-ray emission from solar flares: Summary of Fermi large area telescope detections and analysis of two M-class flares. *Astrophys. J.* **787**, 15. DOI. ADS.

Asai, A., Shimojo, M., Isobe, H., Morimoto, T., Yokoyama, T., Shibasaki, K., Nakajima, H.: 2001, Periodic acceleration of electrons in the 1998 November 10 solar flare. *Astrophys. J. Lett.* **562**, L103. DOI. ADS.

Aschwanden, M.J., Fletcher, L., Schrijver, C.J., Alexander, D.: 1999, Coronal loop oscillations observed with the Transition Region and Coronal Explorer. *Astrophys. J.* **520**, 880. DOI. ADS.

Chamberlin, P.C., Woods, T.N., Eparvier, F.G., Jones, A.R.: 2009, Next generation X-ray sensor (XRS) for the NOAA GOES-R satellite series. In: *SPIE* **CS-7438**. DOI. ADS.

Dennis, B.R., Zarro, D.M.: 1993, The Neupert effect – What can it tell us about the impulsive and gradual phases of solar flares? *Solar Phys.* **146**, 177. DOI. ADS.

Dolla, L., Marqué, C., Seaton, D.B., Van Doorsselaere, T., Dominique, M., Berghmans, D., Cabanas, C., De Groof, A., Schmutz, W., Verdini, A., West, M.J., Zender, J., Zhukov, A.N.: 2012, Time delays in quasi-periodic pulsations observed during the X2.2 solar flare on 2011 February 15. *Astrophys. J. Lett.* **749**, L16. DOI. ADS.

Fletcher, L., Hudson, H.S.: 2008, Impulsive phase flare energy transport by large-scale Alfvén w aves and the electron acceleration problem. *Astrophys. J.* **675**, 1645. DOI. ADS.

Fletcher, L., Dennis, B.R., Hudson, H.S., Krucker, S., Phillips, K., Veronig, A., Battaglia, M., Bone, L., Caspi, A., Chen, Q., Gallagher, P., Grigis, P.T., Ji, H., Liu, W., Milligan, R.O., Temmer, M.: 2011, An observational overview of solar flares. *Space Sci. Rev.* **159**, 19. DOI. ADS.

Gruber, D., Lachowicz, P., Bissaldi, E., Briggs, M.S., Connaughton, V., Greiner, J., van der Horst, A.J., Kanbach, G., Rau, A., Bhat, P.N., Diehl, R., von Kienlin, A., Kippen, R.M., Meegan, C.A., Paciesas, W.S., Preece, R.D., Wilson-Hodge, C.: 2011, Quasi-periodic pulsations in solar flares: new clues from the Fermi Gamma-Ray Burst Monitor. *Astron. Astrophys.* **533**, A61. DOI. ADS.

Hudson, H.S., Warmuth, A.: 2004, Coronal loop oscillations and flare shock waves. *Astrophys. J. Lett.* **614**, L85. DOI. ADS.

Hudson, H.S., Strong, K.T., Dennis, B.R., Zarro, D., Inda, M., Kosugi, T., Sakao, T.: 1994, Impulsive behavior in solar soft X-radiation. *Astrophys. J. Lett.* **422**, L25. DOI. ADS.

Inglis, A.R., Ireland, J., Dominique, M.: 2015, Quasi-periodic pulsations in solar and stellar flares: Re-evaluating their nature in the context of power-law flare Fourier spectra. *Astrophys. J.* **798**, 108. DOI. ADS.

Inglis, A.R., Nakariakov, V.M.: 2009, A multi-periodic oscillatory event in a solar flare. *Astron. Astrophys.* **493**, 259. DOI. ADS.

Inglis, A.R., Nakariakov, V.M., Melnikov, V.F.: 2008, Multi-wavelength spatially resolved analysis of quasi-periodic pulsations in a solar flare. *Astron. Astrophys.* **487**, 1147. DOI. ADS.

Ireland, J., McAteer, R.T.J., Inglis, A.R.: 2015, Coronal Fourier power spectra: Implications for coronal seismology and coronal heating. *Astrophys. J.* **798**, 1. DOI. ADS.

Janssens, T.J., White, K.P. III: 1969, Microwave pulse trains observed before and during a solar flare. *Astrophys. J. Lett.* **158**, L127. DOI. ADS.

Kallunki, J., Pohjolainen, S.: 2012, Radio pulsating structures with coronal loop contraction. *Solar Phys.* **280**, 491. DOI. ADS.

Kupriyanova, E.G., Melnikov, V.F., Nakariakov, V.M., Shibasaki, K.: 2010, Types of microwave quasi-periodic pulsations in single flaring loops. *Solar Phys.* **267**, 329. DOI. ADS.

Mariska, J.T.: 2005, Observations of solar flare Doppler shift oscillations with the Bragg crystal spectrometer on Yohkoh. *Astrophys. J. Lett.* **620**, L67. DOI. ADS.

Mariska, J.T.: 2006, Characteristics of solar flare Doppler-shift oscillations observed with the Bragg crystal spectrometer on Yohkoh. *Astrophys. J.* **639**, 484. DOI. ADS.

McTiernan, J.M., Kane, S.R., Loran, J.M., Lemen, J.R., Acton, L.W., Hara, H., Tsuneta, S., Kosugi, T.: 1993, Temperature and density structure of the 1991 November 2 flare observed by the YOHKOH soft X-ray telescope and hard X-ray telescope. *Astrophys. J. Lett.* **416**, L91+. DOI. ADS.

Meegan, C., Lichti, G., Bhat, P.N., Bissaldi, E., Briggs, M.S., Connaughton, V., Diehl, R., Fishman, G., Greiner, J., Hoover, A.S., van der Horst, A.J., von Kienlin, A., Kippen, R.M., Kouveliotou, C., McBreen, S., Paciesas, W.S., Preece, R., Steinle, H., Wallace, M.S., Wilson, R.B., Wilson-Hodge, C.: 2009, The Fermi Gamma-Ray Burst Monitor. *Astrophys. J.* **702**, 791. DOI. ADS.

Melnikov, V.F., Reznikova, V.E., Shibasaki, K., Nakariakov, V.M.: 2005, Spatially resolved microwave pulsations of a flare loop. *Astron. Astrophys.* **439**, 727. DOI. ADS.

Mrozek, T., Tomczak, M.: 2004, Solar impulsive soft X-ray brightenings and their connection with footpoint hard X-ray emission sources. *Astron. Astrophys.* **415**, 377. DOI. ADS.

Nakariakov, V.M., Melnikov, V.F.: 2009, Quasi-periodic pulsations in solar flares. *Space Sci. Rev.* **149**, 119. DOI. ADS.

Nakariakov, V.M., Melnikov, V.F., Reznikova, V.E.: 2003, Global sausage modes of coronal loops. *Astron. Astrophys.* **412**, L7. DOI. ADS.

Nakariakov, V.M., Verwichte, E.: 2005, Coronal waves and oscillations. *Living Rev. Solar Phys.* **2**, 3. DOI. ADS.

Nakariakov, V.M., Zimovets, I.V.: 2011, Slow magnetoacoustic waves in two-ribbon flares. *Astrophys. J. Lett.* **730**, L27. DOI. ADS.

Neupert, W.M.: 1968, Comparison of solar X-ray line emission with microwave emission during flares. *Astrophys. J. Lett.* **153**, L59. DOI. ADS.

O'Dwyer, B., Del Zanna, G., Mason, H.E., Weber, M.A., Tripathi, D.: 2010, SDO/AIA response to coronal hole, quiet Sun, active region, and flare plasma. *Astron. Astrophys.* **521**, A21. DOI. ADS.

Paciesas, W.S.: 2011, Fermi GBM: Instrument description and science highlights. In: Florinski, V., Heerikhuisen, J., Zank, G.P., Gallagher, D.L. (eds.) *Am. Inst. Phys.* **CS-1366**, 155. DOI. ADS.

Parks, G.K., Winckler, J.R.: 1969, Sixteen-second periodic pulsations observed in the correlated microwave and energetic X-ray emission from a solar flare. *Astrophys. J. Lett.* **155**, L117. DOI. ADS.

Reznikova, V.E., Melnikov, V.F., Su, Y., Huang, G.: 2007, Pulsations of microwave flaring emission at low and high frequencies. *Astron. Rep.* **51**, 588. DOI. ADS.

Schwartz, R.A., Dennis, B., Tolbert, A.K., Murphy, R., Share, G., Fishman, G., Briggs, M., Longo, F., Diehl, R., Wijers, R.: 2010, Fermi GBM and LAT solar flare X ray and γ ray observations. In: *Bull. Am. Astron. Soc.* **41**, #404.06. ADS.

Simões, P.J.A., Graham, D., Fletcher, L.: 2015, Impulsive heating of solar flare ribbons above 10 MK. *Solar Phys.* submitted.

Simões, P.J.A., Fletcher, L., Hudson, H.S., Russell, A.J.B.: 2013, Implosion of coronal loops during the impulsive phase of a solar flare. *Astrophys. J.* **777**, 152. DOI. ADS.

Stepanov, A.V., Kopylova, Y.G., Tsap, Y.T., Shibasaki, K., Melnikov, V.F., Goldvarg, T.B.: 2004, Pulsations of microwave emission and flare plasma diagnostics. *Astron. Lett.* **30**, 480. DOI. ADS.

Švestka, Z.: 1976, *Solar Flares*, Springer, Berlin. ADS.

Torrence, C., Compo, G.P.: 1998, A practical guide to wavelet analysis. *Bull. Am. Meteorol. Soc.*, 61.

Vaughan, S.: 2005, A simple test for periodic signals in red noise. *Astron. Astrophys.* **431**, 391. DOI. ADS.

Vaughan, S.: 2010, A Bayesian test for periodic signals in red noise. *Mon. Not. Roy. Astron. Soc.* **402**, 307. DOI. ADS.

White, S.M., Thomas, R.J., Schwartz, R.A.: 2005, Updated expressions for determining temperatures and emission measures from GOES soft X-ray measurements. *Solar Phys.* **227**, 231. DOI. ADS.

Reprinted from *Solar Physics* Journal, DOI 10.1007/s11207-015-0726-8
SOLAR AND STELLAR FLARES

Triggering an Eruptive Flare by Emerging Flux in a Solar Active-Region Complex

Rohan E. Louis[1] · Bernhard Kliem[2,3] · B. Ravindra[4] · Georgios Chintzoglou[3]

Received: 12 February 2015 / Accepted: 16 June 2015 / Published online: 8 July 2015
© Springer Science+Business Media Dordrecht 2015

Abstract A flare and fast coronal mass ejection originated between solar active regions NOAA 11514 and 11515 on 2012 July 1 (SOL2012-07-01) in response to flux emergence in front of the leading sunspot of the trailing region 11515. Analyzing the evolution of the photospheric magnetic flux and the coronal structure, we find that the flux emergence triggered the eruption by interaction with overlying flux in a non-standard way. The new flux neither had the opposite orientation nor a location near the polarity inversion line, which are favorable for strong reconnection with the arcade flux under which it emerged. Moreover, its flux content remained significantly smaller than that of the arcade (≈ 40 %). However, a loop system rooted in the trailing active region ran in part under the arcade between the active regions, passing over the site of flux emergence. The reconnection with the emerging flux, leading to a series of jet emissions into the loop system, caused a strong but confined rise of the loop system. This lifted the arcade between the two active regions, weakening its downward tension force and thus destabilizing the considerably sheared flux under the arcade. The complex event was also associated with supporting precursor activity in an enhanced network near the active regions, acting on the large-scale overlying flux, and with two simultaneous confined flares within the active regions.

Solar and Stellar Flares: Observations, Simulations, and Synergies
Guest Editors: Lyndsay Fletcher and Petr Heinzel

Electronic supplementary material The online version of this article (doi:10.1007/s11207-015-0726-8) contains supplementary material, which is available to authorized users.

✉ R.E. Louis
rlouis@aip.de

1 Leibniz-Institut für Astrophysik Potsdam (AIP), An der Sternwarte 16, 14482 Potsdam, Germany

2 Institut für Physik und Astronomie, Universität Potsdam, Karl-Liebknecht-Str. 24-25, 14476 Potsdam, Germany

3 School of Physics, Astronomy and Computational Sciences, George Mason University, Fairfax, VA 22030, USA

4 Indian Institute of Astrophysics, Koramangala, Bengaluru 560034, India

Keywords Flares, dynamics · Sunspots, magnetic fields · Chromosphere, active · Corona · Prominences, active

1. Introduction

Eruptive solar flares arise from a disruption of the coronal magnetic field, which evolves into a coronal mass ejection (CME). In many cases a filament or prominence eruption is associated. The flare source region comprises highly sheared magnetic fields in the corona that overlie a photospheric polarity inversion line (PIL) (Martres *et al.*, 1968; Hagyard, Venkatakrishnan, and Smith, 1990; Wang, 2006). These locations inevitably show a filament channel in the chromosphere and often contain a filament or prominence in the corona above (Martin, 1998). It is widely (but not universally) accepted that a filament represents a weakly twisted magnetic flux rope holding the cool material (Mackay *et al.*, 2010). The formation and instability of a flux rope is a key element of storage-and-release eruption models (van Tend and Kuperus, 1978; Forbes and Priest, 1995; Amari *et al.*, 2003; Gibson and Fan, 2006; Mackay and van Ballegooijen, 2006; Kliem and Török, 2006).

Eruptions can be triggered through the emergence of flux as well as through flux cancellation. In the case of the former, the eruption occurs when newly emerged magnetic fields appear in a region of pre-existing flux, often in or in close proximity to a filament channel (*e.g.*, Rust, 1972; Martin *et al.*, 1982; Démoulin *et al.*, 1993; Feynman and Martin, 1995; Li *et al.*, 2000; Sakajiri *et al.*, 2004; Schrijver, 2009; Sun *et al.*, 2012). This can lead to the formation of a magnetic flux rope through bodily emergence (Low, 1996; Lites, 2005), by reconnection within an emerged magnetic arcade (Manchester *et al.*, 2004; Archontis and Hood, 2010), or by reconnection with the pre-existing flux (Kusano *et al.*, 2012). A number of variants for the triggering of eruptions by emerging flux have been proposed (see Section 5 for details). According to nearly all of them, the presence of pre-existing flux, orientated in a direction that facilitates reconnection with the emerging flux, is crucial to the eruption of the coronal flux rope (*e.g.*, Chen and Shibata, 2000; Archontis and Török, 2008; Kusano *et al.*, 2012; Leake, Linton, and Antiochos, 2014). Otherwise, a specific, narrow range of locations (Xu, Fang, and Chen, 2008) or a large amount of emerging flux (Kaneko and Yokoyama, 2014) is required. Lin, Forbes, and Isenberg (2001) argued on general grounds that while the emergence of new flux can cause a loss of equilibrium, there is no simple, universal relation between the orientation of the emerging flux and the likelihood of an eruption, due to the complexity of the equilibrium surface in the multidimensional parameter space of the configuration.

Flux cancellation is also an important mechanism for triggering solar flares (Livi *et al.*, 1989; Sterling *et al.*, 2010; Green, Kliem, and Wallace, 2011; Savcheva *et al.*, 2012; Burtseva and Petrie, 2013). When canceling photospheric magnetic flux patches lie at the base of sheared coronal flux, then the cancellation is associated with magnetic reconnection low in the atmosphere. This gradually forms a filament channel in the photosphere and chromosphere and a flux rope above the PIL where the cancellation occurs (van Ballegooijen and Martens, 1989; Wang and Shi, 1993; Mackay and van Ballegooijen, 2006; Aulanier *et al.*, 2010; Amari *et al.*, 2011). The flux rope gradually rises as more flux is reconnected into it, implying a higher current flowing along the rope. When the rope becomes unstable and rises much faster, a vertical current sheet is formed underneath and becomes the site of fast "flare" reconnection (Lin and Forbes, 2000).

In this article we employ multiwavelength and dual-viewpoint observations to analyze an eruptive flare that originated between two active regions (ARs) that formed a quadrupolar active-region complex. Both the emergence of new flux and flux cancellation occurred under the erupting middle flux lobe. The flux cancellation was weak, leaving the flux emergence as the primary candidate for the triggering of the eruption. However, the flux emergence did not correspond to any of the standard scenarios for such triggering: the direction and emergence site of the new flux were not favorable for strong reconnection with the pre-existing erupting flux, and its magnitude was only moderate. Additionally, the eruption was preceded by a complex sequence of events in the vicinity of the AR complex and accompanied by confined flares within the two ARs. The data analysis yields a suggestion for a non-standard triggering of the eruption by the emerging flux. We also discuss this eruption from a quadrupolar source region in relation to the breakout model (Lynch *et al.*, 2008; Karpen, Antiochos, and DeVore, 2012). A further eruptive flare in the region of flux emergence, following about 19 hr later and most likely due to flux cancellation, was analyzed in Louis *et al.* (2014, hereafter referred to as Lo14).

2. Observations and Data Processing

We used full-disk solar observations from the *Solar Dynamics Observatory* (SDO; Pesnell, Thompson, and Chamberlin, 2012) that include data from the *Helioseismic and Magnetic Imager* (HMI, Schou *et al.*, 2012) and the *Atmospheric Imaging Assembly* (AIA, Lemen *et al.*, 2012) for the two ARs of interest. The HMI data set consists of 4k × 4k-pixel images of continuum intensity and line-of-sight (LOS) magnetograms derived from the photospheric Fe I 617.3 nm line at a spatial sampling of $0.5''$ pixel^{-1}. The HMI data were processed in a similar manner as described in Lo14. The AIA data consist of 4k × 4k-pixel images with a spatial sampling of $0.6''$ pixel^{-1} in a range of UV and EUV bands. In particular, we have analyzed the bands including He II 304 Å, Fe IX 171 Å, Fe XII 193 Å, Fe XIV 211 Å, Fe XVI 335 Å, Fe XVIII 94 Å, Fe XXI 131 Å, and the continuum at 1700 Å. The observations span a duration of 28 hr starting on 2012 July 1 with a cadence of 12 sec for the AIA data and 45 sec for the HMI data.

We also used full-disk Hα images from the *Global Oscillations Network Group* (GONG; Harvey *et al.*, 1996, 2011) recorded at Big Bear Solar Observatory. The 2k × 2k images have a spatial sampling of about $1''$ with a cadence of 5 minutes and cover the passage of the eruptive flare.

Limb views of the event are provided by the SECCHI instrument package (Howard *et al.*, 2008) onboard the STEREO *Behind* spacecraft. We analyzed 195 and 304 Å data from the EUVI-B imager and data from the COR1-B coronagraph.

3. Source Region

NOAA ARs 11514 and 11515 appear at the east limb on June 26, 2012 as young regions still in their emergence phase. By the beginning of July 1, their centers are located at S15E09 and S16E24, respectively. The leading AR 11514 is then already dispersing and spotless, while the following AR 11515 still develops strongly, with considerable rearrangements of its multiple spots. Both regions are essentially bipolar and follow Hale's law, so that they jointly form a quadrupolar configuration. AR 11515 contains far more flux, amounting to 2.6×10^{22} Mx, while the total magnetic flux in AR 11514 is around 6.1×10^{21} Mx. South-west of AR 11514, four areas of enhanced network with alternating polarity are located. Filaments have developed at all three resulting PILs, with the first two forming a joint,

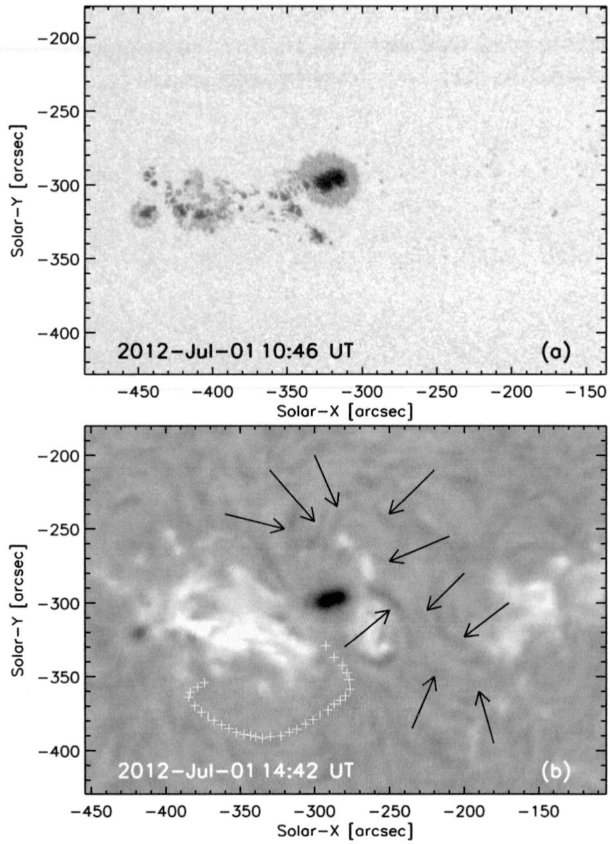

Figure 1 Overview images of the AR complex. (a) HMI continuum intensity; (b) GONG Hα from Big Bear Solar Observatory (Hα images from other observatories near the time of the other panels in Figures 1 and 2 are less sharp). Arrows in the bottom panel mark fibrils that indicate the forming filament channel involved in the eruption, while the plus symbols indicate the overall alignment of Hα fibrils south of AR 11515 with the loop system seen in the AIA 171 and 193 Å images (Figure 2(c – d)), similar to the absorbing material seen in He II 304 Å (Figure 2(b)). An animation of HMI continuum and magnetogram and AIA 171, 304, and 1700 Å images during 2012 July 1 is provided as online material.

forward S-shaped filament, which becomes involved in the eruption studied here. The AR complex and adjacent enhanced network area are displayed at various wavelengths in Figures 1 and 2, and the accompanying animation shows their evolution during July 1. Their appearance in the AIA 211 and 335 Å bands is very similar to the appearance at 193 Å. It is also worth mentioning that AIA 211 and 193 Å images provide a weak suggestion for the existence of a magnetic connection between AR 11515 and AR 11513 located in the other hemisphere at N15E12. Both ARs 11515 and 11513 are rather active during their disk passage, with much of temporally associated, possibly sympathetic activity occurring in the days around July 1 – 2.

Figure 3 shows field lines of a potential field source surface (PFSS) extrapolation of the AR complex immediately before the onset of the eruptive flare. One can clearly see that the two ARs are magnetically coupled: the excess positive flux in AR 11515, largely concentrated in the leading sunspot (LS), mostly connects to the excess negative flux in AR 11514, located in its trailing part. This yields the three-lobe magnetic configuration characteristic of a quadrupolar source region and agrees with the conclusion of Lo14 that the (later splitting) LS has strong magnetic connections with AR 11514. The PFSS extrapolation also indicates magnetic connections of both ARs with the enhanced network area southwest of AR 11514, and northward-pointing field lines starting in the LS, which most likely connect to AR 11513, a region dominated by negative flux. High-reaching field lines are also rooted in both the leading positive polarity of AR 11514 and the trailing negative

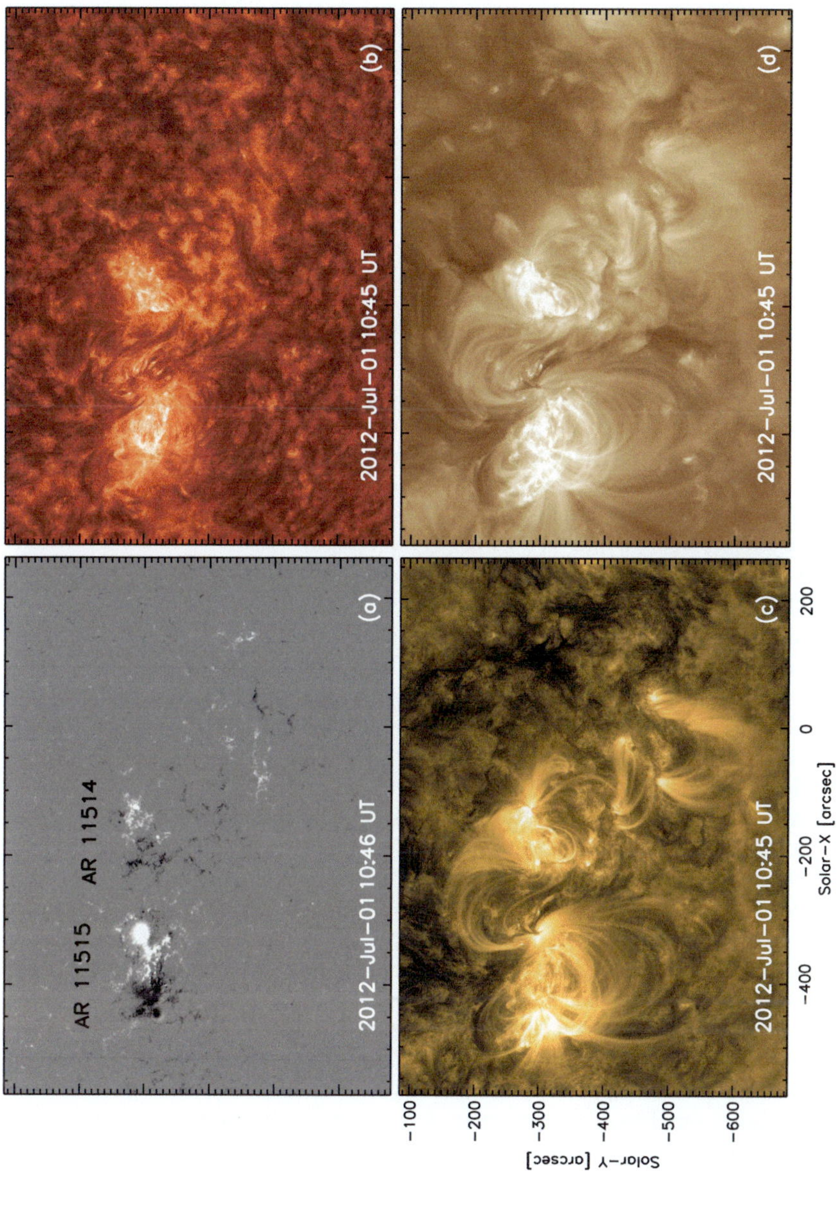

Figure 2 Continuation of Figure 1 showing (a) the HMI LOS magnetogram and the chromospheric and coronal structure in AIA images at (b) 304 Å, (c) 171 Å, and (d) 193 Å. The enhanced network area southwest of the AR complex, which contains the S-shaped filament centered at solar disk coordinates $(x, y) = (0, -480)$, is included here.

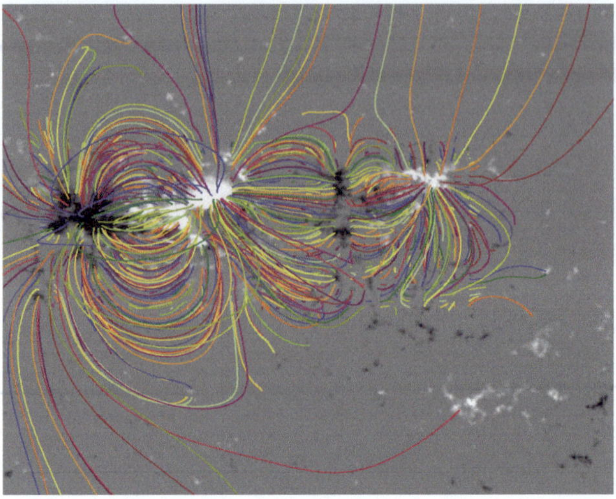

Figure 3 Field lines of a PFSS extrapolation using the HMI LOS magnetogram of ARs 11514 and 11515 and their environment at 15:24:45 UT patched into a synoptic full-Sun magnetogram. The field is computed using the method by Zhao and Hoeksema (1995) with 330 spherical harmonics and the source surface located at 2.5 $R_{\odot}$. The grayscale of the magnetogram display is saturated at $|B| = 500$ G.

polarity of AR 11515. These may indicate the existence of an overarching fourth lobe of flux, as required by the breakout model, or open flux. While the PFSS extrapolation is not conclusive in this regard, the EUV observations described below (see Sections 4.1 – 4.2 and Figures 8 and 10) strongly suggest the existence of overarching flux, albeit with a different orientation. Moreover, the PFSS extrapolation indicates that the amount of overarching flux is considerably smaller than the flux in the three central lobes.

Beginning at about 05 UT on July 1 and continuing into the following day, new flux emerges in AR 11515 just in front (westward) of the LS, *i.e.* under the middle magnetic lobe of the AR complex, near the left edge of the lobe (Figures 4 and 5). This is dominantly negative flux mixed with some small positive flux patches, thus forming a new PIL in front of the LS. It is difficult to discern with certainty where the corresponding positive flux emerges. Part of it must emerge within the preexisting scattered positive flux just outside the northwest periphery of the LS (marked by a white arrow). The amount of this flux clearly increases during the emergence of the negative flux ahead of the LS (Figure 5). Additionally, a moderate positive flux patch grows on the western side just in front of the emerging negative flux (marked by a second white arrow). It is possible, however, that part of the positive flux emerges in the moat of the LS, where it is hard to detect. Any coronal field lines connecting the LS and the emerging negative flux must have a direction rather similar to the overlying flux of the middle lobe of the AR complex. However, the field lines connecting the two polarities within the emerging flux in front of the LS have a considerable north-south component, forming a small-to-moderate angle with the overlying flux. This allows for the occurrence of magnetic reconnection, although only with a moderate likelihood. The minor part of the emerging flux rooted in the moderate positive flux patch on the western edge of the emergence area is directed oppositely to the overlying flux; this flux has a high likelihood for reconnection to occur. The relative amount of emerged flux by the time of the eruptive flare is estimated from the measurement of negative flux in Figure 5 and from the amount of negative flux in the footpoint area of the middle flux lobe in the PFSS extrapolation to be $\sim$40 %.

Another major development in AR 11515 is the splitting of its LS, which becomes apparent after the eruptive flare studied here and is analyzed in Lo14.

In the EUV (see Figure 2 and the corresponding animation), both active regions are bright throughout the observation, with the usual multitude of small-scale brightenings and

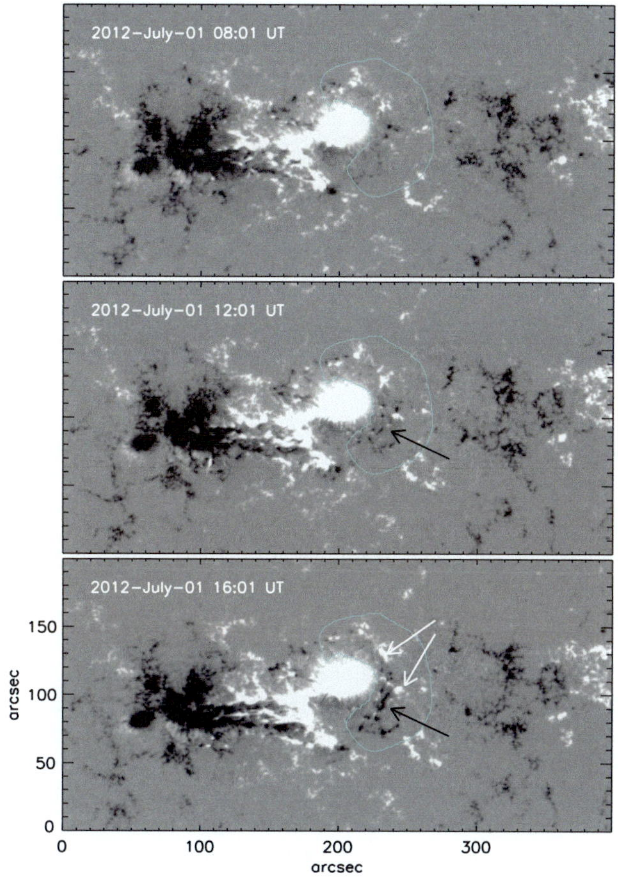

Figure 4 Emergence of new flux ahead of the LS in AR 11515. Mostly negative new flux can be seen, as indicated by the black arrows. Regions of increasing positive flux are marked by white arrows. The grayscale is saturated at $|B| = 500$ G, except around the PIL between the ARs, where the flux has been enhanced by a factor of 4, to visualize the motion of weak negative flux in the trailing part of AR 11514 toward the PIL in the final hours before the eruption. The cyan contour outlines the region of emerging flux.

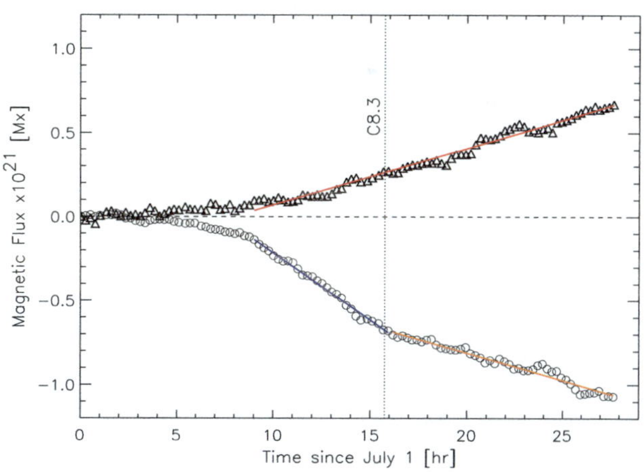

Figure 5 Temporal evolution of magnetic flux in the emergence area ahead of the LS (area enclosed by the cyan contour in Figure 4) in AR 11515 over the course of 2012 July 1. The triangles and circles correspond to positive and negative flux, respectively. The flux shown in the plot is relative to the initial flux at 00:00 UT. The blue and orange lines represent linear fits to the observed values, before and after the occurrence of the C8.3 flare.

changing coronal loops reflecting the ongoing evolution of the photospheric flux distribution. Additionally, the region of new flux emergence is bright and highly variable. Before the eruptive flare, there are far fewer coronal loops between the ARs, and they are mostly

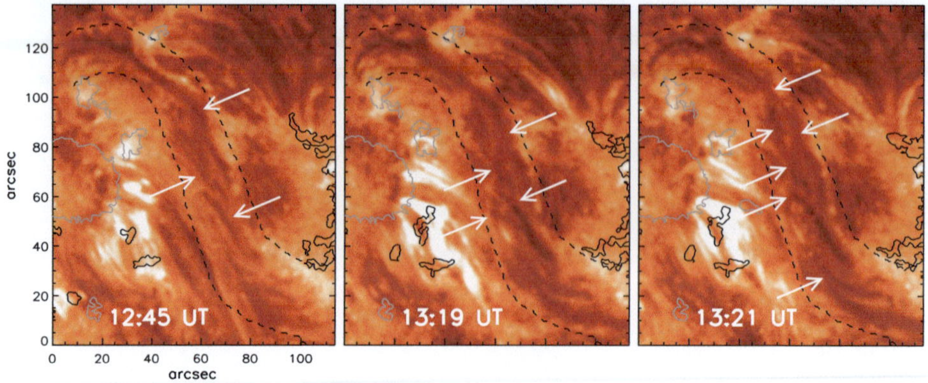

Figure 6 AIA 304 Å images of the forming filament channel between the two ARs showing instances of apparently crossing threads in the middle of the channel (indicated by the arrows). Dashed lines outline the filament channel. Gray (black) contours represent the LOS magnetic field of positive (negative) polarity.

fainter than the loops within the ARs. Many of the inter-region loops are also distinguished by their clear reverse-S shape, indicative of magnetic shear, and by their higher temperature, as they are more pronounced in the "warmer" AIA bands (193, 211, and 335 Å) than in the 171 Å band. The higher temperature is seen already before the new flux emergence and is therefore indicative of the dissipation of enhanced currents in the sheared field.

The existence of considerable shear in the field between the ARs is also obvious from the strong alignment of the He II threads at the PIL of the central field lobe, shown in Figure 6. A filament channel, indicated in the figure, is obviously forming here; it extends with a reverse-S shape from the northern edge of the LS in AR 11515 to the southern edge of the trailing flux in AR 11514. Its shear is left-handed, consistent with the reverse-S shape of the overlying coronal loops. Hα-absorbing material can be seen to be arranged along the forming filament channel in elongated patches (Figure 1). These patches are often changing in shape and position, with the trend to delineate the filament channel more completely by the end of the day.

Flows converging at the PIL are known to be essential for the formation of a filament channel (Martin, 1998). Since the photospheric field is weak in a broad strip around the PIL between the two ARs, an attempt to track magnetic features using the DAVE algorithm (Schuck, 2006) did not yield a reliable map of the photospheric flow field around the PIL. However, the sequence of HMI magnetograms unambiguously shows that the narrow strip of weak, mostly negative flux at the trailing edge of AR 11514 moves toward the PIL as a consequence of a strong expansion of the adjacent network cell from about 10 UT onwards (see Figure 4 and the AIA 1700 Å data included in the online movie).

The detailed arrangement of the threads in the forming filament channel is highly variable. Overall, they indicate a highly sheared, left-handed arcade at all times, but at a number of instances several threads display a crossing pattern that might indicate the formation of a flux rope (see examples in Figure 6). Since this pattern does not persist, the actual formation of a flux rope at this location remains unclear.

Absorbing (filament) material is also observed in the AIA 304 Å band (Figure 2(b)) in the heterogeneous structures extending up to $\sim 20^\circ$ south of AR 11515, which on a large scale are similar to the loops seen in the "warm" bands at 171 Å, 193 Å, and higher (Figure 2(c, d)). This cool material is also seen above the limb in EUVI-B 304 Å images, with variable shape (Figure 7(a)). The Hα images show an arrangement of the fibrils in this area in a large-scale loop pattern that resembles the structure in the AIA 304 Å band (some of them are indicated by crosses in Figure 1(b)).

Finally, we note that right-handed helicity is indicated for the filaments in the enhanced network southwest of AR 11514 by the orientation of the absorbing He II threads, the shape of the brightenings during their later activation (Section 4.1), and by the forward-S shape of the western filament.

4. Eruptive Flare and Associated Pre- and Post-flare Activity

Both ARs 11515 and 11513 display frequent subflaring and flaring activity throughout the days around 2012 July 1, some of which might be sympathetic activity. Therefore, it is difficult to identify with certainty the initial element in a chain of events leading to the eruptive C-class (C8.3) flare in the AR complex formed by ARs 11514 and 11515, which peaks at 15:45 UT. These activities occur in different places in and around the AR complex, indicating a highly complex structure of the magnetic field, even if a possible connection to AR 11513 on the other hemisphere is not taken into account. Only part of this complexity can be inferred from the EUV images in the present article.

The basic sequence of events consists of 1) two relatively similar activations of the loop system containing filament material south of AR 11515, which remain confined, 2) the eruptive C8.3 flare and fast CME, which originate at the PIL between the two ARs but involve a complex pattern of bright ribbons and dimming areas scattered over the AR complex and the neighboring area of enhanced network flux containing two filaments, and 3) related confined flaring activities within the two ARs in the peak and decay phases of the eruptive flare.

4.1. Pre-flare Filament Activations

In the AIA data we count 19 episodes with single or multiple ejections of bright material from AR 11515 on July 1 before the eruptive flare; some of these were observed by EUVI-B as well. Initially, they all originate in the region's LS, but from 05 UT onward, the area of newly emerging flux (NEF) increasingly becomes the source of the jets. A double jet emission from the LS during 03:40 – 04 UT in both north and south directions deserves mentioning, as co-temporal brightenings occur in the S-shaped filament in the enhanced network southwest of the AR complex, as well as in, and adjacent to, a filament channel extending from the S-shaped filament to the west. The existence of magnetic connections to these quite remote areas, which are marked in Figure 8 below, is thus indicated. Part of the connections are also seen in the PFSS extrapolation (Figure 3). Later, these remote areas show the rise of material, associated with brightenings and dimmings, in close relation to the main eruptive flare.

The chain of coupled events can be traced back at least to the activation of the large loop system containing filament material (dark in He II 304 Å) south of AR 11515, which commenced at ~05 UT, *i.e.* with the first jet ejected from the NEF area. AIA 304 and 193 Å images show that the jet is ejected into a large loop connecting the area of the LS and the negative polarity of AR 11515 and extending to about S35. The loop then expands in the south, east, and west directions, reaching ~S50 by 10:30 UT, when it fades in the 304 Å images. A second, somewhat smaller loop, lying below the first one in the EUVI-B limb view and seen only in the 304 Å band, also starts to expand. These loops are relatively faint in the AIA images, where they are most clear in animated image sequences, but are clearly seen in the limb view of STEREO *Behind*. The EUVI-B images (Figure 7(a – d)) show the rise

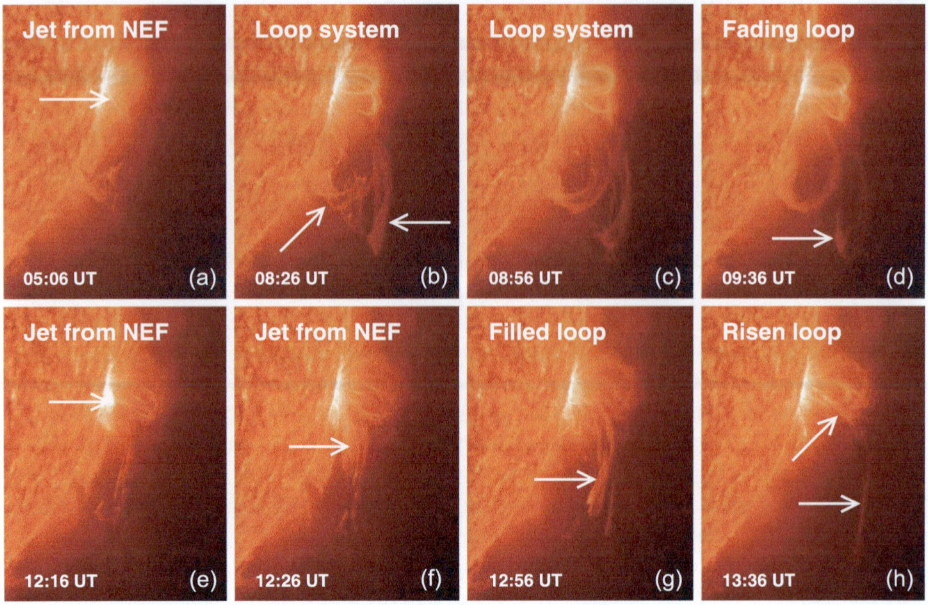

Figure 7 EUVI-B 304 Å snapshots of the two confined loop and filament activations south of AR 11515 during 05 – 10:30 UT (top row) and 12:10 – 13:40 UT (bottom row), both before the eruptive flare. The rise of three loops south of the AR complex, all containing filament material, is seen following the ejection of a jet from the NEF at ∼05 and 12:10 UT, respectively. The loops fade subsequently, mostly due to the draining of the cool material (the lower loop seen in panel (d) fades by the time of panels (e) – (f)).

of a multithreaded loop, filled with He II-emitting plasma and apparently rooted in the AR complex and about 10° farther south, from 05 UT onwards. The threads in the southern part, which in part obviously belong to the smaller loop in the AIA images, show indications of helical structure and undergo various changes until the flux is organized in two loops, which reach maximum heights of ~ 0.15 and ~ 0.25 $R_\odot$ during 9 – 10 UT. Both loops then fade at these positions, the upper one by $\approx 10{:}20$ UT, the lower one about two hours later. This filament activation appears to be confined.

Of the further jet emissions in AR 11515, the southward ones from the NEF area commencing at about 10:50 and 12:10 UT trigger similar effects in the AR complex, which eventually evolve into the eruptive flare. The initially slow expansion of a similar southward-oriented loop (lying inside the previously expanded one in the AIA images) is triggered by the first jet and accelerated by the second. This loop shows a stronger westward expansion. Its western leg sweeps over the S-shaped filament in the enhanced network southwest of the AR complex, reaching it at 13:12 UT, simultaneously with the onset of brightenings in the filament (Figure 8). The loop fades in the AIA 304 Å images by 13:42 UT, when it crosses over the filament roughly in the middle. The jet at 12:10 UT and subsequent loop expansion are also imaged by EUVI-B (Figure 7(e – h)). These 304 Å images show that the jet is ejected into the height range of the previously activated loop (~ 0.15 $R_\odot$). The He II-emitting material forms a new multithreaded loop, which quickly rises to ~ 0.4 $R_\odot$, but stops at this height. The material then slides northward, back to the surface, and the loop fades, disappearing at $\approx 14{:}00$ UT; this activity apparently also remains confined. Nevertheless, the flux above the legs of this loop, which become nearly vertical, from $\sim 25°$ to the surface at the time of the jet ejection to $\sim 70°$ at the fading, must be substantially disturbed. Primarily the western leg,

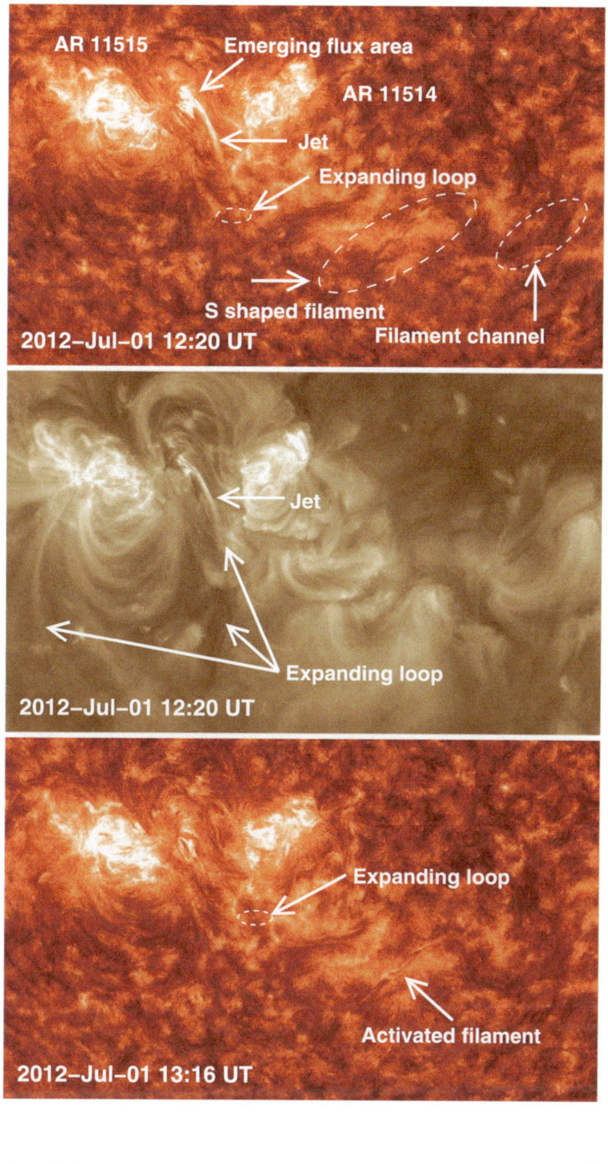

Figure 8 AIA 304 and 193 Å snapshots of the second activation that directly leads to the eruptive flare. Several thin threads of absorbing (filament) material in the expanding loop system are enclosed by small ellipses in the 304 Å images. At 193 Å the legs of the expanding loop system are clearly seen in the still image, while the top part can only be discerned against the structured background in animated image sequences. Two large ellipses outline the positions of the S-shaped filament and the filament channel westward of it, which become involved in the eruption.

running under the middle field lobe of the AR complex, must be seen in the EUVI-B 304 Å data, because this leg is much clearer than the eastern leg in the corresponding AIA data. The loops seen above the AR complex in the EUVI-B 304 Å images extend within the area of the two ARs, but do not pass over the expanding loop (Figure 8); therefore, they are not lifted.

The corresponding EUVI-B 195 Å images show the rise of diffuse, multiple loop structures south of the AR complex from ∼11 UT, also stalled at $\sim 0.4\ R_{\odot}$ by ∼14 UT. These are not displayed here, since they show the rising structures only very faintly.

The second activation of the loop system south of AR 11515 and its strong westward expansion trigger several brightenings in the S-shaped filament in the enhanced network southwest of AR 11514, and fast motions of brightened material along the axis of the filament. At

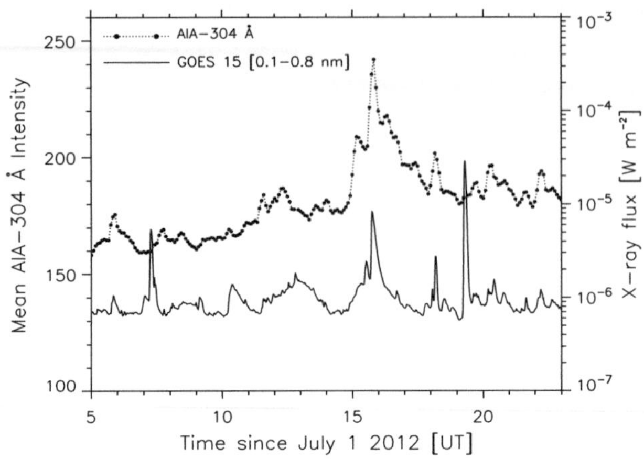

Figure 9 Temporal evolution of the mean AIA 304 Å intensity (dotted line with filled circles) in the active region and the GOES 15 soft X-ray flux in the 0.1 – 0.8 nm band (solid line). Both the C5.4 flare at 07:11 UT and the M2.8 flare at 19:11 UT occurred in AR 11513.

least one thread detaches and moves away from the filament in a perpendicular (southwest) direction, followed by a pair of moderately intense ribbons developing on either side of the filament from 13:48 UT onward (these can still be seen 90 minutes later in Figure 10(a)). Clearly, there is an eruption of flux in the channel that includes this filament, but is mostly lying above the filament. A partial eruption of the S-shaped filament is also obvious from its reduced darkness in the Hα images. The AIA 193 Å images reveal more clearly that the corona above the S-shaped filament strongly expands southward as the ribbons to the sides of the filament spread, suggestive of the opening of the field in this area, which dims subsequently.

Related activities also occur in and adjacent to the filament channel extending westward from the S-shaped filament. These include a very localized ejection of bright material near its western end during 12:45 – 13 UT, precisely at the location of a brightening caused by the earlier activation during 03:40 – 04 UT, and a southward-directed eruption from this filament channel essentially simultaneous with the development of the ribbons of the S-shaped filament. This eruption leads to a deep and extended dimming. The locations of these developments in the area of enhanced network are identified in Figures 8 and 10.

4.2. The Eruptive Flare

Although the pre-eruptive and eruptive activity extends over the whole AR complex and the large neighboring area of enhanced network to the southwest, the core of the eruption clearly is the PIL between ARs 11514 and 11515. The bright ribbons begin to form near the ends of this PIL, where they enclose the initial dimmings, and the ribbons and flare loop arcade extend along this PIL. Subsequently, extended dimmings, interspersed with small-scale brightenings, form in other places, indicating that the overlying flux had a complex topology and that the erupting flux reconnected with overlying flux, exchanging the footpoints.

The ribbons of the eruptive C8.3 flare, which peaks at 15:45 UT (Figure 9), begin to form in the positive flux ahead and northwest of the LS in AR 11515 at 14:30 UT and about 8 minutes later at the eastern edge of the most eastern enhanced network cell south of AR 11514 (Figure 10(a)). This onset time coincides with the time the ribbons at the activated S-shaped filament in the enhanced network have reached their maximum separation. Thus, there is a continuous sequence of events, starting with jet ejections at the NEF, continuing into the activation and rise of a large loop that runs under the middle lobe of the AR

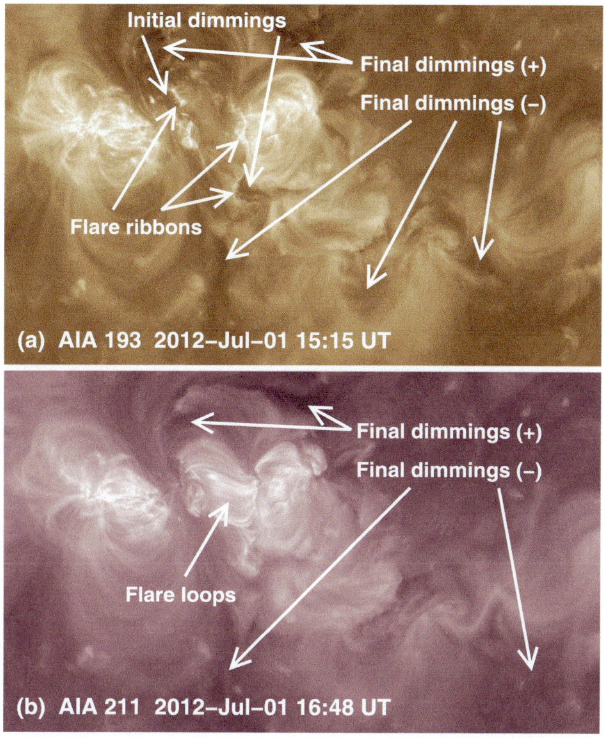

Figure 10 AIA images of the early (top, 193 Å) and late (bottom, 211 Å) evolution of the eruptive C8.3 flare. The magnetic polarity of the dimming areas is indicated.

complex and at the same time sweeping over the area of enhanced network. This is followed by the partial eruption of the S-shaped filament and the adjacent filament channel in the enhanced network, and finally by the destabilization of the flux in the middle lobe of the AR complex. The arc-shaped initial section of the ribbon in the positive-flux area encloses the initial dimming area, which includes about one-third of the LS. On the other side, the initial dimming extends over half of the network cell bounded by the initial ribbon. These are the best candidates for the footpoint regions of the destabilized flux. They are marked in Figure 10(a).

The erupting flux can be discerned only partly and is extremely difficult to visualize, since it does not contain any cool material. Animated images of AIA 193, 211, and 335 Å display a very faint and diffuse loop that propagates westward from the middle field lobe of the AR complex, up to $\sim 20^\circ$ west from the PIL, during 14:15 – 15:15 UT. The loop appears to be rooted in or near the initial positive flare ribbon, *i.e.* in or near the LS. In the southward direction it cannot be traced beyond the edge of AR 11514 against the complex and changing background. This is consistent with the dimming at the initial negative flare ribbon being its southern footpoint. The two initial dimmings deepen synchronously with the westward motion of the loop. The detection in the 193, 211, and 335 Å bands, and the non-detection in the 94 and 131 Å bands, indicate a temperature in the range 1.5 MK $\lesssim T < 7$ MK. The structure appears to be similar to the hot channel detected in other events, mostly above the limb (*e.g.*, Zhang, Cheng, and Ding, 2012), with the lower temperature here corresponding to the moderate intensity of the flare. The hot channel is assumed to show the erupting flux rope.

Within the next half hour, the ribbons quickly extend along the PIL, especially the ribbon in the weaker negative flux, which then covers the whole north-south extent of negative flux

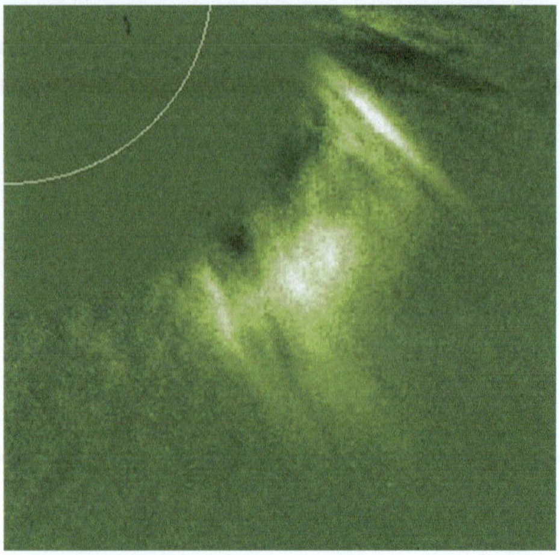

Figure 11 COR1-B running-difference image at 15:30 UT showing the considerable southward angular extent of the erupting flux, consistent with it being rooted in the negative-polarity dimming areas south of the AR complex, which are marked in Figure 10.

at the trailing side of the enhanced network and AR 11514 (this is indicated in Figure 10(a)). The other ribbon extends into positive network north of the LS and propagates toward the center of the LS. Simultaneously, several dimming areas begin to develop. These include an extended area of positive network north of the unstable middle flux lobe (where a prominent set of interconnecting loops existed before the eruption and was strongly disturbed), the erupting S-shaped filament and adjacent filament channel in the enhanced network southwest of AR 11514, and a substantial area of negative network extending south from the initial negative ribbon (Figure 10(b)). These dimming areas last for several hours. The three southern ones formed in the first place during the pre-eruption activities (rise of the large loops south of AR 11515 and rise of flux above the S-shaped filament and adjacent filament channel southwest of AR 11514 from 13:12 UT onward), but extend farther in the course of the eruptive flare. They indicate that the erupting flux reconnects and exchanges footpoints with originally overlying flux rooted in at least some of the dimming areas.

This is supported by the EUVI-B 195 Å images of the eruption, which show a loop expanding radially and southward above the AR complex from about 14:40 UT onward (then at a height of $\sim 0.3\ R_{\odot}$). The loop appears to be rooted in the AR complex and up to 20° south of it (the precise location being masked by foreground and background loops). This is consistent with the dimmings in the positive network regions southwest and south of the AR complex. Since the loop in the EUVI-B 195 Å images is too faint for a visualization in still images, even when using difference images, Figure 11 shows a COR1-B difference image of the CME at an early stage, when it still expands in the same latitude range as the loop imaged by EUVI-B at 195 Å.

The direction of the flux overlying the middle field lobe of the AR complex, implied by the location of the dimmings in AIA images, is roughly from northeast to southwest, *i.e.* rather close to the direction of the flux in the lobe, especially if the sheared loops in the lobe are considered. This is different from the configuration of the breakout model for eruptions.

The arcade of flare loops forms at the section of the PIL between the two ARs, supporting the view that the destabilization of sheared flux in the middle lobe of the AR complex caused the eruption. The usual evolution from considerably sheared first flare loops to more potential ones later on can be seen (Figures 10(b) and 12(b–d)). The first flare loops form after 14:38 UT in AIA 94 and 131 Å images ($T \sim 7-11$ MK).

A fast, southward-propagating CME appears in the LASCO/C2 field of view at 15:35 UT. A second-order fit yields a velocity of 800 km s^{-1} at the heliocentric distance of 3 $R_{\odot}$. The propagation direction is southwestward in the field of view of COR1-B (Figure 11), where the CME appears at 15:00 UT.

The C2.4 flare peaking at 15:32 UT, *i.e.* during the rise of the flare in the AR complex studied here, is a compact and confined, likely sympathetic flare in AR 11513.

4.3. Superimposed Confined Flaring Activity

Two confined flares, one in AR 11515, the other in AR 11514, are closely related to the eruptive flare. Both occur under the closed, main field lobe of their bipolar host AR. The one in AR 11515 marks the peak of the emission of the C8.3 flare at 15:45 UT, *i.e.*, is brighter than the eruptive flare. However, the confined flare is clearly a separate event, occurring in a different lobe of flux, where it remains compact and confined, and developing on a much shorter timescale. It is located at the PIL formed by the westward tongue of negative flux in the middle of the AR. In the AIA 131 Å band, it develops only from 15:37 UT onward, one hour after the onset of the eruptive flare. The other confined flare occurs within 5 – 10 minutes from the first at the northern end of the internal PIL in AR 11514. These confined flares are most sharply imaged in the 171 Å band, as shown in Figure 12; they are of course more prominent, but more diffuse, in the "hot" AIA bands, up to 131 Å. Both may be triggered by changes in their overlying flux caused by the rise of the flux in the eruptive flare. In particular, material is seen rising from a position very near the second flare immediately before its onset.

A third confined flare, of X-ray class C2.7, occurs in the NEF during the decay of the eruptive flare, *i.e.* under the flare loop arcade (see also Figure 12). Its loops are oriented north-south, as is part of the field rooted in the newly emerged flux ahead of the LS (Section 3). A detailed inspection of the 171 Å data suggests that the NEF brightens due to reconnection with loops rooted under the flare loop arcade, which show brief brightenings and changes in shape (see the 171 Å data in the animation accompanying Figures 1 and 2). Most likely, the continuously rising NEF pushes against the re-closed field of the middle lobe in the AR complex, and the temporal association with the decay phase of the eruptive flare is accidental.

5. Discussion: Initiation of the Eruption

The eruption occurs at the PIL between ARs 11514 and 11515, under the middle lobe of a three-lobe magnetic configuration, where the field is considerably sheared and a filament channel is forming. Moreover, new flux is emerging under the lobe near its edge. The amount of new flux emerged by the time of the eruption is only $\sim$40 % of the flux in the middle lobe. The observations are inconclusive with regard to the question whether a flux rope was formed (or is forming) above the PIL, or whether the flux has the structure of a sheared arcade.

Both basic possibilities for the triggering of eruptions appear consistent with the data: the equilibrium of sheared (current-carrying) flux can be destroyed either by an increase of the current beyond the limit of stability or by a weakening of the external flux (rooted in photospheric sources other than the sheared flux) by a sufficient amount, such that it can no longer stabilize the upward-directed Lorentz self-force of the given current. The first option is supported by the flow toward the PIL from the trailing side of AR 11514. This transports negative flux to the PIL, some of which cancels there with positive network flux, increasing the shear and current above the PIL. However, by the amount of flux involved

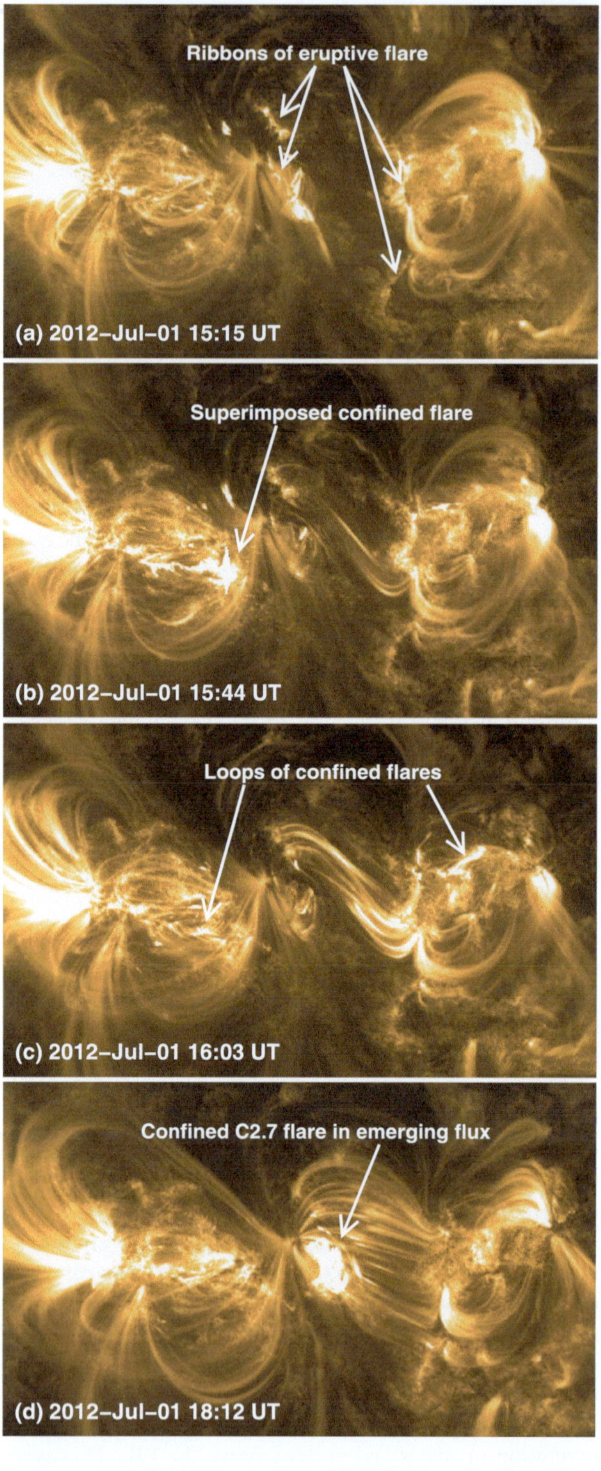

Figure 12 AIA 171 Å images of the additional confined flares. (a) Early phase of the eruptive flare before the confined flares; (b) peak emission of the combined eruptive and first additional confined flares; (c) loops of the two confined flares in AR 11515 and AR 11514 after their peak times; (d) peak emission of the confined C2.7 flare in the NEF under the arcade of the eruptive flare.

(around 1.5×10^{18} Mx, $\approx$450 times smaller than the amount of emerging flux), this is a very weak effect compared to the multitude of other pre-flare activities that affect the large-scale field structure of the AR complex. Therefore, we consider such "internal triggering" at the PIL to be *far* less likely than an external triggering by changes of the ambient field.

A classical model for external triggering in a quadrupolar source region is the breakout model (Lynch *et al.*, 2008; Karpen, Antiochos, and DeVore, 2012). This model relies on the weakening of the ambient flux in the middle lobe by reconnection with an overlying fourth lobe of essentially opposite direction. We do not find signatures that such a lobe exists and contains a sufficient amount of flux. The PFSS extrapolation is inconclusive with regard to the existence and direction of such a field lobe, but indicates that not much flux could be contained if it exists. The dimmings developing in association with the eruption indicate that the overlying flux is directed relative to the middle field lobe at an angle smaller than 90°, basically opposite to that in the model. Therefore, it is unlikely that the breakout model can explain the onset of the eruption.

The event yields many strong indications that the episode of newly emerging flux under the middle flux lobe triggers interactions with the large-scale field in the AR complex, which lead to an upward expansion of the middle lobe, weakening its stabilizing downward tension force. The most likely scenario in our opinion is that the rise of the large loop system south of AR 11515, which is rooted near the area of NEF and the LS and initially runs at a low elevation, *i.e.* essentially under the middle flux lobe, lifts the lobe quite strongly (see, in particular, Figure 7(f–h)). Additionally, the expansion of the loop system probably triggers the eruption of the S-shaped filament in the enhanced network southwest of AR 11514, and of the filament channel extending westward from the filament. These secondary eruptions perturb the large-scale field structure in the AR complex, as is obvious from the development of dimmings, and very likely have a share in lifting the middle flux lobe. This is indicated by their close temporal association with the main eruption, which commences while they are in progress. Each of the relevant pre-eruption activities commences with, or is amplified by, a jet emission from the NEF. The jets are emitted *into* the large loop system south of AR 11515, which thus has become magnetically connected with the NEF through reconnection. The loop system must carry currents already before the eruption, since the PFSS extrapolation does not yield any low-lying field lines at the location of the loop system. There is no doubt that the currents, and consequently the equilibrium of the loop system, are strongly perturbed when the loop system reconnects with the NEF.

Thus, we trace the origin of the eruption to the NEF, one of the standard options for eruption triggering (*e.g.*, Schrijver, 2009). The particular trigger mechanism in the present event, however, is partly different from the mechanisms considered so far. Both observations (Feynman and Martin, 1995) and modeling find the triggering of eruptions by emerging flux to be most efficient when the new flux is directed oppositely to the existing flux, such that reconnection commences easily. This is true for emergence both near the PIL and near the edge of the arcade field overlying the PIL, although the latter case was found to be restricted in the range of distances to the PIL (Xu, Fang, and Chen, 2008). The reconnection destabilizes a pre-existing flux rope (Chen and Shibata, 2000; Xu, Chen, and Fang, 2005) or transforms the inner part of a pre-existing sheared arcade into an unstable flux rope (Kusano *et al.*, 2012; Kaneko and Yokoyama, 2014). Alternatively, the emerging flux itself forms a flux rope, and the reconnection of this flux rope with a pre-existing arcade is essential for the removal of the rope's own stabilizing envelope, such that the rope will erupt (Archontis and Török, 2008; Archontis and Hood, 2012; Leake, Linton, and Antiochos, 2014). None of these mechanisms appears to be at work in the eruption studied here. The dominant orientation of the NEF is between westward and southward, not much different from the orientation of the pre-existing flux in the middle field lobe, so that the mechanisms based on the occurrence of reconnection with antiparallel or nearly antiparallel field, elaborated in the above articles, are not applicable. Additionally, the amount of flux emerged by the time of the eruption is considerably smaller than the pre-existing flux, clearly insufficient for the alternative arcade compression mechanism studied by Kaneko and Yokoyama (2014).

Our suggestion of destabilization by lifting the overlying flux bears some similarity to the findings in Wang and Sheeley (1999). These authors used a superposition of potential (vacuum) fields to model the effect of new flux emergence and obtained a lifting or sidewards deflection of flux overlying filaments that erupted. This is conceptually different from the lifting of the middle flux lobe by the rise of the activated loop system in our suggested scenario, which must be driven by Lorentz forces, but has a similar effect on the stability of the sheared flux in the filament channel.

The NEF in the present event does reconnect with overlying, pre-existing flux, but this flux is not part of the later erupting middle lobe of the quadrupolar AR complex, but rather part of a side lobe (AR 11515). This flux, seen as a large loop system south of AR 11515, and rooted at both ends in this AR, runs under one side of the middle lobe. Its direction is not much different from the field direction in the middle lobe, and no signs of reconnection with the middle lobe (any brightenings other than the jet ejection from the NEF into the loop system) are seen in either AIA or EUVI-B images. This indicates that the rise of the loop system and the consequent lifting of the flux in the middle lobe is the dominant effect that destabilizes the lobe.

Although the magnetic topology of the eruption may be a relatively rare occurrence, it need not be a singular case. Configurations with both interconnecting loops to a preceding active region and internal loops in an active region emerging from the leading sunspot are common. New flux is often seen to emerge in the vicinity of existing flux concentrations.

The observations do not yield a conclusive hint on whether or not a flux rope already exists by the time the middle lobe is lifted. If none existed before, a rope must form in response to the lifting because otherwise there would be no eruption – there is no ideal MHD instability of a sheared arcade (*e.g.*, Roumeliotis, Sturrock, and Antiochos, 1994; Mikic and Linker, 1994; Amari *et al.*, 1996). One can expect that the lifting of a sufficiently sheared arcade by external forces induces Lorentz forces in the lifted flux that point toward the middle of the arcade and lead to a horizontal constriction that steepens a vertical current sheet in the center of the arcade. This is similar to the enforced rise of an arcade over an unstable flux rope, which leads to a constriction of the flux under the rope, current sheet steepening, and eventually to flare reconnection. Here the reconnection would form a flux rope if none existed before and then continue in the standard manner of flare reconnection, amplified by the eruption of the formed rope.

Following perturbations by the NEF, part of the loop system running south of AR 11515 and under the middle lobe begins to rise. However, all loops find a new equilibrium height (Figure 7), whereas the sheared flux under the middle lobe erupts into a CME. This difference is most likely related to a difference in the nature of the two structure's force-free equilibria. We suggest the following explanation. The current in the sheared flux rooted near the PIL is probably stabilized by the overlying flux of the middle lobe rooted to a large part away from the PIL at the edges of the adjacent ARs. This is the classic force-free equilibrium of flux carrying a net current in external, essentially potential flux, as proposed by van Tend and Kuperus (1978) and elaborated by Titov and Démoulin (1999). This equilibrium admits an instability (torus instability) when the sheared flux transforms, at least partly, into a flux rope above the PIL (van Tend and Kuperus, 1978; Kliem and Török, 2006). It is difficult to see how such an equilibrium could be established for the loop system south of

AR 11515 because there is hardly any significant amount of flux rooted in the area south of the loop system (Figure 2). Therefore, the equilibrium of the loop system, which also carries electric currents (as argued above), is most likely provided by a return current such that the net current along the loop system vanishes. Such force-free equilibria exist for arcades and flux ropes; the flux rope can have the shape of a loop. While the arcades are stable, flux ropes may admit an instability of the helical kink mode. However, the existence of instability for such flux ropes is not yet established (Török and Kliem, 2003; Aulanier, Démoulin, and Grappin, 2005). If instability occurs, a minimum twist of about one field line turn is required. The EUV images of the loop system do not provide any evidence supporting this substantial requirement (see Figures 2 and 8). Therefore, the loop system is probably in a very stable force-free equilibrium. Its reconnection with the NEF must change its currents, such that it acquires a new equilibrium at a higher position without erupting.

6. Conclusions

The present investigation of an eruptive C8.3 flare and fast CME on 2012 July 1 in the AR complex consisting of ARs 11514 and 11515 reveals a new scenario of eruption triggered by emerging flux. The emergence occurs under the middle flux lobe of a quadrupolar configuration (between the two ARs) far away from the PIL. Most of the emerging flux is roughly aligned with the arcade-like large-scale flux in this lobe, not favorable for reconnection with it. Moreover, the amount of emerging flux remains moderate, reaching only $\sim$40 % of the pre-existing flux of the lobe by the time of the eruption. A relatively particular circumstance is given by the fact that a large loop system of the trailing AR 11515, rooted in the leading spot of the region, passes under the middle lobe, which is also largely rooted in this spot. The new flux emerges in front of the spot, under the loop system. It reconnects with the loop system, ejecting a series of jets into the loop system, which then rises considerably (changing its inclination to the surface from 25° to 70°), but does not erupt by itself. The rise of the loop system lifts the flux of the middle lobe, reducing its stabilizing downward tension. This triggers the eruption of the strongly sheared flux in the center of the lobe.

The eruption is also noteworthy for its location. The structure of the magnetogram would hardly suggest that an eruption occurs in the extended area of very weak field between the two ARs, one of them already decaying. Instead, one would expect an eruption to occur in the complex field of the developing AR, which also harbors the emergence of new flux in the immediate vicinity of its leading spot (where an eruption, analyzed in Lo14, indeed occurs 19 hours later). However, the chromospheric and coronal structures indicate magnetic shear at the PIL between the ARs, correctly pointing at the site of the eruption studied here.

Acknowledgements We gratefully acknowledge constructive comments by the referee, which were helpful in improving the clarity of this article. R.E.L. is grateful for the financial assistance from the German Science Foundation (DFG) under grant DE 787/3-1 and the European Commission's FP7 Capacities Programme under the Grant Agreement number 312495. G.C. and B.K. acknowledge support by the NSF under Grant No. 1249270. B.K. also acknowledges support by the DFG. HMI data are courtesy of NASA/SDO and the HMI science team. They are provided by the Joint Science Operations Center – Science Data Processing at Stanford University. EUVI-B and COR1-B images are supplied courtesy of the STEREO Sun Earth Connection Coronal and Heliospheric Investigation (SECCHI) team. This work utilizes data obtained by the Global Oscillation Network Group (GONG) Program, managed by the National Solar Observatory, which is operated by AURA, Inc. under a cooperative agreement with the National Science Foundation. The data were acquired by instruments operated by the Big Bear Solar Observatory, High Altitude Observatory, Learmonth Solar Observatory, Udaipur Solar Observatory, Instituto de Astrofísica de Canarias, and Cerro Tololo Interamerican Observatory. We have used the SOHO/LASCO CME catalog, generated and maintained at the CDAW Data Center by NASA and The Catholic University of America in cooperation with the Naval Research Laboratory. SOHO is a project of international cooperation between ESA and NASA.

References

Amari, T., Luciani, J.F., Aly, J.J., Tagger, M.: 1996, Plasmoid formation in a single sheared arcade and application to coronal mass ejections. *Astron. Astrophys.* **306**, 913. ADS.

Amari, T., Luciani, J.F., Aly, J.J., Mikic, Z., Linker, J.: 2003, Coronal mass ejection: Initiation, magnetic helicity, and flux ropes. II. Turbulent diffusion-driven evolution. *Astrophys. J.* **595**, 1231. DOI. ADS.

Amari, T., Aly, J.-J., Luciani, J.-F., Mikic, Z., Linker, J.: 2011, Coronal mass ejection initiation by converging photospheric flows: Toward a realistic model. *Astrophys. J. Lett.* **742**, L27. DOI. ADS.

Archontis, V., Hood, A.W.: 2010, Flux emergence and coronal eruption. *Astron. Astrophys.* **514**, A56. DOI. ADS.

Archontis, V., Hood, A.W.: 2012, Magnetic flux emergence: A precursor of solar plasma expulsion. *Astron. Astrophys.* **537**, A62. DOI. ADS.

Archontis, V., Török, T.: 2008, Eruption of magnetic flux ropes during flux emergence. *Astron. Astrophys.* **492**, L35. DOI. ADS.

Aulanier, G., Démoulin, P., Grappin, R.: 2005, Equilibrium and observational properties of line-tied twisted flux tubes. *Astron. Astrophys.* **430**, 1067. DOI. ADS.

Aulanier, G., Török, T., Démoulin, P., DeLuca, E.E.: 2010, Formation of torus-unstable flux ropes and electric currents in erupting sigmoids. *Astrophys. J.* **708**, 314. DOI. ADS.

Burtseva, O., Petrie, G.: 2013, Magnetic flux changes and cancellation associated with X-class and M-class flares. *Solar Phys.* **283**, 429. DOI. ADS.

Chen, P.F., Shibata, K.: 2000, An emerging flux trigger mechanism for coronal mass ejections. *Astrophys. J.* **545**, 524. DOI. ADS.

Démoulin, P., van Driel-Gesztelyi, L., Schmieder, B., Hémoux, J.C., Csepura, G., Hagyard, M.J.: 1993, Evidence for magnetic reconnection in solar flares. *Astron. Astrophys.* **271**, 292. ADS.

Feynman, J., Martin, S.F.: 1995, The initiation of coronal mass ejections by newly emerging magnetic flux. *J. Geophys. Res.* **100**, 3355. DOI. ADS.

Forbes, T.G., Priest, E.R.: 1995, Photospheric magnetic field evolution and eruptive flares. *Astrophys. J.* **446**, 377. DOI. ADS.

Gibson, S.E., Fan, Y.: 2006, The partial expulsion of a magnetic flux rope. *Astrophys. J. Lett.* **637**, L65. DOI. ADS.

Green, L.M., Kliem, B., Wallace, A.J.: 2011, Photospheric flux cancellation and associated flux rope formation and eruption. *Astron. Astrophys.* **526**, A2. DOI. ADS.

Hagyard, M.J., Venkatakrishnan, P., Smith, J.B. Jr.: 1990, Nonpotential magnetic fields at sites of gamma-ray flares. *Astrophys. J. Suppl.* **73**, 159. DOI. ADS.

Harvey, J.W., Hill, F., Hubbard, R.P., Kennedy, J.R., Leibacher, J.W., Pintar, J.A., Gilman, P.A., Noyes, R.W., Title, A.M., Toomre, J., Ulrich, R.K., Bhatnagar, A., Kennewell, J.A., Marquette, W., Patron, J., Saa, O., Yasukawa, E.: 1996, The Global Oscillation Network Group (GONG) project. *Science* **272**, 1284. DOI. ADS.

Harvey, J.W., Bolding, J., Clark, R., Hauth, D., Hill, F., Kroll, R., Luis, G., Mills, N., Purdy, T., Henney, C., Holland, D., Winter, J.: 2011, Full-disk solar H-alpha images from GONG. In: *AAS/Solar Physics Division Abstracts* **#42**, 1745. ADS.

Howard, R.A., Moses, J.D., Vourlidas, A., Newmark, J.S., Socker, D.G., Plunkett, S.P., Korendyke, C.M., Cook, J.W., Hurley, A., Davila, J.M., Thompson, W.T., St Cyr, O.C., Mentzell, E., Mehalick, K., Lemen, J.R., Wuelser, J.P., Duncan, D.W., Tarbell, T.D., Wolfson, C.J., Moore, A., Harrison, R.A., Waltham, N.R., Lang, J., Davis, C.J., Eyles, C.J., Mapson-Menard, H., Simnett, G.M., Halain, J.P., Defise, J.M., Mazy, E., Rochus, P., Mercier, R., Ravet, M.F., Delmotte, F., Auchere, F., Delaboudiniere, J.P., Bothmer, V., Deutsch, W., Wang, D., Rich, N., Cooper, S., Stephens, V., Maahs, G., Baugh, R., McMullin, D., Carter, T.: 2008, Sun Earth Connection Coronal and Heliospheric Investigation (SECCHI). *Space Sci. Rev.* **136**, 67. DOI. ADS.

Kaneko, T., Yokoyama, T.: 2014, Simulation study of solar plasma eruptions caused by interactions between emerging flux and coronal arcade fields. *Astrophys. J.* **796**, 44. DOI. ADS.

Karpen, J.T., Antiochos, S.K., DeVore, C.R.: 2012, The mechanisms for the onset and explosive eruption of coronal mass ejections and eruptive flares. *Astrophys. J.* **760**, 81. DOI. ADS.

Kliem, B., Török, T.: 2006, Torus instability. *Phys. Rev. Lett.* **96**(25), 255002. DOI. ADS.

Kusano, K., Bamba, Y., Yamamoto, T.T., Iida, Y., Toriumi, S., Asai, A.: 2012, Magnetic field structures triggering solar flares and coronal mass ejections. *Astrophys. J.* **760**, 31. DOI. ADS.

Leake, J.E., Linton, M.G., Antiochos, S.K.: 2014, Simulations of emerging magnetic flux. II. The formation of unstable coronal flux ropes and the initiation of coronal mass ejections. *Astrophys. J.* **787**, 46. DOI. ADS.

Lemen, J.R., Title, A.M., Akin, D.J., Boerner, P.F., Chou, C., Drake, J.F., Duncan, D.W., Edwards, C.G., Friedlaender, F.M., Heyman, G.F., Hurlburt, N.E., Katz, N.L., Kushner, G.D., Levay, M., Lindgren, R.W., Mathur, D.P., McFeaters, E.L., Mitchell, S., Rehse, R.A., Schrijver, C.J., Springer, L.A., Stern, R.A., Tarbell, T.D., Wuelser, J.-P., Wolfson, C.J., Yanari, C., Bookbinder, J.A., Cheimets, P.N., Caldwell, D., Deluca, E.E., Gates, R., Golub, L., Park, S., Podgorski, W.A., Bush, R.I., Scherrer, P.H., Gummin, M.A., Smith, P., Auker, G., Jerram, P., Pool, P., Soufli, R., Windt, D.L., Beardsley, S., Clapp, M., Lang, J., Waltham, N.: 2012, The Atmospheric Imaging Assembly (AIA) on the Solar Dynamics Observatory (SDO). *Solar Phys.* **275**, 17. DOI. ADS.

Li, H., Sakurai, T., Ichimoto, K., UeNo, S.: 2000, Magnetic field evolution leading to solar flares I. Cases with low magnetic shear and flux emergence. *Publ. Astron. Soc. Japan* **52**, 465. ADS.

Lin, J., Forbes, T.G.: 2000, Effects of reconnection on the coronal mass ejection process. *J. Geophys. Res.* **105**, 2375. DOI. ADS.

Lin, J., Forbes, T.G., Isenberg, P.A.: 2001, Prominence eruptions and coronal mass ejections triggered by newly emerging flux. *J. Geophys. Res.* **106**, 25053. DOI. ADS.

Lites, B.W.: 2005, Magnetic flux ropes in the solar photosphere: The vector magnetic field under active region filaments. *Astrophys. J.* **622**, 1275. DOI. ADS.

Livi, S.H.B., Martin, S., Wang, H., Ai, G.: 1989, The association of flares to cancelling magnetic features on the Sun. *Solar Phys.* **121**, 197. DOI. ADS.

Louis, R.E., Puschmann, K.G., Kliem, B., Balthasar, H., Denker, C.: 2014, Sunspot splitting triggering an eruptive flare. *Astron. Astrophys.* **562**, A110. DOI. ADS.

Low, B.C.: 1996, Solar activity and the corona. *Solar Phys.* **167**, 217. DOI. ADS.

Lynch, B.J., Antiochos, S.K., DeVore, C.R., Luhmann, J.G., Zurbuchen, T.H.: 2008, Topological evolution of a fast magnetic breakout CME in three dimensions. *Astrophys. J.* **683**, 1192. DOI. ADS.

Mackay, D.H., van Ballegooijen, A.A.: 2006, Models of the large-scale corona. I. Formation, evolution, and liftoff of magnetic flux ropes. *Astrophys. J.* **641**, 577. DOI. ADS.

Mackay, D.H., Karpen, J.T., Ballester, J.L., Schmieder, B., Aulanier, G.: 2010, Physics of solar prominences: II – Magnetic structure and dynamics. *Space Sci. Rev.* **151**, 333. DOI. ADS.

Manchester, W. IV, Gombosi, T., DeZeeuw, D., Fan, Y.: 2004, Eruption of a buoyantly emerging magnetic flux rope. *Astrophys. J.* **610**, 588. DOI. ADS.

Martin, S.F.: 1998, Conditions for the formation and maintenance of filaments (invited review). *Solar Phys.* **182**, 107. DOI. ADS.

Martin, S.F., Dezso, L., Antalova, A., Kucera, A., Harvey, K.L.: 1982, Emerging magnetic flux, flares and filaments – FBS interval 16–23 June 1980. *Adv. Space Res.* **2**, 39. DOI. ADS.

Martres, M.-J., Michard, R., Soru-Iscovici, I., Tsap, T.T.: 1968, Étude de la localisation des éruptions dans la structure magnétique évolutive des régions actives solaires. *Solar Phys.* **5**, 187. DOI. ADS.

Mikic, Z., Linker, J.A.: 1994, Disruption of coronal magnetic field arcades. *Astrophys. J.* **430**, 898. DOI. ADS.

Pesnell, W.D., Thompson, B.J., Chamberlin, P.C.: 2012, The Solar Dynamics Observatory (SDO). *Solar Phys.* **275**, 3. DOI. ADS.

Roumeliotis, G., Sturrock, P.A., Antiochos, S.K.: 1994, A numerical study of the sudden eruption of sheared magnetic fields. *Astrophys. J.* **423**, 847. DOI. ADS.

Rust, D.M.: 1972, Flares and changing magnetic fields. *Solar Phys.* **25**, 141. DOI. ADS.

Sakajiri, T., Brooks, D.H., Yamamoto, T., Shiota, D., Isobe, H., Akiyama, S., Ueno, S., Kitai, R., Shibata, K.: 2004, A study of a tiny two-ribbon flare driven by emerging flux. *Astrophys. J.* **616**, 578. DOI. ADS.

Savcheva, A.S., Green, L.M., van Ballegooijen, A.A., DeLuca, E.E.: 2012, Photospheric flux cancellation and the build-up of sigmoidal flux ropes on the Sun. *Astrophys. J.* **759**, 105. DOI. ADS.

Schou, J., Scherrer, P.H., Bush, R.I., Wachter, R., Couvidat, S., Rabello-Soares, M.C., Bogart, R.S., Hoeksema, J.T., Liu, Y., Duvall, T.L., Akin, D.J., Allard, B.A., Miles, J.W., Rairden, R., Shine, R.A., Tarbell, T.D., Title, A.M., Wolfson, C.J., Elmore, D.F., Norton, A.A., Tomczyk, S.: 2012, Design and ground calibration of the Helioseismic and Magnetic Imager (HMI) instrument on the Solar Dynamics Observatory (SDO). *Solar Phys.* **275**, 229. DOI. ADS.

Schrijver, C.J.: 2009, Driving major solar flares and eruptions: A review. *Adv. Space Res.* **43**, 739. DOI. ADS.

Schuck, P.W.: 2006, Tracking magnetic footpoints with the magnetic induction equation. *Astrophys. J.* **646**, 1358. DOI. ADS.

Sterling, A.C., Chifor, C., Mason, H.E., Moore, R.L., Young, P.R.: 2010, Evidence for magnetic flux cancelation leading to an ejective solar eruption observed by Hinode, TRACE, STEREO, and SoHO/MDI. *Astron. Astrophys.* **521**, A49. DOI. ADS.

Sun, X., Hoeksema, J.T., Liu, Y., Wiegelmann, T., Hayashi, K., Chen, Q., Thalmann, J.: 2012, Evolution of magnetic field and energy in a major eruptive active region based on SDO/HMI observation. *Astrophys. J.* **748**, 77. DOI. ADS.

Titov, V.S., Démoulin, P.: 1999, Basic topology of twisted magnetic configurations in solar flares. *Astron. Astrophys.* **351**, 707. ADS.

Török, T., Kliem, B.: 2003, The evolution of twisting coronal magnetic flux tubes. *Astron. Astrophys.* **406**, 1043. DOI. ADS.

van Ballegooijen, A.A., Martens, P.C.H.: 1989, Formation and eruption of solar prominences. *Astrophys. J.* **343**, 971. DOI. ADS.
van Tend, W., Kuperus, M.: 1978, The development of coronal electric current systems in active regions and their relation to filaments and flares. *Solar Phys.* **59**, 115. DOI. ADS.
Wang, H.: 2006, Rapid changes of photospheric magnetic fields around flaring magnetic neutral lines. *Astrophys. J.* **649**, 490. DOI. ADS.
Wang, J., Shi, Z.: 1993, The flare-associated magnetic changes in an active region. II – Flux emergence and cancellation. *Solar Phys.* **143**, 119. DOI. ADS.
Wang, Y.-M., Sheeley, N.R. Jr.: 1999, Filament eruptions near emerging bipoles. *Astrophys. J. Lett.* **510**, L157. DOI. ADS.
Xu, X.-Y., Chen, P.-F., Fang, C.: 2005, A parametric survey of the CME triggering process by numerical simulations. *Chin. J. Astron. Astrophys.* **5**, 636. DOI. ADS.
Xu, X.-Y., Fang, C., Chen, P.-F.: 2008, A statistical study on the filament eruption caused by new emerging flux. *Chin. Astron. Astrophys.* **32**, 56. DOI. ADS.
Zhang, J., Cheng, X., Ding, M.-D.: 2012, Observation of an evolving magnetic flux rope before and during a solar eruption. *Nat. Commun.* **3**, 747. DOI. ADS.
Zhao, X., Hoeksema, J.T.: 1995, Prediction of the interplanetary magnetic field strength. *J. Geophys. Res.* **100**, 19. DOI. ADS.

Reprinted from *Solar Physics* Journal, DOI 10.1007/s11207-015-0752-6

The Origin of Superflares on G-Type Dwarf Stars of Various Ages

M.M. Katsova[1] · **M.A. Livshits**[2]

Received: 2 December 2014 / Accepted: 28 July 2015 / Published online: 7 August 2015
© Springer Science+Business Media Dordrecht 2015

Abstract We analyze new observations of superflares on G-stars discovered in the optical and near-IR ranges with the *Kepler* mission. An evolution of solar-type activity is discussed. We give an estimate of the maximal total energy, $E_{\mathrm{tot}} = 10^{34}$ erg, of a flare that could have occurred on the young Sun at its age of 1 Gyr when the cycle was formed. We believe that the main source of the flare optical continuum is a low-temperature condensation forming in the course of the response of the chromosphere to an impulsive heating. For a superflare on the young Sun, we adopt an accelerated electron flux, $F_e\ (E > 20\ \mathrm{keV}) = 3 \times 10^{11}\ \mathrm{erg\,cm^{-2}\,s^{-1}}$, that is limited by the return current and obtain that the area of the optical continuum source on a G-star is $S \approx 10^{19}\ \mathrm{cm}^2$. This value is close to the area of the Hα ribbons in the largest solar flares, while the area of bright patches of a white-light flare on the contemporary Sun is lower by about two orders of magnitude. At the same electron flux and hard electron spectrum, a stellar flare of similar energy should be accompanied by a microwave source of about 2 mJy at frequencies 10 – 100 GHz at a distance of 100 pc. We discuss the possible detection of the flare-produced lithium in the course of spallation reactions. The detection of the flare microwave source and the emission in the Li resonant line could demonstrate how effective particle acceleration can be on stars in the lower part of the main sequence.

Keywords Flares · Solar · Stellar · Particle acceleration · Magnetic fields

1. Introduction

The first observations of powerful flares of the optical continuum in G stars testify that huge numbers of particles, in particular, electrons, are accelerated during these events. These

Solar and Stellar Flares: Observations, Simulations, and Synergies
Guest Editors: Lyndsay Fletcher and Petr Heinzel

✉ M.M. Katsova
maria@sai.msu.ru

[1] Sternberg State Astronomical Institute, Lomonosov Moscow State University, 119991 Moscow, Russia

[2] Pushkov Institute of Terrestrial Magnetism, Ionosphere, and Radio Wave Propagation, Russian Academy of Sciences, 142190 Troitsk, Moscow, Russia

results of the *Kepler* mission were presented by Maehara *et al.* (2012), who reported on total bolometric energies of the flare emission from 10^{33} to 10^{36} erg.

Maehara *et al.* (2012) presented observations of 148 G stars that produced 365 superflares with a total energy $> 3 \times 10^{33}$ erg registered during 120 days. The *Kepler* mission monitoring was made in the V and R bands and in two time-resolution regimes, short-cadence, 1 min, and long-cadence, 30 min. The first results were based on a set of data with the lower time resolution.

The energies of white-light stellar flares were recalculated in bolometric values in black-body approximation. Maehara *et al.* (2012) estimated the occurrence frequency of flares with energies from 10^{33} to 10^{36} erg to be somewhat higher for the cooler G stars ($T_{\rm eff} = 5100-5600$ K) than that for the hotter G stars ($T_{\rm eff} = 5600-6000$ K). Superflares were found to occur more often on G stars rotating with periods from a few hours to several days; the frequencies of these events on stars with rotational periods > 10 days are several times smaller.

The *Kepler* data of monitoring during 500 days were analyzed in more detail by Shibayama *et al.* (2013) and Balona (2015). The results by Shibayama *et al.* (2013) were based on a longer series of observations and confirmed those obtained by Maehara *et al.* (2012). They provide a reliable flare frequency and evidence that superflares occur quite rarely on G stars. Balona (2015) found that flare energies are strongly correlated with stellar luminosity and radius.

The detection of such powerful non-stationary processes on normal late-type stars raises the question about the principal possibility of these phenomena that may require extreme physical conditions such as strong magnetic fields, close companions (a star or a planet), comet impact etc. But here we do not go beyond the standard knowledge about the physics of flares. The most effective particle acceleration takes place during some of the largest flares on the Sun. These events are known well and are studied in detail. By analyzing the white-light solar flare on 14 July 2000, Livshits and Belov (2004) assumed that the source of this effective acceleration is situated in close proximity to a sunspot. This was confirmed by the analysis of other powerful events such as those in August 1972 (Svestka, 1976) and on 28 October 2003, but most clearly, it was seen during the flare on 20 January 2005 above sunspot umbrae (Grechnev *et al.*, 2008). These high-energy phenomena are widely discussed (see, for instance, the review by Ramaty, 1986; Fletcher *et al.*, 2011). Before the *Kepler* observations, it was hard to imagine that the acceleration in moderate magnetic fields, observed on normal G stars, could be much more efficient than in large solar flares.

We assume that the basic property of superflares is an efficient acceleration of electrons and protons. These particles are the cause of both line emission and optical continuum in solar flares. In superflares, this continuum emission is extremely intense. The dynamic response of the solar chromosphere to impact by accelerated electrons leads to the formation of the hard X-ray radiation, line and continuum emission, in particular, in optical wavelengths. The explosive evaporation of the chromosphere occurs in many large impulsive flares on the Sun and red dwarf stars. We believe that this process also takes place in superflares.

The aim of this work is to understand this strong optical continuum emission in the framework of the current knowledge about flares on the Sun and red dwarfs.

The solution of this basic problem requires information about G stars with superflares. As a rule, these stars are younger than the present Sun. The current experience of flare studies relates to phenomena observed on the Sun and other G stars of comparable age. Therefore, one needs to make an educated guess where superflares occur and how they are associated with other active phenomena on these stars.

When the Sun was younger than 1 Gyr, its non-stationary processes may have differed significantly from solar events in the contemporary epoch. The general character of activity of late-type stars depends on their age. The activity level of open cluster members is noticeably higher than in the present Sun. The problem of the evolution of stellar activity is widely discussed after the article by Barnes (2003). A method for estimating the stellar age from their activity level (a single-parametric gyrochronology) was suggested by Mamajek and Hillenbrand (2008). For a set of phenomena as those observed on the Sun today, including formation of the regular cycle, we use the term "solar-type activity".

One can expect that this type of activity was formed on a main-sequence G star to the age slightly older than 1 Gyr (see, for example, Katsova and Livshits 2011; 2014). On young stars of ages from 50 to 500 Myr, the character of the magnetic fields and non-stationary processes differ from the solar ones. The basic distinction is due to a greater impact of the large-scale magnetic fields on these objects, in particular, to the longer duration of powerful flares. In this context, the *Kepler* observations of superflares can shed light on age-related changes of flare activity from young T Tau-type stars to the young Sun.

So far, observations have demonstrated that the characteristics of these events, apart from their total energy, do not differ significantly from the largest solar analogs. Indeed, their duration in the optical is from a few minutes to one or several hours, the temporal profiles are similar to many X-class flares on the Sun. This was the reason why the first theoretical analysis by Shibayama *et al.* (2013) and some other works used simple scaling relationships. A disadvantage of this approach is that to explain the observed powerful phenomena, one needs to assume that the area occupied by spots is close to 100 % of the stellar surface. The relation between spot areas and the flare energy was investigated recently by Aulanier *et al.* (2013). Of course, because observed values of stellar spots on young G star do not exceed 30 % (Berdyugina, 2005), the simple scaling relationship by Shibayama *et al.* (2013) cannot be applied to explain superflares of the largest energies. It is necessary here to analyze the possibility that strong optical emission is generated, which probably is more effective than the emission on the contemporary Sun.

We consider a few outstanding problems. The first of them relates to the question whether it is possible to explain the strong optical continuum observed during superflares in the framework of the gas-dynamic response of the chromosphere to an impact of accelerated particles. The second question is whether superflares are associated with the efficient acceleration of particles. This can be verified through their manifestations in flares. We estimate the microwave radiation of superflares and discuss the possibility of detecting it. Then, we discuss under which conditions a nuclear spallation reaction can occur during superflares and estimate the resulting expected enhancement of the resonant Li lines at 6708 Å.

The question posed in the title of the article by Shibata *et al.* (2013) is likewise of interest: "Can superflares occur on our Sun?". The answer requires analyzing the relation between the development of flares and the magnetic fields, which have recently been observed on G stars (Marsden *et al.*, 2014). We formulate the problem as follows: Are superflares similar to solar phenomena? Are there hypothetical possibilities to explain events with such power, which cannot in principle occur on the Sun, while they are observed on the youngest stars?

2. Evaluation of the Accelerated Electron Fluxes for Flares Registered with the *Kepler* Mission

First of all, we discuss the problem of formation of the strong flare optical continuum. It was previously thought that white-light flares occur only very rarely on the Sun. However, the *Yohkoh* observations (Matthews *et al.*, 2003) and the data obtained with TRACE (Hudson,

Wolfson, and Metcalf, 2006; Fletcher *et al.*, 2007) in the blue spectral range have demonstrated a greater prevalence of these phenomena.

Great progress has been achieved in the past years in the study of the X-ray radiation of flares. The results of the RHESSI observations in particular deserve to be mentioned here, which permitted to obtain comprehensive information about the accelerated electron fluxes. These particles can be responsible for the heating and radiation by the flares in different spectral ranges. The sudden enhancement of the optical continuum has also been observed during flares on red dwarf stars (Gershberg, 2005). Modern observations of stellar flares give evidence for both the common properties and some differences of physical processes on the Sun and red dwarfs (Schmitt *et al.*, 2008; Kowalski *et al.*, 2013).

The *Kepler* observations of late-type stars were performed in the optical range 4300–8300 Å. Here we use the total energies of the optical flare emission presented by Shibayama *et al.* (2013). We assume that the basic source of the optical continuum of large solar flares is a low-temperature condensation, which is formed as a result of the gas-dynamical response of the chromosphere to the impulsive heating by the accelerated electron flux. As is well known, the first numerical modeling of this process was carried out by Kostyuk and Pikelner (1975). This approach was further developed by Livshits *et al.* (1981) in the article entitled "The optical continuum of solar and stellar flares". The corresponding model of impulsive flares on red dwarf stars was developed by Katsova, Boiko, and Livshits (1997). An exhaustive physical justification of individual aspects of the process of explosive evaporation was given by Fisher, Canfield, and McClymont (1985). We recall that the gas-dynamical response includes the primary heating of a small area of the chromosphere, formation of two shock waves propagating upward and downward, and the formation of the low-temperature condensation between the shock front and downward-moving temperature jump. This is accompanied by an upward outflow of hot plasma, and as a result, hot gas gradually fills the coronal loops. Over the past years many articles have appeared that interpret individual events on the Sun and the stars, and now we can assume that this process is actually implemented in impulsive flares (see, for instance, Antonucci, 1989; Liu *et al.*, 2006; Martínez Oliveros *et al.*, 2012).

When modeling processes of a superflare with an extremely high total energy, two main problems have to be taken into account. First, at high fluxes of the accelerated electrons, a return current arises that prevents energy from penetrating into deep chromospheric layers. The gas-dynamical modeling with return current was performed by Karlicky and Henoux (1992). This complicated problem was further considered by Matthews, Brown, and van Driel-Gesztelyi (1998), Zharkova (2000), and Lee, Büchner, and Elkina (2008). Calculations showed that this electron flux is so high possibly because of processes in the outer atmosphere of the Sun (Boiko and Livshits, 1999). In our calculations we assume that the optical continuum emission source arises in powerful events when the energy of directly precipitating electrons F_e reaches $3 \times 10^{11}\ \mathrm{erg\,cm^{-2}\,s^{-1}}$. Here, for definiteness, we assume that the low-energy cutoff of these electrons is 20 keV in X-ray spectra of the largest flares.

Second, such strong electron fluxes during impulsive events are a source of perturbations that can penetrate as deeply as the region of the temperature minimum, *i.e.*, to the upper boundary of the photosphere. When we modeled solar impulsive flares, we assumed that the radiative losses were determined mainly by the hydrogen line emission. In superflares, physical conditions in a low-temperature condensation are as follows: the distribution of the hydrogen atoms over the energy levels reaches equilibrium, and the energy losses in the hydrogen lines are not enough to compensate for the heating inside the condensation. Therefore, we have to consider motions of the plasma, the excitation and ionization of the hydrogen in a homogeneous layer behind the downward-moving shock front in solar and

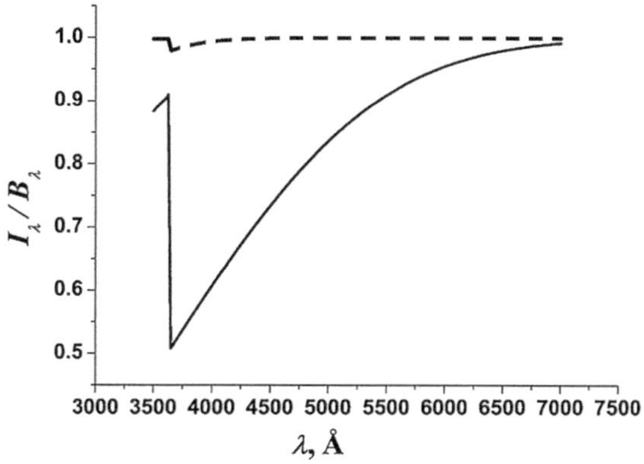

Figure 1 Theoretical spectrum of the optical continuum of a homogeneous 10 km-layer of hydrogen plasma with a temperature of 11 600 K. The ratio of intensity of the optical radiation of this layer to the corresponding black-body radiation, I_λ / B_λ (T), is given for densities 7×10^{15} cm^{-3} and 3×10^{16} cm^{-3} for the lower and the upper (dashed) lines correspondingly. The optical thickness for both cases are 1.05 and 5.79 at $\lambda = 4170$ Å.

stellar flares. The optical depth of this condensation in the optical continuum is close to one for the most powerful events.

Here the new advances in the elementary processes in partially ionized plasma are taken into account. The transition from the optically thin to optically thick plasma in the optical continuum of a low-temperature condensation is studied in detail by Morchenko, Bychkov, and Livshits (2015) for the atmosphere of a red dwarf (when the influence of the photospheric radiation can be neglected). We apply the two-temperature approximation where the electrons and ions are heated differently behind the shock front. In our gas-dynamics model, the condensation is presented as a homogeneous planar layer with a fixed thickness of 10 km. The electron temperature is 11 600 K (1 eV). At high densities the optical spectrum represents an intensified optical continuum, against which the lines are seen as small maxima of their centers. As an example, the results of modeling the optical continuum are given in Figure 1: the lower curve corresponds to a density of 7×10^{15} cm^{-3}, and the upper curve is for 3×10^{16} cm^{-3}. It is clear that the intensity ratio before and behind the Balmer limit decreases, which means that the Balmer jump is diminished. The optical thickness at $\lambda = 4170$ Å is 1.05 and 5.79 for both cases correspondingly. This means that the emission of this 10 km-layer at $T = 11\,600$ K and $n = 3 \times 10^{16}$ cm^{-3} almost becomes black-body radiation.

This approach permits us to evaluate the contribution of the radiation of the gas-dynamical condensation to the optical continuum of solar and stellar flares of different energies. Observations of superflares allow us to evaluate the area of the optical emission source in the framework of the gas-dynamics model. In powerful events, the radiation approaches that of a black body, and the total emitted energy is

$$E_{\text{flare}} = \sigma T^4 S \Delta t, \tag{1}$$

where S is the area of the optical source and Δt is the duration of the radiation.

As we show below, the evolution of the magnetic fields on a G star with the $R = R_\odot$ can lead to flares with a maximal total energy 10^{34} erg. Our point is that we make no distinction between the blue and the red continua and adopt the temperature of the emission as 10 000 K. Then from the expression (1) for a flare lasting $\Delta t = 0.5$ hour, we obtain $S = 1.0 \times 10^{19}$ cm^2. This value is slightly higher than the Hα ribbon area of the greatest solar flares.

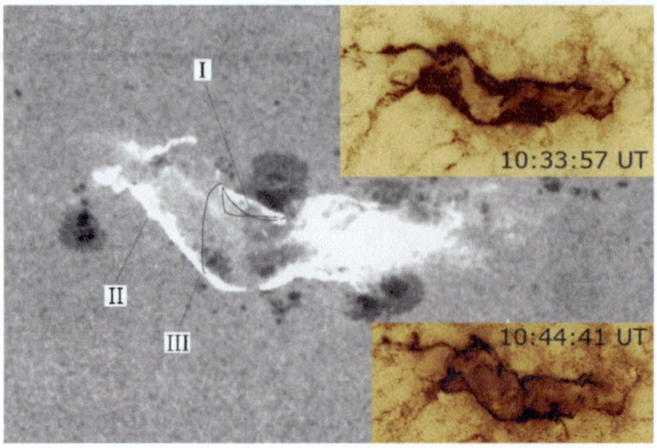

Figure 2 A schema of the position of sources I, II, and III of the optical radiation of superflares. Two TRACE images of the flare on 14 July 2000 obtained close to the γ-ray burst maximum at 10:27 UT, in white light and at 195 Å, are superimposed. Two ribbons at 195 Å emitted weakly at the loop top are seen (the brightness of the western part is reduced). The TRACE images at 1600 Å are presented for 10:33:57 UT (the upper right corner) and 10:44:41 UT (in the bottom right corner). The source I is better seen at 10:33:57 UT. The loop connecting sources I and III was gradually stretched eastward during 10 min that can be due to the lifting of its top. The optical continuum of superflares is assumed to form in the flare ribbons and intensifies in sources I, II, and III.

It is known that the brightest patches of white-light flares on the Sun occupy an area of about of 2×10^{16} cm^{-2}. Now it is confirmed by current data about the area of the brightest footpoints in the G-band of the continuum (Krucker *et al.*, 2011). In large solar flares, the area of the white-light radiation is 100 times smaller than the area of the Hα ribbon emission in the same events. Such a large optical continuum emission area provides evidence that accelerated particles within the loop system continue to maintain this radiation in the loop footpoints for some time longer immediately after one or more episodes of an impulsive acceleration. We discuss this in more detail in the next section.

3. On a Model of the Optical Continuum Source of Superflares

Now we propose a model of the optical continuum source for a superflare with a total energy of 10^{34} erg that can occur on the young Sun. This model is based on an analysis of contemporary observations of the largest solar flares. In fact, the study in this direction started with the Bastille Day flare on 14 July 2000, because it became possible to obtain both EUV-line and X-ray images. Livshits and Belov (2004) showed that features of the radiation in different spectral ranges are associated with the extremely efficient particle acceleration occurring at low heights in the immediate vicinity of spot penumbrae. In particular, Livshits and Belov (2004) schematically presented coronal flare loops and ribbons against the photospheric layers of the active region (this is shown in Figure 2). The hard X-ray image and the temporal profile of the coronal emission of the hardest source are also given. Tentatively, three types of hard X-ray sources were revealed. The first one is the region of the main acceleration (I); this is a single or a few very low-lying loops located above the polarity inversion line (PIL) crossing the large spot. The system of coronal loops begins to form above the PIL over the entire active region, and some of footpoints of loops in this post-eruptive arcade are sources of the enhanced X-ray radiation (II). The coronal loop, which connects the region of the main particle acceleration with another ribbon, is often a source of strong microwave radiation, because many particles of relativistic energies are captured therein. The footpoint of this loop (III) is also bright in the hard X-ray and γ-ranges.

We note that in previous analyses of data on the Bastille Day flare the X-ray sources I, II, and III were identified as intense inhomogeneities at 50 – 200 keV-images of ribbons (Fletcher and Hudson, 2001). These hard X-ray sources can be detected from other observations as well. For the Bastille Day flare it was obtained that the maximum of the radiation of the region I even of the cool coronal plasma ($T \sim 10^6$ K) at 171 Å coincided with a peak of the γ-ray burst (Aschwanden and Alexander, 2001). This confirms the conclusion that the high-energy events took place in region I, where electrons are effectively accelerated. A portion of the accelerated electrons is captured into a coronal loop system. The enhanced intensity both the hard X-ray and EUV emission in some points of another ribbon can be considered as the site of electron precipitation (Fletcher and Hudson, 2001).

In the context of superflares, we reanalyzed TRACE data in the EUV ranges for this Bastille Day flare and found that the hard X-ray sources I, II, and III are the footpoints of the corresponding coronal loops, *e.g.* they coincide with the EUV sources. This coincidence of the sources is better seen just after the maximum of the hard X-ray burst. These TRACE data permit us to study the evolution of coronal loops linking source I with sources II and III. For example, Figure 2 shows a change of shape and height of a loop between sources I and III.

Coincidences of the bright points in the ribbons in the EUV- and in hard X-ray ranges are seen in many powerful events, for example, on 28 October 2003, and even in the flare on 6 December 2006 located near the limb in not quite favorable conditions for observation (Krucker *et al.*, 2011).

The idea by Livshits and Belov (2004) about efficient particle acceleration above spots was reliably confirmed by Grechnev *et al.* (2008) by an analysis of the whole set of observational data for the flare on 20 January 2005. In fact, many features of the largest solar flares, including motions, radiation in different spectral ranges, and appearance of elementary particles are caused by this efficient acceleration of electrons and protons.

Before discussing a model of superflares, we briefly summarize the knowledge on the appearance of the flare optical continuum. Short brightness enhancement was registered quite rarely in small regions of about a few arcsec. For instance, these white-light (WL) flares were seen distinctly in images obtained at the Gyula Observing Station of the Debrecen Heliophysical Observatory (with a filter of about 100 Å effective half-width centered at 5460 Å) (Kovács, 1977; Dezső *et al.*, 1980; Babin, Dyatel, and Livshits, 1985). These WL flares were located in close vicinity of spots or directly above them, and their brightness exceeded the brightness of the surrounding photosphere by up to 20 – 30 %.

Most of the WL flares are characterized by fewer contrasts in the visible range. Zirin and Neidig (1981) proposed that this "blue continuum" including the Balmer jump and hydrogen lines originates from a thin layer of the hydrogen plasma above the chromosphere. Such a weak continuum is registered not only in bright patches near spots, but also in the Hα ribbon of powerful flares observed with TRACE. An analysis of the observations showed that the optical depth of WL flare source is close 0.01. This value was obtained by Potts *et al.* (2010) in the red spectral range, but it is typical also for the *Yohkoh* aspect camera in the blue range.

The gas-dynamics modeling demonstrated that this radiation is associated with the downward-moving low-temperature condensation and is the basis for interpreting observations of line spectra of impulsive solar and stellar flares. The chromospheric response to an impulsive heating can be different depending on the energy and hardness of the electron beam. This explains a certain spread in properties of specific events, in particular, the appearance of the optical continuum in the condensation occurs rarely and only during the largest flares. We emphasize that the explosive evaporation occurs in a different way in small areas during the same flare (Sadykov *et al.*, 2014).

However, the entire solar WL flare observations contain a few points that cannot be explained in the framework of the gas-dynamics model. First of all, this refers to classical observations about the high contrasts of WL flares in the red spectral range. An analysis of their spectral data requires adopting the optical continuum source that is situated in the upper photosphere. The opacity of layers beneath the temperature minimum is due to the negative hydrogen ion, H^-, and the hydrogen atoms. Any changes of physical conditions in these layers will be reflected in the intensity of red and IR spectral ranges.

The red continuum is observed directly in spectra of powerful impulsive flares on red dwarf stars. Kowalski *et al.* (2013) presented observations of line and continuum emission from simultaneous high-cadence spectra covering near-ultraviolet and optical wavelengths (3400 – 9200 Å) and photometry for twenty M-dwarf flares. This analysis allowed them to trace the behavior of the Balmer continuum at $\lambda < 3646$ Å and features of the line profiles in weak and large flares. Observations in the blue spectral range agree well with both the two-temperature gas-dynamical modeling of an impulsive flare on the red dwarf (Katsova, Boiko, and Livshits, 1997) and the new single-temperature radiative hydrodynamic flare models by Allred *et al.* (2005). The most important result of the analysis of these observations by Kowalski *et al.* (2013) is the detection of two components of the flare optical continuum: blue continuum at the flare maximum that linearly decreases with wavelength from $\lambda =$ 4000 – 4800 Å, which indicates the hot, black-body emission with typical temperatures of $T_{BB} \sim 9000 - 14\,000$ K; the red continuum gives evidence of a $T_{BB} \sim 8000$ K component at the beginning of the gradual decay phase during eight flares, most of which are impulsive events. These observations reveal some "mysterious" relationship between properties of the blue and red continua.

The optical continuum, not related to the radiation of the low-temperature condensation, can apparently appear on G-type stars at high energies of non-stationary processes more often than during the large flares on the contemporary Sun. We note that the required heating of dense layers can have different causes. This may be associated with signatures of propagation of high-energy particles, in particular, electrons with $E > 500$ keV, which can outstrip the main stream of the lower-energy particles, penetrate deeper and affect a larger area than the main flux of electrons with $E \approx 20$ keV. In principle, the heating of the upper photosphere can be due to the return current (Matthews, Brown, and van Driel-Gesztelyi, 1998).

What can be imagined about a model of the optical source of a superflare on G-type stars based on this experience of the study of solar phenomena? One clear result is the following: the continuum emission arises in loops with a footpoint area comparable with the area of Hα ribbons of large solar flares. Most likely, three types of sources, I, II and III, similar to those presented in Figure 2, will stand out here. The emission of source III is due to gas-dynamical response to the particle flux, which fills up the coronal loop. We note that here particles can be both captured into an arcade and accelerated directly in the vertical current sheet, as usually occurs in weak solar flares (Kopp-Pneuman model). The optical emission sources of ribbons are low-temperature chromospheric condensations in footpoints of flare coronal loops.

In addition to this blue continuum and line emission, the optical continuum of the photospheric layers of the active region arises in low-lying loops near large spots where particles are accelerated (source I). This region close to the temperature minimum emits in a broad spectral interval from IR to the visible range – the red continuum with a low effective temperature. There are a few sources of small additional heating of the upper photosphere of G-type stars similar to those in red dwarfs. One of them can be associated with penetration of high-energetic particles from the acceleration site to the photosphere.

An impact of the hardest particles on the temperature minimum region can also be substantial in source III. Therefore the ratio between the intensity of the blue and red continua can differ from one event to another. In addition, in some specific superflares, pairs I and III can occur and may be spaced in time.

It is possible to imagine the temporal development of a superflare. The start will most likely be a giant plasma ejection that either erupts from the stars as a CME or lifts up into the corona. At the beginning of the event, strong absorption arises due to the negative hydrogen ion, H^-, in source I, sometimes referred to as a black-light flare (van Driel-Gesztelyi *et al.*, 1994; Ding *et al.*, 2003). Then series of elementary episodes of particle acceleration develops, accompanied by peaks of hard X-ray emission. Each of acceleration episodes provides the red continuum, and the radiation of the condensation arises within a few seconds. As a result, the flare optical continuum extends from the IR- to **UV** ranges.

We note that the total energy of the optical radiation of a superflare is the sum of the energies of elementary events, and each of them is characterized by the optical radiation in a broad spectral range. The radiation during the post-eruptive phase also contributes to the total flare energy of the superflare. Of course, the study of the photospheric response to an impulsive heating may enable us to concretize these predictions.

Returning to problems of superflares, we note that their contrast on the stellar surface has to be close to that observed on the Sun. This means that at the same total energy 10^{34} erg, the observed flux ratio, $\Delta F/F = 10^{-3}$, corresponds to the same contrast as that in patches of WL flares on the Sun. However, on the Sun, these patches can be observed only rather closely to the particle acceleration site (source I). In superflares, most of the energetic particles can reach another footpoint of the loop and penetrate there into deeper layers of the atmosphere before the main stream of particles with lower energy would arrive there. This may enhance the low-temperature optical continuum.

We do not discuss the radiative gas-dynamical simulations of processes in powerful impulsive flares in full detail. However, we have dealt with these problems earlier since 1980 (Livshits *et al.*, 1981), and some years ago we carried out new gas-dynamics modeling based on the algorithm by Caunt and Korpi (2001). Using the same approach for radiative losses, we have obtained values close to our previous results (Livshits *et al.*, 1981). Advance in such a modeling for superflares is difficult because there is no final decision on the question of how the return current limits the particle energy from penetrating the chromosphere.

However, before radiative gas-dynamical simulations of superflares, we studied the problem of how the appearance of strong optical radiation can be explained in principle in the framework of the gas-dynamics model and which consequences follow from this consideration. Unlike most other gas-dynamics models, we applied the single-fluid two-temperature approximation, which allowed us to derive more exact physical characteristics of a flare source in the first $1-2$ seconds after injection of the accelerated particle beam into the chromosphere and upper photosphere. This two-temperature model better describes the relationship between high-energy phenomena arising during the interaction of accelerated particles with the dense plasma. This model allows tracing the connection between nuclear spallation reaction, microwave radiation, and subsequent formation of sources of the thermal X-ray and the optical emission. When the heating beam is hard, two sources of the optical continuum arise: the downward-moving chromospheric condensation, and a region of enhanced heating in the upper photosphere. Furthermore, we used just one result of our gas-dynamics calculations: that the flare continuum emission arises on the Sun (or the young G star) when the energy of the accelerated electrons is 3×10^{11} erg cm^{-2} s^{-1} and the differential spectrum $\gamma \leq 3$. This minimal energy flux of the electron provides the formation of the optical continuum that conventionally is identified with the blue component found by Kowalski *et al.* (2013) for flares on red dwarfs.

This chromospheric source is characterized by the maximal radiation area. But when accelerated electrons have a hard spectrum, the heating of the upper photosphere occurs. On the Sun, two types of white-light flares are considered: i) hydrogen recombination continua, *e.g.* Balmer continuum similar to Figure 1 (see recent observations with IRIS by Heinzel and Kleint, 2014), and ii) photospheric white-light radiation due to photospheric temperature enhancement (*e.g.* backwarming), see review by Ding (2007). During superflares, both sources of continuum emission probably arise. Their relative contribution to the total radiation varies from one superflare to the next.

It is important that this result allows us to evaluate the flux of accelerated electrons penetrating the chromosphere. The energy flux is fixed, while the spectrum of electrons with $E > 20$ keV can have different power-law indices. However, in this case, the mean energy of the beam is around 23 – 26 keV, and for the highest electron flux 3×10^{11} erg cm^{-2} s^{-1}, we obtain an electron flux 7.5×10^{18} electron cm^{-2} s^{-1}, and the total number of electrons that enter the chromosphere is 7.5×10^{37} electron s^{-1} in this superflare. Apparently, injecting accelerated electrons with $E > 20$ keV into the coronal part of the source will have the same order of magnitude. For M5 – X1 solar flares, this value is lower by a factor ~ 100.

This is a parameter determined by the X-ray and radio emission of flares. The hardness of this electron beam, the magnetic fields, the temperature, and the density of the plasma in the radiation sources are important here.

In the case of the hard spectrum of accelerated electrons, this model has several consequences, which can be verified in current observations. This point is discussed in the next two sections.

4. The Microwave Radiation of Flares

The optical observations of superflares with $E = 10^{34}$ erg give information about the total number of electrons with $E > 20$ keV $N \approx 10^{38}$ electron s^{-1}. It is clear that this energy probably is provided by primary injection of accelerated electrons from the acceleration site into the coronal loop. If the electron spectrum is soft, then gas-dynamical processes will heat the corona.

The soft X-ray radiation that arises during such an event is high enough to be observable at stellar distances. Even if the electron spectrum is hard, the X-ray radiation at the range above 20 keV from stellar flares is not yet available for observations. But in the case of the hard spectrum, the same total number of accelerated electrons will result in effects that can be registered and be a test to verify our assumption that the powerful optical radiation of superflares is associated with effective particle acceleration.

We also evaluate a level of the microwave radiation of superflares. First, the gyrosynchrotron radiation is frequently studied in large solar flares. The spectrum of the microwave burst has a maximum at frequencies of around 10 GHz, but in solar flares larger than M3 (with possible particle acceleration), this maximum is shifted toward the higher frequencies. The highest fluxes at 35 GHz are within a wide range $I_{35} = 1 - 100\,000$ sfu (1 sfu $= 10^{-22}$ W m^{-2} Hz^{-1}). The IZMIRAN data base for observations in 1976 – 2006 with different radio telescopes shows that this maximal flux I_{35} at this frequency weakly depends on the soft X-ray flux, I_R, for faint flares (where the thermal radiation is still significant), while the I_{35} value is proportional to $I_R^{1.5}$ for flares with particle acceleration.

We are interested here in the greatest solar flares. The high-energy phenomena in some of them were studied in detail in the past solar cycle. It is clear that microwave bursts arise if electron beams contain a large amount of microwave-emitting electrons with E $>$ 0.5 MeV. The largest events with $E \geq 10^{32}$ erg are accompanied by 3 – 20 GHz bursts with a peak flux of up to 10^5 sfu. A similar flux was registered for the flare on 20 January 2005, where footpoints were situated above the sunspots of the opposite polarity (Grechnev *et al.*, 2008). Because this large flare occurred at 60 W, this case is not best suited for extrapolating the magnetic field into the corona. We have selected another large flare on 28 October 2003 near the disk center. Unfortunately, RHESSI was disabled right during the maximum of this flare > X17, but its X-ray radiation was registered by SONG/KORONAS-F, HEND/*Mars-Odyssey* and some radio telescopes. Using the code for computing the gyrosynchrotron radiation (Fleishmen and Kuznetsov, 2010; Nita *et al.*, 2014), we calculated the spectrum of the microwave radiation of the accelerated electrons with the single slope $\gamma = -2.5$. The magnetic field in the corona was reconstructed from a SOHO/MDI magnetogram. The microwave emission arose in the loop of length of 2×10^{10} cm and width of 3×10^9 cm. The magnetic field strength in the loop top was 175 G. In this case, it is impossible to achieve consent between observed and calculated spectrum (the flare on 20 January 2005 is shown in Figure 11 in Grechnev *et al.*, 2008, and our modeling by another code also confirms this).

Although the two flares have similar microwave spectra, the soft X-ray radiation of the flare on 28 October 2003 exceeded the corresponding value for the flare on 20 January 2005, and the highest microwave fluxes of the second flare were somewhat higher, by a factor of 1.5 – 2. Therefore, we report the version of computations of the emission in the reconstructed shape of the flare loop for the 28 October event, but with corrected values of the accelerated electron flux. The result is presented in Figure 3. The calculation was carried out for the following parameters: the density of electrons with $E > 20$ keV was 3×10^7 cm^{-3}, the density of the thermal electrons was $n_0 = 5 \times 10^9$ cm^{-3} in the loop and 1×10^9 cm^{-3} in the surrounding corona. The density of thermal electrons decreased according to the exponential law for $T = 2 \times 10^7$ K and 1×10^6 K in the loop and surrounding corona correspondingly. We note that the total number of accelerated electrons is 3.5×10^{35}. These conditions probably describe the situation in the greatest hard X-ray flares on the Sun when the highest microwave flux reaches 10^5 sfu at 10 – 100 GHz.

Now we can evaluate the microwave flux in superflares on G-type main sequence stars. Here certain information about solar phenomena, namely, the statistical proportional dependence of I_{35} on $I_R^{1.5}$ for flares with particle acceleration can be helpful. The I_R value for flares is closely related to their total energy. If we adopt the total energy of the largest solar flares to be 10^{32} erg and use the same law for variations of the microwave emission in the form of I_{35} is proportional to the total flare energy $E_{\text{tot}}^{1.5}$, then we obtain that the microwave flux of flares with $E = 10^{34}$ erg at the distance of 1 AU is about 10^8 sfu. This means that such a flare on a star at a distance of 100 pc will produce a microwave flux above 2 mJy.

A microwave flux of 80 mJy was registered at the peak of the optical flare on the young star V773 Tau with NAOJ Nobeyama at 98, 64, 43, 23 GHz (Iizuka *et al.*, 2014). However, this phenomenon relates rather to non-solar analogs. It would be of importance to carry out the monitoring with VLA and ALMA of some of G-type stars with a high occurrence of superflares.

5. Lithium Production During Impulsive Flares

We again consider phenomena associated with the efficient particle acceleration, when many electrons but also protons can be accelerated up to high energies, in particular, protons can reach energies of up to dozens of MeV. The high-energy protons interact with nuclei of the ambient photosphere, and a series of spallation reactions like the $\alpha - \alpha$ process is developed.

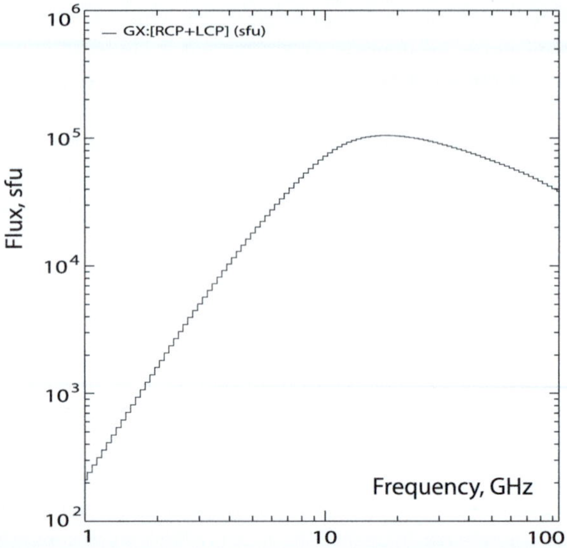

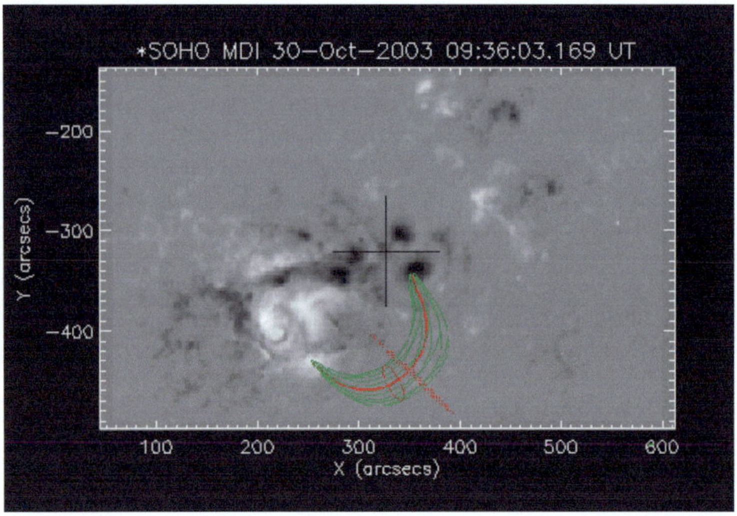

Figure 3 The calculated microwave spectrum for the solar flare on 28 October 2003 (upper panel). The calculation parameters are given in the text. The lower panel presents the location of the loop in an MDI image.

A review of the nuclear processes in solar flares was presented by Ramaty (1986). From observation of the ^{7}Li line at 0.478 MeV, Livshits (1997) gave an estimate of the amount of lithium produced in a large impulsive flare on the Sun. He concluded that about 10^{30} Li nuclei can be produced in the largest impulsive flares on the Sun. It was also shown that the diffusion of new Li ions is slow under the conditions of turbulent convection, and the produced excess of the Li atoms can be observable during several hours after the flare. Ramaty *et al.* (2000) later found a nuclear reaction where accelerated ^{3}He reacts with ^{4}He, which can lead to production of the ^{6}Li nuclei during flares.

There is no direct observational evidence for this effect in solar flares. But Montes and Ramsey (1998) reliably found the enhancement of the resonant Li 6708 Å line during a long-duration optical stellar flare. The Li line equivalent width followed the temporal Hα-profile of this flare. The idea of the production of Li by spallation reactions is confirmed by the observations of the rise of the ^{6}Li/^{7}Li ratio in this flare. This flare occurred on an active K0 star with a rotational period of 9.98 days and lasted for three days. These data were reviewed by Cutispoto (2002) in the context of the interrelation of lithium and other light elements and stellar activity. It is argued that one should expect that the ^{6}Li and ^{7}Li isotopes arise during large flares, but the abundance of ^{7}Li nuclei are not seen to change against the abundance of Li atoms originating from the stellar interiors, where the proton-proton reactions occur, while the ^{6}Li atoms arising in the flare are manifested and lead to a change in the ^{6}Li/^{7}Li ratio.

A new approach to the study of the question whether new Li has been produced by nuclear reactions during flares was proposed by Flores Soriano, Strassmeier, and Weber (2015). They observed in spectra of the young active K2 dwarf LQ Hya (with a rotation period $P_{\rm rot} = 1.6^{\rm d}$) "that the strength of the Li line and other temperature sensitive lines is rotationally modulated and clearly enhanced in the hemisphere of the star with the highest magnetic activity", *i.e.*, on the hemisphere with the larger spot area. Generally speaking, this modulation is associated with an increase in average temperature. But on the same hemisphere, the observed spectral features were interpreted by Flores Soriano, Strassmeier, and Weber (2015) as the appearance of new ^{6}Li nuclei. Because "sporadic flares of moderate intensity were also detected" at the same hemisphere in the hydrogen and helium lines, the authors assumed that these Li atoms are produced during flares.

One can expect a significant amount of lithium being produced during superflares. The set of spectral observations of the equivalent width of the resonant Li 6708 Å line, $W_{\rm Li}$, can help to reveal a certain part of temporal variations with the rotation period associated with starspots. The detection of another, irregular variability of $W_{\rm Li}$ will indicate a flare. Both the duration and shape of the temporal profile of the equivalent width variations differ significantly. These observations require long-term spectral monitoring with large telescopes.

6. On the Evolution of Activity of Late-Type Stars

Where can superflares with the total energy $E > 10^{34}$ erg occur? To address this question, we briefly discuss some results of the study of activity of low-mass G0 – K5 stars.

As it well known, a protostar spins faster during a stage of the gravitational contraction, and the primary relict magnetic field is enhanced. Then the rotation of a star is decelerated, and some activity processes are developed, as follows from observations of star-members of open clusters of ages of 30 – 600 Myr.

The dependence of the X-ray radiation of late-type main-sequence stars on the rotation period shows that activity of fast-rotating stars is saturated, the coronal activity index, $R_x = \log L_x/L_{\rm bol}$ is close to -3 and weakly depends on the rotation period (Pallavicini *et al.*, 1981; Wright *et al.*, 2011). Stars with moderate activity up to the solar level demonstrate the clear relationship $\log L_x \sim P_{\rm rot}^{-2} \sim v^2$, where v is the rotation rate (Pallavicini *et al.*, 1981; Reiners, Schüssler, and Passegger, 2014). The rotation rate is a direct indicator of both stellar age and activity (Skumanich, 1972). The further analysis of the "rotation–age" and "activity–rotation" relationships allowed Mamajek and Hillenbrand (2008) to propose a method of gyrochronology for estimating a star's age from its activity level. The transition from young stars with activity saturation to less active, older stars occurs when their rotation

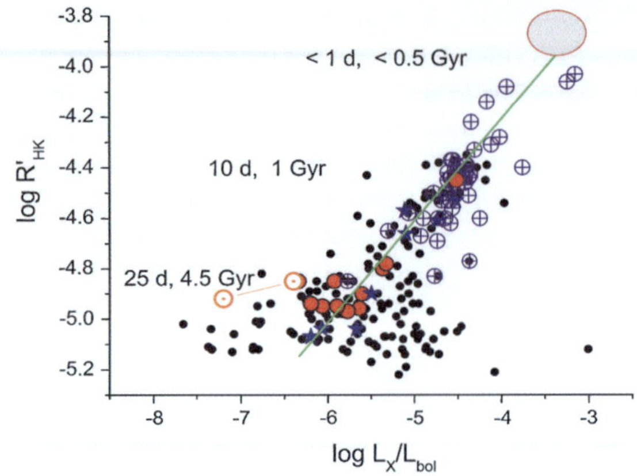

Figure 4 The "chromosphere–corona" diagram of the chromospheric and coronal activity indices for late–type stars. Stars with detected lithium abundances are marked as violet crosses inside the circles (similar to Figue 9 in Mishenina *et al.*, 2012). The stars of the basic data set are represented by black dots. According to quality of detection of the cycle, the stars with "*Excellent*" cycles are plotted as red circles, the stars with "*Good*" cycles are indicated as blue asterisks, the Sun at maximum and minimum is denoted by its own sign, connected by a straight line. The location of stars with saturated activity is conditionally designated with oval. The straight line corresponds to a single-parametric gyrochronology.

slows down to periods of 10 days. Periods of 7 – 12 days are characteristic both for the first types of stars and for another group of intermediate-age stars.

To investigate the activity of stars older than 600 Myr, we considered data on chromospheric activity, namely, the chromospheric activity index, $\log R'_{\mathrm{HK}}$, that is, the ratio of the emission components of the H and K Ca II line to the optical continuum at 4000 Å. We analyzed the results of the HK Project of the monitoring of long-term chromospheric variations of more than 100 late-type stars in the solar vicinity since 1965 (Baliunas *et al.*, 1995). In addition, we added these data to observations obtained during the Planet Search Program (Wright *et al.*, 2004; Arriagada, 2011) and built a "chromosphere – corona" diagram for the chromospheric, $\log R'_{\mathrm{HK}}$, and coronal activity indices $\log L_x/L_{\mathrm{bol}}$ (Figure 4). Then we included observations of stars with high proper motions (Maldonado *et al.*, 2010) and stars with detectable lithium abundance (Mishenina *et al.*, 2012). The location of the young stars with activity saturation is indicated by an oval according to the data of Martínez-Arnáiz *et al.* (2011).

At present it appears that dating stellar ages from data of open clusters is more reliable (Meibom *et al.*, 2015, and references therein). This analysis allowed us to improve the chromosphere – corona diagram. Figure 4 shows that the basic group of stars is concentrated near the straight line in the diagram that connects young stars of ages of hundred million years to the older stars whose age is comparable to that of the Sun (4.5 Gyr) or more. The distribution of young stars with activity saturation clearly continues in the distribution of the stars with high lithium abundance. Strongly spotted stars, most of them are the BY Dra-type stars, are situated in a compact group on the chromosphere – corona diagram around the values $\log R'_{\mathrm{HK}} = -4.5$ and $R_x = -4.5$. The stars whose activity most closely matches what is observed in the Sun today are located below this point, along the straight line. There is reason to believe that activity of all stars from this level to activity of the contemporary Sun is caused by local magnetic fields with a definite governing role of the large-scale magnetic field. In particular, it relates to such known active stars as HD 1835 (BE Cet), HD 20630 (κ^1 Cet), and HD 152391 (V 2292 Oph). The relative spot area of the G star BE Cet reaches

3.3 % (Alekseev, 2001). The G7 star V 2292 Oph is characterized by faster rotation than for other stars with *Excellent* (*well-detected*) cycles. In recent years, it has been noted that the cyclic activity with a period of about 11 years became less regular in this star. This allows us to consider it as an example of a star with a *Good* (*established*) cycle. (The classification of stellar cycle types was introduced in the HK Project by Baliunas *et al.*, 1995). The rotation periods of these stars are 8, 9, and 11 days, respectively. There is reason to consider that the age of main-sequence stars rotating with periods of about 10 days is slightly older than 1 Gyr. We call such a main-sequence G star "a young sun", and for active processes on stars of ages from young Suns to the contemporary Sun, use the term "solar-type activity". We note that this term involves the existence of a cycle on a given star. In fact, the diagram shows two basic groups of stars with activity close to the saturation level and to the solar-type one (sees Figure 4 in Reiners, Schüssler, and Passegger, 2014, where these groups are shown separately). These problems of the evolution of activity were discussed in more detail by Katsova (2012). We note that the activity of stars situated below the straight line that describes single-parametric gyrochronology can evolve through another path (Katsova and Livshits, 2011).

On the stars with activity saturation and stars where Li is reliably detected, the rotational modulation of the chromospheric and coronal emission is clearly pronounced. This is due to the concentration of active regions around certain fixed longitudes. Unlike for stars with solar-type activity, large-scale magnetic fields have strong effects here.

The clearest result of the analysis of the *Kepler* superflares is that these phenomena mainly occur on fast-rotating stars. For instance, if we plot the number of superflares versus the rotational periods of stars, the main maximum at periods of $1-5$ days and the second one at $12-13$ days emerge clearly. The corresponding histogram showing the minimum of the flare number is given by Katsova and Livshits (2014, Figure 5 therein). Thus, the superflares occur on G stars of the first group with activity saturation or close to this level. Only a small fraction of superflares occur on G-stars with solar-type activity and a strong spottedness.

7. Magnetic Fields of G-Stars and the Maximal Flare Energies

Until now, variations in the magnetic field strength with the axial rotation phase together with spectral data provide an opportunity to simulate the distribution of inhomogeneities over the surface of several active G and K main-sequence stars. This method of Zeeman Doppler mapping allows us in the first approximation to separate the contributions of spots and of the large-scale magnetic field.

New data on the magnetic fields of F, G, and K stars were recently obtained in the frameworks of the *"Bcool Collaboration"* (Marsden *et al.*, 2014). The spectropolarimetric observations of 170 late-type stars were carried out with large telescopes. The mean surface longitudinal magnetic field, B_l, was detected in about 40 % of the Bcool sample stars. This value represents the B_l signal averaged over all observations of the star.

We analyzed these observations for G stars of different ages alone. We chose all G stars with values B_l exceeding 3σ from Table 3 of Marsden *et al.* (2014). From that list, we excluded some of the rapidly rotating and hence very young stars with strong magnetic fields. Thus, the final list includes 28 G stars whose rotation periods are > 5 days. The result is presented in Figure 2 in Katsova and Livshits (2014). We obtained the average value of the modulus of the field $|B_l| = 4.72 \pm 0.53$ G for G stars. The obtained average value corresponds to a rotation rate of about 4 $\mathrm{km\,s^{-1}}$, *i.e.*, twice faster than the rotation of the modern Sun. This allows us to adopt this value of the field as an average value for a G star with a rotation period of 10 days, which corresponds to an age of about 1 Gyr.

We compared this value with data on the general magnetic field of the Sun as a star. Values averaged over the Carrington rotation near the maximum of the cycle do not exceed 0.5 G (see, for example, Kotov, Haneychuk, and Tsap, 1999 for data on the magnetic field of the Sun as a star in 1991). Thus, the average magnetic field strength of young G stars is higher at least by an order of magnitude than that for the Sun at maximum in the modern era.

We note that this evaluation is supported by the fact that the magnitude of $|B_l|$ for stars whose parameters are most similar to those of the young Sun, is 8.5 G for V 2292 Oph and 7.7 G for κ^1 Cet (other data for this star $|B_l| = 7.0$, do Nascimento *et al.*, 2014).

Now there is certain information on the structure of the large-scale magnetic fields of late-type stars. In particular, it was found that slowly rotating stars have magnetic fields with the same structure as that on the Sun, which are more-or-less regularly changing during the 22-year magnetic cycle. In other words, the global dipole does exist virtually all the time, and only during the polarity reversal is there a shift in latitude of the magnetic equator, which can be considered as a manifestation of a large-scale toroidal component. Active regions are located rather chaotically, active longitudes are allocated poorly, and toroidal fields are characterized by comparatively small scales (local fields). At the same time, the large-scale toroidal component is clearly pronounced on G stars with rotation periods shorter than 12 days (Petit *et al.*, 2008). Although it prevails over large-scale poloidal magnetic fields, the difference of the magnetic field strength is small. So far, it is difficult to relate these conclusions with new data on spots of the rapidly rotating G stars.

Thus, we conclude that the general nature of the magnetic fields on G and K stars is similar to that observed on the Sun today. Namely, there is a large-scale magnetic field, most clearly manifested in the polar regions and in the local fields at low latitudes. A magnetic field strength of $1-3$ kG was found in spots covering up to 10 % of the stellar surface (see Saar, 1996; Gershberg, 2005).

The development of non-stationary processes in solar active regions is associated not with the field energy of a given magnetic configuration, but with its distinction from the potential energy. It is characterized by the free energy. It increases with an emergence of a new magnetic flux, and it is probably spent in flares and CMEs. It was found that on the Sun the free energy is accumulated at altitudes of about 2000 to 5000 km (Sun *et al.*, 2012) and $10-15$ % of the free energy of a given active region can be spent on a large flare in it. For four large flares this follows from results by Jing *et al.* (2009), and for dozens of other flares, this result can be derived from Falconer *et al.* (2011). Both these conclusions are based on extrapolating the photospheric magnetic field into the corona (Metcalf *et al.*, 2008, where authors applied nonlinear force-free field extrapolation algorithms) and also agree with our calculation in Livshits *et al.* (2015).

A non-stationary process starting immediately above the dense chromosphere layers quickly extends to the coronal part of an active region. Accelerated particles fill up one or a few coronal loops. Basic sources of the hard X-ray radiation are located in footpoints of the loops. We note that a detailed discussion of the location of hard X-ray sources, ribbons in the chromosphere, and white-light emission is given by Krucker *et al.* (2011), and one can only agree with their statement that "future observations should focus on the three-dimensional structure of the sources."

The measured average magnetic field of a star is composed of the large-scale field and the field of spots. Like the Sun, the field of individual spots reaches a strength of several kG, but spottedness of the stellar surface, filling-factor, in rapidly rotating stars exceeds that of the Sun by 1 – 3 orders of magnitude. Finally, the contribution of small-scale and large-scale fields to the measured average magnetic field of a star is about the same (Petit *et al.*, 2008; Marsden *et al.*, 2014). According to measurements of the fields on individual stars and on a large group of stars, we can conclude that the magnetic fields on G-type dwarfs have approximately the same general character as that of the Sun, but the strength of the field in the vicinity of neutral lines, separating the oppositely directed magnetic field, is one order of magnitude larger than that on the Sun.

This evaluation corresponds to an increase in the field energy above the active region on G-type dwarfs by two orders of magnitude as compared to large active regions on the Sun. Repeating the arguments similar to those used above for the Sun, we obtain an upper limit of the total flare energy on the young Sun (with an axial rotation period of 10 days that corresponds to an age of about 1 Gyr) equal to 10^{34} erg. This evaluation is obtained by applying the magnetic virial theorem (Livshits *et al.*, 2015).

As for flares of higher energies, from 10^{34} erg to 10^{36} erg, there is evidence for occurrence of these events on stars with a radius larger than one solar radius. This correlation has been revealed by Balona (2015). If the radius of the G star increases, then the characteristics of activity, including flares, changes significantly.

Of course, superflares can have a nature that differs from non-stationary processes on the Sun. In particular, they may result from an interaction between planets and the host star; these problems were discussed in Rubenstein and Schaefer (2000), Cuntz, Saar, and Musielak (2000), and Ip, Kopp, and Hu (2004).

8. Concluding Remarks

Before the detection of superflares on stars, only one event was registered on a red dwarf that differed from the solar analogs and indicated a possible realization of nuclear spallation reactions. The simultaneous X-ray and optical observations of a giant flare on the red dwarf CN Leo by Schmitt *et al.* (2008) and Fuhrmeister *et al.* (2008) demonstrated the definite difference of this phenomenon from typical impulsive flares on red dwarfs and on the Sun. In particular, instead of motion of low-temperature condensations downward to the photosphere, an explosion of the coronal and chromospheric plasma was registered in the first seconds of the flare onset.

In the context of finding super-powerful phenomena on G stars, we tried to explain where and how such events can occur. These observations were carried out in the optical. Until now, an interpretation of these superflares was associated with direct solar analogs (Shibata *et al.*, 2013). In light of the fact that these phenomena are caused by particle acceleration, we provided observational evidence of the origin of these superflares. We assumed that the optical continuum mainly arises during an explosive evaporation of the chromosphere and estimated an area of the optical flare as high as 10^{19} cm^2 for a total flare energy of 10^{34} ergs. This means that all flare ribbons emit in the optical continuum, differently from what occurs on the Sun. This radiation is enhanced in regions above the Hα ribbon where particles are accelerated, and also in the second footpoint of the main flare loop.

If this electron flux is hard (slope γ is 2 – 3), then the microwave emission can reach 2 mJy at a distance of 100 pc, which can be detected by modern radio facilities. Secondly, protons above 30 MeV and α-particles when interacting with ambient particles produce light elements, in particular, lithium, due to a series of spallation reactions (Ramaty, 1986; Livshits, 1997; Ramaty *et al.*, 2000). There is great interest in detecting an enhancement of the resonant Li I line in superflares.

One of the important results of the *Kepler* observations is a statistics of the flare occurrence frequency on G stars: superflares are observed more often on the faster-rotating stars. We have analyzed in more detail which stars produce these powerful phenomena from the point of view of the evolution of solar-type activity. Among the active late-type stars there are young objects with saturated activity, which weakly depends on the rotation rate. There is another group of main-sequence stars where the chromospheric and coronal activity is determined by the axial rotation. Many arguments indicate that phenomena on stars with a rotation period > 10 days can be considered as solar-type activity. In particular, here the interconnection between large-scale and local magnetic fields changes, while their role in the formation of the activity varies as well. We note that the young Sun with a rotation period of about 10 days is about 1 Gyr old. Namely, the solar-type activity has already formed by this age.

As is known, on the Sun the free energy of the magnetic field accumulates in the chromosphere and is then released in non-stationary processes. Powerful events with effective particle acceleration occur in regions where the polarity inversion line (PIL) passes sites with a strong magnetic field (near spots). It is clear that the total flare energy in a given active region cannot exceed the free energy therein. Because the free energy is released in several steps, every large flare "devours" $10-15$ % of the deposited free energy in this region. These estimates derived from data on the full vector of the magnetic field agree well with the energies of largest solar flares, 10^{32} ergs.

We have analyzed new observations of magnetic fields on G stars with the rotation periods > 7 days. The mean longitudinal magnetic field is 10 times stronger than that on the Sun at the maximum of the cycle. This allows us to estimate that the highest energy of flares on these G stars is not more than 10^{34} ergs.

In conclusion, we would like to emphasize that interpreting the observations of superflares on G stars helps understanding extreme processes on the Sun. In addition, the detection of the flare microwave source and the emission in the Li resonant line could demonstrate how effective particle acceleration can be on stars in the lower part of the main sequence.

Acknowledgements We are grateful to the Organizing Committee for the possibility of presenting our results on "Solar and Stellar Flares" at the conference in Prague, June 2014. We are thankful to V.V. Grechnev and T.I. Kaltman for help in analyzing the microwave data, and to A.B. Belov and B.A. Nizamov for discussions. We are deeply grateful to the referee for the helpful remarks.

This work is supported by Russian Foundation of Basic Research under grants No 14-02-00922, 15-02-06271 and by the grant from the Russian Federation President supporting Scientific Schools 1675.2014.2.

References

Alekseev, I.Yu.: 2001, Spotted low-mass stars. *Astroprint. Odessa*, 137 pp. (in Russian).
Allred, J.C., Hawley, S.L., Abbett, W.P., Carlsson, M.: 2005, *Astrophys. J.* **630**, 573.
Antonucci, E.: 1989, *Solar Phys.* **121**, 31.
Arriagada, P.: 2011, *Astrophys. J.* **734**, 70.
Aschwanden, M.J., Alexander, D.: 2001, *Solar Phys.* **204**, 93.
Aulanier, G., Démoulin, P., Schrijver, C.J., Janvier, M., Pariat, E., Schmieder, B.: 2013, *Astron. Astrophys.* **549**, A66.
Babin, A.N., Dyatel, N.P., Livshits, M.A.: 1985, *Soviet Astron.* **29**, 676.
Baliunas, S.L., Donahue, R.A., Soon, W.H., Horne, J.H., Frazer, J., Woodard-Eklund, L., *et al.*: 1995, *Astrophys. J.* **438**, 269.
Balona, L.: 2015, *Mon. Not. Roy. Astron. Soc.* **447**, 2714.
Barnes, S.A.: 2003, *Astrophys. J.* **586**, 464.
Berdyugina, S.V.: 2005, *Living Rev. Solar Phys.* **2**, 8.
Boiko, A.Ya., Livshits, M.A.: 1999, *Astron. Rep.* **43**, 759.

Caunt, S.E., Korpi, M.J.: 2001, *Astron. Astrophys.* **369**, 706.
Cuntz, M., Saar, S.H., Musielak, Z.E.: 2000, *Astrophys. J.* **533**, L151.
Cutispoto, G.: 2002, *Astron. Nachr.* **323**, 325.
Dezső, L., Gesztelyi, L., Kondás, L., Kovács, A., Rostás, S.: 1980, *Solar Phys.* **67**, 317.
Ding, M.D.: 2007. In: Heinzel, P., Dorotovič, I., Rutten, R.J. (eds.) *The Physics of Chromospheric Plasmas*, *ASP Conf. Ser.* **368**, 417.
Ding, M.D., Liu, Y., Yeh, C.-T., Li, J.P.: 2003, *Astron. Astrophys.* **403**, 1151.
do Nascimento, J.D., Petit, P., Castro, M., Porto de Mello, G.F., Jeffers, S.V., Marsden, S., *et al.*: 2014. In: Petit, P., Jardine, M.M., Spruit, H.C. (eds.) *Magnetic Fields Throughout Stellar Evolution*, *Proc. IAU Symp.* **302**, 142.
Falconer, D., Barghouty, A.F., Khazanov, I., Moore, R.: 2011, *Space Weather* **9**, S04003.
Fuhrmeister, B., Liefke, C., Schmitt, J.H.M.M., Reiners, A.: 2008, *Astron. Astrophys.* **487**, 283.
Fisher, G.H., Canfield, R.C., McClymont, A.N.: 1985, *Astrophys. J.* **289**, 434.
Fleishmen, G.D., Kuznetsov, A.: 2010, *Astrophys. J.* **721**, 1127.
Fletcher, L., Hannah, I.G., Hudson, H.S., Metcalf, T.R.: 2007, *Astrophys. J.* **656**, 1187.
Fletcher, L., Hudson, H.: 2001, *Solar Phys.* **204**, 71.
Fletcher, L., Dennis, B.R., Hudson, H.S., Krucker, S., Phillips, K., Veronig, A., *et al.*: 2011, *Space Sci. Rev.* **159**, 19.
Flores Soriano, M., Strassmeier, K.G., Weber, M.: 2015, *Astron. Astrophys.* **575**, A57.
Gershberg, R.E.: 2005, *Solar-Type Activity in Main-Sequence Stars*, Springer, Berlin.
Grechnev, V.V., Kurt, V.G., Chertok, I.M., Uralov, A.M., Nakajima, H., Altyntsev, A.T., *et al.*: 2008, *Solar Phys.* **252**, 149.
Heinzel, P., Kleint, L.: 2014, *Astrophys. J. Let.* **794**(2), L23.
Hudson, H.S., Wolfson, C.J., Metcalf, T.R.: 2006, *Solar Phys.* **234**, 79.
Iizuka, R., Matsumura, T., Tsuboi, Y., Ogawa, S., Akiyama, M., Higa, M., *et al.*: 2014, Suzaku-MAXI 2014: expanding the frontiers of the X-ray universe. In: Ishida, M., Petre, R., Mitsuda, K. (eds.) *Proc. of the Conf.*, *Ehime University, Japan*, **154**.
Ip, W.-H., Kopp, A., Hu, J.-H.: 2004, *Astrophys. J.* **602**, L53.
Jing, J., Chen, P.F., Wiegelmann, T., Xu, Y., Park, S.-H., Wang, H.: 2009, *Astrophys. J.* **696**, 84.
Karlicky, M., Henoux, J.-C.: 1992, *Astron. Astrophys.* **264**, 679.
Katsova, M.M.: 2012. In: *The Sun: New Challenges*, *Astrophys. Space Sci. Proc.* **30**, Springer, Berlin, 19.
Katsova, M.M., Livshits, M.A.: 2011, *Astron. Rep.* **55**, 1123.
Katsova, M.M., Livshits, M.A.: 2014, *Geomagn. Aeron.* **54**, 982.
Katsova, M.A., Boiko, A.Ya., Livshits, M.A.: 1997, *Astron. Astrophys.* **321**, 549.
Kotov, V.A., Haneychuk, V.I., Tsap, T.T.: 1999, *Astron. Rep.* **43**, 185.
Kostyuk, N.D., Pikelner, S.B.: 1975, *Soviet Astron.* **18**, 590.
Kovács, A.: 1977, *Publ. Debr. Heliophys. Obs., Heliogr. Ser.* **3**, 207.
Kowalski, A.F., Hawley, S.L., Wisniewski, J.P., Osten, R.A., Hilton, E.J., Holtzman, J.A., *et al.*: 2013, *Astrophys. J. Suppl.* **207**, 15.
Krucker, S., Hudson, H.S., Jeffrey, N.L.S., Battaglia, M., Kontar, E.P., Benz, A.O., *et al.*: 2011, *Astrophys. J.* **739**, 96.
Lee, K.W., Büchner, J., Elkina, N.: 2008, *Astron. Astrophys.* **478**, 889.
Liu, W., Liu, S., Jiang, Y.W., Petrosian, V.: 2006, *Astrophys. J.* **649**, 1124.
Livshits, M.A.: 1997, *Solar Phys.* **173**, 377.
Livshits, M.A., Badalyan, O.G., Kosovichev, A.G., Katsova, M.M.: 1981, *Solar Phys.* **73**, 269.
Livshits, M.A., Belov, A.V.: 2004, *Astron. Rep.* **48**, 665.
Livshits, M.A., Rudenko, G.V., Katsova, M.M., Myshyakov, I.I.: 2015, *Adv. Space Res.* **55**(3), 920.
Maehara, H., Shibayama, T., Notsu, S., Notsu, Y., Nagao, T., Kusaba, S., *et al.*: 2012, *Nature* **485**, 478.
Maldonado, J., Martinez-Arnaiz, R., Eiroa, C., Montes, D., Montesinos, B.: 2010, *Astron. Astrophys.* **521**, A12.
Mamajek, E., Hillenbrand, L.A.: 2008, *Astrophys. J.* **687**, 1264.
Martínez Oliveros, J.-C., Hudson, H.S., Hurford, G.J., Krucker, S., Lin, R.P., Lindsey, C.: 2012, *Astrophys. J. Lett.* **753**, L26.
Marsden, S.C., Petit, P., Jeffers, S.V., Morin, J., Fares, R., Reiners, A., *et al.*: 2014, *Mon. Not. Roy. Astron. Soc.* **444**, 3517.
Martínez-Arnáiz, R., López-Santiago, J., Crespo-Chacón, I., Montes, D.: 2011, *Mon. Not. Roy. Astron. Soc.* **414**, 2629.
Matthews, S.A., Brown, J.C., van Driel-Gesztelyi, L.: 1998, *Astron. Astrophys.* **340**, 277.
Matthews, S.A., van Driel-Gesztelyi, L., Hudson, H.S., Nitta, N.V.: 2003, *Astron. Astrophys.* **409**, 1107.
Metcalf, T.R., De Rosa, M.L., Schrijver, C.J., Barnes, G., van Ballegooijen, A.A., Wiegelmann, T., *et al.*: 2008, *Solar Phys.* **247**, 269.

Meibom, S., Barnes, S.A., Platais, I., Gilliland, R.L., Latham David, W., Mathieu, R.: 2015, *Nature* **517**(7536), 589.
Mishenina, T.V., Soubiran, C., Kovtyukh, V.V., Katsova, M.M., Livshits, M.A.: 2012, *Astron. Astrophys.* **547**, A106.
Montes, D., Ramsey, L.W.: 1998, *Astron. Astrophys.* **340**, L5.
Morchenko, E., Bychkov, K.V., Livshits, M.A.: 2015, *Astrophys. Space Sci.* **357**, 119.
Nita, G.M., Fleishman, G.D., Kuznetsov, A., Kontar, E.P., Gary, D.E.: 2014, *Astrophys. J.* **789**, 152.
Pallavicini, R., Golub, L., Rosner, R., Vaiana, G.S., Ayres, T., Linsky, J.L.: 1981, *Astrophys. J.* **248**, 279.
Petit, P., Dintrans, B., Solanki, S.K., Donati, J.-F., Aurière, M., Lignières, F., *et al.*: 2008, *Mon. Not. Roy. Astron. Soc.* **388**, 80.
Potts, H., Hudson, H., Fletcher, L., Diver, D.: 2010, *Astrophys. J.* **722**, 1514.
Ramaty, R.: 1986. In: Sturrock, P.A., Holzer, T.E., Mihalas, D.M., Ulrich, R.K. (eds.) *Physics of the Sun* **2**, Reidel, Dordrecht, 291.
Ramaty, R., Tatischeff, V., Thibaud, J.P., Kozlovsky, B., Mandzhavidze, N.: 2000, *Astrophys. J. Lett.* **534**, L207.
Reiners, A., Schüssler, M., Passegger, V.M.: 2014, *Astrophys. J.* **794**, 144.
Rubenstein, E.P., Schaefer, B.E.: 2000, *Astrophys. J.* **529**, 1031.
Saar, S.H.: 1996. In: Uchida, Y., Kosugi, T., Hudson, H.S. (eds.) *Magnetodynamic Phenomena in the Solar Atmosphere – Prototype of Stellar Magnetic Activity*, Reidel, Dordrecht, 367.
Sadykov, V.M., Vargas Dominguez, S., Kosovichev, A.G., Sharykin, I.N., Struminsky, A.B., Zimovets, I.V.: 2014, *Astrophys. J.* **805**, 167.
Schmitt, J.H.M.M., Reale, F., Liefke, C., Wolter, U., Fuhrmeister, B., Reiners, A., Peres, G.: 2008, *Astron. Astrophys.* **481**, 799.
Shibata, K., Isobe, H., Hillier, A., Choudhuri, A.R., Maehara, H., Ishii, T.T., *et al.*: 2013, *Publ. Astron. Soc. Japan* **65**, 49.
Shibayama, T., Maehara, H., Notsu, S., Notsu, Y., Nagao, T., Honda, S., Ishii, T.T., *et al.*: 2013, *Astrophys. J. Suppl.* **209**, 5.
Skumanich, A.: 1972, *Astrophys. J.* **171**, 565.
Sun, X., Hoeksema, J.T., Liu, Y., Wiegelmann, T., Hayashi, K., Chen, Q., Thalmann, J.: 2012, *Astrophys. J.* **748**, 77.
Svestka, Z.: 1976, *Solar Flares*, Reidel, Dordrecht, 415 pp.
van Driel-Gesztelyi, L., Hudson, H.S., Anwar, B., Hiei, E.: 1994, *Solar Phys.* **152**, 145.
Wright, J.T., Marcy, G.W., Butler, R.P., Vogt, S.S.: 2004, *Astrophys. J. Suppl.* **152**, 261.
Wright, N.J., Drake, J.J., Mamajek, E.E., Henry, G.W.: 2011, *Astrophys. J.* **743**, 48.
Zharkova, V.V.: 2000. In: Ramaty, R., Mandzhavidze, N. (eds.) *High Energy Solar Physics Workshop – Anticipating HESSI*, *ASP Conf. Series* **206**, 264.
Zirin, H., Neidig, D.F.: 1981, *Astrophys. J. Lett.* **248**, L45.

Reprinted from *Solar Physics* Journal, DOI 10.1007/s11207-014-0644-1

X-ray Flare Spectra from the DIOGENESS Spectrometer and Its Concept Applied to ChemiX on the Interhelioprobe Spacecraft

Janusz Sylwester[1] · Zbigniew Kordylewski[1] · Stefan Płocieniak[1] · Marek Siarkowski[1] · Mirosław Kowaliński[1] · Stanisław Nowak[1] · Witold Trzebiński[1] · Marek Stęślicki[1] · Barbara Sylwester[1] · Eugeniusz Stańczyk[1] · Ryszard Zawerbny[1] · Żaneta Szaforz[1] · Kenneth J.H. Phillips[2] · František Fárník[3] · Anatolyi Stepanov[4]

Received: 10 September 2014 / Accepted: 16 December 2014 / Published online: 20 January 2015
© The Author(s) 2015. This article is published with open access at Springerlink.com

Abstract The DIOGENESS X-ray crystal spectrometer on the CORONAS-F spacecraft operated only for a single month (25 August to 17 September) in 2001, but in its short lifetime obtained one hundred and forty high-resolution spectra of eight solar flares with GOES importance ranging from C9 to X5. The instrument included four scanning flat crystals with wavelength ranges covering the regions of Si XIII (6.65 Å), S XV (5.04 Å), and Ca XIX (3.18 Å) X-ray lines and associated dielectronic satellites. Two crystals covering the Ca XIX lines were oriented in a "dopplerometer" manner, *i.e.* such that spatial and spectral displacements, both of which commonly occur in flares, can be separated. We describe the DIOGENESS spectrometer and the spectra obtained during flares that include lines not hitherto seen from spacecraft instruments. An instrument with a very similar concept is currently being built for the two Russian *Interhelioprobe* spacecraft that are scheduled for launch in 2020 and 2022 and will make a near-encounter (perihelion $\sim$0.3 AU) with the Sun in its orbit. We outline the results that are likely to be obtained.

Solar and Stellar Flares: Observations, Simulations, and Synergies
Guest Editors: Lyndsay Fletcher and Petr Heinzel

E. Stańczyk and R. Zawerbny are deceased.

✉ J. Sylwester
js@cbk.pan.wroc.pl

K.J.H. Phillips
kennethjhphillips@yahoo.com

F. Fárník
ffarnik@asu.cas.cz

A. Stepanov
as@izmiran.ru

[1] Space Research Centre, Polish Academy of Sciences, Kopernika 11, 51-622 Wrocław, Poland

[2] Earth Sciences Department, Natural History Museum, London SW7 5BD, UK

[3] Astronomical Institute, The Czech Academy of Sciences, 25165 Ondřejov, Czech Republic

[4] Institute of Terrestrial Magnetism and Radiowave Propagation (IZMIRAN), Troisk, Moscow, Russia

Keywords Flares, Spectrum · Spectral Line, Intensity and Diagnostics · Spectrum, X-ray

1. Introduction

Scanning X-ray flat crystal spectrometers on rockets and spacecraft have been used to observe flares and non-flaring active regions for many years, and the resulting spectra have helped our understanding of the energetics and physical characteristics of the emitting regions and the mechanism of flare energy release. Instruments with an accurate absolute intensity calibration have also enabled element abundances in the corona to be found. The rapidly developing initial phases of flares, in which high-temperature plasma is ejected from localised sources such as the footpoints of magnetic flux tubes, requires X-ray spectrometers that can acquire and store data on very short time-scales. This has led to the application of bent crystal spectrometers with position-sensitive detectors for solar-dedicated spacecraft. Early examples include those on the *Solar Maximum Mission* (BCS or *Bent Crystal Spectrometer*: Acton *et al.*, 1980), *Yohkoh* (BCS or *Bragg Crystal Spectrometer*: Culhane *et al.*, 1991), and CORONAS-F (RESIK or *REntgenovsky Spektrometr s Izognutymi Kristalami*: Sylwester *et al.*, 2005).

Bent crystal spectrometers have the advantage of obtaining spectra over their wavelength ranges instantaneously in a particular data-gathering interval, typically a few seconds, and also of generally having a higher sensitivity than flat crystal spectrometers. However, to observe the initial phases of flares, spectra from bent crystal spectrometers, especially those without a fine collimator that reduces their sensitivity, are subject to a confusion of spectral and spatial blurring without being able to untangle the two unless independent measurements are available. This difficulty may be overcome with the simultaneous observation of flares with a pair of crystals oriented in such a way that the spatial displacements occur to longer wavelengths for one of them and towards shorter wavelengths for the other. This "dopplerometer" mode was proposed for an instrument initially launched on a *Vertical* rocket in the Russian space programme by the Space Research Centre of the Polish Academy of Sciences in 1981. A later version, called DIOGENESS (*DIagnostics Of Global ENErgy Sources and Sinks*) was built by SRC to be included in the Russian *CORONAS-I* spacecraft in the mid-1990s and, following the failure of the instrument, for the subsequent *CORONAS-F* spacecraft, which was operational from 2001 to 2006. Four scanning flat crystals were included to observe the vicinity of intense lines of high-temperature ions – Si XIII (6.65 Å), S XV (5.04 Å), and Ca XIX (3.18 Å) – which are prominent in flare and some active region spectra, with two of the crystals observing the Ca XIX lines in a dopplerometer arrangement.

The *CORONAS-F* spacecraft was launched on 31 July 2001 and worked successfully until the end of 2006, with RESIK and DIOGENESS included in the instrument package. While RESIK obtained spectra until May 2003, DIOGENESS operated for only a few weeks because a fault in the scanning drive mechanism occurred on 17 September 2001. However, eight flares with GOES importance up to X5.5 were observed and one hundred and forty spectra were obtained in four wavelength channels. Detailed analysis was delayed for some years while the more extensive RESIK data set were examined. In the meantime, another version of DIOGENESS, called *ChemiX*, is being built to be included in the instrument package on two Russian *Interhelioprobe* spacecraft, scheduled for launch in 2020 and 2022, which will eventually orbit the Sun in highly elliptical paths, with 30° inclination to the ecliptic plane, bringing them to within 0.3 astronomical units from the Sun. It is appropriate, therefore, to describe the DIOGENESS and *ChemiX* instruments in some detail at this time, even though the two earlier versions were launched some years ago. In particular, it is of interest to describe the DIOGENESS spectra, which have particularly good resolution and allow the flare dynamics to be determined as well as investigation of lines not previously noted in solar spectra.

In this article, we describe the DIOGENESS instrument and some details of the *CORONAS-F* spacecraft and operations (Section 2), the flare spectra obtained in its operating lifetime, including a list of the lines observed (Section 3), and describe the forthcoming *ChemiX* instrument that is scheduled to fly on the *Interhelioprobe* missions (Section 4).

2. CORONAS-F and the DIOGENESS Instrument

2.1. Spacecraft and Performance

CORONAS-F was the second of three Russian spacecraft designed to observe various solar phenomena in the period 1994 – 2009, with thirteen instruments studying global oscillations, variations in ultraviolet, X-ray, and γ-ray, emission, and particle emission. The spacecraft was launched into a near-circular orbit around Earth with an altitude range of 501 – 549 km. The orbital plane was inclined at $82^{\circ}.5$ to the equator, and its initial period was 94.9 minutes. Intervals of uninterrupted solar observation were possible for up to 20 days on two occasions during a year, while at other times the spacecraft night-time periods lasted no more than 35 minutes. Solar X-ray observations were also interrupted by passages through the auroral oval particle zones near the Earth's magnetic poles and the South Atlantic Anomaly. Spacecraft pointing at the Sun was achieved with a 15-arcminute precision.

The instrument package is described by Oraevsky and Sobelman (2002). The RESIK and DIOGENESS instruments were built by teams led by the Space Research Centre, Polish Academy of Sciences, in Wrocław, Poland. RESIK has been described by Sylwester *et al.* (2005). It was a bent crystal spectrometer with four channels observing X-ray flare and non-flaring active regions in the range 3.3 – 6.1 Å, which included emission lines of highly ionized Si, S, Cl, Ar, and K ions. An absolute flux calibration of ~ 20 % precision was achieved for the instrument, allowing element abundances to be determined. These estimates have been discussed in several publications: see Sylwester *et al.* (2014) and references therein. RESIK operated from the time of spacecraft launch until May 2003, when the instrument power supply failed.

2.2. DIOGENESS: Instrument and Performance

The DIOGENESS instrument, in contrast to RESIK, had a lifetime of only a few weeks, but the instrumental concept, including that of the dopplerometer, was adequately verified. The construction of the instrument was outlined by Płocieniak *et al.* (2002) and Siarkowski *et al.* (2002). DIOGENESS had four channels covering wavelength ranges in the vicinity of intense lines of highly ionized Si, S, and Ca lines visible in solar flares. Two quartz crystals covered the Ca XIX lines (channels 1 and 4), and ADP and beryl crystals (channels 2 and 3) covered the S XV and Si XIII lines, respectively. Table 1 lists details of the wavelength ranges and crystals. The mono-crystals were flat and mounted on a common rotatable shaft, so a scanning motion back and forth of the shaft allowed X-rays with various wavelengths λ to be diffracted according to Bragg's diffraction law,

$$n\lambda = 2d \sin\theta, \tag{1}$$

Table 1 DIOGENESS instrument channels.

Channel	1	2	3	4
Crystal	Quartz	ADP	Beryl	Quartz
Diffracting plane	$10\bar{1}1$	101	1010	$10\bar{1}1$
$2d$ spacing (Å)[a]	6.6855	10.5657	15.9585	6.6875
Central wavelength (Å)	3.1781	5.0374	6.6488	3.1781
Principal lines in range	Ca XIX	S XV	Si XIII	Ca XIX
Min. of wavelength range (Å)	3.1436	4.9807	6.1126	2.9601
Max. of wavelength range (Å)	3.3915	5.3721	6.7335	3.2123
Crystal reflectivity (μrad)[a]	91	91	15	90
Rocking curve (FWHM, arcsec)[a]	24.1	68.1	94.1	25.6

[a]Prelaunch measurement.

where d is the crystal lattice spacing, θ the angle of incidence, and n the diffraction order. Only first-order ($n = 1$) spectra were recorded. The angular range of the crystal shaft was 140 arcmin, giving the wavelength ranges specified in Table 1. The spectral resolution of each crystal is defined by the rocking curve, whose values (FWHM) are given in arcsec in Table 1. The instrument was uncollimated, but as flares generally have small angular extents (typically a few arcmin), the emission region can be considered to be a point source, so a collimator is generally not necessary. The diffracted X-rays are registered by double proportional counters with beryllium entrance windows of 145 μm thickness and are filled with argon gas with a small admixture of carbon dioxide quenching gas at $\sim$0.5 atmospheric pressure. An Fe^{55} source (energy 5.9 keV) illuminates the other, non-solar proportional counter section, so that the detector's electronic system energy gain could be checked. A multi-slit collimator with soft X-ray detectors scanning in the direction of the spectrometer's dispersion was also included to give spatial context information, but unfortunately, this part of the instrument failed early in the mission.

For channels 2 and 3, the S XV and Si XIII lines were repeatedly scanned over the period of a flare in the direction of alternately increasing and decreasing wavelengths. Figure 1 shows the arrangement for channels 1 and 4, which viewed the Ca XIX lines with identical quartz crystals cut from the same parent mono-crystal at an angle α between the diffracting planes of the crystals, the so-called dopplerometer mode. Solar X-rays are incident on each crystal in the manner illustrated in the figure (left panel). As the two crystals on their common shaft are rotated, solar X-ray lines are successively recorded by each channel from the point-like flare on the Sun, as shown in the right panel of Figure 1. The angle α was selected such that the Ca XIX resonance line w emitted by flare emission with zero radial velocity was observed simultaneously, its theoretical value being 56°.7416. (In practice, owing to co-alignment uncertainty, the angle α was set to 56°.8000.) In Figure 1, the two occasions when the vicinity of the Ca XIX w line was scanned simultaneously in the sense of increasing wavelengths (red scan in the figure) and decreasing wavelengths (blue scan) are shown. The horizontal axis indicates the stepper drive's address rather than wavelength. For zero radial velocity of the flare source, the Ca XIX w line would be coincident for each scan, but because of the presence of an approach velocity (such wavelength shifts have been commonly attributed to "chromospheric evaporation" or upward convection of the emitting plasma), the lines are separated in each scan. The linearity of the scanning mechanism was checked during the spacecraft mission using strain-gauge systems.

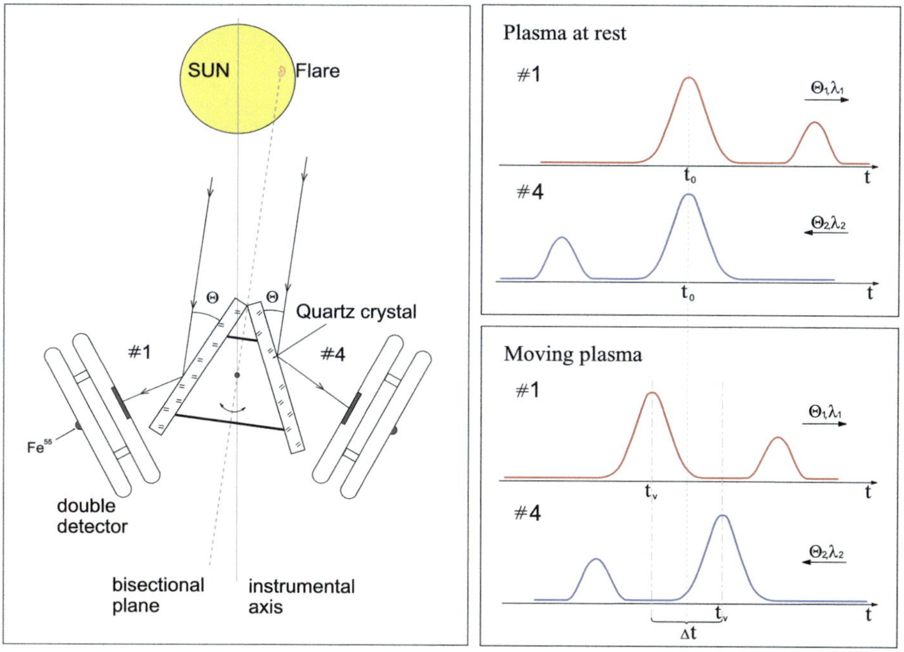

Figure 1 Left panel: Principle of the DIOGENESS dopplerometer. Solar X-rays are incident on two identical quartz crystals oriented as shown. The diffracted rays around the wavelengths of the Ca XIX lines near 3.177 Å are detected by double proportional counters (labelled 1 and 4 in the figure). The detector energy gain is checked with Fe^{55} sources at the rear of each detector. Right panel: Resulting Ca XIX spectra, showing the more intense resonance (w) and less intense intercombination (x, y) lines (see Table 3). In the upper panel the flare plasma is assumed to be at rest with respect to the instrument, and the Ca XIX w (resonance) lines are coincident (for detector 1 wavelength λ increases to the right, for detector 4 it increases to the left). In the lower panel the flare plasma is moving with approaching velocity (as is usually the case with flare impulsive phases), so the line wavelengths are displaced to shorter wavelengths in opposite directions. Any shifts due to spatial movement across the line of sight will be seen as wavelength shifts in identical directions.

The effective areas of the four DIOGENESS channels (see Figure 2) were measured before spacecraft launch and are the product of detector efficiency, filter transmission, crystal integrated reflectivity, and projected area of the detector window as seen from the Sun.

3. DIOGENESS Observations

3.1. Flare Spectra

Eight intense flares were observed during the lifetime of DIOGENESS; see Table 2 for details of these flares including times, standard IAU flare notation, GOES and Hα importance, and heliographic location. One hundred and forty spectra were obtained in the four channels for these events.

The X5 flare on 25 August 2001 was observed to great advantage by DIOGENESS, from the earliest pre-flare state including a small precursor flare (16:17 UT) and the impulsive phase at around 16:30 – 16:32 UT, also seen with the *Yohkoh Hard X-ray Telescope* (HXT), until late in the decay phase, a total of nearly two hours. Figure 3 shows the light curves

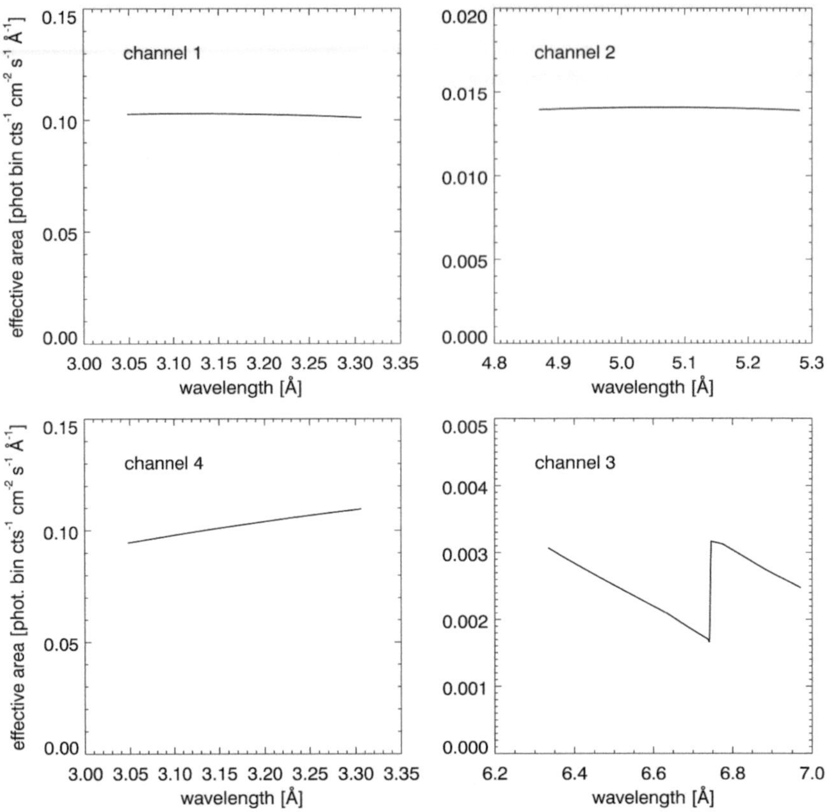

Figure 2 Effective areas of the four channels of DIOGENESS. The units are photon bin count^{-1} cm^{-2} sec^{-1} Å^{-1}. The channel numbers are indicated in the legend.

Table 2 Flares observed by DIOGENESS.

Date (2001)	UT of peak	Flare notation	GOES importance	Hα importance	Location[a]
Aug. 25	16:45	SOL2001-08-25T16:45	X5.3	3B	S17E34
Aug. 30	17:57	SOL2001-08-30T17:57	M1.5	2N	S21W28
Sept. 2	06:02	SOL2001-09-02T06:02	M1.3	1F	S17W66
Sept. 2	13:48	SOL2001-09-02T13:48	M3.0	2N	S21W65
Sept. 3	01:58	SOL2001-09-03T01:58	C9.0	–	S17E90
Sept. 3	17:16	SOL2001-09-03T17:16	M1.1	–	N10W06
Sept. 3	18:41	SOL2001-09-03T18:41	M2.5	–	S26E90
Sept. 16	03:53	SOL2001-09-16T03:53	M5.6	2N	S29W54

[a]No Hα records for the Sept. 3 flares; locations from Yohkoh HXT images.

from the two GOES channels and the total emission in DIOGENESS channel 1 (Ca XIX channel). The DIOGENESS light curve shows the Ca XIX line emission because the Bragg diffraction condition (Equation (1)) was successively satisfied for the Ca XIX emission lines

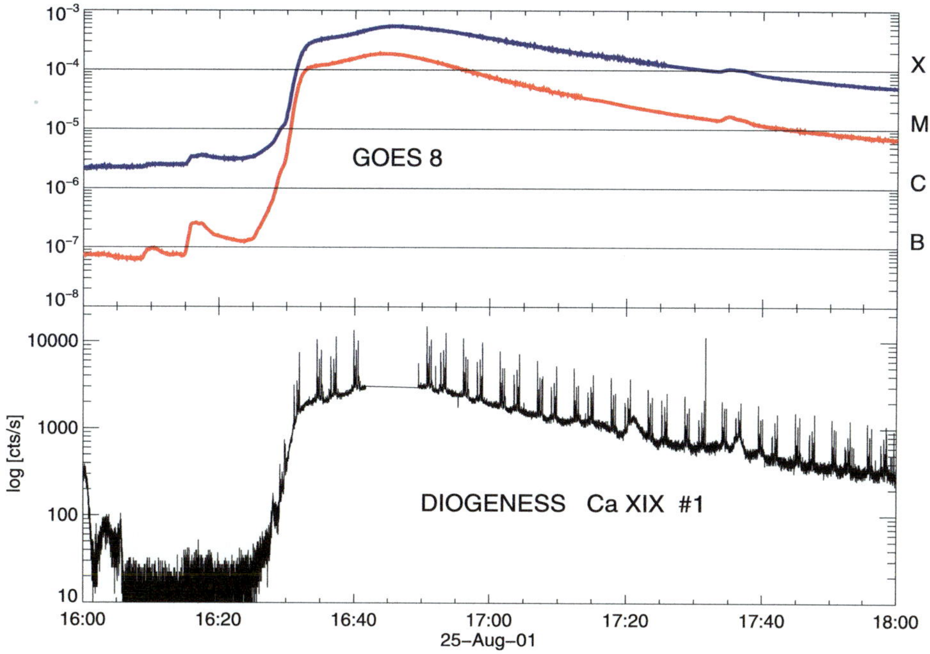

Figure 3 Top: Light curves of the two channels of GOES (red – 0.5 – Å; blue – 1 – 8 Å), irradiance scale in W m^{-2} on the left axis (GOES importance scale on the right axis) for the 25 August 2001 flare. Lower panel: DIOGENESS Ca XIX count rate (logarithmic scale), showing the scans through the Ca XIX lines as the crystals rocked back and forth (time of scan 140 s). A passage of the *CORONAS-F* spacecraft through the auroral oval radiation belts before the flare occurred around 16:00 – 16:10 UT.

a total of ∼30 scans, *i.e.* approximately 140 s per scan. At the impulsive phase, the HXT saw emission up to its highest-energy channel (53 – 93 keV): see the light curves at four energy bands in Figure 4 (left-hand panel). The HXT images show the double-footpoint emission characteristic of large flares with soft X-ray emission (from the *Yohkoh Soft X-ray Telescope*) in between (see right-hand panel of Figure 4).

In Figure 5, channel 4 spectra over the Ca XIX lines are indicated during the 25 August 2001 flare with forward and backward scans (wavelength increasing and decreasing with time, respectively).

3.2. Plasma Dynamics in the 25 August 2001 Flare

Velocities of the emitting plasma during the 25 August 2001 flare were found from DIOGENESS Si XIII, S XV, and Ca XIX spectra, with the Ca XIX spectra from the dopplerometer part of the instrument. The left panel of Figure 6 shows the Ca XIX spectra in channels 1 and 4 plotted in stepper motor addresses. The main lines apparent are the resonance (w), intercombination (x, y), and forbidden (z) lines of Ca XIX in order of wavelength (see Table 3). The channel 1 spectrum (in blue in Figure 6) has wavelength increasing to the left, the channel 4 spectrum (in red) with channel 1 to the right. The scan times of the spectra are almost coincident. The deduced velocities for all channels are given as a function of time in Figure 6 (right panel). They indicate that there were high approach velocities at

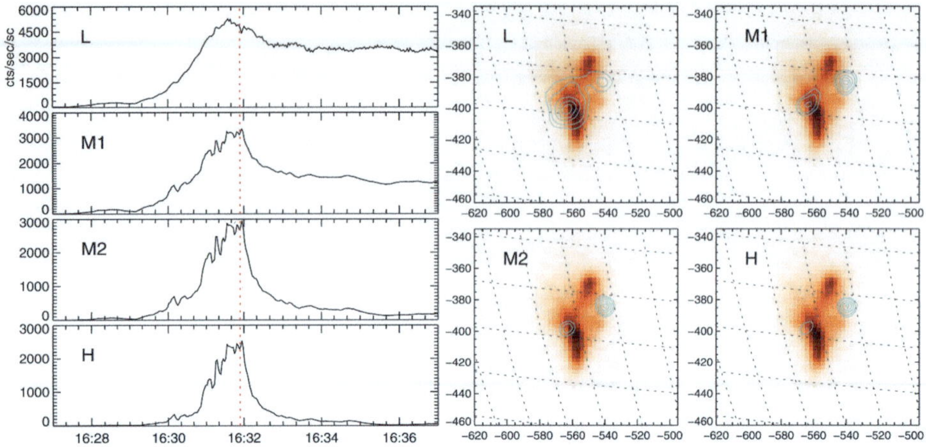

Figure 4 Left: Light curves over the impulsive phase of the 25 August 2001 flare from the *Yohkoh* HXT instrument in its four channels: L – 14 – 23 keV, M1 – 23 – 33 keV, M2 – 33 – 53 keV, H – 53 – 93 keV. Right: *Yohkoh* SXT images (red intensity scale) and HXT images (blue contours) in the four channels of HXT indicated at the top of each image.

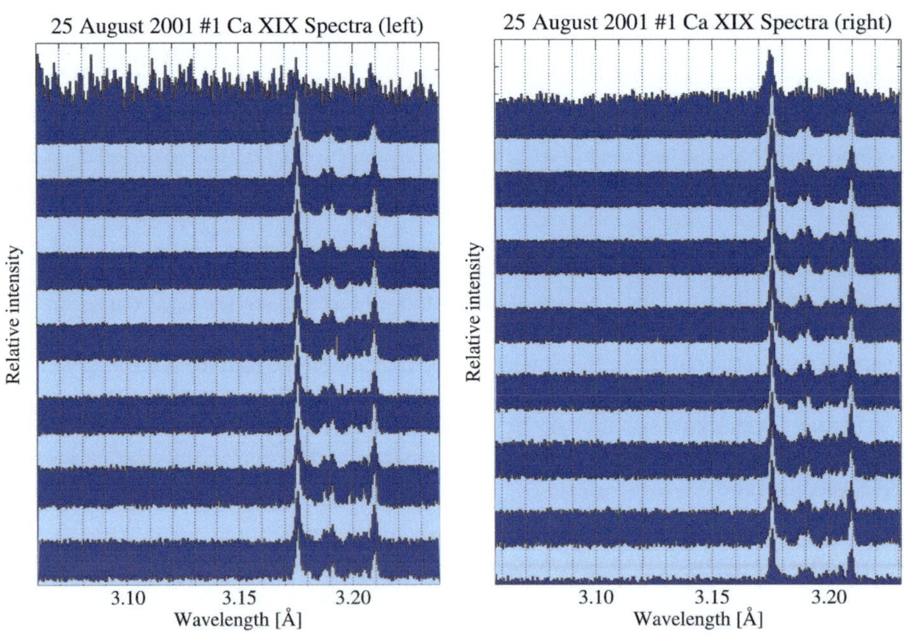

Figure 5 Left: Forward (increasing wavelength) scans for DIOGENESS channel 4 (Ca XIX lines) during the 25 August 2001 flare stacked with increasing times from top to bottom (time range 16:29 – 18:00 UT). Right: Corresponding backward (decreasing wavelength) scans.

the time (approximately 27 – 16:29 UT) at the start of the hard X-ray impulses recorded by the *Yohkoh* HXT M1, M2, and H channels (see light curves and images in Figure 4). The highest velocity, 165 km s^{-1}, is recorded by the S XV (temperature $\sim$10 MK) line emission. The velocity of the higher temperature ($\sim$20 MK) Ca XIX line emission indicates a velocity

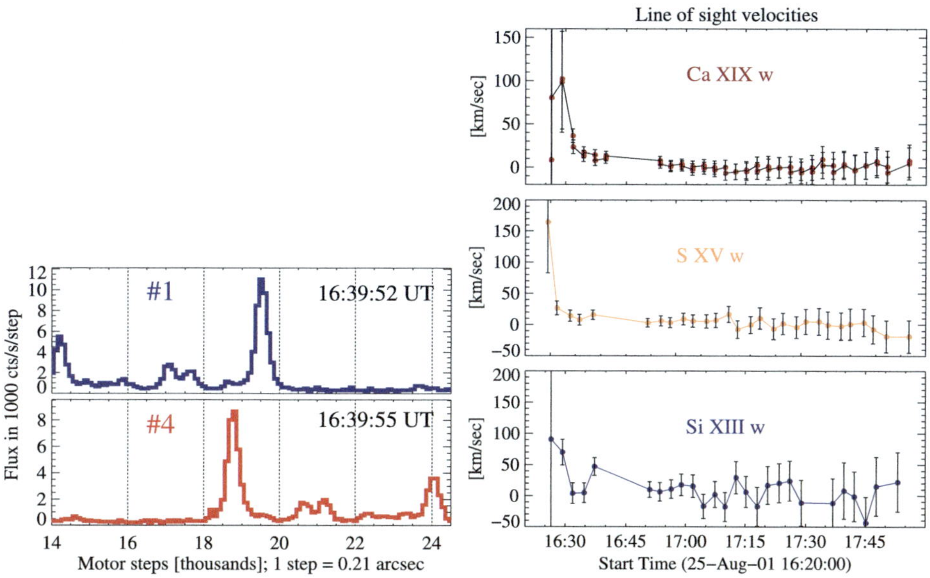

Figure 6 Left: Ca XIX spectra from channels 1 and 4 (reversed), plotted as a function of stepper motor addresses, for two times (indicated) in the 25 August 2001 flare. Right: Line-of-sight velocities of X-ray flare plasma as deduced from Doppler shifts of Ca XIX, S XV, and Si XIII lines during the flare of 25 August 2001. See Płocieniak *et al.* (2002) for details.

of about 100 $\mathrm{km\,s^{-1}}$, as does ($\approx$90 $\mathrm{km\,s^{-1}}$) the lower temperature ($\sim$7 MK) Si XIII line emission.

These velocities are smaller than those previously observed for large disk flares. Doschek *et al.* (1980), who discussed several X-class flares seen with X-ray spectrometers on the *P78-1* spacecraft, found approach velocities of about 400 $\mathrm{km\,s^{-1}}$ for disk flares, as did Tanaka *et al.* (1982) for a very large disk flare seen with an X-ray spectrometer on the *Hinotori* spacecraft. It is possible that plasma motions during the 25 August 2001 flare observed by DIOGENESS were directed at a substantial angle to the line of sight (although the flare was on the disk, it was 34° east of the central meridian), which would account for the lower observed approach velocities in its rise phase.

3.3. Spectral Line Identifications

Figure 7 shows an averaged spectrum from DIOGENESS channels 1 and 4 during the 25 August 2001 flare, peak time in soft X-rays at 16:45 UT (Table 2), with flux on a logarithmic scale to show the weak line emission either side of the Ca XIX line emission to better advantage. Table 3 gives the wavelengths of identifiable line features for the 3.05 – 3.30 Å range. Channels 1 and 4 spectra cover the region of the Ca XIX resonance (w) line and include other Ca XIX lines and dielectronic satellites of the Li-like (Ca XVIII) and lower stages of ionization. These lines are familiar from previously flown X-ray spectrometers on the *P78-1*, SMM, *Hinotori*, and *Yohkoh* spacecraft. The ratios of the principal satellites (k and the $d13$, $d15$ lines) to the Ca XIX w line have been used extensively to find flare and non-flaring active region temperatures and emission measures, based on the theory of Gabriel (1972) (whose line notation we use here) and Bely-Dubau *et al.* (1982). Ca XVIII satellite j is indistinguishable from the Ca XIX line z. The ratio of the inner-shell excitation satellite q to

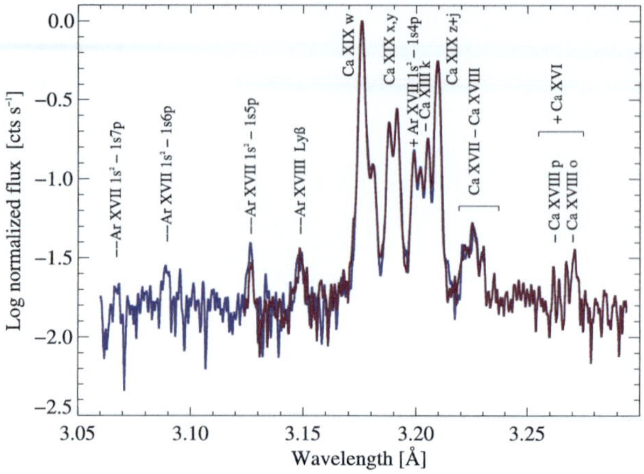

Figure 7 Averaged DIOGENESS spectrum from channels 1 and 4 during the 25 August 2001 flare showing Ar XVII and Ca XIX lines and dielectronic satellites of Ca XVIII, Ca XVII, and Ca XVI.

line w depends on the ion fraction $N(\mathrm{Ca}^{+17})/N(\mathrm{Ca}^{+18})$ and therefore in principle is a means of finding departures of ionization equilibrium in the emitting plasma. However, satellite q is blended with the Ar XVII line $w4$ (this notation being used for the transition $1s^2 - 1s4p$).

In the DIOGENESS spectrum shown in Figure 7, the Rydberg series of Ar XVII lines up to the $w7$ line can be discerned. The Ca XIX and Ar XVII spectra up to the $w10$ line have been observed during strong flares with crystal spectrometers on the *P78-1* spacecraft (Doschek, Feldman, and Seely, 1985; Seely and Feldman, 1985), with lines $w9$ and $w10$ blended with Ca XIX dielectronic satellites on the long-wavelength side of the Ca XX Ly-α doublet. Spectra from the Alcator C-Mod tokamak plasmas, recently reported by Rice *et al.* (2014), also show Ar XVII high-n lines up to the ionization limit at 3.009 Å. All these lines, which are shortward of 3.06 Å, are outside the DIOGENESS range. We note that the estimated temperatures (~ 30 MK) and electron densities ($\sim 10^{15}$ cm^{-3}) in the Alcator plasmas are both higher than those typical of large solar flare plasmas (~ 25 MK and $\sim 10^{11} - 10^{12}$ cm^{-3}, respectively), but this has little effect on the appearance of the spectral lines, whose the principal excitation mechanism is electron collisional excitation from the ground state of ions in both cases.

In Table 3, we follow Rice *et al.* (2014) and Seely and Doschek (1989) in identifying lines in the DIOGENESS spectrum on the long-wavelength side of the Ca XIX forbidden line z as Ca XVII satellites including the strong inner-shell-excited line β, as well as Ca XVIII satellites u and v, which are both dielectronically and collisionally excited (Gabriel, 1972). In addition, a feature possibly composed of several lines between 3.261 Å and 3.272 Å, not apparently identified before, is probably caused by a group of Ca XVI lines by analogy with equivalent lines of Fe XXII seen in solar flare spectra with the *SMM Bent Crystal Spectrometer* (Phillips *et al.*, 1983). The Ca XVIII two-electron-jump satellites o and p (transitions $1s^2\,2p\,^2P_{3/2} - 1s\,2s^2\,^2S_{1/2}$, $1s^2\,2p\,^2P_{1/2} - 1s\,2s^2\,^2S_{1/2}$) may also contribute to this line feature: their wavelengths are estimated by Bely-Dubau *et al.* (1982) to be 3.2688 Å and 3.2636 Å, respectively.

In DIOGENESS channel 2 spectra, the predominant lines are the resonance, intercombination, and forbidden lines of S XV and the Ly-β ($1s - 3p$) line of Si XIV, prominent also in RESIK spectra. The x and y lines are partially blended as are S XIV satellites $d13$, $d15$ with S XV line w and S XIV satellites j, k with S XV line z. There is a possible long-wavelength "shoulder" of emission to the Si XIV line, which may be due to the Si XIII $w6$ line. In channel 3 spectra, the Si XIII $w - z$ lines are the most prominent with Si XII satellites $d13 + d15$,

Table 3 Spectral line identifications in DIOGENESS spectra.

Measured line λ (Å)	Line identification Ion	λ (Å)	Transition	Name[a]	Source[b]
Channels 1 and 4					
3.068	Ar XVII	3.068	$1s^2\,{}^1S_0 - 1s7p\,{}^1P_1$	$w7$	R, S
3.089					
3.095	Ar XVII	3.095	$1s^2\,{}^1S_0 - 1s6p\,{}^1P_1$	$w6$	K, S
3.127	Ar XVII	3.128	$1s^2\,{}^1S_0 - 1s5p\,{}^1P_1$	$w5$	K, S
3.148					
3.177	Ca XIX	3.177	$1s^2\,{}^1S_0 - 1s2p\,{}^1P_1$	w	K
3.181	Ca XVIII	3.1809	$1s^2 3p\,{}^2P_{3/2} - 1s2p({}^1P)3p\,{}^2D_{5/2}$	$d13$	K
3.188	Ca XIX	3.1889	$1s^2\,{}^1S_0 - 1s2p\,{}^3P_2$	x	K
3.191	Ca XIX	3.1925	$1s^2\,{}^1S_0 - 1s2p\,{}^3P_1$	y	K
3.199	Ca XVIII	3.2003	$1s^2\,2s\,{}^2S_{1/2} - 1s(2s2p\,{}^3P)\,{}^2P_{3/2}$	q	K
	Ar XVII	3.200	$1s^2\,{}^1S_0 - 1s4p\,{}^1P_1$	$w4$	K
3.203	Ca XVIII	3.2031	$1s^2\,2s\,{}^2S_{1/2} - 1s(2s2p\,{}^3P)\,{}^2P_{1/2}$	r	K
3.206	Ca XVIII	3.2058	$1s^2\,2p\,{}^2P_{1/2} - 1s2p^2\,{}^2D_{3/2}$	k	K
3.210	Ca XIX	3.2111	$1s^2\,{}^1S_0 - 1s2s\,{}^3S_1$	z	K
	Ca XVIII	3.2097	$1s^2 2p\,{}^2P_{3/2} - 1s2p^2\,{}^2D_{5/2}$	j	K
3.222	Ca XVII	3.2217	$1s^2\,2s^2\,{}^1S_0 - 1s\,2s^2\,2p\,{}^1P_1$	β	R
3.225	Ca XVII	?	$1s^2 2s^2 - 1s2s^2 2p$		Ph
3.225	Ca XVIII	3.2266	$1s^2\,2s\,{}^2S_{1/2} - 1s2s2p\,{}^4P_{3/2}$	u	R, D
3.230	Ca XVIII	3.2277	$1s^2\,2s\,{}^2S_{1/2} - 1s2s2p\,{}^4P_{1/2}$	v	R, D
3.230	Ca XVII	?	$1s^2 2s^2 - 1s2s^2 2p$		Ph
3.272	Ca XVI	?	$1s^2 2s^2 2p - 1s2s2p^2$		Ph
Channel 2					
5.037	S XV	5.0385	$1s^2\,{}^1S_0 - 1s2p\,{}^1P_1$	w	K
5.047	S XIV	5.0468	$1s^2 3p\,{}^2P_{3/2} - 1s2p({}^1P)3p\,{}^2D_{5/2}$	$d13$	K
5.062	S XV	5.06?	$1s^2\,{}^1S_0 - 1s2p\,{}^3P_2$	x	K
5.064	S XV	5.0662	$1s^2\,{}^1S_0 - 1s2p\,{}^3P_1$	y	K
5.083	S XIV	5.0861	$1s^2 2s\,{}^2S_{1/2} - 1s(2s2p\,{}^3P)\,{}^2P_{3/2}$	q	K
5.099	S XIV	5.0964	$1s^2 2p\,{}^2P_{1/2} - 1s2p^2\,{}^2D_{3/2}$	k	K
	S XV	5.101	$1s^2\,{}^1S_0 - 1s2s\,{}^3S_1$	z	K
	S XIV	5.1010	$1s^2 2p\,{}^2P_{3/2} - 1s2p^2\,{}^2D_{5/2}$	j	K
	Si XIV	5.217	$1s\,{}^2S_{1/2} - 2p\,{}^2P_{1/2,3/2}$	Ly-β	
Channel 3					
6.410	Mg XII ?	6.497	$1s\,{}^2S_{1/2} - 8p\,{}^2P_{1/2,3/2}$	Ly-η	K
6.490	Mg XII ?	6.497	$1s\,{}^2S_{1/2} - 6p\,{}^2P_{1/2,3/2}$	Ly-ϵ	K
6.513					
6.545					
6.574	Mg XII ?	6.580	$1s\,{}^2S_{1/2} - 5p\,{}^2P_{1/2,3/2}$	Ly-δ	K
6.641	Si XIII	6.6477	$1s^2\,{}^1S_0 - 1s2p\,{}^1P_1$	w	K
6.681					
6.686	Si XIII	6.6879	$1s^2\,{}^1S_0 - 1s2p\,{}^3P_1$	y	K
6.713	Si XII	6.718	$1s^2 2s\,{}^2S_{1/2} - 1s(2s2p\,{}^3P)\,{}^2P_{3/2}$	q	K
6.737	Si XIII	6.740	$1s^2\,{}^1S_0 - 1s2s\,{}^3S_1$	z	K

[a]Gabriel (1972) for He-like ion $1s^2 - 1s2l$ transitions and Li-like ion satellites.

[b]D = Doschek *et al.* (1985); K = Kelly (1987); R = Rice *et al.* (2014); Ph = Phillips *et al.* (1983); S = Seely and Feldman (1985).

Table 4 ChemiX Spectral Channels.

Channel No.	Crystal	Diffracting plane	2*d* (Å)	Wavelength Range (Å)	Av. spectral resolution (mÅ per pixel)
Spectrometer channels					
1	Si	111	6.271	1.500 – 2.713	1.46
2	Quartz	$10\bar{1}0$	8.514	2.700 – 4.304	1.95
3	KDP	011	10.185	4.290 – 5.228	1.43
4	KAP	001	26.64	5.200 – 8.800	4.73
Dopplerometer channels					
1	LiF	022	2.848	1.830 – 1.925	0.10
2	Si	111	6.271	3.150 – 3.245	0.10
3	Si	111	6.271	3.90 – 4.080	0.19

q, and k identifiable. Possible very weak line features to the short-wavelength side of the Si XIII line w occur. The Mg XII Rydberg sequence $1s - np$ occur in this region and were seen with the crystal spectrometers on OSO-8 (Parkinson *et al.*, 1978). They may account for the 6.410 Å ($n = 8$), 6.490 Å ($n = 6$), and 6.580 Å ($n = 5$) lines in the DIOGENESS spectra, although the $n = 7$ Mg XII line at 6.448 Å is not apparent. The $n = 4$ Mg XII line blends with the Si XIII z line. Note that the anomaly in the effective area due to the beryl crystal reflectivity at about 6.75 Å (see Figure 2) distorts the intensity of the Si XIII z line.

In time, the line identifications given in Table 3 will be refined with better atomic data. In particular, we plan to run atomic codes to find the wavelengths of the satellites due to Ca XVI, as was done for the case of the equivalent Fe ions (Phillips *et al.*, 1983).

4. The ChemiX Instrument for Interhelioprobe

ChemiX (*CHEMical composition In X-rays*) is a bent crystal spectrometer under construction that will observe the soft X-ray spectra of solar flares, continuing the objectives of previous spectrometers such as the *Flat Crystal Spectrometer* and *Bent Crystal Spectrometer* on *Solar Maximum Mission* (Acton *et al.*, 1980), SOLEX and SOLFLEX on *P78-1* (Doschek, Kreplin, and Feldman, 1979), and RESIK and DIOGENESS on *CORONAS-F*. We briefly describe it here, while a more detailed account is in preparation (Siarkowski *et al.*). ChemiX will have four "spectrometer" channels viewing spectral lines of Fe XXV, Ca XIX, S XV, Si XIII and Si XIV, and three pairs of crystals in a dopplerometer arrangement like that on DIOGENESS and its predecessors, viewing lines of Fe XXV, Ca XIX, and Ar XVII. There will be important improvements over both the RESIK and DIOGENESS instruments in that the detectors will be X-ray-sensitive, cooled (to < -20 °C) charge-coupled devices (CCDs), and there will be a two-dimensional collimator with ~ 3 arcmin (FWHM) field of view that will enable the instruments to view single active regions and so avoid possible spatial and spectral confusion for the spectrometer channels. A projected 1-s time resolution will allow the highly dynamic impulsive phases of flares to be observed, in particular changes to spectral line profiles and shifts indicative of plasma turbulence and motions. Table 4 gives details of the Bragg-diffracting crystals to be used and the wavelength ranges in each of the seven channels. There will be full spectral coverage in the range 1.5 – 9 Å with a spectral resolution

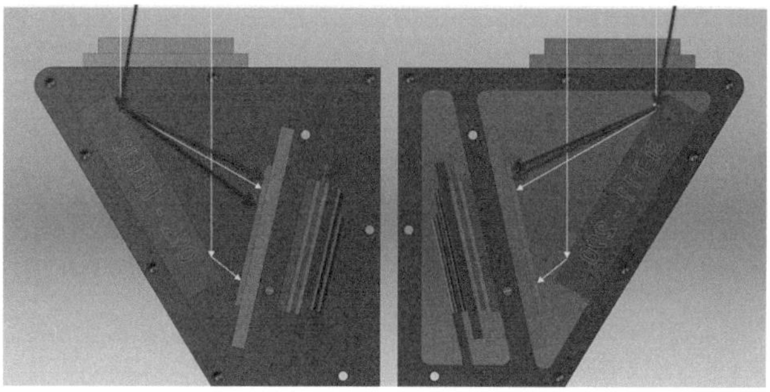

Figure 8 Scheme of ChemiX. Solar X-rays are incident from the top of the figure (blue, yellow, and red arrows), are diffracted off the two bent crystals (Si 111 in both cases), and are then incident on the CCDs. The yellow rays are for a source in the centre of the collimator field of view (FOV), the blue rays from a source moved across the FOV (tangential motion), and the red rays for a source approaching the instrument (*i.e.*, subject to a Doppler shift).

that is expected to be four to ten times better than that of RESIK because of the use of CCDs over gas proportional counters. The signal-to-noise ratio will also be much improved, and because of the use of crystals with low atomic number, there will be very little instrumental background due to crystal fluorescence, enabling the solar continuum to be measured. Figures 8 and 9 give the scheme for ChemiX, which is at present under construction.

Other subsystems of ChemiX include an energetic particle (electrons, protons, and alpha-particles) detector, so that the instrument will have its own monitor for safety concerns with the electronics; a soft X-ray pin-hole camera with CCD detector, providing X-ray context images with one-arcmin spatial resolution that will help to ascertain which active region a flare originated from; and a movable target-pointing platform that can lock the spectrometer and CCDs on to a particular flare or other region within seconds of command from a flare trigger. Spectral atlases consisting of full spectral scans over the entire 1.5 – 9 Å range will be taken while the instrument is directed at particular sources with a user-chosen intensity level (*e.g.* a total of 10 000 photon counts over the spectrum). Spectral line and continuum fluxes will be obtained with unprecedented precision, especially when the spacecraft is at perihelion, about 0.3 AU from the Sun. Thermal filters over the instrument entrance aperture will prevent solar ultraviolet radiation from entering.

5. Conclusions

The DIOGENESS instrument on *CORONAS-F* operated for only a few weeks in 2001, but several flares with GOES importance of up to X5 were recorded. Its scanning crystal spectrometers observed the neighbourhood of the He-like Ca, S, and Si (Ca XIX, S XV, and Si XIII) X-ray emission lines at 3.17 Å, 5.04 Å, and 6.65 Å, with the two quartz crystals of channels 1 and 4 in a dopplerometer mode, *i.e.* arranged in such a way that the Doppler line shifts at flare impulsive phases could be separated from spatial shifts. The concept was validated for the X5 flare on 25 August 2001, in which velocities of up to 150 $\mathrm{km\,s^{-1}}$ were observed. The strong X-ray emission during this flare enabled many lines to be distinguished and identified. The Ca XIX lines predominate in channels 1 and 4 spectra, with dielectronic satellites of Ca XVIII, Ca XVII, and Ca XVI also present. The spectra are similar to those recorded with the X-ray spectrometer on the *P78-1* spacecraft (Doschek, Feldman, and Seely, 1985; Seely and Feldman, 1985), and resemble to a high degree those obtained from

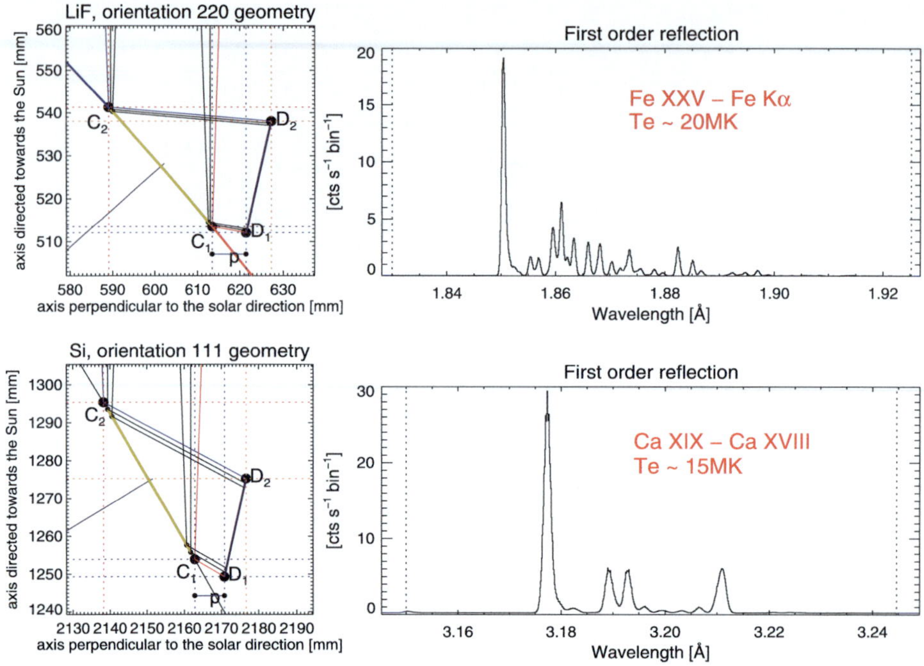

Figure 9 Scheme of crystal diffraction for dopplerometer channels 1 (Fe XXV, top) and 2 (Ca XIX, lower) and resulting spectra in ChemiX (simulated for a GOES M5 flare using the CHIANTI database and software package). Left: Solar X-rays are incident from the top of the figure on the crystal (light green), are diffracted by the crystals (C_1–C_2), and then are incident on the CCD detectors (D_1–D_2). Right: Simulated spectra calculated for a Sun–spacecraft distance of 1 AU using the CHIANTI database – top panel is the Fe XXV spectrum with Fe XXIV dielectronic satellites, lower panel is the Ca XIX spectrum with relatively much weaker Ca XVIII satellites. Temperatures (in MK) are indicated in the figure legend.

the Alcator C-Mod tokamak (Rice *et al.*, 2014), including the Rydberg series of Ar XVII $1s^2 - 1snp$ lines, which are seen up to $n = 7$ in DIOGENESS spectra. In channel 2 and 3 spectra, lines of S XV, Si XIII, and Si XIV predominate, with dielectronic satellites of the Li-like stages occurring.

The ChemiX instrument on the forthcoming Russian *Interhelioprobe* spacecraft, which will be launched into elliptical orbits around the Sun in 2020 and 2022, will have four channels viewing the Fe XXV, Ca XIX, S XV, Si XIII, and Si XIV X-ray emission lines, with a full coverage of the 1.5–9 Å range. The diffracting crystals will be arranged in a dopplerometer manner as for DIOGENESS, but will be bent and non-moving, not flat and scanning. This will be a considerable improvement in that the time resolution should be much higher. The use of cooled CCD detectors will also give a much better spectral resolution than did the RESIK spectrometer on *CORONAS-F*, which had position-sensitive proportional counters. A collimator will allow the emission from individual active regions to be observed.

Acknowledgements We acknowledge support from the Polish National Science Centre grants 2011/01/B/ST9/05861, 4 T12E 045 29, 2011/01/M/ST9/06096, 2314/7.PR/2012/2 and 2011/01/M/ST9/05878. The research leading to these results has received funding from the European Union's Seventh Programme for Research, Technological Development and Demonstration under Grant Agreement No. 284461 (eHEROES project). We also thank the Rutherford Appleton Laboratory (UK) for providing the beryl crystal of DIOGENESS, which was a spare crystal from constructing the *Solar Maximum Mission Flat Crystal Spectrome-*

ter. The CHIANTI atomic database and code is a collaborative project involving the George Mason University, the University of Michigan (USA), and the University of Cambridge (UK).

Open Access This article is distributed under the terms of the Creative Commons Attribution License which permits any use, distribution, and reproduction in any medium, provided the original author(s) and the source are credited.

References

Acton, L.W., Finch, M.L., Gilbreth, C.W., Culhane, J.L., Bentley, R.D., Bowles, J.A., Guttridge, P., Gabriel, A.H., Firth, J.G., Hayes, R.W.: 1980, The soft X-ray polychromator for the Solar Maximum Mission. *Solar Phys.* **65**, 53. DOI.

Bely-Dubau, F., Faucher, P., Steenman-Clark, L., Dubau, J., Loulergue, M., Gabriel, A.H., Antonucci, E., Volonte, S., Rapley, C.G.: 1982, Dielectronic satellite spectra for highly-charged helium-like ions. VII – Calcium spectra: Theory and comparison with SMM observations. *Mon. Not. Roy. Astron. Soc.* **201**, 1155.

Culhane, J.L., Hiei, E., Doschek, G.A., Cruise, A.M., Ogawara, Y., Uchida, Y., Bentley, R.D., Brown, C.M., Lang, J., Watanabe, T., Bowles, J.A., Deslattes, R.D., Feldman, U., Fludra, A., Guttridge, P., Henins, A., Lapington, J., Magraw, J., Mariska, J.T., Payne, J., Phillips, K.J.H., Sheather, P., Slater, K., Tanaka, K., Towndrow, E., Trow, M.W., Yamaguchi, A.: 1991, The Bragg crystal spectrometer for SOLAR-A. *Solar Phys.* **136**, 89. DOI.

Doschek, G.A., Feldman, U., Kreplin, R.W., Cohen, L.: 1980, High-resolution X-ray spectra of solar flares. III – General spectral properties of X1 – X5 type flares. *Astrophys. J.* **239**, 725. DOI.

Doschek, G.A., Feldman, U., Seely, J.F.: 1985, Element abundances from solar flare spectra. *Mon. Not. Roy. Astron. Soc.* **217**, 317.

Doschek, G.A., Kreplin, R.W., Feldman, U.: 1979, High-resolution solar flare X-ray spectra. *Astrophys. J. Lett.* **233**, L157. DOI.

Gabriel, A.H.: 1972, Dielectronic satellite spectra for highly-charged helium-like ionlines. *Mon. Not. Roy. Astron. Soc.* **160**, 99.

Oraevsky, V.N., Sobelman, I.I.: 2002, Comprehensive studies of solar activity on the CORONAS-F satellite. *Astron. Lett.* **28**, 401. DOI.

Parkinson, J.H., Wolff, R.S., Kestenbaum, H.L., Ku, W.H.M., Lemen, J.R., Long, K.S., Novick, R., Suozzo, R.J., Weisskopf, M.C.: 1978, Silicon X-ray line emission from solar flares and active regions. *Solar Phys.* **60**, 123. DOI.

Phillips, K.J.H., Lemen, J.R., Cowan, R.D., Doschek, G.A., Leibacher, J.W.: 1983, Inner-shell transitions in Fe XIX – XXII in the X-ray spectra of solar flares and Tokamaks. *Astrophys. J.* **265**, 1120. DOI.

Płocieniak, S., Sylwester, J., Kordylewski, Z., Sylwester, B.: 2002, Determination of wavelengths and line shifts based on X-ray spectra from Diogeness. In: Wilson, A. (ed.) *Solar Variability: From Core to Outer Frontiers*, *ESA* **SP-506**, 963.

Rice, J.E., Reinke, M.L., Ashbourn, J.M.A., Gao, C., Victora, M.M., Chilenski, M.A., Delgado-Aparicio, L., Howard, N.T., Hubbard, A.E., Hughes, J.W., Irby, J.H.: 2014, X-ray observations of Ca^{19+}, Ca^{18+} and satellites from Alcator C-Mod tokamak plasmas. *J. Phys. B, At. Mol. Opt. Phys.* **47**(7), 075701. DOI.

Seely, J.F., Doschek, G.A.: 1989, Measurement of wavelengths for inner-shell transitions in CA XVII – XIX. *Astrophys. J.* **338**, 567. DOI.

Seely, J.F., Feldman, U.: 1985, Precision measurement of transition energies and Lamb shifts in two-electron argon and potassium in solar-flare spectra. *Phys. Rev. Lett.* **54**, 1016. DOI.

Siarkowski, M., Sylwester, J., Plocieniak, S., Kordylewski, Z.: 2002, Flare X-ray spectra observed by Diogenes scanning Bragg spectrometer. In: Wilson, A. (ed.) *Solar Variability: From Core to Outer Frontiers*, *ESA* **SP-506**, 753.

Sylwester, B., Sylwester, J., Phillips, K.J.H., Kępa, A., Mrozek, T.: 2014, Solar flare composition and thermodynamics from RESIK X-ray spectra. *Astrophys. J.* **787**, 122. DOI.

Sylwester, J., Gaicki, I., Kordylewski, Z., Kowaliński, M., Nowak, S., Płocieniak, S., Siarkowski, M., Sylwester, B., Trzebiński, W., Bakała, J., Culhane, J.L., Whyndham, M., Bentley, R.D., Guttridge, P.R., Phillips, K.J.H., Lang, J., Brown, C.M., Doschek, G.A., Kuznetsov, V.D., Oraevsky, V.N., Stepanov, A.I., Lisin, D.V.: 2005, RESIK: a bent crystal X-ray spectrometer for studies of solar coronal plasma composition. *Solar Phys.* **226**, 45. DOI.

Tanaka, K., Watanabe, T., Nishi, K., Akita, K.: 1982, High-resolution solar flare X-ray spectra obtained with rotating spectrometers on the HINOTORI satellite. *Astrophys. J. Lett.* **254**, L59. DOI.

MIX
Papier aus verantwortungsvollen Quellen
Paper from responsible sources
FSC® C105338

If you have any concerns about our products,
you can contact us on
ProductSafety@springernature.com

In case Publisher is established outside the EU,
the EU authorized representative is:
Springer Nature Customer Service Center GmbH
Europaplatz 3, 69115 Heidelberg, Germany

Printed by Libri Plureos GmbH
in Hamburg, Germany